Biochemistry of the Essential Ultratrace Elements

BIOCHEMISTRY OF THE ELEMENTS

Series Editor: **Earl Frieden**
Florida State University
Tallahassee, Florida

Volume 1	BIOCHEMISTRY OF NONHEME IRON Anatoly Bezkorovainy
Volume 2	BIOCHEMISTRY OF SELENIUM Raymond J. Shamberger
Volume 3	BIOCHEMISTRY OF THE ESSENTIAL ULTRATRACE ELEMENTS Edited by Earl Frieden

A Continuation Order Plan is available for this series. A continuation order will bring delivery of each new volume immediately upon publication. Volumes are billed only upon actual shipment. For further information please contact the publisher.

Biochemistry of the Essential Ultratrace Elements

Edited by
Earl Frieden
The Florida State University
Tallahasse, Florida

PLENUM PRESS • NEW YORK AND LONDON

Library of Congress Cataloging in Publication Data

Main entry under title:

Biochemistry of the essential ultratrace elements.

(Biochemistry of the elements; v. 3)
Bibliography: p.
Includes index.
1. Trace elements—Metabolism. 2. Trace elements in the body. 3. Biological chemistry. I. Frieden, Earl. II. Series.
QP534.B54 1984   574.1'9214   84-17973
ISBN 0-306-41682-4

© 1984 Plenum Press, New York
A Division of Plenum Publishing Corporation
233 Spring Street, New York, N.Y. 10013

All rights reserved

No part of this book may be reproduced, stored in a retrieval system, or transmitted in any form or by any means, electronic, mechanical, photocopying, microfilming, recording, or otherwise, without written permission from the Publisher

Printed in the United States of America

Contributors

Nicholas M. Alexander • Division of Clinical Pathology, Department of Pathology, University of California School of Medicine, San Diego, California 92103

Richard A. Anderson • Beltsville Human Nutrition Research Center, U.S. Department of Agriculture, Beltsville, Maryland 20705

Janet S. Borel • Beltsville Human Nutrition Research Center, U.S. Department of Agriculture, Beltsville, Maryland 20705

Edith M. Carlisle • Division of Nutrition, School of Public Health, University of California, Los Angeles, California 90024

W. M. Dugger • Department of Botany and Plant Sciences, University of California, Riverside, California 92521

Earl Frieden • Department of Chemistry, Florida State University, Tallahassee, Florida 32306

Leon L. Hopkins • Department of Food and Nutrition, Texas Tech University, Lubbock, Texas 79409

Lucille S. Hurley • Department of Nutrition, University of California, Davis, California 95616

Carl L. Keen • Department of Nutrition, University of California, Davis, California 95616

M. Kirchgessner • Institute of Nutrition Physiology, Technical University München D-8050 Freising-Weihenstephan, F.R.G.

Bo Lönnerdal • Department of Nutrition, University of California, Davis, California 95616

Carol Lovatt • Department of Botany and Plant Sciences, University of California, Riverside, California 92521

Harold H. Messer • Department of Oral Biology, University of Minnesota, Minneapolis, Minnesota 55455

David B. Milne • U.S. Department of Agriculture, Agricultural Research Service, Grand Forks Human Nutrition Research Center, Grand Forks, North Dakota 58202

Forrest H. Neilsen • U.S. Department of Agriculture, Agricultural Research Service, Grand Forks Human Nutrition Research Center, and Department of Biochemistry, University of North Dakota, Grand Forks, North Dakota 58202

K. V. Rajagopalan • Department of Biochemistry, Duke University Medical Center, Durham, North Carolina 27710

A. M. Reichlmayr-Lais • Institute of Nutrition Physiology, Technical University München D-8050 Freising-Weihenstephan, West Germany

Gerhard N. Schrauzer • Department of Chemistry, University of California at San Diego, La Jolla, California 92093

Raymond J. Shamberger • The Cleveland Clinic, 9500 Euclid Avenue, Cleveland, Ohio 44106

Harry A. Smith • Department of Chemistry, Florida State University, Tallahassee, Florida 32306

Barbara J. Stoecker • Department of Food and Nutrition, Texas Tech University, Lubbock, Texas 79409

Eric O. Uthus • U.S. Department of Agriculture, Agricultural Research Service, Grand Forks Human Nutrition Research Center and Department of Biochemistry, University of North Dakota, Grand Forks, North Dakota 58202

Roland S. Young • Consulting Chemical Engineer, 605, 1178 Beach Drive, Victoria, B.C. V8S 2M9, Canada

Preface

The remarkable development of *molecular biology* has had its counterpart in an impressive growth of a segment of biology that might be described as *atomic biology*. The past several decades have witnessed an explosive growth in our knowledge of the many elements that are essential for life and maintenance of plants and animals. These essential elements include the bulk elements (hydrogen, carbon, nitrogen, oxygen, and sulfur), the macrominerals (sodium, potassium, calcium, magnesium, chloride, and phosphorus), and the trace elements. This last group includes the ultratrace elements and iron, zinc, and copper. Only the ultratrace elements are featured in this book.

Iron has attracted so much research that two volumes are devoted to this metal—*The Biochemistry of Non-Heme Iron* by A. Bezkoravainy, Plenum Press, 1980, and *The Biochemistry of Heme Iron* (in preparation). Copper and zinc are also represented by a separate volume in this series.

The present volume begins with a discussion of essentiality as applied to the elements and a survey of the entire spectrum of possible required elements. This is followed by an outline of the history of the research on trace elements that led to their identification as essential. Each of the remaining chapters is devoted to an essential ultratrace element and is written by an expert on that particular element. Individual chapters include fluorine, iodine, manganese, cobalt, chromium, molybdenum, selenium, vanadium, silicon, nickel, boron, arsenic, tin, cadmium, and lead. The inclusion of the last several elements may represent a premature designation as essential. Nonetheless, a separate independent chapter for each of these elements should be useful, if only because of the extensive research now in progress. This text hopes to bring together our present knowledge of the essential ultratrace elements and to serve as a guide to further biochemical research on these elements and others not yet so designated.

EARL FRIEDEN
Professor of Chemistry
Florida State University
Tallahassee, Florida

Contents

1. **A Survey of the Essential Biochemical Elements**   1
   Earl Frieden

   1.1 Essentiality   1
       1.1.1 Stimulatory Metals   3
       1.1.2 Evolution of the Essential Trace Elements   4
   1.2 A Survey of the Biochemistry of the Elements   5
       1.2.1 The Nonmetals: Bromine   8
       1.2.2 The Metals   9
       1.2.3 Pretransition Metals   9
       1.2.4 The Alkali Metals; Lithium, Rubidium   9
       1.2.5 The Alkaline Earth Metals; Beryllium, Strontium, Barium   11
       1.2.6 Remaining Pre-transition Metals   11
       1.2.7 The Transition Metals   11
       1.2.8 The Post-transition Metals   12
       1.2.9 Lanthanides and Actinides   13
   1.3 Mechanism of Action of the Essential Ultratrace Elements   13
   1.4 Summary   14
       General References   15
       Specific References   15

2. **The Discovery of the Essential Trace Elements: An Outline of the History of Biological Trace Element Research**   17
   Gerhard N. Schrauzer

   2.1 Introduction   17
   2.2 Classification of the Bioelements   18
   2.3 The Concept of Essentiality   19

2.4	Trace Element Discoveries from 1925 to 1956	20
	2.4.1 Copper	20
	2.4.2 Manganese	20
	2.4.3 Zinc	21
	2.4.4 Cobalt	21
	2.4.5 Molybdenum	22
2.5	Discoveries from 1956 to 1978: The Era of Klaus Schwarz	22
	2.5.1 Selenium	22
	2.5.2 Chromium	24
	2.5.3 Tin	24
	2.5.4 Vanadium	25
	2.5.5 Fluoride	25
	2.5.6 Silicon	25
	2.5.7 Nickel	26
	2.5.8 Lead, Cadmium, and Arsenic	26
2.6	Current Trends and Problems	27
	References	28

3. Iodine 33

Nicholas M. Alexander

3.1	Introduction	33
3.2	Chemistry of Iodine and Iodoamino Acids	34
	3.2.1 Properties of Iodine	34
	3.2.2 Iodine Isotopes	34
	3.2.3 Important Chemical and Biochemical Reactions of Iodine	35
	3.2.4 Iodotyrosines and Iodothyronines	37
3.3	Iodine Metabolism	39
	3.3.1 Iodine Absorption, Evolution, and Thyroid Hormone Biosynthesis	39
	3.3.2 Thyroid Hormones in Blood	42
	3.3.3 Thyroid Hormones in Peripheral Tissues	43
	3.3.4 Iodine Deficiency	44
	3.3.5 Iodine Toxicity	45
3.4	Mechanism of Action of Thyroid Hormones	46
	3.4.1 General and Cellular Effects	46
	3.4.2 Structure–Activity Relationships	47
3.5	Summary	49
	References	50

4. Fluorine — 55
H. H. Messer

4.1 Introduction	55
4.2 Fluoride in Cells and Tissues	56
4.2.1 Mineralized Tissues	56
4.2.2 Cells and Soft Tissues	59
4.2.3 Extracellular Fluid	60
4.3 Fluoride Deficiency and Function	61
4.3.1 Deficiency	61
4.3.2 Functions of Fluoride	63
4.4 Metabolism and Toxicity of Fluoride	67
4.4.1 Metabolism	67
4.4.2 Toxicity	75
4.5 Summary	83
References	83

5. Manganese — 89
Carl L. Keen, Bo Lönnerdal, and Lucille S. Hurley

5.1 Introduction	89
5.2 Manganese Concentration in Animal Tissues	89
5.3 Metabolism of Manganese	92
5.3.1 Absorption	92
5.3.2 Transport and Tissue Distribution	94
5.3.3 Excretion	96
5.4 Biochemistry of Manganese	97
5.4.1 Manganese Chemistry	97
5.4.2 Manganese as a Cofactor and in Metalloenzymes	98
5.4.3 Manganese and Carbohydrate Metabolism	104
5.4.4 Manganese and Lipid Metabolism	108
5.4.5 Manganese and Brain Function	110
5.5 Manganese Nutrition	112
5.5.1 Manganese Deficiency	112
5.5.2 Genetic Interactions and Manganese Metabolism	115
5.5.3 Human Requirements	117
5.5.4 Manganese Content of Foods	117
5.6 Manganese Toxicity	117
5.7 Manganese in Relation to Immunocompetence and Cancer	121
5.8 Summary	122
References	123

6. Cobalt — 133

Roland S. Young

6.1	Introduction and History	133
6.2	Cobalt and Its Compounds in Cells and Tissues	134
	6.2.1 Cobalt in Soils	135
	6.2.2 Cobalt in Plants	135
	6.2.3 Cobalt in Animals	137
6.3	Cobalt Deficiency and Function	138
	6.3.1 Cobalt in Animal Nutrition	139
	6.3.2 Cobalt in Human Nutrition	139
6.4	Metabolism and Toxicity of Cobalt	141
	6.4.1 Effect of Cobalt on Plants	141
	6.4.2 Effect of Cobalt on Animals	143
	6.4.3 Effect of Cobalt on Microorganisms	143
	6.4.4 Toxicity of Cobalt	144
6.5	Conclusion	146
	References	146

7. Molybdenum — 149

K. V. Rajagopalan

7.1	Introduction and History	149
7.2	Molybdenum and Its Compounds in Cells and Tissues	150
	7.2.1 Molybdenum-Containing Enzymes	151
	7.2.2 The Molybdenum Cofactor	162
	7.2.3 General Aspects of Molybdenum Biochemistry	165
7.3	Nutritional Aspects of Molybdenum	167
	7.3.1 Molybdenum in the Diet	167
	7.3.2 Molybdenum Deficiency	168
	7.3.3 Molybdenum Toxicity	169
7.4	Conclusion	170
7.5	Summary	170
	References	171

8. Chromium — 175

Janet S. Borel and Richard A. Anderson

8.1	Introduction	175
8.2	Chromium: Physical and Chemical Properties	175

8.3	Biologically Active Chromium	176
8.4	Absorption and Transport of Chromium	177
8.5	Chromium Occurrence in Blood, Tissues, and Hair	178
	8.5.1 Blood	179
	8.5.2 Tissues	180
	8.5.3 Hair	181
8.6	Chromium Excretion	182
8.7	Functions of Chromium and Signs of Chromium Deficiency	184
8.8	Factors Affecting Chromium Metabolism	187
8.9	Chromium and Stress	188
8.10	Dietary Requirements of Chromium	189
8.11	Effects of Chromium Supplementation	190
8.12	Toxicity of Chromium	193
8.13	Summary	194
	References	195

9. Selenium 201

Raymond J. Shamberger

9.1	Introduction and History	201
9.2	Selenium and its Compounds in Cells and Tissues	204
	9.2.1 Low Molecular Weight Compounds	204
	9.2.2 Macromolecular Weight Compounds	207
9.3	Selenium Deficiency and Function	213
	9.3.1 Dietary Liver Necrosis and Factor 3	213
	9.3.2 Hepatosis Dietetica	216
	9.3.3 Nutritional Muscular Dystrophy	216
	9.3.4 Exudative Diathesis	217
	9.3.5 Pancreatic Regeneration	218
	9.3.6 Mulberry Heart Disease	218
	9.3.7 Reproductive Problems	218
	9.3.8 Myopathy of the Gizzard	220
	9.3.9 Growth	220
	9.3.10 Selenium-Responsive Unthriftiness of Sheep and Cattle	220
	9.3.11 Periodontal Diseases of Ewes	221
	9.3.12 Encephalomalacia	221
9.4	Metabolism and Toxicity of Selenium	222
	9.4.1 Absorption	222
	9.4.2 Excretion	224
	9.4.3 Placental Transfer	225
	9.4.4 Mechanism of the Antioxidant Action of Selenium	225

	9.4.5	Interactions of Selenium with Other Substances	228
	9.4.6	Toxicity of Selenium	231
9.5	Summary		236
	References		236

10. Vanadium 239

Barbara J. Stoecker and Leon L. Hopkins

10.1	Introduction and History		239
	10.1.1	Discovery and History	239
	10.1.2	Occurrence and Distribution	239
	10.1.3	Nuclear and Chemical Characteristics	240
	10.1.4	Essentiality	240
10.2	Vanadium in Tissues		240
	10.2.1	Vanadium in Plants and Plant Products	240
	10.2.2	Vanadium in Tunicates, Crustaceans, Shellfish, and Fish	241
	10.2.3	Human Intakes of Vanadium	242
	10.2.4	Vanadium Levels in Human Beings	242
10.3	Vanadium Deficiency and Function		243
	10.3.1	Growth	243
	10.3.2	Reproduction	243
	10.3.3	Nutritional Edema	243
	10.3.4	Manic-Depressive Illness	244
	10.3.5	Dental Caries	244
	10.3.6	Inotropic Effects of Vanadium	245
	10.3.7	Vanadium and Renal Function	245
	10.3.8	Vanadium and Glucose Metabolism	246
	10.3.9	Vanadium and Lipid Metabolism	246
	10.3.10	Vanadium and ATPases	247
	10.3.11	Additional Effects of Vanadium	248
10.4	Vanadium Metabolism		248
	10.4.1	Absorption of Vanadium	248
	10.4.2	Tissue Distribution of Vanadium	249
	10.4.3	Effects of Hormones on Vanadium Metabolism	249
	10.4.4	Excretion of Vanadium	250
10.5	Vanadium Toxicity		250
	10.5.1	Factors Affecting Toxicity of Vanadium	250
	10.5.2	Toxicity in Chicks, Rats, and Sheep	250
	10.5.3	Toxicity in Human Beings	251
10.6	Summary		252
	References		252

11. Silicon   257
Edith Muriel Carlisle

11.1	Introduction	257
	11.1.1 Discovery and History	257
	11.1.2 Occurrence and Distribution	258
	11.1.3 Chemistry	258
	11.1.4 Essentiality	259
11.2	Silicon in Tissues	259
	11.2.1 Primitive Organisms	259
	11.2.2 Higher Plants	260
	11.2.3 Animals and Man	262
11.3	Silicon Deficiency and Functions	264
	11.3.1 Growth and Development	264
	11.3.2 Calcification	265
	11.3.3 Bone Formation	267
	11.3.4 Cartilage and Connective Tissue Formation	269
	11.3.5 Connective Tissue Matrix	270
	11.3.6 Enzyme Activity	271
	11.3.7 Connective Tissue Cellular Component	272
	11.3.8 Structural Component	272
	11.3.9 Aging	275
11.4	Metabolism	277
	11.4.1 Absorption	277
	11.4.2 Transport	278
	11.4.3 Excretion	279
	11.4.4 Interaction with Molybdenum	279
	11.4.5 Enzyme Interaction	279
11.5	Toxicity	281
	11.5.1 Pneumoconioses in Man	281
	11.5.2 Silicosis	282
	11.5.3 Asbestosis	283
	11.5.4 Renal Toxicity	284
11.6	Summary	287
	References	288

12. Nickel   293
Forrest H. Nielsen

12.1	Introduction and History	293
12.2	Nickel and Its Compounds in Cells and Tissues	293
12.3	Nickel Deficiency	296

12.4	Nickel Function	300
12.5	Biological Interactions between Nickel and Other Trace Elements	301
12.6	Nickel Metabolism and Toxicity	304
12.7	Summary	306
	References	307

13. Tin 309

David B. Milne

13.1	Introduction	309
13.2	Tin in Cells and Tissues	309
	13.2.1 Chemical Properties	309
	13.2.2 Distribution in Mammalian Tissues	311
13.3	Deficiency and Function	311
13.4	Metabolism and Toxicity	312
	13.4.1 Inorganic Tin	312
	13.4.2 Organotin Compounds	315
13.5	Summary	316
	References	316

14. Arsenic 319

Forrest H. Nielsen and Eric O. Uthus

14.1	Introduction and History	319
14.2	Arsenic and Its Compounds in Cells and Tissues	320
14.3	Arsenic Deficiency and Interaction with Other Nutrients	323
14.4	Arsenic Function	327
14.5	Arsenic Metabolism	328
14.6	Arsenic Toxicity	334
14.7	Summary	335
	References	336

15. Cadmium 341

Harry A. Smith

15.1	Introduction	341
	15.1.1 Historical Perspectives and Properties of Cadmium	341
	15.1.2 Metallothionein and Its Interactions with Cadmium	342
15.2	Chemistry of Cadmium: Biological Perspectives	345
	15.2.1 General Chemical Properties of Cadmium	345
	15.2.2 Biological Implications	345

15.3	Evidence for the Possible Essentiality of Cadmium	346
15.4	Metabolism of Cadmium	348
	15.4.1 Absorption of Cadmium	348
	15.4.2 Transport of Cadmium in Blood	350
	15.4.3 Organ, Tissue, and Subcellular Distribution of Cadmium	351
	15.4.4 Cadmium Excretion	352
15.5	Biochemical Effects of Cadmium	354
	15.5.1 Nucleic Acid and Protein Synthesis	354
	15.5.2 Induction of Thionein by Cadmium	357
	15.5.3 Other Biochemical Effects of Cadmium	360
15.6	Summary	362
	References	363

16. Lead   367

A. M. Reichlmayr-Lais and M. Kirchgessner

16.1	Introduction and History	367
16.2	Metabolism of Lead	368
	16.2.1 Occurrence and Intake	368
	16.2.2 Absorption of Lead	369
	16.2.3 Excretion of Lead	370
	16.2.4 Transport and Distribution	370
	16.2.5 Interactions	373
16.3	Lead Deficiency	375
16.4	Toxicity of Lead	376
	16.4.1 Hematologic Effects of Toxic Lead Doses	377
	16.4.2 Neurotoxic Effects of Lead	379
	16.4.3 Renal Effects of Toxic Lead	381
	16.4.4 Intranuclear Inclusion Body	382
	16.4.5 Mutagenic, Mitogenic, and Teratogenic Effects of Lead	382
16.5	Conclusions	383
	References	384

17. Boron   389

Carol J. Lovatt and W. M. Dugger

17.1	Boron in Biology	389
	17.1.1 Introduction	389
	17.1.2 Criteria for Essentiality	389
	17.1.3 Effect of Boron on the Growth of Organisms	390
	17.1.4 Plant Evolution and an Essential Role for Boron	392

	17.1.5	Boron Toxicity	393
	17.1.6	Problems Associated with Studies of Boron Metabolism	394
	17.1.7	10_B (n, α) ^7Li Nuclear Reaction	394
	17.1.8	Therapeutic Uses for Boron and Organoborates in Medicine	395
17.2	Carbohydrate Metabolism		396
	17.2.1	Boron Complexes	397
	17.2.2	Sugar Translocation	398
	17.2.3	Photosynthesis	398
	17.2.4	Respiration	399
	17.2.5	Starch	399
	17.2.6	Cellulose and Cell Wall Glucans	400
	17.2.7	Phenols	401
	17.2.8	Lignin	402
	17.2.9	Boron in Enzymic Reactions	402
	17.2.10	Pollen Germination	404
	17.2.11	Conclusions	405
17.3	Hormone Action		405
17.4	Membrane Structure and Function		409
17.5	Nucleic Acid Biosynthesis		410
17.6	Summary		414
	References		415

Index 423

A Survey of the Essential Biochemical Elements

Earl Frieden

1.1 Essentiality

The simplest definition of an essential element is that it is an element required for the maintenance of life; its absence results in death or a severe malfunction of the organism. Experimentally, this rigorous criterion cannot always be satisfied and this has led to a broader definition of essentiality. An element is considered essential when a deficient intake produces an impairment of function and when restoration of physiological levels of that element prevents or relieves the deficiency. The organism can neither grow nor complete its life cycle without the element in question. The element should have a direct influence on the organism and be involved in its metabolism. The effect of the essential element cannot be wholly replaced by any other element. The importance of modifying any of these specific criteria in identifying the essentiality of a particular element has been emphasized by Luckey and Venugopal (1977).

Some years ago, the late G. C. Cotzias (1967) defined several biochemical criteria for an essential element. The element is present in tissues of different animals at comparable concentrations. Its withdrawal produces similar physiological or structural abnormalities regardless of species. Its presence reverses or prevents these abnormalities. Finally, these abnormalities are accompanied by specific biochemical changes that can be prevented or remedied when the deficiency is prevented or remedied.

The essential trace elements provide a classic example of required nutrients

---

Earl Frieden • Department of Chemistry, Florida State University, Tallahassee, Florida 32306.

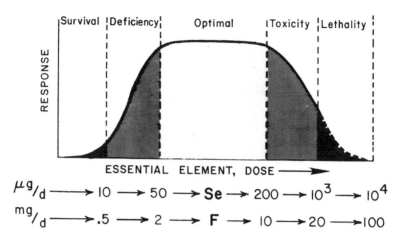

Figure 1-1. Dose response range of an essential element. Estimates of specific requirements in terms of micrograms per day for selenium or milligrams per day for fluorine are included.

as described by Bertrand as early as 1912. An organism passes through several stages as the concentration of an essential nutrient progresses from deficiency to excess (Figure 1-1), In absolute deficiency, death may result. With limited intake, the organism survives but may show marginal insufficiency. With increasing nutrient a plateau representing optimal function is reached. As the nutrient is given in excess, first marginal toxicity, then mortal toxicity are attained. While this curve may vary quantitatively for each essential nutrient, the basic pattern holds for virtually all the elements. This is illustrated in Figure 1-1 for selenium and fluoride. For these two elements there is barely a ten-fold range between intake per day for survival and that for the appearance of toxic effects. Mertz (1981) has emphasized two conclusions from Bertrand's model. Each has a range of safe exposures that maintain optimal tissue concentrations and functions. Also, every trace element has a toxic range when its safe exposure is exceeded.

The pioneer in this field, Klaus Schwarz (1977), has emphasized that high toxicity does not preclude biological essentiality. Selenium and fluoride are two examples where relatively high toxicity has delayed but not prevented their ultimate recognition as essential ultratrace elements. Schwarz (1977) also pointed out that to exclude an element from being essential requires that it has been shown not to be essential at a precise, accurately determined level. Thus recognition of essentiality is singularly dependent on the sensitivity and accuracy of the analytical methods applied to the specific element in question.

A Survey of the Essential Biochemical Elements

1.1.1 Stimulatory Metals

Some metal ions and other nutrients may exhibit stimulatory effects when subjected to a dose-response study. This can result in a spurious impression of a positive response to the addition of the metal in the diet, as illustrated in Figure 1-2. There can be growth stimulation even below the basal requirement for a given nutrient, or stimulation can be superimposed on the classical dose response curve Figure 1-1. Either response can be misread to represent an apparent essential activity of the nutrient.

There are numerous examples of stimulatory metals (called hormetins by Luckey and Venugopal, 1977). Elements in every group of the periodic table (Figure 1-3) have been found to be stimulatory to animal growth and survival, including lithium, titanium, gallium, germanium, rubidium, zirconium, antimony, barium, gold, mercury, and many elements in the lanthanide series. The

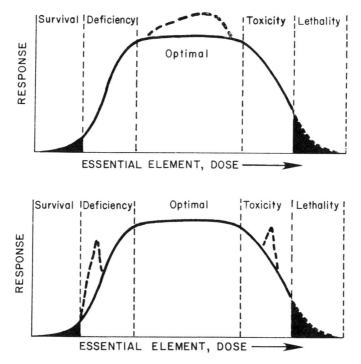

Figure 1-2. The effect of a stimulatory element (or nutrient) on the dose–response curve of an essential element. (Adapted from Luckey and Venugopal, 1977.)

effect of stimulatory substances can only be resolved by careful study of the complete range of nutrient effects. Clear proof of stimulation may be quite difficult to obtain if there is a sharp peak of response over a relatively narrow concentration range. Only a comprehensive dose-response study of the nutrient will permit the distinction of a true essentiality from a stimulatory action. The discussion in Chapter 15 by H. A. Smith (p. 346) emphasizes the difficulties in distinguishing an essential role from a stimulatory role for the metal ion, cadmium.

An interesting stimulatory effect on the growth of plants was described by Steenbjerg (1951). A decrease in the concentration of a particular essential element results in a beneficial effect on growth in plants.

Certain relatively toxic elements or compounds when used at low doses can act as therapeutic agents with beneficial effects. Examples are the effect of gold salts on arthritis and platinum complexes on certain cancers. Stimulatory activity can also be harmful. Cancer induction is observed with a variety of metals in the fourth and fifth periods. Toxicity of metals, mainly by blocking of an essential metabolite, also can arise from apparent stimulatory activity.

1.1.2 Evolution of the Essential Trace Elements

The close relationship between elemental abundance and the requirement for specific elements has been noted by a number of scholars interested in the origin and evolution of chemical life, including T. Egami (1974), J. P. Ferris, E. Frieden (1974), and J. H. McClendon (1976). Ferris and Usher (1983) have written that

> the elements abundant in living systems today generally reflect their relative amounts in the ocean (Table 1-1) and, to a lesser extent in the earth's crust. Since it is very likely that life originated in the presence of water, the elements required for life were probably established from their concentrations in the primitive oceans four billion years ago.

However, this close correlation is mitigated by several factors. The first relates to the enrichment of living organisms in their content of carbon, nitrogen, phosphorus, and sulfur. This is accounted for by the fact that life probably originated at the edge of the sea or in sediments where the concentration of phosphates and

Table 1-1. The Abundance of the Elements in Oceans

$>10^6$ nM	H, O, Na, Cl, Mg, S, K, Ca, C, N
10^6–10^2 nM	Br, B, Si, Sr, F, Li, P, Rb, I, Ba
10^2–1 nM	Mo, Zn, Al, V, Fe, Ni, Ti, U, Cu, Cr, Mn, Cs, Se, Sb, Cd, Co, W

Source: Adapted from Egami, 1974.

other minerals was much higher than in the primitive oceans. Perhaps these sediments also contained higher concentrations of the other three possible essential elements (As, Pb, Sn), which are too dilute in the oceans to be included in Table 1-1.

Another factor to be considered is that the distribution of the elements does *not* reflect an earlier era when the earth's atmosphere was not strongly oxidizing. Then iron would have been present in the more soluble Fe(II) oxidation state. This would explain why iron is so prevalent in later biochemical systems despite its high dilution in sea water.

A third factor is the natural selection process, which led to the choice of a few elements that had superior reactivity. For example, divalent manganese, nickel, cobalt, and zinc may have had similar functions, as illustrated by the observation that these metals can replace the zinc ions in zinc metalloenzymes. The reconstituted enzyme is usually less efficient whenever a metal other than zinc is bound at the active site, indicating that zinc possesses the optimal properties for that metalloenzyme.

1.2 A Survey of the Biochemistry of the Elements

The 29 essential elements can be classified into two major groups as depicted in Figure 1-3:

1. The bulk elements, all of which are required in gram quantities, include the six basic structural elements—hydrogen, carbon, nitrogen, oxygen,

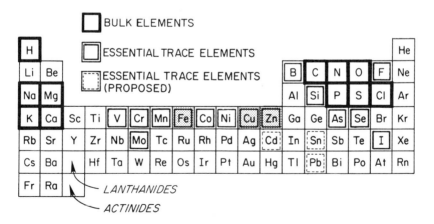

Figure 1-3. Present status of the essentiality of the elements and the periodic table. The 11 bulk elements, 3 essential trace elements (Fe, Zn, Cu), 12 essential ultratrace elements, and the 3 proposed essential ultratrace elements are identified. (Adapted from Valkovic, 1978.)

phosphorus, and sulfur—and the five macrominerals: sodium, potassium, magnesium, calcium, and chloride ion.
2. The trace elements, required in milligram or less than milligram amounts: iron, zinc, copper, and the 15 elements considered in Chapters 3–17 of this text.

Three trace elements—iron, zinc, and copper—stand out as a distinctive subgroup with many ramifications. These three metals have been recognized as essential for over 50 years and are required in milligram quantities to implement multiple catalytic and other functions. They have attracted so much research that each has earned a separate volume in this series (Bezkorovainy, 1980). The remaining trace elements, a more numerous but less ubiquitous group, are utilized in submilligram or even microgram quantities. They constitute the essential ultratrace elements that are the subject of this book.

The concluding part of this chapter is devoted to a survey of all the chemical elements in regard to their potential of each being recognized as an essential element. Attention is directed to those elements that are especially prominent in the biosphere but do not satisfy rigorous requirements for essentiality in studies to date. However, it should be noted that they can be accumulated by certain plants or animals. Those justifying further discussion include aluminum, barium, bismuth, bromine, germanium, lithium, rubidium, strontium, titanium, and tungsten. Several of these elements are as likely candidates for essential ultratrace elements as B, As, Cd, Pb, or Sn. We emphasize that frequently the properties of any of these elements do not permit a sharp line of distinction as to essentiality or even to stimulatory effects.

The prevailing view (Da Silva, 1978) is that these 10 non-essential elements have been acquired by contact with the environment through chemical or physical

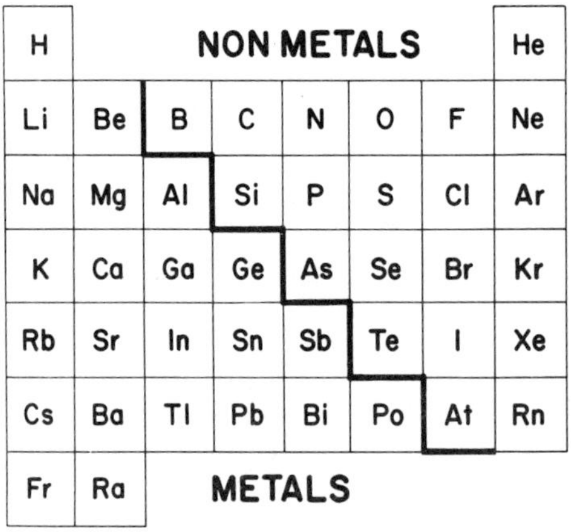

H		NON METALS					He
Li	Be	B	C	N	O	F	Ne
Na	Mg	Al	Si	P	S	Cl	Ar
K	Ca	Ga	Ge	As	Se	Br	Kr
Rb	Sr	In	Sn	Sb	Te	I	Xe
Cs	Ba	Tl	Pb	Bi	Po	At	Rn
Fr	Ra	METALS					

Figure 1-4. A portion of the periodic table for the elements showing the division between metals and nonmetals.

A Survey of the Essential Biochemical Elements

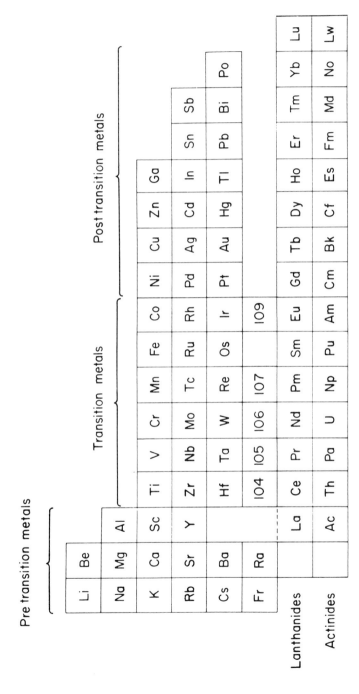

Figure 1-5. The classification of the different metals based on their relationship to the transition metals. (From Choppin and Johnsen.)

processes. Chemical uptake may represent competition with "normal" elements for the binding sites in biological systems. If the abnormal condition in which this competition takes place persists, the biological system will have started to "use" the element. Thus what appears to be a new element dependency might be initiated.

To simplify our discussion of the prospects of the remaining elements as essential, we will employ a classification of the elements used by Choppin and Johnsen (1972). The broadest division of the elements is into two groups: metals and nonmetals (Figure 1-4). The 80 metallic elements will be discussed later and their classification or function in biological systems further resolved according to their relation to the transition metals (Figure 1-5). The rare and noble gases and the metalloids are mentioned separately from the other nonmetals.

1.2.1 The Nonmetals: Bromine

All nonmetals of an atomic weight less than bromine are considered to be essential. Some prominent nonmetals (C, N, O, P, S, Cl) are major constituents of most cells. The rare gases are sufficiently inert to be excluded as participants in any critical biological activity. In addition to the essential metalloids—B, Si, and As—discussed in Chapters 11, 17, and 14, antimony, astatine, germanium, and tellurium are not essential. Antimony has no known function in living organisms, though it can be stimulatory. It is not highly toxic, and trivalent antimony compounds are effective in the treatment of schistosomiasis. Tellurium and germanium are not essential and astatine is a synthetic element with high radioactivity.

Fluoride and iodide (Chapters 3 and 4) are the two halogens (Group VIIA) generally accepted as essential trace elements. Bromine is perhaps the remaining possible nonmetal that may eventually qualify as an essential element. Bromine is one of the most abundant (1–9 ppm in human tissue) and ubiquitous of the trace elements in the biosphere. However, bromine has not been shown to perform any essential function in bacteria, plants, or animals. Bromide can replace chloride in the growth of several algae. It partially replaces chloride and has a stimulatory effect on the growth of chicks. In rats, however, no effect on growth or reproductive performances was observed in rats fed 0.5 ppm bromide. Bromide can substitute for chloride and for iodide in the thyroid gland but is not used for the biosynthesis of thyroid hormones, even though the tri- and tetrabromothyronines are active as thyroid hormones. Many organic bromides, mostly bromophenols, are found in algae and in microorganisms and presumably serve as selective antibiotics. Corals can contain up to 4% bromide in their solids, mostly as 3-mono- and 3,5-di- bromotyrosine.

1.2.2 The Metals

Eighty of the elements can be classified as metals, although the degree to which they exhibit the chemical and physical properties of metals varies considerably. These 80 or so metallic elements can be further divided into four main groups within the periodic table (Figure 1-5).

1. The pretransition elements, which include Groups I and II as well as Al, Sc, and Y. These elements lose s electrons (plus one p electron for Al and one d electron for Sc and Y) to form ions that have noble gas electronic structures.
2. The transition elements in which a d orbital is partially filled.
3. The posttransition elements that include Group III (except Al), Sn and Pb of Group IV, Sb and Bi of Group V, Po of Group VI, and Cu, Ag, Au, Zn, Cd, and Hg. All these elements have filled d subshells where they form ions in the oxidation state of their periodic group.
4. The lanthanide and actinide series of elements, in which f orbitals are involved in the electronic structure.

Many of these elements are familiar to us in their metallic state (e.g., aluminum, chromium, and iron). Others are unlikely to be encountered in their metallic states (e.g., the lanthanides). Still others exist in such small amounts and for such brief times that they have never been prepared in a metallic state; these are the synthetic radioactive elements, francium and others with an atomic number greater than 99.

While many of the heaviest metals are rarely found on earth, those that are available share their most characteristic chemical property, the tendency to oxidize (i.e., to form cations and stable oxides). This common property arises from loss of electrons from outer s orbitals, which accounts for the similar chemical properties of the metallic elements.

1.2.3 Pretransition Metals

The pretransition metals include the alkali metals (Group IA) and the alkaline earth metals (Group IIA) as well as Group IIIA elements. We will consider these pretransition metals in the same order mentioned earlier.

1.2.4 The Alkali Metals; Lithium, Rubidium

The alkali metals (Group IA) have a simple electronic structure with one electron in an s orbital. This leads to charge of $+1$ and the formation of highly polar compounds. Their relatively high solubility in water fits their role as the predominant ions of all living organisms. Thus K^+ is the major intracellular cation and Na^+ the principal extracellular cation, balancing Cl^- as the dominant anion.

Besides Na^+ and K^+, Li is the most likely to be an essential element of the Group IA elements. However, the high charge-to-mass ratios of the lighter metals, especially Li and Be, mitigate against their use in essential reactions of mammalian life. Many vital reactions require light metals, but most of these involve the essential macrominerals of the third and fourth periods of the table of elements or are transition elements.

Interest in Li^+ as a potential essential element has increased since Cade (1949) reported that Li^+ in relatively large doses ameliorated the manic phase of manic-depressive psychosis. This effect has been confirmed but no rationale has been offered other than altered electrolyte distribution in the brain, which is restored to normal by Li^+ treatment. Salt imbalance may also play a role in the protective influence of Li^+ on atherosclerotic heart disease in man. Li^+ shares some of the properties of extracellular Na^+ and intracellular K^+ but is metabolized somewhat differently from other cations. Substituting LiCl for NaCl reduces hypertension in hypertensive patients.

A variety of other significant physiological events may prove to be dependent on Li^+ in the diet. Li^+ was reported to be essential for the slow respiration of nuclear membranes. Li^+ stimulated the uptake of glucose in isolated rat diaphragms. An extensive toxicology has emerged from the relative high doses of Li^+ therapy. Noteworthy are the interference with glucose metabolism in neural tissues, teratogenicity of both mice and rats, and hypothyroidism with goiter development. An extensive summary of the biological activity of lithium has been presented by Birch (1978).

These clues have now received crucial support from Anke *et al.* (1981), who found over a 3-month period that lithium-deficient goats showed 28% less weight gain than controls. Reduced fertility was observed in female goats lacking lithium. Thus lithium may be the most likely of the remaining elements to achieve essentiality.

Another element in this group, rubidium, also shows some properties that are suggestive of essentiality. Rubidium is widely distributed both in sea water (120 ppb) and in the earth's crust (100 ppm). Normal human adults contain about 300 mg in all tissues, more than most of the other ultratrace elements. The metabolism of rubidium and cesium are closely related to that of potassium, and they show an interchangeability with potassium in a variety of biological systems with little evidence for any toxicity. In a recent paper, Lombeck, *et al.* (1980) reported a homeostatic control of rubidium levels in the blood of children (12 ± 3 ppm), which they interpreted as suggestive of an essential role for rubidium. Rubidium is a good example of an element that cannot be excluded from being essential but has not been shown to be essential at a certain, accurately determined level (Schwarz, 1977).

1.2.5 The Alkaline Earth Metals; Strontium, Barium, Beryllium

The alkaline earth metals (Group IIA) have two electrons in a s orbital. These metals become divalent and contribute to specific ionic equilibrium in living cells. Calcium and magnesium ions perform many essential functions in the transmission of nervous impulses, initiating muscular contraction, forming bones, and activating enzyme systems.

The possibility that strontium or barium are essential has not been confirmed (or denied) despite the claim of Rygh (1949). He reported that the omission of either strontium or barium from the diet resulted in depressed growth and reduced calcification of bones and teeth in rats and guinea pigs. Both Sr and Ba can spare calcium and are relatively nontoxic. They also show some stimulatory action.

Beryllium is peculiar in many ways when compared with the other Group IIA elements. Beryllium's small size, intense ionic field, and limited covalency of 4 contribute to this difference. The propensity of beryllium to form covalent bonds with nonmetals results in kinetics too slow to be useful in most biological systems. Beryllium is the most permeable and toxic element of Group IIA. Among its deleterious effects is the inhibition of a broad spectrum of enzymes, including those involved in DNA biosynthesis. Thus no essential or stimulatory function has been observed or is likely to be found for beryllium.

1.2.6 Remaining Pre-transition Metals

None of the remaining pre-transition metals (Group III) is considered to be an essential element. Neither aluminum or scandium are essential or stimulatory. Poor absorption of these metals has also precluded evidence for serious toxicity. However, interest in aluminum toxicity has greatly increased since the symptoms of aluminum intoxication have been reported to be similar to the progressive senile dementia observed in Alzheimer's disease.

1.2.7 The Transition Metals

The transition metals are so named because these elements have d orbitals that are increasingly filled with electrons, leading to a gradual transition in metallic properties. They are also especially effective in forming stable complexes with nitrogen, oxygen, and sulfur, all of which are basic constituents of proteins. Iron occupies a unique position among transition metals in the vertebrates as the metal of choice in hemoglobin and other heme proteins (Calvin, 1969).

Except for titanium, all the first-level transition metals are considered essential, including V, Cr, Mn, Fe, and Co. Several of the neighboring posttransition metals—Cu, Zn, and Ni—are important trace elements. The abundance of several of these elements in the early oceans, particularly iron, zinc, and molybdenum, has stimulated speculation about the origins and the evolution of the dependence of organisms on the transition metals (Egami, 1975; Ferris, 1973; Frieden, 1974; McClendon, 1976).

Titanium inhibits numerous hydrolytic enzymes, and as titanate, can be stimulatory, but there is no evidence that it is required. Zirconium is relatively nontoxic and inactive; 5 ppm in the drinking water of rats did not affect their health or longevity. Of the heavier transition elements, only molybdenum has been proved essential. The remaining transition metals are poorly absorbed in cells and are relatively nontoxic. The metal ion of tungsten has several properties of interest. There is no evidence for its essentiality but tungsten has been found to stimulate growth in plants. It is frequently antagonistic to molybdenum and specifically decreases the activity of several oxidative enzymes that require molybdenum (e.g., xanthine oxidase). Tungsten has been reported to be a biologically active metal as a component of the enzyme formate dehydrogenase in several bacteria (Ljungdahl, 1976).

1.2.8 The Post-transition Metals

This group of metals has filled d subshells when they give up electrons. This gives them extremely high reactivities and complex forming abilities. The most prominent essential trace elements in this group are Zn and Cu, which are coenzymes for a large number of crucial enzymes. Probable essential ultratrace elements are Ni, Cd, Sn, and Pb, which are discussed in detail in Chapters 12, 15, 13, and 16.

Several of the other post-transition metals have played a significant role in the history of our civilization but are still not considered essential. This includes gold, mercury, platinum, and silver. These elements owe their toxicity to their affinity for sulfhydryl groups in proteins. Medical interest in gold has centered around its therapeutic use in arthritis. Early literature suggests a stimulatory role for gold. Silver competes effectively with copper and selenium in many biological reactions. For example, Ag^+ displaces Cu^{2+} in the biosynthesis of ceruloplasmin, the copper transport protein of vertebrate plasma, which is also involved in iron mobilization. Mercury occurs widely and its useful properties in society were recognized centuries ago. The inorganic form is readily metabolized to more permeable and toxic forms, monomethyl- and dimethyl-mercury. Combination of these various forms of mercury effect disastrous pathological changes in nerve and brain tissue. In all these studies, there has been no evidence to suggest an essential role for this element, but it can be stimulatory in certain biological systems.

1.2.9 Lanthanides and Actinides

Lanthanides and actinides are two groups of heavy elements that are so radioactive and toxic as to be precluded from serving a useful part in any biochemical system. Moreover, the actinides are found only in extremely small quantities. The lanthanides are more common and metallic than usually appreciated. Normally, neither the lanthanides nor actinides occur in animal or plant tissues. Only traces of lanthanides (<0.5 ppm) are detected in the bones of animals exposed to them. There has been some interest in lanthanum as a "super calcium" in calcium-sensitive *in vitro* systems. It is not anticipated that further studies will provide evidence for any essential role by either of these series of elements.

1.3 Mechanism of Action of the Essential Ultratrace Elements

The essential ultratrace elements are universally required for survival. They normally occur and function in cells at extremely low concentrations, usually far less than one micromolar and as low as 10^{-9} M. Our understanding of the biological events that link an ultratrace element to its specific vital function(s) is still fragmentary. Obviously, the amplification machinery of the organisms—enzymes, carrier proteins, hormones, key structural sites—is involved. No single pattern will encompass what is known to date about these metal ions. The dominant theme has been that trace elements are essential because they serve as required coenzymes for irreplaceable metal-ion-activated enzymes or metalloenzymes. It has been estimated that one-fourth to one-third of all known enzymes involves a metal ion as a required participant.

In the metalloenzymes, a fixed number of specific metal atoms (usually Fe, Zn, Cu, Mn, Mo, Co, Se, Ni, etc.) are firmly associated with a specific protein. This combination produces a unique enzyme with a unique catalytic function. There are a number of variations on this theme. The metalloenzyme, superoxide dismutase, is isolated from mammalian cytoplasm with two atoms of copper and zinc per molecule. To retain catalytic activity, the zinc ion can be replaced by almost any transition metal ion, but no active replacement for copper is possible. Vallee (1980) has described how cobalt and various other metal ions can substitute for native zinc atoms in several zinc enzymes (e.g., carboxypeptidase, alkaline phosphatase) with retention of enzyme activity.

Cobalt functions in a complicated system that is not yet understood. Cobalt is an essential component of vitamin B_{12}, the coenzyme for methyl transferase that is necessary for thymidine synthesis and, ultimately, DNA biosynthesis and the transcription process itself. Evolution of iron enzymes and vital iron proteins has taken several major paths, one of which involves prophyrin groups and one of which involves constituent amino acids (cysteinc, histidine, tyrosine) of specific proteins. At one time the effects of selenium deficiency were thought to be due primarily to an absence of adequate amounts of glutathione peroxidase leading to excessive hemolysis. More recently, interconversions of methionine and cysteine have been related to this element (Shamberger, 1983).

The two essential nonmetals, iodine and fluorine, have different histories. Iodine is needed for the biosynthesis of thyroid hormones, which in turn greatly affect development and metabolism in all vertebrates. The primary role of fluoride may be in the structural integrity of mineralized tissues. Fluoride's anticaries effect is not regarded as implying the existence of a fundamental fluoride deficiency disease.

As the elements become more ultratrace, establishing deficiency syndromes and identifying their possible coenzymic functions become increasingly difficult. One important advantage is that the metal ion provides a natural label of that part of the protein that is directly involved in the reaction, the active site of the enzyme. The unique spectroscopic properties of metalloenzymes have also been especially valuable in describing events at the active site. The availability and use of a convenient radioactive isotope of the metal becomes almost mandatory. Refinement of atomic absorption analytical instrumentation is another asset in addressing this problem. Certainly the elucidation of the mode of action of the essential ultratrace elements will be a major challenge for future research in biochemistry.

1.4 Summary

The chemical elements have been surveyed for their essentiality or potential essentiality. The difference between essential and stimulatory elements is defined and explained. How the essential elements, both nonmetal and metal, have evolved was also considered. Among the nonmetals only bromine is a possible candidate for essentiality. Numerous additional metals of the pretransition, transition, and posttransition series may quality as essential. Those metals justifying further discussion as to possible essentiality include aluminum, barium, bismuth, germanium, lithium, rubidium, strontium, titanium, and tungsten. Chapter 1 concludes with a brief reflection on the mechanism of action of the essential ultratrace elements.

General References

Bowen, H. J. M., 1979. *Environmental Chemistry of the Elements*, Academic Press, New York.
Calvin, M., 1969. *Chemical Evolution*, Clarendon Press, Oxford, England.
Choppin, G. R., and Johnsen, R. H., 1972. *Introductory Chemistry*, Addison-Wesley, Reading, Mass.
Frieden, E., 1972. The chemical elements of life, *Sci. Am.* 227:52–62.
Luckey, T. D., and Venugopal, B., 1977. *Metal Toxicity in Mammals*, Vol. 1, Plenum Press, New York.
Mertz, W., 1981. The essential trace elements, *Science* 213:1332–1338.
Underwood, E. J., 1977. *Trace Elements in Human and Animal Nutrition*, 4th ed, Academic Press, New York.
Williams, R. J. P., and Da Silva, J. R. R. F., (eds.) 1978. *New Trends in Bio-inorganic Chemistry*, Academic Press, New York.

Specific References

Anke, M., Grün, M., Groppel, B., and Kronemann, H., 1981. The biological importance of lithium, in *Mengen- und Spurenelemente*, Arbeitstagung, Leipzig, pp. 217–239.
Bertrand, G., 1912. *Proc. Int. Congr. Appl. Chem.*, 8th Vol. 28:30.
Bezkorovainy, A., 1980. *Biochemistry of Nonheme Iron*, Plenum Press, New York.
Birch, N. J., 1978. Lithium in medicine, in *New Trends in Bio-inorganic Chemistry*, Williams and Da Silva (eds.), Academic Press, New York, pp. 389–436.
Cade, J. F. J., 1949. Lithium salts in the treatment of psychotic excitement, *Med. J. Aust.* 36:349–352.
Cotzias, G. C., 1967. *Proc. First Annual Conf. Trace Subst. in Environmental Health*, D. H. Hemphill (ed.), Columbia, Mo., pp. 5–19.
Da Silva, J. R. R. F., 1978. Interaction of the chemical elements with biological systems, in *New Trends in Bio-inorganic Chemistry*, Williams and Da Silva (eds.), Academic Press, New York, pp. 449–484.
Egami, F., 1974. Minor elements and evolution. *J. Mol. Evol.* 4:113–120.
Egami, F., 1975. Origin and early evolution of transition element enzymes, *J. Biochem.* 77:1165–1169.
Ferris, J. P., and Usher, D. A., 1973. *Origins of life in Biochemistry*, G. Zubay (ed.), Addison-Wesley, Reading, Mass., pp. 1191–1241.
Frieden, E., 1974. The evolution of metals as essential elements in *Advances in Experimental Medicine and Biology*, Vol. 48, *Protein-Metal Interactions*, M. Friedman (ed.), pp. 1–32.
Ljungdahl, L. G., 1976. Tungsten, a biologically active metal, *Trends Biochem. Sci.* 1:63–64.
Lombeck, I., Kasperek, K., Feinendegen, L. E., and Bremer, H. J., 1980. Rubidium—a possible essential trace element, *Biol. Trace Element Res.* 2:193–198.
McClendon, J. H., 1976. Elemental abundance as a factor in the origins of mineral nutrient requirements, *J. Mol. Evol.* 8:175–195.
Rygh, O., 1949. Trace elements importance of strontium, barium and zinc in the animal organism, *Bull. Soc. Chim. Biol.* 31:1052.
Schwarz, K., 1977. Essentiality vs. toxicity of metals, in *Clinical Chemistry and Chemical Toxicology of Metals*, S. S. Brown (ed.), Elsevier/N.H., New York, pp. 3–22.
Shamberger, R., 1983. *The Biochemistry of Selenium*, Plenum Press, New York.
Steenbjerg, F., 1951. Yield curves and chemical plant analysis. *Plant and Soil* 3:97–109.
Valkovic, V. 1978. *Trace Substances in Environmental Health*, Vol. XII, D. D. Hemphill (ed.), University of Missouri, Columbia, pp. 75–88.
Vallee, B., 1980. Zinc and other active site metals as probes of local conformation and function of enzymes, *Carlsberg Res. Com.* 45:423–441.

2
The Discovery of the Essential Trace Elements: An Outline of the History of Biological Trace Element Research*

Gerhard N. Schrauzer

2.1 Introduction

After decades of intense research activity on *organic* nutrients, enzymes, vitamins, and hormones, investigators from different disciplines in recent years have directed their attention increasingly to the long neglected *inorganic* nutrients, and in particular the trace elements. Recent extraordinary developments have created a need for a concise account of the major advancements in this field for specialists as well as for members of neighboring disciplines. The present monograph is intended to fill this need. Responding to the invitation of the editor to contribute an introductory chapter to this volume, I have prepared a brief account of the history of biological trace element research, concentrating on individual discoveries of the essentiality of elements for animals and man. It is dedicated to the memory of three men who deserve prominent positions in the Hall of Fame of biological trace element research: to Jules Raulin (1836–1896), for his contribution to the development of the concept of "essentiality"; to Klaus Schwarz (1914–1978), uniquely successful discoverer of new essential trace elements; and to Eric J. Underwood (1905–1980), researcher, author, and until his death one of the unifying philosophical leaders in the field (Carles, 1972; Schrauzer, 1979, 1980).

Gerhard N. Schrauzer • University of California at San Diego, La Jolla, California 92093.

2.2 Classification of the Bioelements

All living things are composed to about 99% of 12 common elements from the first 20 of the periodic table. These are the "bulk" or "constituent" elements, whose occurrence in living matter is dictated primarily by their high abundance on the earth's crust, in the oceans, and in the atmosphere. However, all organisms in addition contain small amounts of heavier elements, which may be somewhat arbitrarily subdivided into trace and ultramicro trace elements. In all, 81 of the 92 naturally occurring elements were detected in the human body (Vohora, 1982). Of these, 15 are presently established as essential (Fe, I, Cu, Zn, Mn, Co, Cr, Mo, Ni, V, Se, As, F, Si, Li) and at least four others are serious candidates of essentiality (Cd, Pb, Sn, Rb). In Figure 1, the 12 bulk elements, the essential,

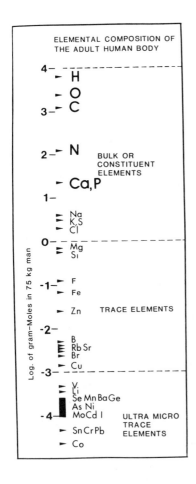

Figure 2-1. Elemental composition of the human body on a logarithmic scale.

and some nonessential trace and ultramicro trace elements, are arranged as a function of their molar abundance in the human body. Most of the unlisted remaining trace elements are considered nonessential. They are present in the body chiefly as a result of environmental exposure. However, future research could prove that some of these elements have physiological functions as well.

2.3 The Concept of Essentiality

Although biological trace element research could be said to have started in the late seventeenth century, when the importance of iron in the treatment of anemias was recognized, or at least during the first half of the eighteenth century, when iodine deficiency was associated with goiter, these empirical discoveries did not contribute to the further development of the field.

Systematic biological trace element research began with J. Raulin's discovery of the essentiality of zinc for the growth of the common mold *Aspergillus niger* in 1869 (Raulin, 1870). Raulin, a pupil of Pasteur and later a professor of chemistry at Lyon, France, demonstrated that *Asp. niger* could grow only if zinc was added to the culture medium, at amounts so small that it could be overlooked easily or considered as a mere contaminant. Raulin demonstrated that the requirement for zinc was *absolute,* that zinc was irreplaceable and thus essential.

The importance of the concept of essentiality was immediately recognized by the plant physiologists of the period. Employing sand culture and hydroponic growth techniques, they in turn began to establish the full trace mineral requirements for all plants. This task was completed only comparatively recently (see, e.g., Arnon, 1958). While bacteriologists also recognized the importance of trace elements for the growth of microorganisms in the years immediately following Raulin, human and animal nutritionists, with the exception of a few outsiders, seemed to have taken little notice of these developments. In consequence, iron and iodine remained the only two trace elements considered as essential for animals and man until well into the twentieth century.

During the twentieth century, two major periods of activity in biological trace element research may be discerned: a first, or "classical," period from 1925 to 1956, and a second, "modern" period, from 1957 to the present. During the first period, almost all trace element discoveries were made more or less by accident, or in response to emergency situations such as localized outbreaks of unexplained diseases of livestock. During the second period, which was dominated by Klaus Schwarz, trace element research was for the first time beginning to be conducted systematically. Experimental methods were devised that enabled researchers to produce specific trace element deficiency states in laboratory animals by maintaining them on specially formulated synthetic or semisynthetic diets. This could not have been accomplished prior to the commercial availability of purified amino acids and the development of sensitive and precise analytical techniques, and, in retrospect, was undoubtedly responsible for the delayed development of the field.

2.4 Trace Element Discoveries from 1925 to 1956

2.4.1 Copper

As late as the early to mid-1920s a new trace element, copper, was suggested, on the basis of empirical evidence, to be of value in the diet of rats (Bodansky, 1921; McHargue, 1925, 1926). Copper deficiency was subsequently shown to inhibit hematopoiesis in the rat (Hart *et al.*, 1928) and in exclusively milk-fed human infants (Josephs, 1931). However, it was later discovered that copper is required for the formation of aortic elastin (O'Dell *et al.*, 1961), and thus is of crucial importance for heart functioning. Following these findings, chronic copper deficiency, or a relative copper deficiency induced by high zinc intakes, has been suggested to be a major etiological factor in human ischemic heart disease (Klevay, 1975). Copper-deficient laboratory animals have since been found to be hypercholesterolemic and hyperuricemic and to exhibit glucose intolerance and abnormalities of cardiac function. They also show abnormal connective tissues and lipid deposits in the arteries. Deficient animals may die suddenly with a ruptured heart, caused by thinning of the aortic wall. These findings have ominous significance in the light of recent copper estimates in typical human diets in the United States; 75% of the diets examined furnished less than 2 mg of copper per day, the amount thought to be required by adults (Klevay, 1982).

2.4.2 Manganese

Following the discovery of the essentiality of copper, Kemmerer and Todd, associates of Hart at the University of Wisconsin, Madison, succeeded in demonstrating the essentiality of manganese in 1931. Manganese deficiency can be induced in a variety of laboratory animal species. Specific deficiency syndromes are unknown in humans but occur spontaneously in pigs and poultry (Underwood, 1971). Newborn Mn-deficient animals typically exhibit retarded growth, skeletal deformities, ataxia, and convulsions. The skeletal abnormalities, caused by a defect in the formation of the organic bone matrix, are attributed to dysfunctions of two Mn-dependent enzymes, the polymerase effecting the conversion of uridine diphosphate (UDP)-*N*-acetylgalactosamine to the polysaccharide, and to a galactotransferase that brings about the transfer of trisaccharide to the protein moiety (Leach, 1967, 1971). Impairment of bone development and growth in Mn-deficient animals also affects the skull; improper otolith development is believed to be the major cause of neonatal ataxia (Erway *et al.*, 1966).

2.4.3 Zinc

Three years after the discovery of the essentiality of manganese, the Wisconsin group demonstrated that zinc was also an essential element for mammals (Todd, Elvehjem, and Hart, 1934). However, their discovery elicited embarrassingly little interest. It took another 20 years until zinc deficiency was shown to be responsible for parakeratosis in swine (Tucker and Salmon, 1955) and for Laennec's cirrhosis, a type of cirrhosis formerly thought to be due to alcoholism (Vallee *et al.*, 1956, 1957). And even though beneficial effects of zinc containing unguents were known to physicians in the nineteenth century, the accelerating effects of zinc on wound healing were objectively described only less than 20 years ago (Pories *et al.*, 1967).

Zinc is required by nearly 100 different mammalian enzymes. It is essential for the maintenance of growth, development, cell division, protein, and DNA synthesis. In spite of its obviously central importance, its roles in human and animal nutrition are only now beginning to be recognized. Nutritional zinc deficiency caused by malnutrition, or conditioned deficiency induced by special dietary habits such as the consumption of unleavened bread, causes dwarfism and hypogonadism in boys in Iran and Egypt (Prasad *et al.*, 1961, 1963). Marginal zinc deficiency produces a variety of mild or vague symptoms, including inappetence, smell and taste dysfunctions. These have since been shown to occur in a surprisingly high percentage of American adults (Henkin, 1982). At the moment, marginal zinc deficiency is not considered clinically significant; nor is there, to this author's knowledge, as yet much concern about possible long-term effects of marginal zinc deficiency in man.

2.4.4 Cobalt

Prior to World War II, Australian workers associated a lack of cobalt in the pastures with a condition in cattle known as "wasting" or "coast disease" (Underwood and Filmer, 1935; Marston, 1935; Lines, 1935). While mildly affected animals show no symptoms other than a nonspecific unthriftiness, it is fatal in the acute form, leading to rapid wasting, anemia, and death. The disease can be prevented by the addition of traces of inorganic cobalt salts to the feedstock or pastures but was eventually recognized to be caused by vitamin B_{12} rather than by cobalt deficiency (Smith *et al.*, 1951). The discovery of the presence of one atom of tightly bound cobalt in the molecule of vitamin B_{12} added new dimensions to trace element research, especially after vitamin B_{12} coenzyme was shown to be a compound with a direct cobalt–carbon bond (Lenhert and Crowfoot-Hodgkin, 1961). The corrinoid coenzyme is one of the most efficient biocatalysts known, effecting unusual molecular rearrangements in a number of enzymes.

2.4.5 Molybdenum

Evidence for a physiological role for molybdenum in mammals was established after World War II through the discovery of the molybdenum-dependent enzyme, xanthine oxidase (Bray *et al.*, 1955). However, the nutritional molybdenum requirement of most species is so low that deficiency states cannot be readily generated in experimental animals. In chicks, a low-Mo state could eventually be induced by adding tungstate, a molybdenum antagonist, to molybdenum-deficient diets. The Mo-deficient chicks exhibited a diminished ability to oxidize xanthine to uric acid, as expected (Higgins *et al.*, 1956), as well as increased mortality. In humans, neither molybdenum deficiency nor toxicity presently give cause for concern. However, the element was already known in the 1930s to cause a scouring disease, "teart," in cattle. Animals grazing on Mo-rich pastures develop severe diarrhea and rough, discolored coats. Since the animals rapidly lose weight, the condition is fatal unless diagnosed in time. First observed in England (Ferguson *et al.*, 1938), teart was subsequently reported from cattle-raising areas in the United States, Ireland, and New Zealand. The discovery that the toxic effects of molybdenum could be counteracted by the administration of copper (Brouwer *et al.*, 1938) ultimately revealed that molybdenum causes a severe diminution of copper uptake, attributed to the formation of insoluble copper thiomolybdates in the digestive tract.

2.5 Discoveries from 1957 to 1978: The Era of Klaus Schwarz

2.5.1 Selenium

Long known and feared only as a severe poison, selenium was shown in 1957 to be the integral part of a powerfully antihepatonecrotic factor isolated from pork kidney, and thus became a nutritionally essential element (Schwarz and Foltz, 1957). While still working at what was then the Kaiser Wilhelm Institute of Medical Research at Heidelberg, Schwarz identified dietary liver necrosis as an independent disease entity, clearly different from cirrhosis of the liver, with which it often had been confused (Schwarz, 1944). The condition was known to occur in rats that were fed low-protein diets. Schwarz showed that sulfur amino acids and/or vitamin E provided some protection. However, after joining the NIH as Chief, Section of Experimental Diseases, National Institute of Arthritis and Metabolic Diseases in 1949, Schwarz discovered a powerfully antinecrotic third factor (Factor 3), first in alcohol extracts of casein and later in liver and in pork kidney (Schwarz, 1951). From 1 ton of the latter he was eventually able to obtain 1.3 mg of the highly concentrated fraction in which selenium was detected. Factor 3 could never be isolated in pure form. It presumably consisted of one or more organoselenium compounds, derived from selenomethionine and possibly other naturally occurring selenoamino acids and their oxidation products. Structural identification proved to be unnecessary, however, as sodium selenite and other inorganic and organic selenium compounds were found to possess sufficiently high antinecrogenic activities to render them useful as feed additives.

In 1957, exudative diathesis in chicks was shown to be a selenium deficiency disease (Patterson *et al.,* 1957; Schwarz *et al.,* 1957). After annually causing millions of dollars of damage to the poultry industry, exudative diathesis became readily preventable through the addition of 0.1 ppm of selenium to the feed. Identification of "white muscle disease," a muscular dystrophy in lambs and calves of New Zealand and Oregon, was similarly shown to be caused by selenium deficiency (Muth *et al.,* 1958; McLean *et al.,* 1959).

Selenium was shown in 1972 to be the functional component of the enzyme glutathione peroxidase (Rotruck *et al.,* 1972). This enzyme assures the maintenance of the structural integrity of liver and other cell membranes by protecting them from the destructive effects of oxygen radicals. Oxygen radicals are formed during lipid metabolism, but also on exposure of organs and tissues to ionizing radiation, as well as by certain drugs. Only a few of the many other protective functions of selenium, the fascinating aspects of its functional similarities with the tocopherols, or selenium-dependent bacterial enzymes can be mentioned here. Selenium is required for the maintenance of fertility, the functioning of the eye, the heart, and the immune system. It also exerts protective effects against certain heavy metals, notably mercury and cadmium. Recently, it has also been identified as a nutritional cancer-protecting agent, and for this function alone could be regarded as the most important of all trace elements (Schrauzer, 1979*b*). However, its value for human health was even further enhanced through recent reports from the People's Republic of China, where it was recognized that selenium deficiency was the cause of "Keshan disease," an endemic and often fatal cardiomyopathy affecting mainly young women and children. A massive selenium supplementation program was initiated in 1974 in all affected provinces.

By 1980 the disease was all but eliminated (Chen *et al.,* 1980). Further epidemiological studies in the low-selenium regions of China have since revealed that the Kaschin–Beck syndrome, an inflammatory, painful disease related to arthritis, which is locally known as "big joint disease," is also probably caused by selenium deficiency. In addition, however, molybdenum deficiency and/or NO_3^- or NO_2^- poisoning have been implicated (Zhu, 1980).

2.5.2 Chromium

The discovery of the essentiality of selenium led Schwarz to emphasize repeatedly that "toxicity is no counterargument against biological essentiality" (Schwarz, 1975, 1977). Correctly sensing that even the most toxic element could have a narrow "concentration window" within which it could be essential, Schwarz began to direct his efforts to the systematic search for new essentiality trace elements. In 1959, while still at NIH, traces of chromium were shown to be required for health as part of a still somewhat elusive "glucose tolerance factor" (Schwarz and Mertz, 1959). Apart from its role in glucose metabolism, chromium is presumably also involved in lipid and cholesterol metabolism.

2.5.3 Tin

During his work with chromium, Schwarz realized that for further trace element discoveries to be successful, radical modifications of the existing experimental techniques would be required. Accordingly, he formulated semisynthetic or synthetic amino acid diets of exactly known composition to minimize trace element contaminations. To eliminate other potential sources of contamination, the water had to be specially purified and was stored and dispensed in specially cleaned plastic containers. The air was passed through special filters and the then commonly used metal cages were replaced by plastic isolators (Smith and Schwarz, 1967). Now able to maintain animals under "trace-element sterile conditions" for extended periods of time, Schwarz entered the most productive period of his career. Working at his "Laboratory for Experimental Metabolic Diseases" at the VA Hospital, Long Beach, from 1963 to his death in 1978, he first concentrated on the element tin, which he felt had all the prerequisites of being an essential element. In 1970, Schwarz *et al.* reported the first positive growth responses of tin in young rats at levels from 0.5 to 2 ppm in the diet (Schwarz, Milne, and Vinyard, 1970). It is possible that tin, by virtue of its existence in the oxidation states of $+II$ and $+IV$ and its ability to form complexes with a variety of ligands, acts as an electron transfer catalyst *in vivo;* however, more work is required to establish its essentiality and nutritional roles.

2.5.4 Vanadium

Applying the isolator technique, Schwarz and Milne (1971) were able to observe definite growth effects in rats receiving diets to which 0.05–0.1 ppm vanadium in the form of sodium orthovanadate had been added. Vanadium in addition stimulated the pigmentation of incisors. Simultaneously, Hopkins and Mohr (1971) demonstrated that vanadium-deficient chicks and rats developed higher plasma cholesterol levels than the controls. Vanadium appears to act as a biocatalyst of oxidation of certain substrates. It is possible that this explains its cholesterol-lowering effect, which in humans was demonstrated already before the nutritional requirement of this element was known (Lewis, 1959). However, vanadium also appears to inhibit cholesterol biosynthesis *in vivo* (Curran, 1964).

2.5.5 Fluoride

While fluoride was long known to be a normal constituent of bone and to have anticariogenic properties, this in itself does not yet constitute proof of the nutritional essentiality of this element. To demonstrate a nutritional requirement of fluoride, Schwarz and Milne (1972a) first had to prepare an amino acid diet with a specially low fluoride content. This was eventually achieved, after the calcium phosphate was identified as the major source of fluoride and replaced by a specially synthesized phosphate. The addition of 1.5–2.0 ppm of F^- in the form of NaF to this diet produced a definite enhancement of longitudinal growth of young rats during the first 4 weeks of postnatal life. In other experiments, fluoride-deficient female mice developed diminished fertility and became anemic (Messer *et al.*, 1972a,b). As to the toxicity of fluoride, it is remarkable that diets containing even 300–500 ppm fluoride may be administered to laboratory animals for prolonged periods of time without causing apparent harm (Schwarz, 1975). Fluoride in the water, on the other hand, produces a characteristic mottling of the enamel of permanent teeth at concentrations as low as 1.5 ppm. At 1 ppm, the concentration at which fluoride is added to some drinking water supplies, mottling does not occur, while the anticariogenic effects are optimal.

2.5.6 Silicon

In 1972, Schwarz and Milne (1972b) provided definite evidence for a nutritional essentiality of silicon in the rat. Silicon, the second most abundant element in the lithosphere, is nevertheless a biological "trace" element, for only traces of it are needed and taken up by mammals. Schwarz's silicon-deficient rats showed impaired growth, a disturbance of bone formation that was particularly noticeable in the skull, and a diminution in the pigmentation of the incisors.

Ironically, silicon's essentiality was discovered in the same year by a worker at the same institution: At UCLA's Department of Public Health, E. M. Carlisle showed that silicon accumulates in the actively calcifying zone of bones of chicks; subsequently, Carlisle also demonstrated a nutritional requirement for this element (Carlisle, 1972). Schwarz, though also affiliated with UCLA but separated by a distance of 35 miles from Carlisle's laboratory, at the time had no knowledge of this work.

2.5.7 Nickel

Other researchers in the United States and elsewhere were now also beginning to contribute trace element discoveries. In 1975, a nutritional requirement for nickel was announced by three separate research groups (Nielsen and Ollerich, 1974; Anke *et al.*, 1974; Schnegg and Kirchgessner, 1975). In rats, inadequate dietary supply of nickel reduced growth and lowered the erythrocyte count and hematocrit and hemoglobin level in blood. Nickel appears to be required for proper iron utilization. However, a lack of nickel also impairs copper and zinc metabolism and lowers the activity of glucose-6-phosphate dehydrogenase and malate dehydrogenase in rat liver homogenates (Kirchgessner and Schnegg, 1976). 2,5,8 Lead, Cadmium, and Arseneic.

2.5.8 Lead, Cadmium, and Arsenic

In the mid-1970s, Schwarz and others directed their attention to the study of the nutritional roles of lead, cadmium, and arsenic, three elements that until then had always been regarded only as poisonous. Schwarz could still report initial evidence for the biological essentiality of all three of them (Schwarz, 1977), but his death prevented completion of these studies. Although not all criteria of essentiality of Cd and Pb have been met, his observations were independently confirmed and extended by Anke *et al.* (1977) and Kirchgessner and Reichlmayr-Lais (1981). With arsenic, positive growth responses in young rats were observed (Schwarz, 1977), but at that time convincing evidence for the essentiality of arsenic had already been produced by two other groups (Anke *et al.*, 1976; Nielsen *et al.*, 1975). Nielsen's arsenic-deficient rats exhibited rough fur, increased osmotic fragility of the erythrocytes, and abnormally enlarged spleens containing excessive amounts of iron. Anke's As-deficient goats and pigs showed decreased fertility, low birth rates, and retarded growth. Lactating arsenic-deficient goats were also observed to die suddenly with myocardial damage. The discovery of the essentiality of arsenic and of the possible essentialities of lead and cadmium should of course not detract from the established fact that higher concentrations of these elements pose definite health hazards.

2.6 Current Trends and Problems

Since Schwarz's death an increasing number of scientists from a variety of disciplines have become actively interested in trace elements, and a new generation of workers is beginning to make exciting contributions. Several new essential trace element discoveries have been announced in recent years. Among the alkali elements, the essentiality of lithium may now be considered as virtually certain (Anke *et al.*, 1981), while evidence for a physiological function of rubidium is accumulating (Lombeck *et al.*, 1980). Lithium-deficient goats exhibited 28% lower weight gains than controls during three months of observation. Female goats showed diminished fertility and higher postpartum mortalities. While no defined Rb-deficiency syndrome has been identified thus far, this element is a normal constituent of all tissues and body fluids.

At the present level of activity, the last remaining essential trace elements will probably be discovered within the next two decades. It will subsequently become necessary to investigate the health effects of all nonessential trace elements, since some of them may have important controlling functions by virtue of their interaction with other essential trace elements.

However, there are also interactions among the essential elements, giving rise to complex synergistic and antagonistic effects. The copper–zinc antagonism, which has been briefly mentioned in Section 2.4.1, represents but one example, and interactions between three or more elements are also possible. The molybdenum–copper antagonism, which was discussed in Section 2.4.5, provides one example of a ternary interaction, since the formation of copper thiomolybdates in the digestive system also depends on the sulfur content of the diet.

Between the 15 known essential elements, as many as 105 binary and 455 ternary interactions exist, of which only a few have been studied thus far. Will it be possible to study all of them? Perhaps; but this will require some planning effort on national and international levels, to prevent a possibly chaotic situation with much duplication of effort and squandering of the generally diminishing funds. The future of biological trace element research will depend on the willingness of its protagonists to communicate and collaborate with each other and their ability to convince governmental agencies and the public of the importance of this work, which it undoubtedly has, even though it may not be immediately obvious. This planning of the future and dissemination of information should begin immediately. Several national and international organizations exist that could provide the administrative framework for these efforts.

In recent years the exchange of scientific information between the workers in the field has also markedly improved. In July 1969, trace element researchers from all over the world gathered in Aberdeen, Scotland, for a first symposium, "Trace Element Metabolism in Animals" (TEMA 1). Since then, TEMA symposia have been held every four years. Their scope has also been expanded to include human trace elements ("TEMA" now stands for "Trace Elements in Man and Animals"). There also exists an International Association of Bioinorganic Scientists (IABS), founded by the author in 1975, with Klaus Schwarz serving as its first vice president. This nonprofit organization publishes the journal *Biological Trace Element Research* and holds regular conferences on selected topics (i.e., on "Inorganic and Nutritional Aspects of Cancer and Other Diseases"), of which three have been held thus far. The IABS also awards the "Klaus Schwarz Medal" to leading workers in biological trace element research; first recipients were Milton Scott (1980), E. J. Underwood (posthumously, 1981), E. Cecil Smith (1982) and Manfred Anke (1983). In 1984, two medals were awarded to the Keshan Disease Research Groups of the Chinese Academy of Medical Sciences, Beijing, and of Xian Medical College. Last but not least, the Society of Environmental Geochemistry and Health (SEGH) should be mentioned, whose aim is to provide a unifying forum for research workers in the health, environmental, and geochemical sciences.

Acknowledgment

Preparation of this review was supported by Grant CHE 79-50003 from the National Science Foundation.

REFERENCES

Anke, M., Grün, M., Dittrich, G., Groppel, B., and Hennig, A., 1974. Low nickel rations for growth and reproduction in pigs. In: *Trace Metabolism in Animals,* W. G. Hoeckstra, *et al.,* (eds.), University Park Press, Baltimore, pp. 716–718.

Anke, M., Grün, M., Groppel, B., and Kronemann, H., 1981. The biological importance of lithium, in Mengen- und Spurenelemente, Arbeitstagung Leipzig 1981, pp. 217–239.

Anke, M., Hennig, A., Groppel, B., Partschefeld, M., and Grün, M., 1978. The biochemical role of cadmium, in *Proceedings of the Third International Symposium on Trace Element Metabolism in Man and Animals,* M. Kirchgessner (ed.), Freising-Weihenstephan, Germany, pp. 540–548.

Anke, M., Hennig, A., Grün, M., Partschefeld, M., Groppel, B., and Lüdke, 1976. Arsen- ein neues essentielles Spurenelement, *Arch. Tierernähr.* 26:742–743.

Anke, M., Hennig, A., Grün, M., Patschefeld, M., Groppel, B., and Lüdke, H., 1977, Nickel- ein essentielles Spurenelement, *Arch. Tierernähr.* 27:25–38.

Arnon, D. I., 1958. The role of micronutrients in plant nutrition with special reference to photosynthesis and nitrogen assimilation, in *Trace Elements. Proceedings of the Conference Held at Ohio Agricultural Experimental Station,* Wooster, Ohio, Oct. 14–16, 1957, C. A. Lamb, O. G. Bentley, and J. M. Beattie (eds.), Academic, New York, pp. 1–32.

Bodansky, M., 1921. The zinc and copper content of the human brain, *J. Biol. Chem.* 48:361–364.

Bray, R. C., Avis, P. G., and Bergel, F., 1955. The chemistry of xanthine oxidase from cow milk, *Congr. Internatl. Biochem. Resumes communs.* 3eme *Congress, Brussels,* 30.

Brouwer, F., Frens, A. M., Reitsma, P., and Kalisvaast, C., 1938. *Verslag Landbouwk. Onderzoek.* 44C:267, quoted by E. J. Underwood, in *Trace Elements in Human and Animal Nutrition*, 3rd ed., Academic, New York, 1971, pp. 130–140.
Carles, J. B., 1972. Raulin, Jules, in *Dictionary of Scientific Biography*, Vol. XI, C. C. Gillespie (ed.), Scribner's, New York, pp. 310–311.
Carlisle, M., 1972, Silicon, an essential element for the chick, *Fed. Proc.* 31:700.
Chen, X., Yang, G., Chen, J., Chen, X., Wen, Zh., and Ge, K., 1980. Studies on the relations of selenium and Keshan disease, *Biol. Trace Element Res.* 2:91–107.
Curran, G. L., 1964. Effect of certain transition group elements on hepatic synthesis of cholesterol in the rat, *J. Biol. Chem.* 210:765–768.
DeRenzo, E. C. E., Kaleita, E., Heytler, P. G., Oleson, J. J., Hutchings, B. L., Williams, J. H., 1953. Identification of the xanthine oxidase factor as molybdenum, *Arch. Biochem. Biophys.* 45:247.
Erway, K., Hurley, L. S., and Fraser, A., 1966. Neurologic defect. Manganese in phenocopy and prevention of genetic abnormality in inner ear, *Science* 152:1766–1768.
Ferguson, W. S., Lewis, A. H., and Watson, S. J., 1938. Action of molybdenum in nutrition of milking cattle, *Nature (London)* 141:553.
Hart, E. B., Steenbock, H., Waddell, J., and Elvehjem, C. A., 1928. Iron in Nutrition. VII. Copper as a supplement to iron for hemoglobin building in the rat, *J. Biol. Chem.* 77:797–812.
Henkin, R. I., 1982. A New Approach to Zinc Deficiency, 1982, Abstracts, 3rd Int. Conf. Inorganic and Nutritional Aspects of Cancer and Other Diseases, La Jolla, 1982, pp. 2–3.
Higgins, E. S., Richert, D. A., and Westerfield, W. W., 1956. Molybdenum deficiency and tungstate inhibition studies, *J. Nutr.* 59:539–559.
Hopkins, L. L., Jr., and Mohr, H. W., 1971. Effect of vanadium deficiency on plasma cholesterol of chicks, *Fed. Proc.* 30:462.
Josephs, H. W., 1932. Studies on iron metabolism and the influence of copper, *J. Biol. Chem.* 96:559–571.
Josephs, H. W., 1931. Treatment of anemia of infancy with iron and copper, *Bull. Johns Hopkins Hosp.* 49:246–258.
Kemmerer, A. R., and Todd, W. R., 1931. The effect of diet on manganese content of milk, *J. Biol. Chem.* 94:317–321.
Kirchgessner, M., and Reichlmayr-Lais, A., 1981. Changes of iron concentration and iron binding capacity in serum resulting from alimentary lead deficiency, *Biol. Trace Element Res.* 3:279.
Klevay, L. M., 1975. Coronary heart disease: The zinc/copper hypothesis, *Am. J. Clin. Nutr.* 28:764–774.
Klevay, L. M., 1982. Copper and Ischemic Heart Disease. Abstracts, 3rd Int. Conf. Inorganic and Nutritional Aspects of Cancer and Other Diseases, La Jolla, 1982, p. 1.
Leach, R. M., 1967. Role of manganese in synthesis of mucopolysaccharides, *Fed. Proc.* 26:118–120.
Leach, R. M., 1971. Role of manganese in mucopolysaccharide metabolism, *Fed. Proc.* 30:991–994.
Lenhert, P. G., and Crowfoot-Hodgkin, D., 1961. Structure of the 5,6-dimethylbenzimidazolyl-cobamide coenzyme, *Nature (London)* 192:937–938.
Lewis, C. E., 1959. The biological actions of vanadium. I. Effects upon serum cholesterol levels in man, *AMA Arch. Ind. Health* 19:419–425.
Lines, E. W., 1935. *J. Council Sci. Ind. Res.* 8:117.
Lombeck, I., Kasperek, K., Feinendegen, L. E., and Bremer, H. J., 1980. Rubidium—a possible essential trace element, *Biol. Trace Element Res.* 2:193–198.
Marston, H. R., 1935. *J. Council Sci. Ind. Res.* 8:111, quoted by Underwood, E. J., in *Trace Elements in Human and Animal Nutrition*, 3rd ed., Academic, New York, 1972, pp. 141–168.
McHargue, J. S., 1925. The association of copper with substances containing the fat soluble A vitamin, *Am. J. Physiol.* 72:583–594.

McHargue, J. S., 1926. Further evidence that small quantities of copper, manganese and zinc are factors in the metabolism of animals, *Am. J. Physiol.* 77:245–255.

McLean, J. W., Thompson, G. G., and Claxon, J. H., 1959. Growth responses to selenium in lambs, *Nature (London)* 184:251.

Messer, H. H., Armstrong, W. D., and Singer, L., 1972a. Fertility impairment in mice following low fluoride intake, *Science* 177:893–894.

Messer, H. H., Wong, K., Wegner, M., *et al.*, 1972b. Effect of reduced fluoride intake of mice on hematocrit values, *Nature (London), New Biol.* 240:218–220.

Muth, O. H., Oldfield, J. E., Rennert, L. F., and Schubert, J. R., 1958, Effects of selenium and vitamin E on white muscle disease, *Science* 128:1090.

Nielsen, F. H., Givad, S. H., and Myron, D. R., 1975. Evidence for a possible requirement for arsenic by the rat, *Fed. Proc.* 34:923–925.

Nielsen, F. H., Ollerich, D. A., 1974. Nickel: A new trace element, *Fed. Proc.* 33:1767–1769.

O'Dell, B. L., Hardwick, B. C., Reynolds, G., and Savage, J. E., 1961. Connective tissue defect in the chick resulting from copper deficiency, *Proc. Soc. Exp. Biol. Med.* 108:402–404.

Patterson, E. L., Milstrey, R., and Stokstad, E. L. R., 1957. Effect of selenium in preventing exudative diathesis in chicks, *Proc. Soc. Exp. Biol. Med.* 95:617–620.

Pories, W. J., Henzel, J. H., Rob, C. G., and Strain, W. H., 1967. Acceleration of wound healing, *Lancet* 1:121–124.

Prasad, A. S., Halsted, J. A., and Nadimi, M., 1961. Syndrome of iron deficiency anemia, hepatosplenomegaly, dwarfism, hypogonadism, and geophagia, *Am. J. Med.* 31:532–546.

Prasad, A. S., Miale, A., Farid, Z., Schulert, A., and Sandstead, H. H., 1963. Zinc metabolism in normals and patients with the syndrome of iron deficiency, anemia, hypogonadism and dwarfism, *J. Lab. Clin. Med.* 61:537–549.

Raulin, J., 1870. Études chimiques sur la végetation, *Ann. Sci. Nat. Bot.*, 5th ser. 51:93–299.

Rotruck, J. T., Hoekstra, W. G., Pope, A. L., Ganther, H., Swanson, A. B., and Hafeman, D. G., 1972. Relation of selenium to GSH peroxidase, *Fed. Proc.* 31:691 (Abstract 2684).

Schnegg, A., and Kirchgessner, M., 1975. Zur Essentialität von Nickel für das tierische Wachstum, *Ztschr. Tierphysiol., Tierernaehrung u. Futtermittelkunde* 36:63–74.

Schrauzer, G. N., 1979a. Klaus Schwarz, 1914–1978, Commemoration of a leader in trace element research, in *Trace Metals in Health and Disease*, N. Kharash (ed.), Raven Press, New York, pp. 251–261.

Schrauzer, G.N., 1979b. Trace elements in carcinogenesis, *Advances in Nutritional Research*, Vol. 2, H. Draper (ed.), Plenum, New York, pp. 219–244 and references therein.

Schrauzer, G. N., 1980. In Memoriam Eric John Underwood, *Biol. Trace Elements Res.* 2:231–234.

Schwarz, K., 1944, Über einen ernaehrungsbedingten, tödlichen Leberschaden und seine Verhütung durch Leberschutzstoffe. *Z. Physiolog. Chem.* 281:101–108.

Schwarz, K., 1951. Production of dietary necrotic liver degeneration using American Torula yeast, *Proc. Soc. Exp. Biol. Med.* 80:319–323.

Schwarz, K., 1975, Neuere Erkenntnisse über den essentiellen Charakter einiger Spurenelemente, in *Spurenelemente in der Entwicklung von Mensch und Tier*, K. Betke, and F. Bindlingmaier (eds.), Urban und Schwarzenberg, München, pp. 1–30.

Schwarz, K., 1977, Essentiality versus toxicity of metals, in *Clinical Chemistry and Chemical Toxicology of Metals*, S. S. Brown (ed.), Elsevier/North Holland Biomedical Press, New York, Amsterdam, pp. 3–22.

Schwarz, K., and Foltz, C. M., 1957. Selenium as an integral part of factor 3 against necrotic liver degeneration, *J. Am. Chem. Soc.* 79:3292.

Schwarz, K., Bieri, J. G., Briggs, G. M., and Scott, M. L., 1957. Prevention of exudative diathesis in chicks by factor 3 selenium, *Proc. Soc. Exp. Biol. Med.* 95:621–625.

Schwarz, K., and Mertz, W., 1959. Chromium(III) and the glucose tolerance factor, *Arch. Biochem. Biophys.* 85:292–295.

Schwarz, K., and Milne, D. B., 1971, Growth effects of vanadium in the rat, *Science 174:425*.

Schwarz, K., and Milne, D. B., 1972a, Fluorine requirement for growth in the rat, *Bioinorg. Chem.* 1:331–358.

Schwarz, K., and Milne, D. B., 1972b. Growth promoting effects of silicon in rats. *Nature (London)* 239:333–334.

Schwarz, K., Milne, D. B., and Vinyard, E., 1970. Growth effects of tin compounds in rats maintained in a trace element-controlled environment, *Biochem. Biophys. Res. Commun.* 40:22–29.

Smith, J. C., and Schwarz, K., 1967. A controlled environment system for new trace element deficiencies, *J. Nutr.* 93:182–188.

Smith, S. E., Koch, B. A., and Turk, K. L., 1951. The response of cobalt-deficient lambs to liver extract and vitamin B_{12}, *J. Nutr.* 44:455–459.

Todd, W. R., Elvehjem, C. A., and Hart, E. B., 1934. Zinc in the nutrition of the rat, *Am. J. Physiol.* 107:146–156.

Tucker, H. F., and Salmon, W. D., 1955. Parakeratosis or zinc deficiency disease in the pig, *Proc. Soc. Exp. Biol. Med.* 88:613.

Vallee, B. L., Wacker, W. E. C., Bartholomay, A. F., and Robin, E. D., 1956. Zinc metabolism in hepatic dysfunction. I. Serum zinc concentrations in Laennec's cirrhosis and their validation by sequential analysis, *New Engl. J. Med.* 255:403–408.

Vallee, B. L., Wacker, W. E. C., Bartholomay, A. F., and Hoch, F. L., 1957. Zinc metabolism in hepatic dysfunction. II. Correlation of metabolic patterns with biochemical findings, *New Engl. J. Med.* 257:1055–1065.

Underwood, E. J., and Filmer, J. F., 1935. *Aust. Vet. J.* 11:84–88; sec Underwood, E. J., in *Trace Elements in Human and Animal Nutrition*, 3rd ed., Academic, New York, 1971, pp. 123–156.

Vohora, S. B., 1982. *Earth Elements and Man*, Department of Philosophy of Medicine, Inst. of Hist. of Medicine and Medical Research, Hamdard Nagar, New Delhi-110062, India.

Zhu, Mei-Nuan, 1980. An exploration of the chemical mechanism for Kashin–Beck's disease and environmental factors—multivariate functional relationship of molybdenum with biogeochemical environment, *Huan Ching K'o Hsueh* 1(3):31–37 *Chem. Abstr.* 21:53876 (1980).

Iodine 3

Nicholas M. Alexander

3.1 Introduction

Iodine was discovered in 1812 by Courtois in Paris while preparing saltpeter for the manufacture of gunpowder during the Napoleonic wars. Aqueous extracts of kelp mixed with hot sulfuric acid in a copper retort yielded a "vapor of a superb violet color that condensed in the form of a brilliant crystalline plate" (Parkes, 1967; Pitt-Rivers, 1978). In the following year Gay-Lussac established the elementary nature of iodine and named it from the Greek ιοεδής (violet).

Although iodine is widely distributed it only represents 10^{-5}% of the earth's crust and is the sixty-fourth most abundant element (Ahrens, 1965; Encyclopedia Brittanica, 1978). It exists naturally as iodate and iodide salts, or covalently bound to tyrosine and thyronine (thyroid hormones), but it is not found as elemental iodine (I_2). It is present in kelp (0.1–0.3%), seawater (0.05 ppm), sea plants and animals, some land plants and animals (thyroid gland), cod liver oil, mineral springs, and salt brine from some petroleum wells and is combined with silver and lead ores. Most of the commercial iodine is obtained from caliche, a nitrate rock primarily found in Chile that contains about 0.2% iodine as iodate salts. The halogen is also prepared from seaweed and salt brine from oil wells (Parkes, 1967).

Iodine is consumed in the United States at the rate of about 8 million pounds per year (Layman, 1982) and is used for food additives, dyes, X-ray contrast media, catalysts, sanitizers, photographic film, water treatment, and pharmaceuticals. It has been postulated that early inhabitants of Asia and Europe recognized the therapeutic value of ingesting seaweed or burnt sponge to reduce goiters, although this has been disputed (Pitt-Rivers, 1978; Werner, 1978).

Nicholas M. Alexander • Division of Clinical Pathology, Department of Pathology, University of California School of Medicine, San Diego, California 92103.

Table 3-1. Properties of Iodine

Atomic number	53
Atomic weight	126.9
Melting point	113°
Boiling point	184°
Vapor pressure	0.03 mm (0°C)
	3.08 mm (55°C)
Specific gravity	4.93
Color of gas	Violet
Solubility (g/100 ml solvent)	0.029 (water 20°)
	20.5 (ethanol 15°)

Source: Data compiled from Parkes, 1967, and *Handbook of Chemistry and Physics,* 44th ed., The Chemical Rubber Publishing Co., Cleveland.

3.2 Chemistry of Iodine and the Iodoamino Acids

3.2.1 Properties of Iodine

Some properties of elemental iodine are listed in Table 3-1. At ambient temperatures it is a dark, blue-black, crystalline solid that slowly sublimes, as seen by its low vapor pressure at 0° and 55°. It forms brown aqueous solutions and is more soluble in alcohol than in water; it is violet when dissolved in chloroform, carbon disulfide, benzene, and other hydrocarbons. The brown color is due to chemical combination of iodine with solvent, whereas no combination with solvent occurs in violet solutions (Sneed *et al.,* 1961). The addition of starch to brown iodine solutions turns them blue because I_2 is adsorbed by starch, presumably by complexing with the interior of the helical polysaccharide molecule. This starch-iodine color has been advantageously exploited as an indicator in iodometric titrations. The solubility of I_2 in water is considerably enhanced by the presence of I^-, which forms a brown I_3^- complex that is similar in its chemical reactivity to I_2.

3.2.2 Iodine Isotopes

More than 20 isotopes of iodine have been described (Barnes *et al.,* 1978). Some of them and their properties are listed in Table 3-2. ^{131}I is the most widely used radioisotope in medicine, primarily for diagnosis and therapy of thyroid diseases, while ^{125}I has been extensively exploited in biological and medical research for radioimmunoassays, metabolic studies, and structure–function investigations. Some short-lived isotopes (^{121}I, ^{123}I, ^{130}I, ^{132}I) may be useful in the future. ^{128}I was the first isotope employed in biology, but its extremely short

Table 3-2. Useful Isotopes of Iodine in Biology and Medicine

Mass number	Half-life	Mode of decay
121	1.8 hours	β^+, e^-
123	13 hours	γ, e^-, EC[a]
125	56 days	γ, EC
127	Stable	None
128	25 minutes	β,γ
129	1.7×10^6 years	β,γ
130	12.5 hours	β,γ
131	8 days	β,γ
132	2.4 hours	β,γ

[a] EC refers to electron, or K, capture.
Source: Data compiled from the *Handbook of Chemistry and Physics*, 44th ed., The Chemical Rubber Publishing Co., Cleveland.

half-life has restricted its usefulness. ^{129}I is a useful standard because of its very long half-life, and it provides excellent analytical sensitivity after its conversion to ^{130}I by neutron activation.

3.2.3 Important Chemical and Biochemical Reactions of Iodine

Redox and Substitution Reactions

An important reaction that was widely utilized in laboratory medicine is the iodide-catalyzed ceric–arsenite redox reaction (Sandell and Kolthoff, 1937). Ce^{4+} is reduced to Ce^{3+} and arsenite is oxidized to arsenate while I^- and I_2 undergo cyclic interconversion (Figure 3-1). The reaction is proportional to the iodide concentration and is sufficiently sensitive to detect and quantitate nanogram quantities of iodide. It was extensively used to analyze thyroid hormones in blood prior to radioimmunoassay (RIA) methodology. It is still useful for analyzing iodoamino acids that are resolved by ion-exchange chromatography (Rolland et al., 1970; Sorimachi and Ui, 1974), since the iodoamino acids react directly with ceric–arsenite (Barker, 1971).

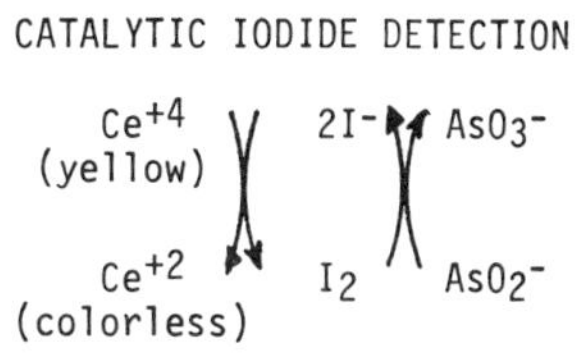

Figure 3-1. Cerate–arsenite reaction catalyzed by iodine (Sandell–Kolthoff reaction).

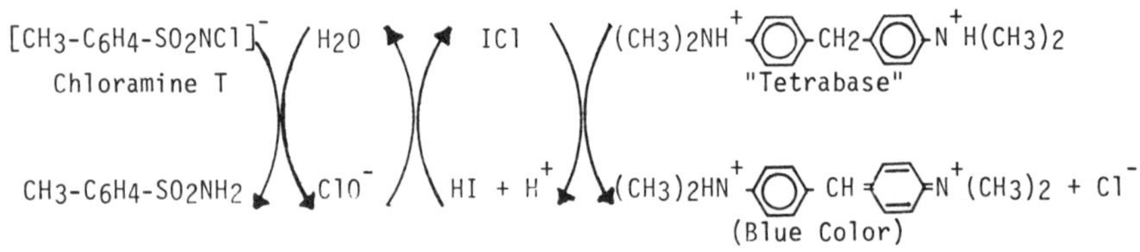

$$[CH_3\text{-}C_6H_4\text{-}SO_2NCl]^- \quad H_2O \quad ICl \quad (CH_3)_2NH^+\text{-}\bigcirc\text{-}CH_2\text{-}\bigcirc\text{-}N^+H(CH_3)_2$$
$$\text{Chloramine T} \qquad\qquad\qquad\qquad\qquad\qquad\qquad \text{"Tetrabase"}$$
$$CH_3\text{-}C_6H_4\text{-}SO_2NH_2 \quad Cl^- \quad HI + H^+ \quad (CH_3)_2HN^+\text{=}\bigcirc\text{=}CH\text{-}\bigcirc\text{=}N^+(CH_3)_2 + Cl^-$$
$$\text{(Blue Color)}$$

Figure 3-2. "Tetrabase" reaction catalyzed by iodine. Iodine catalyzes the oxidation of 4,4′-bis(dimethylamino)diphenylmethane by chloramine-T.

Another useful catalytic iodine reaction is the so-called tetrabase reaction (Feigl and Anger, 1972). Iodide catalyzes the oxidation of "tetrabase" (4,4′-bis(dimethylamino)-diphenylmethane) by chloramine-T to a blue quinoidal compound (Figure 3-2). It has been used with high-performance liquid chromatography to measure nanogram quantities of iodoamino acids (Lankmayer et al., 1981). Iodine must be released from the iodoamino acids with zinc, because they do not react directly in the tetrabase reaction. Parenthetically, it is notable that iodide (0.1–1.0 μM) can be determined with an ion-selective electrode.

Aromatic iodinations are electrophilic substitution reactions, and some evidence suggests that I^+ is an iodinating intermediate in uncatalyzed reactions, but only minute amounts of I^+ are generated in these equilibria (Berliner, 1966; Mayberry, 1972):

$$I_2 \rightleftharpoons I^+ + I^- \qquad (K = 10^{-20})$$

$$I_2 + H_2O \rightleftharpoons H_2OI^+ + I^- \qquad (K = 10^{-11})$$

H_2OI^+ represents protonated hypoiodous acid and is not hydrated iodinium cation.

The oxidation potential of iodine in N-iodosuccinimide and ICl is very similar to I^+, since these positive halogen compounds oxidatively cleave tyrosyl peptide bonds and transform phloretic acid to a dienone lactone, whereas neither I_2 nor I_3^- are effective (Junek et al., 1969; Alexander, 1973, 1974).

Peroxidase-Catalyzed Iodination

Thyroid peroxidase oxidizes iodide with H_2O_2 to "active iodine" for iodotyrosine and iodothyronine synthesis (Alexander, 1959, 1961, 1976, 1980). It is significant that "active iodine" is similar in oxidation potential to ICl and N-iodosuccinimide (Alexander, 1973, 1974a, 1974b, 1976). Extrathyroidal peroxidases catalyze iodination of tyrosine, but thyroid hormone synthesis does not occur in these tissues because thyroglobulin is absent. It is of interest that leucocyte peroxidase with H_2O_2 and iodide effectively kills bacteria, viruses, and fungi (Klebanoff, 1967; Belding *et al.*, 1970).

Antiseptic Iodine Solutions

Tincture of iodine (I_2 dissolved in alcohol) is utilized as an antiseptic, as are other water-soluble complexes of iodine (iodophors) that release free iodine in solution. One such example is povidone-iodine ("Betadine"), a complex of iodine and polyvinylpyrrolidone that reacts with a wide variety of biological molecules (Alexander, 1981, 1983).

Radioiodination of Biological Molecules

Radioiodine labeling of biological molecules for metabolic, structure–function, and radioimmunoassay studies has practically become routine in biochemical research laboratories. A popular method is to oxidize radioiodide with excess chloramine-T to ICl (Figure 3-2), which then iodinates tyrosine residues in peptides and proteins (Greenwood *et al.*, 1963; McConahey and Dixon, 1980). Histidine is iodinated at a slower rate than tyrosine.

To minimize potential oxidation of essential groups in proteins (e.g., cysteine, tryptophan, methionine) that occurs with chloramine-T (Alexander, 1973, 1974a), peroxidase-catalyzed radioiodinations have been employed (Morrison, 1980). In addition, a water-insoluble, stable, mild iodinating reagent, 1,3,4,6-tetrachloro-3α, 6α-diphenylglycoluril ("iodogen") has been developed for the same reason (Fraker and Speck, 1978). Iodogen is easily separated from the iodinated product and it provides the capability for maximum radioiodine incorporation.

Nucleic acids have been labeled *in vitro* with radioiodide and thallic trichloride (Commerford, 1980). Cytosine residues are converted to 5-iodocytosine, a modification that does not significantly alter the biological properties of the nucleic acids.

3.2.4 Iodotyrosines and Iodothyronines

Iodine is an essential constituent of the iodothyronines and iodotyrosines whose structures are shown in Figure 3-3. Thyroxine (T_4) was isolated from thyroid hydrolysates in 1914 (Kendall, 1919) and its correct structure was determined by chemical degradation and synthesis (Harington and Barger, 1927). The other iodoamino acids were detected in the thyroid gland and plasma some years later (Pitt-Rivers, 1978).

Iodoamino acids have been isolated and measured by various chromato-

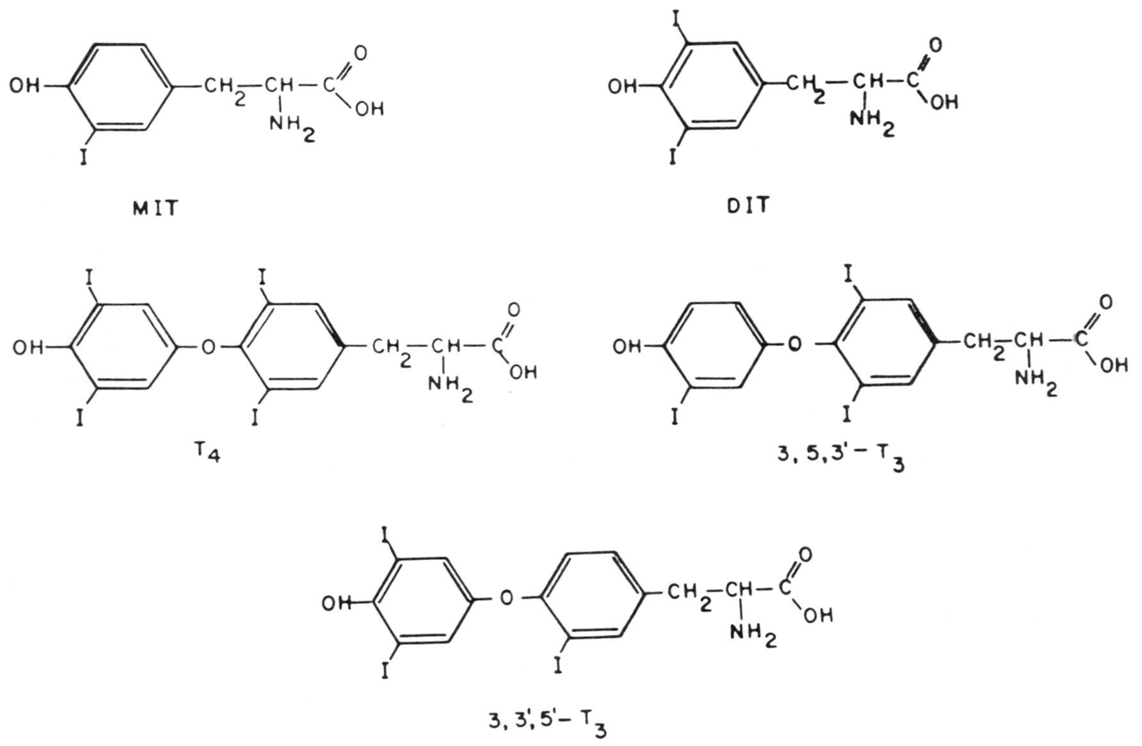

Figure 3-3. Chemical structures of the iodotyrosines and iodothyronines. MIT is 3-monoiodotyrosine and is also abbreviated as ITyr. DIT is 3,5-diiodotyrosine and is also abbreviated as I_2Tyr. T_4 is thyroxine, or 3,5,3',5'-tetraiodothyronine, and 3,3',5'-triiodothyronine is reverse T_3 and abbreviated as rT_3. T_3 refers to 3,5,3'-triiodothyronine.

graphic techniques (Cahnmann, 1972), and more recently, high-performance liquid chromatographic methods have been particularly useful (Hearn et al., 1978; Nachtmann et al., 1978; Alexander and Nishimoto, 1979; Alexander, 1984), especially for resolving the enantiomeric isomers of both T_4 and T_3 (Lankmayr et al., 1981; Hay et al., 1981). All the naturally occurring iodoamino acids are L(S)-enantiomers.

But the most important assay methods for T_3 and T_4 are competitive radioassay procedures that accurately and precisely (CV 5–8%) detect picogram quantities of hormones (Larsen, 1978; Chopra and Crandall, 1975, 1978; Alexander and Jennings, 1974b, 1974c, 1976). It is indeed impressive that radioimmunoassay (RIA) methodology can detect T_4 in only 3 µl of dried whole blood, a procedure used for screening newborn infants for congenital hypothyroidism. Sensitive fluoro- and enzyme-immunoassays for T_4 have also been developed recently that utilize stable reagents and obviate the requirement for radioisotopes.

Iodotyrosines and iodothyronines in thyroglobulin (unhydrolyzed) may be conveniently analyzed by ultraviolet spectrophotometry (Covelli et al., 1971). This procedure exploits pH dependent changes in the spectra and absorptivities of each of the iodoamino acids, since dissociation of the phenolic group is significantly enhanced by iodine substitutions in the ortho positions. However, the sensitivity of this method is several orders of magnitude less than RIA.

3.3 Iodine Metabolism

3.3.1 Iodine Absorption, Evolution, and Thyroid Hormone Biosynthesis

Iodine, the heaviest element in living organisms, was first identified as a constituent of thyroid glands in 1896 by Baumann (Pitt-Rivers, 1978). This trace element is efficiently concentrated by thyroid glands in fish, amphibians, reptiles, birds, and mammals for the unique purpose of synthesizing thyroid hormones. The tremendous avidity of the gland for iodine is readily apparent from the fact that a 25-g thyroid gland contains more than 80% of the total body iodine in an adult human being. The thyroid normally contains about 10 mg iodine, nearly all of which is covalently linked to thyroglobulin, a large glycoprotein (MW 660,000) that serves as the substrate for thyroid hormone synthesis.

It has been speculated that the evolution of thyroid hormones resulted from the ingestion of iodinated proteins by primitive chordates, and that natural selection favored development of the iodoamino acids into hormonal signals. Iodoproteins containing MIT and DIT are found in many invertebrates (Gorbman, 1978; Roche and Michel, 1951), and indeed, thyroxine synthesis has been reported in the jellyfish Aurelia aurita (Spangenberg, 1974). Spontaneous, or peroxidase-catalyzed iodination of tyrosine presumably occurred in the early evolutionary marine environment.* Iodide would be favored over the other halogens, because of its more favorable oxidation potential, although chloro- and bromo-peroxidases have been discovered in living organisms.

Ingested inorganic iodide is efficiently absorbed from the gastrointestinal tract into the bloodstream and is actively transported into the thyroid cell. Other tissues (gastric mucosa, salivary glands, mammary glands, ovaries, placenta, and skin) also concentrate iodine, but major competition for the halogen comes from the kidney, and urinary excretion of iodide normally exceeds 40 µg/day. Dietary iodate (used as a stabilizer in bread) is reduced to iodide by glutathione (Taurog et al., 1966). Thyroid hormones are absorbed unchanged and thus T_4 medication can be administered orally. Free MIT and DIT are deiodinated by tissue deiodinases.

Iodide ion is oxidized by thyroid peroxidase to "active iodine,"** which iodinates tyrosyl residues in thyroglobulin, a large glycoprotein (MW 660,000) that serves as the substrate for iodination and coupling of peptide-linked iodotyrosines to T_4 and T_3 (Figure 3-4). The coupling reaction is also catalyzed by thyroid peroxidase (Taurog, 1978; Alexander, 1980) and results in the conversion of an alanine side chain to dehydroalanine (Gavaret et al., 1980). Thyroglobulin consists of two polypeptide subunits (MW 330,000) that undergo posttranslational glycosylation and iodination prior to secretion into the lumen of the thyroid follicle (Figure 3-4). The iodoprotein is resorbed into the thyroid cell as colloid droplets that are proteolytically hydrolyzed to release free thyroid hormones for

---

* Primitive forms of life originated in a marine environment that contained all the halogens (Cl—2%, and F, Br, and I 1.3, 6.7, and 0.06 parts per million, respectively). Thus the Swedish chemist J. J. Berzelius (1779–1848) named them halogens from the Greek αλs (sea salt) and γεννάω (I produce). Iodine is more readily oxidized than the other elements, as seen from their standard oxidizing potentials ($X_2 + 2e^- \rightleftharpoons 2X^-$): iodine, +0.54 V; bromine, +1.09 V; chlorine, +1.36 V; and fluorine, +2.85 V (Parkes, 1967), and in fact, fluorine is the most electronegative element known. Astatine (atomic number = 85) is the heaviest element in the group, but it does not exist in nature and was discovered after bombarding bismuth with helium in the cyclotron (Sneed et al., 1954).

** "Active iodine" may be one of three possibilities: (1) molecular iodine (I_2); (2) peroxidase-bound iodinium (E-I^+), possibly as sulfenyl iodide (E-SI); and (3) an enzyme-bound iodine free radical (E-I). Several lines of evidence do not support I_2 (Taurog, 1978; Nunez, 1980) and a free radical would appear to be too unspecific. Hence iodinium appears to be the most likely possibility (Alexander, 1973, 1974a,b, 1976). It should be noted that excess iodide will convert all these possible intermediates to I_3^-, which serves as the basis for a spectrophotometric assay for peroxidase catalyzed oxidation of iodide (Alexander, 1962). A recent analysis of existing data suggests that peroxidase catalyzed iodination of tyrosine with H_2O and iodide involves enzyme-activated HOI (H. B. Dunford and I. M. Ralston, 1983, Biochem. Biophys. Res. Comm. 116:639.

The source of H_2O_2 in the thyroid is not definitively established, but it probably results from pyridine nucleotide oxidation (Taurog, 1978). Superoxide anion does not appear to be an intermediate for peroxide production (Alexander, unpublished).

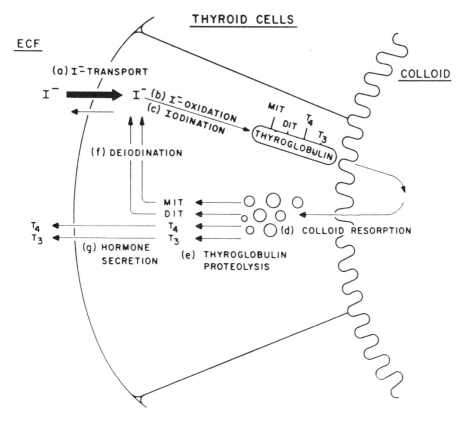

Figure 3-4. Iodine metabolism in the thyroid gland (from Tong, 1971, with permission). ECF represents the extracellular fluid, or plasma. Inorganic iodide is (a) actively transported into the thyroid cell, (b) oxidized by H_2O_2 and thyroid peroxidase to "active iodine," which (c) iodinates tyrosyl residues in thyroglobulin. Peroxidase also catalyzes the coupling of DIT to T_4 and MIT with DIT to T_3. Thyroglobulin is secreted into the lumen of the thyroid follicle and stored as colloid. Glycosylation and iodination are posttranslational events. Iodination occurs intracellularly or in the lumen at the surface of the membrane. Stored thyroglobulin is (d) resorbed into the cell as colloid droplets, and (e) proteolytically hydrolyzed to yield free iodotyrosines and iodothyronines. MIT and DIT are (f) deiodinated to tyrosine and I^- and recycled. T_4 and T_3 are (g) secreted into the plasma for distribution to the peripheral tissues.

secretion into the circulation and distribution to the peripheral tissues. Iodotyrosines released by proteolysis are deiodinated to inorganic iodide and tyrosine for recycling and incorporation into thyroglobulin.

One mole of thyroglobulin containing 1% iodine (52 atoms) consists of 10–12 residues each of MIT and DIT, 3–4 residues of T_4, and less than one residue of T_3. The percent iodine varies and is a function of iodine in the diet.

About 70% of the dry weight of a thyroid gland is thyroglobulin, and it normally stores enough iodine for about three months supply of thyroid hormones.

In man a minimum daily intake of 80 μg iodide is essential, since the daily turnover rates for T_4 and T_3 are 90 and 30 μg, respectively (Nicoloff, 1978; Chopra, 1981). T_4 contains 65.4% iodine and has a biological half-life of 7 days, compared to one day for T_3 (58.5% iodine). Thus 10% of the extrathyroidal T_4 pool (900 μg) is metabolized daily, while 75% of the T_3 pool (40 μg) disappears. Consequently, a daily intake of at least 200 μg iodine has been recommended to ensure adequate hormone synthesis. In the United States it is estimated that three times this amount is consumed as iodized table salt that contains 1 part iodide per 10,000 parts salt (Oddie *et al.*, 1970).

Thyroid-stimulating hormone (TSH) (MW 300,000-glycoprotein with two subunits) from the pituitary gland stimulates thyroid metabolism (via cyclic AMP) and the synthesis of thyroid hormones. TSH release from the pituitary is inhibited (feedback control) by thyroid hormones and its release is stimulated (neural control) by thyrotropin-releasing hormone (TRH), a tripeptide (pyroglutamyl-histidyl-prolinamide) that is synthesized in the hypothalamus and transported to the pituitary through special blood vessels. TRH secretion from the hypothalamus is in turn inhibited by T_3; thus these interacting factors of the thyroid-pituitary-hypothalamus axis maintain homeostasis with regard to the thyroid hormones.

3.3.2 Thyroid Hormones in Blood

The thyroid hormones are transported in blood almost entirely (>99.7%) bound to plasma proteins. In man the normal concentration of plasma T_4 ranges from 45 to 120 μg/L and for T_3 it is about 0.65–2.2 μg/L. Approximately 75% of T_4 and T_3 is noncovalently bound to thyroxine-binding globulin (TBG), 15% to prealbumin, and the remainder to albumin, whereas there are considerable variations in plasma-binding proteins in other animals, as, for example, in chickens, which do not contain TBG.

The plasma inorganic iodide concentration is 2–10 μg/L with normal daily intake of iodide (200–500 μg). Iodotyrosines are usually absent, or present only in trace quantities (Refetoff, 1979), but they are found in abnormal serum iodoproteins that are secreted by some thyroid neoplasms, and in the general circulation of newborn rats (Dratman, 1978). In addition, MIT covalently linked to serum albumin has been observed after topical application of povidone-iodine (Alexander, 1981).*

---

* Other compounds in serum of extrathyroidal origin (deiodination of T_4) having little, or no, biological activity include 3,3'-diiodothyronine (2–13 ng/dl), rT_3 (25–80 ng/dl), and 3,5,3',5'-tetraiodothyroacetic acid (80–160 ng/dl).

The hormone-binding proteins are synthesized in the liver, with females having slightly higher levels of TBG (and T_4) than males. TBG binds with an affinity constant of 10^{10} M^{-1} for T_4 and 10^9 M^{-1} for T_3, while prealbumin and albumin, respectively, bind the hormones 100 and 100,000 times less effectively. As a result of this tight binding only 0.03% of plasma T_4 and 0.3% T_3 are free. Thus protein-bound iodine was used as a diagnostic test for thyroid function. Organic solvents extract T_3 and T_4 from plasma proteins and butanol extractable iodine was also utilized as a thyroid function test. However, these procedures are subject to serious interferences from other nonhormone-iodinated compounds, such as radioopaque X-ray dyes.

Thyroid hormones in blood are now accurately and precisely (CV 5–8%) determined by sensitive and highly specific radioimmunoassays (RIA). Indeed, thyroid hormone measurements are the most frequently performed endocrine tests because newborn children in many countries are screened for congenital hypothyroidism, a disease that occurs in one of every 4000 births and leads to mental retardation if not treated with T_4. Other thyroid function tests have been developed to measure free hormone concentrations in serum to account for alterations in TBG levels. Although T_4, rT_3, and trace quantities of T_3 are found in amniotic fluid during pregnancy (Chopra and Crandall, 1975), they do not reliably predict neonatal thyroid status (Hollingsworth and Alexander, 1983).*

3.3.3 Thyroid Hormones in Peripheral Tissues

The thyroid gland primarily secretes T_4, which rapidly equilibrates with liver and kidney and slowly with other tissues, such as muscle, skin, and so on. It is believed that T_4 serves as a prohormone and is converted to the more active T_3 by a 5′-deiodinase enzyme in the peripheral tissues (T_3 has five times more biological activity than T_4). Because alkyl analogs of T_3 and T_4 are not metabolized and have similar biological activities, the prohormone concept has been challenged (Frieden, 1981).

In addition, T_4 is converted by deiodinase tissue enzymes to reverse T_3 (rT_3), an inactive metabolite. It is estimated that about 40% of the peripheral T_4 is metabolized to T_3 (primarily in liver, kidney, and gut), and an equal amount is converted to rT_3. The remaining (20%) T_4 is mostly metabolized to glucuronide and sulfate conjugates for excretion into urine and feces. T_3 and rT_3 are further deiodinated to all the possible iodothyronine isomers, inorganic iodide, and thyronine.

Most data support the concept that only free T_4 and T_3 can enter cells.

---

* Milk and cerebrospinal fluid also contain small amounts of T_3 and T_4, but determination of the hormones in these fluids is of no diagnostic value (Refetoff, 1979).

Inasmuch as the small fraction of free hormones is in dynamic equilibria with the bound iodothyronines, plasma binding proteins serve as "buffers" to regulate variations in the availability of hormones from the thyroid gland by maintaining free hormone concentrations within normal physiological limits. The extrathyroidal plasma and tissue hormone reservoirs contain 10 days supply of T_4 and 1.5 days supply of T_3.

3.3.4 Iodine Deficiency

Insufficient iodine in the diet is the most common cause of hypothyroidism (myxedema). Although iodized table salt is a convenient mechanism for providing sufficient dietary iodine, endemic goiter resulting from iodine deficiency is still a major public health problem in underdeveloped countries of South America, Asia, and Africa. The intramuscular injection of long-acting iodized poppy seed oil has been utilized successfully as a temporary substitute for iodized salt in some of these goiter areas (Pretell *et al.*, 1969; Stanbury *et al.*, 1974).

Hypothyroidism may develop even with an adequate intake of iodine because of inhibition of hormone synthesis or diminished hormone secretion from the thyroid. Hormone synthesis may be inhibited because of rare genetic thyroid enzyme deficiencies (primary hypothyroidism), and much less frequently because of pituitary (TSH) insufficiency (secondary hypothyroidism), or hypothalamic (TRH) insufficiency (tertiary hypothyroidism). In addition, inhibition of thyroid function occurs after exposure to antithyroid compounds (Green, 1978). These compounds (1) inhibit intrathyroidal iodine metabolism or (2) inhibit peripheral thyroid hormone disposal and metabolism. Thiocyanate, perchlorate ion, thionamides, aniline derivatives, phenols, 3-amino-1,2,4-triazole (a herbicide used in cranberry farming) (Alexander, 1959b) prevent the synthesis of thyroid hormones. Some goitrogens occur in edible plants, such as cabbage, rutabaga, and so on (Underwood, 1977; Prasad, 1978). Lithium used to treat psychiatric patients inhibits hormone release from the gland, and other drugs interfere with peripheral hormone metabolism.

Some of the thionamide drugs (especially propylthiouracil and methimazole) are effective therapeutic agents for treating hyperthyroidism, while perchlorate and thiocyanate are used to diagnose iodide transport defects in the gland. Excess iodide ingestion leads to a temporary block of hormone synthesis (Wolff-Chaikoff effect) from which the gland "escapes," but chronic iodide ingestion occasionally causes hypothyroidism. Serum TSH levels are markedly increased in primary hypothyroidism because the pituitary is responding to the low levels of circulating thyroid hormones.

The clinical symptoms of hypothyroidism include apathy, fatigue, slow heart rate, dry skin, menorrhagia, and intolerance to cold because insufficient quantities of hormones are available to the peripheral tissues. The opposite symptoms are seen with hyperthyroidism, an autoimmune disease with a strong genetic predisposition. Patients with hyperthyroidism usually have thyroid stimulating immunoglobulins circulating in the blood plasma.

Autoimmune thyroiditis (Hashimoto's thyroiditis) is the most common cause of goiter and hypothyroidism in children over 6 years of age in North America. This disease appears to be an autoimmune phenomenon because these patients have circulating antibodies directed against thyroid microsomes and thyroglobulin.

3.3.5 Iodine Toxicity

Although iodine is essential for the synthesis of thyroid hormones, the halogen also displays a pharmacological activity toward the thyroid gland that is dependent on dose and duration of iodine consumption. As mentioned earlier, a large dose of iodide temporarily inhibits organic iodine formation (Wolff-Chaikoff effect), possibly by binding the active form of iodine as I_3^-, but the gland adapts to excess iodine by decreasing its transport into the cell. Nevertheless, iodide-induced hypothyroidism occasionally occurs when large quantities (several milligrams daily) of iodine are consumed for an extended period of time.

It seems paradoxical that hyperthyroidism is induced by the administration of iodine, but a marked increase in the incidence of hyperthyroidism was observed after the consumption of iodine-rich bread (stabilized with iodate) in Tasmania (Braverman, 1978). Most of the people affected by this type of iodide-induced thyrotoxicosis appeared to have underlying Graves' disease (hyperthyroidism), although it occurred in normal individuals too. It is also notable that even though 20 mg of iodine are consumed in some parts of Japan (high seaweed diets), the incidence of goiter, or hypothyroidism, is low. Thus although excess iodine ingestion may result in either hypo- or hyperthyroidism, iodide-induced thyroid disease is rare.

Another form of iodine toxicity is radioiodine ingestion after a nuclear power plant accident or from nuclear explosion fallout. To block uptake of radioactive iodine by the thyroid gland in such cases, it has been recommended that potassium iodide ($K^{127}I$) be made available for distribution to the exposed population (von Hippel, Wolfe, and La Cheen, 1982), but this view has been challenged with respect to the potential exposure hazard following a nuclear accident (Yalow, 1982).

3.4 Mechanism of Action of Thyroid Hormones

3.4.1 General and Cellular Effects

Thyroid hormones are not considered essential for life, but the severe manifestations resulting from a deficiency of hormones (cretinism, stunted growth, myxedema coma, altered brain electrical activity, failure of tadpole metamorphosis) or from an excess of hormones (thyroid storm that is potentially fatal and exophthalmos) are indicative of the profound effects on the health and well-being of an organism. Thus cells require a regulated supply of hormones to maintain normal cellular function.

There have been many investigations *in vivo* and *in vitro* attempting to find a unified mechanism of action of thyroid hormones, and the reader is referred to several reviews that address this question (Barker, 1971; Jorgensen, 1978a,b; Dratman, 1978; Oppenheimer, 1979; Sterling, 1979; Frieden, 1981a,b). Thyroid hormones influence the basal metabolic rate (BMR), heart rate, body weight, amphibian metamorphosis, induction of enzyme synthesis such as α-glycerolphosphate dehydrogenase, and serum levels of cholesterol and tyrosine. Consequently, these biological responses have been exploited to assess the biological activity of the thyroid hormones and hormone analogs. By measuring oxygen consumption of tissues after the injection of thyroxine into thyroidectomized rats, it was found that heart, gastric mucosa, liver, kidney, diaphragm, pancreas, salivary gland, and skeletal muscle responded, whereas spleen, brain, and testes did not respond (Barker, 1971). In addition, it was noted that there was a 6- to 24-hour latent period before the response was observed following administration of thyroxine. Anuran metamorphosis and disappearance of the tadpole tail have also been an important *in vivo* assay for hormonal response.

In addition to these effects in intact animals, several *in vitro* tests have been utilized for measuring thyroid hormone activity. These include binding affinity to plasma transport proteins and cellular components, cell growth of rat pituitary cells, amino acid uptake by rat thymocytes, (Na^+-K^+)-ATPase activity in liver plasma membranes, and binding by lymphocytes. All the subcellular components (cytosol, mitochondria, nucleoli, plasma membrane, and nuclei) bind T_3 and T_4, but most of the attention has been directed at the limited-capacity, high-affinity nonhistone protein (MW 48,000–70,000) in the nuclei (Oppenheimer, 1979; Samuels and Tsai, 1973), which appears to be the nuclear receptor for initiating the genetic expression by thyroid hormones.

It is of interest that thyroxine stimulates early strobilation (conversion of sessile hydroid individuals into swimming individuals) in jellyfish (Spangenberg, 1974). This is the only example of a physiological role for thyroid hormones in an invertebrate species.

Another interesting biological effect of thyroid hormones is their possible relationship to the neoplastic transformation of cells. It was observed that mouse embryo fibroblasts are completely resistant to the neoplastic transforming action of X-rays if they are incubated in media depleted of thyroid hormone for 1 week (Guernsey et al., 1981). It was concluded that thyroid hormone induced the synthesis of a protein that was essential for the X-ray-mediated transformation.

3.4.2 Structure–Activity Relationships

From extensive studies with thyroxine analogs, X-ray crystallography, NMR spectroscopy, and theoretical molecular orbital calculations, the essential chemical and conformational features of the thyroid hormones have been elucidated (Jorgensen, 1978a,b; Cody, 1980). In Figure 3-5 a three-dimensional representation of the T_4 molecule is depicted. It consists of a lipophilic core containing two mutually perpendicular benzene rings (designated as α and β) that are connected by an oxygen bridging atom at an angle of about 120°. The diphenyl ether oxygen atom may be replaced by sulfur or carbon and still retain activity. The alanine side chain may be replaced with a minimum of a two-carbon atom chain containing anionic groups (acetic, propionic). The side chain and the outer ring (β) may lie on the same side (cisoid) or on opposite sides (transoid) of the inner ring (α) plane. The nonpolar groups at positions 3,5 and 3′,5′ on the benzene rings are occupied by iodine in the native molecule, but they may be replaced by other nonpolar alkyl substituents (methyl, isopropyl), and indeed, 3,5-dimethyl-3′ isopropylthyronine possesses significant thyromimetic activity. The activity is enhanced by a halogen at 3′ (distal) as in T_3, and reduced when

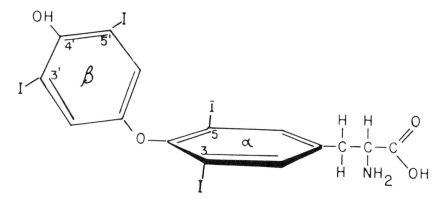

Figure 3-5. Molecular conformation of the thyroxine molecule (from Barker, 1971, with permission). The α-benzene ring is perpendicular to the page while the β-ring is in the plane of the page linked at an angle of 120° in ether linkage with an oxygen bridging atom.

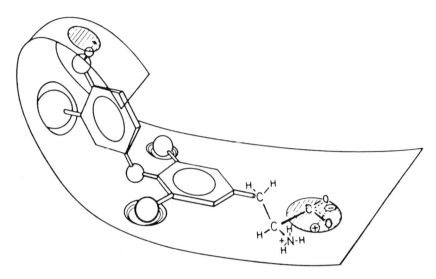

Figure 3-6. A pictorial representation of T_3 binding to the nuclear receptor (from Jorgensen, 1978b, with permission).

5' is occupied by halogen or alkyl substitution. Lack of occupancy at 5 position (α ring) has a profound diminution on activity as seen in rT_3, a deiodinated metabolite of T_4. A 4'-hydroxyl, or functionally similar group, para to the bridging oxygen atom is essential.

It is generally believed that the initiation of thyroid hormone action is the binding of T_3 to a nonhistone receptor protein in the nucleus of thyroid hormone-responsive cells that promotes a cascade of biochemical responses. A pictorial representation of T_3 binding by the nuclear receptor is shown in Figure 3-6, in which the alanine side chain forms an ion-pair bond with a positively charged region on the receptor, and the aromatic rings and their 3,5,3' substituents form hydrophobic associations with the receptor. The 4'-hydroxyl group contributes a hydrogen bond at another receptor site. Steric restraints limit the size of the 5'-substituent so that T_4 is bound less well than T_3.

This image of the nuclear receptor is supported by several observations, including the fact that T_3 is bound 10 times more effectively than T_4, and nuclear binding correlates quite well with thyroid hormone-responsive tissues (Oppenheimer, 1979). However, some results are not consistent with the nuclear receptor hypothesis, since it has been observed that nuclei in hormone-responsive tissues bind inactive D-T_3 as well as L-T_3, and the affinity for L-3,5-dimethyl-3'-isopropylthyronine by the nuclear receptor is very low, even though this alkylthyronine possesses significant thyromimetic activity (Frieden, 1981a).

Other extranuclear mechanisms of thyroid hormone action have been suggested, for example, mitochondrial activation, induction of synthesis of the (Na^+-K^+)-ATPase pump, tyrosine metabolic pathways, adrenergic receptor pathway, plasma membrane effects, and a combination of these with nuclear receptor–T_3 interaction (Sterling, 1979; Dratman, 1978). Of particular interest is the observation that the administration of physiological concentrations of T_3 produced significant increases in cGMP in the liver and tail fins of premetamorphic bullfrog tadpoles (Sidlowski and Frieden, 1982). Additional proposals of thyroid hormone action have included heavy-atom iodine perturbation effects in exciting electrons into a triplet state, electron donating character of iodine, and formation of charge transfer complexes between iodothyronines and other biological molecules. But cogent arguments based on data with noniodine-containing thyromimetic analogs do not support these possibilities (Frieden, 1981a, 1981b). Thus thyroid hormone binding to the nuclear receptor as the initiating event in hormone action appears to be the most attractive possibility at this time, but a final conclusion on the unifying mechanism of action of thyroid hormones at the molecular level awaits further research.

3.5 Summary

Iodine is a trace element that is essential for the synthesis of thyroid hormones in vertebrates, although iodoproteins are present in invertebrates. Several radioactive isotopes of iodine are available but ^{125}I and ^{131}I have been the most important in medical and biological research. Iodine deficiency is still the major cause of hypothyroidism in the world, while hyperthyroidism and some forms of hypothyroidism are primarily due to autoimmune reactions.

The thyroid gland contains more than 80% of the total body iodine and has the unique capability of synthesizing T_4 and T_3 in vertebrates. Iodide is actively transported into the thyroid cell and oxidized with H_2O_2 and thyroid peroxidase for incorporation into tyrosyl residues in thyroglobulin. The peroxidase also oxidatively catalyzes the coupling of peptide-linked iodotyrosines to iodothyronines. Thyroglobulin is stored as colloid in the lumen of thyroid follicles and is resorbed and proteolytically hydrolyzed to T_3 and T_4. The hormones are secreted into the plasma for transport to the peripheral tissues, where 80% of the peripheral T_3 is formed from T_4 deiodination. It appears that thyroxine is a prohormone for T_3, which has four to five times more activity than T_4. Thyroid hormone synthesis and secretion is controlled by an interacting thyroid–pituitary–hypothalamus axis.

Extensive physical and chemical studies have elucidated the structure–activity relationships of the thyroid hormones. The hormones consist of a lipophilic core with two mutually perpendicular benzene rings connected by a diphenyl ether oxygen atom at an angle of 120°. The 4'-hydroxyl group is essential for activity while the bulky iodine atoms serve to constrain the lipophilic core. Iodine atoms may be replaced by hydrophobic alkyl groups because halogen-free 3,5-dimethyl-3'-isopropylthyronine (DIMIT) possesses significant thyromimetic activity.

Cell nuclei contain a nonhistone, limited-capacity, high-affinity receptor for T_3 that is believed to initiate the biological response to thyroid hormones. Nuclear binding of T_3 correlates quite well with thyroid hormone-responsive tissues, although some data are inconsistent with the nuclear receptor hypothesis. For example, biologically inactive D-T_3 is bound as well as the active isomer L-T_3, and DIMIT has significant biological activity but is bound very poorly. In addition, high-affinity binders for thyroid hormones in other subcellular components (mitochondria, plasma membranes, microsomes, nucleoli) have been reported. A unifying mechanism of action has not been conclusively elucidated, although the nuclear receptor hypothesis seems most attractive.

References

Ahrens, L. H., 1965. *Distribution of the Elements in Our Planet*, McGraw-Hill, New York, p. 97.
Alexander, N. M., 1959a. Iodide peroxidase in rat thyroid and salivary glands and its inhibition by antithyroid compounds, *J. Biol. Chem.* 234:1530–1533.
Alexander, N. M., 1959b. Antithyroid action of 3-amino-1,2,4-triazole, *J. Biol. Chem.* 234:148–150.
Alexander, N. M., 1961. The mechanism of iodination reactions in thyroid glands, *Endocrinology* 68:671–679.
Alexander, N. M., 1962. A spectrophotometric assay for iodide oxidation by thyroid peroxidase, *Anal. Biochem.* 4:341–345.
Alexander, N. M., 1973. Oxidation and oxidative cleavage of tryptophanyl peptide bonds during iodination, *Biochem. Biophys. Res. Comm.* 54:614–621.
Alexander, N. M., 1974a. Oxidative cleavage of tryptophanyl peptide bonds during chemical- and peroxidase-catalyzed iodinations, *J. Biol. Chem.* 249:1946–1952.
Alexander, N. M., and Jennings, J. F., 1974b. Analysis for total serum thyroxine by equilibrium competitive protein binding on small, resuable Sephadex columns. *Clin. Chem.* 20:553–559.
Alexander, N. M., and Jennings, J. F., 1974c. Radioimmunoassay of serum triiodothyronine on small, reusable Sephadex columns. *Clin. Chem.* 20:1353–1361.
Alexander, N. M., 1976. Evidence for the oxidation of iodide to I^+ by H_2O_2 and peroxidase, in *Thyroid Research*, J. Robbins and L. E. Braverman (eds.), American Elsevier, New York, pp. 134–138.
Alexander, N. M., 1977. Purification of bovine thyroid peroxidase, *Endocrinology* 100:1610–1620.
Alexander, N. M., and Nishimoto, M., 1979. Rapid analysis for iodotyrosines and iodothyronines in thyroglobulin by 1979 reversed-phase liquid chromatography, *Clin. Chem.* 25:1957–1960.
Alexander, N. M., 1980. Thyroid peroxidase-catalyzed coupling of 3,5-diiodotyrosine (DIT) to thyroxine (T_4): DIT-peptide and activator requirements, in *Thyroid Research*, Vol. VIII, J. R. Stockigt and S. Nagataki (eds.), Australian Academy of Science, Canberra, pp. 117–120.
Alexander, N. M., and Nishimoto, M., 1981. Protein-linked iodotyrosines in serum after topical application of povidone-iodine (Betadine), *J. Clin. Endocr. Metab.* 53:105–108.

Alexander, N. M., 1983. Reaction of povidone-iodine with amino acids and other important biological compounds, in *Proc. Intl. Symposium on Povidone,* G. A. Digenis and J. Ansell (eds.), Univ. of Kentucky, Lexington, Kentucky, pp. 274–288.

Alexander, N. M., 1984. Analysis of iodothyronines and iodotyrosines in biological samples by higher performance liquid chromatography, in CRC *Handbook for the Use of HPLC for the Separation of Amino Acids, Peptides, and Proteins,* W. S. Hancock (ed.), CRC Press, Boca Raton, Fl., pp. 291–301.

Barker, S. B., 1971. Chemistry, cellular and subcellular effects of thyroid hormones, in *The Thyroid,* S. C. Werner and S. H. Ingbar (eds.), Harper & Row, New York, pp. 79–92.

Barnes, H. V., Rhodes, B. A., and Wagner, H. N., Jr., 1978. Radiation physics, in *The Thyroid,* S. C. Werner and S. H. Ingbar (eds.), Harper & Row, New York, pp. 257–273.

Belding, M. E., Klebanoff, S. J., and Ray, C. G., 1970. Peroxidase-mediated virucidal systems, *Science* 167:195–196.

Berliner, E., 1966. The current state of positive halogenating agents, *J. Chem. Educ.* 43:124–133.

Braverman, L. E., 1978, Disorders of iodine excess and deficiency, in *The Thyroid,* S. C. Werner and S. H. Ingbar (eds.), Harper & Row, New York, pp. 528–636.

Cahnmann, H. J., 1972, Iodoamino acids, in *Methods in Investigative and Diagnostic Endocrinology,* Vol. 1, J. E. Rall and I. J. Kopin (eds.), (S. A. Berson, general ed.), American Elsevier, New York, pp. 27–51.

Chopra, I. J., and Crandall, B. F., 1975. Thyroid hormones and thyrotropin in amniotic fluid, *New Engl. J. Med.* 293:740–743.

Chopra, I. J., 1978. Nature, source and biologic significance of thyroid hormones in blood, in *The Thyroid,* S. C. Werner and S. H. Ingbar (eds.), Harper & Row, New York, pp. 100–114.

Chopra, I. J., 1981. *Triiodothyronine in Health and Disease,* Monographs on Endocrinology, Vol. 18. Springer-Verlag, New York.

Cody, V., 1980. Thyroid hormone interactions: Molecular conformation, protein binding, and hormone action, *Endocr. Rev.* 1:140–166.

Commerford, S. L., 1980. In vitro iodination of nucleic acids, in *Methods in Enzymology,* Vol. 70, H. Van Vunakis, and J. J. Langone, (eds.), Academic Press, New York, pp. 247–252.

Covelli, I., vanZyl, A., and Edelhoch, H., 1971. Spectrophotometric determination of monoiodotyrosine, diiodotyrosine and thyroxine in iodoproteins, *Anal. Biochem.* 42:82–90.

Dratman, M. B., 1978. The mechanism of thyroxine action, in *Hormanal Proteins and Peptides,* Vol. 6, C. H. Li (ed.), pp. 205–271.

Encyclopedia Brittanica, Macropaedia, Vol. 6, 1978. Helen Hemingway Benton Publisher, Chicago, p. 702.

Feigl, F., and Anger, V., 1972. *Spot Tests in Inorganic Analysis,* American Elseveir, New York, pp. 253–254.

Fraker, P. J., and Speck, J. C., 1978. Protein and cell membrane iodinations with a sparingly soluble chloramide, 1,3,4,6-tetrachloro-3,6-diphenylglycoluril, *Biochem. Biophys. Res. Comm.* 80:849–857.

Frieden, E., 1981a. Iodine and the thyroid hormones, *Trends Biochem. Sci.* 6:50–53.

Frieden, E., 1981b. The dual role of thyroid hormones in vertebrate development and calorigenesis, in *Metamorphosis,* L. I. Gilbert and E. Frieden (eds.), Plenum, New York, pp. 545–563.

Gavaret, J. M., Nunez, J., and Cahnmann, H. J., 1980. Formation of dehydroalanine residues during thyroid hormone synthesis in thyroglobulin, *J. Biol. Chem.* 255:5281–5285.

Gorbman, A., 1978. Evolution of thyroid function, in *Hormonal Proteins and Peptides,* Vol. 6, C. H. Li (ed.), pp. 383–389.

Green, W. L., 1978. Mechanism of action of antithyroid compounds, in *The Thyroid,* S. C. Werner and S. H. Ingbar (eds.), Harper & Row, New York, pp. 77–87.

Greenwood, F. C., Hunter, W. M., and Glover, J. S., 1963. *Biochem. J.* 89:114–123.

Guernsey, D. L., Borek, C., and Edelman, I. S., 1981. Crucial role of thyroid hormone in x-ray induced neoplastic transformation in cell culture, *Proc. Nat. Acad. Sci.* 78:5708–5711.

Harington, C. R., and Barger, G., 1927. XXIII Chemistry of thyroxine. III. Constitution and synthesis of thyroxine, *Biochem. J.* 21:169–183.

Hay, I. D., Annesley, T. M., Jiang, N. S., and Gorman, C. A., 1981. Simultaneous determination of D- and L-thyroxine in human serum by liquid chromatography with electrochemical detection, *J. Chromatog.* 226:383–390.

Hearn, M. T. W., and Hancock, W. S., 1979. High pressure liquid chromatography of thyromimetic iodoamino acids, *J. Liq. Chromatogr.* 2:217–237.

Hearn, M. T. W., Hancock, W. S., and Bishop, C. A., 1978. High-pressure liquid chromatography of amino acids, peptides and proteins. V. Separation of thyroidal iodo-amino acids by hydrophilic ion-paired reversed-phase high performance liquid chromatography, *J. Chromatog.* 157:337–344.

Hollingsworth, D. R., and Alexander, N. M., 1983. Failure of amniotic fluid hormones to reliably predict neonatal outcome in pregnancies complicated by anencephaly or hyperthyroidism, *J. Clin. End. Metab.* 57:349–355.

Jorgensen, E. C., 1978a. Thyroid hormones and analogs. I. Synthesis, Physical properties and theoretical calculations, in *Hormonal Proteins and Peptides*, Vol. 6, C. H. Li (ed.), pp. 57–105.

Jorgensen, E. C., 1978b. Thyroid hormones and analogs, II. Structure–activity relationships, in *Hormonal Proteins and Peptides*, Vol. 6, C. H. Li (ed.), pp. 107–204.

Junek, H., Kirk, K. L., and Cohen, L. A., 1969. The oxidative cleavage of tyrosyl-peptide bonds during iodination, *Biochemistry* 8:1844–1848.

Kendall, E. C., 1919. Isolation of the iodine compound which occurs in the thyroid, *J. Biol. Chem.* 39:125–147.

Klebanoff, S. J., 1967. Iodination of bacteria: A bactericidal mechanism, *J. Exp. Med.* 126:1063–1076.

Kleinberg, J., Argersinger, W. J., and Griswold, E., 1960. *Inorganic Chemistry*, Heath, Boston, pp. 458–459.

Lankmayr, E. P., Maichin, B., and Knapp, G., 1981. Catalytic detection principle for high-performance liquid chromatography: Determination of enantiomeric iodinated thyronines in blood serum, *J. Chromatogr.* 224:239, 248.

Krinsky, M., and Alexander, N. M., 1971. Thyroid Peroxidase: Nature of the heme binding to apoperoxidase, *J. Biol. Chem.* 246:4755–4758.

Larsen, P. R., 1978. Thyroid hormone concentrations, in *The Thyroid*, S. C. Werner and S. H. Ingbar (eds.), Harper & Row, New York, pp. 321–337.

Layman, P., 1982. *Chemical and Engineering News*, June 14, pp. 12–13.

Mayberry, W. E., 1972. Iodine chemistry, in *Methods in Investigative and Diagnostic Endocrinology*, Vol. 1, J. E. Rall and I. J. Kopin (eds.), (S. A. Berson, general ed.), American Elsevier, New York, pp. 3–26.

McConahey, P. J., and Dixon, F. J., 1980. Radioiodination of proteins by the use of the chloramine-T method, in *Methods in Enzymology*, Vol. 70, H. Van Vunakis and J. J. Langone (eds.), Academic Press, New York, pp. 210–213.

Morrison, M., 1980. Lactoperoxidase-catalyzed iodination as a tool for investigation of proteins, in *Methods in Enzymology*, Vol. 70, H. Van Vunakis and J. J. Langone (eds.), Academic Press, New York, pp. 214–220.

Nachtmann, F., Knapp, G., and Spitzy, H., 1978. Catalytic detection principle for high-performance liquid chromatography. *J. Chromatog.* 149:693–702.

Nicoloff, J. T., 1978. Thyroid hormone transport and metabolism: Pathophysiologic implications, in *The Thyroid*, S. C. Werner and S. H. Ingbar (eds.), Harper & Row, New York, pp. 88–99.

Nunez, J., 1980. Iodination and thyroid hormone synthesis, in *The Thyroid Gland*, M. DeVisscher (ed.), Raven Press, New York, pp. 39–49.

Oddie, T. H., Fisher, D. A., McConahey, W. M., and Thompson, C. S., 1970. Iodine intake in the United States: A reassessment, *J. Clin. Endocr. Met.* 30:659–665.
Oppenheimer, J. H., 1979. Thyroid hormone action at the cellular level, *Science* 203:971–979.
Parkes, G. D., 1967. *Mellor's Modern Inorganic Chemistry*, Wiley, New York, pp. 522–580.
Pitt-Rivers, R., and Tata, J., 1959. *The Thyroid Hormones*, Pergamon Press, New York, pp. xi–xiii, 1–17.
Pitt-Rivers, R., 1978. The thyroid hormones: Historical aspects, in *Hormonal Proteins and Peptides*, Vol. 6, C. H. Li (ed.), pp. 391–422.
Prasad, A. S., 1978. *Trace Elements and Iron in Human Metabolism*, Plenum, New York, pp. 63–75.
Pretell, E. A., Moncloa, F., Salinas, R., Kawano, A., Guerra-Garcia, R., Gutierrez, L., Beteta, J., Pretell, J. and Wan, M., 1969. Prophylaxis and treatment of endemic goiter in Peru with iodized oil, *J. Clin. End. Metab.* 29:1586–95.
Refetoff, S., 1979. Thyroid function tests, in *Endocrinology*, L. J. De Groot, G. F. Cahill, Jr., L. Martini, D. H. Nelson, W. D. Odell, J. T. Potts, Jr., E. Steinberger, and A. I. Winegrad (eds.), Grune & Stratton, New York, pp. 387–428.
Roche, J., and Michel, R., 1951. Natural and artificial iodoproteins, in *Advances in Protein Chemistry*, M. L. Anson, J. T. Edsall, and K. Bailey (eds.), Academic Press, New York, pp. 253–297.
Rolland, M., Aquaron, R., and Lissitzky, S., 1970. Thyroglobulin iodoamino acids estimation after digestion with pronase and leucylaminopeptidase, *Anal. Biochem.* 33:307–317.
Samuels, H. H., and Tsai, J. S., 1973. Thyroid hormone action in cell culture: Demonstration of nuclear receptors in intact cells and isolated nuclei, *Proc. Nat. Acad. Sci. USA* 70:3488–3492.
Sandell, E. B., and Kolthoff, T. M., 1937. Microdetermination of iodine by a catalytic method, *Mikrochim. Acta* 1:9–25.
Sidlowski, J. J., and Frieden, E., 1982. Triiodothyronine induces an increase in cyclic GMP in bullfrog tadpole tissues, *Biosci. Rep.* 2:569–573.
Sneed, M. C., Maynard, J. L., and Brasted, R. C., 1961. *Comprehensive Inorganic Chemistry*, Vol. III, The Halogens, Van Nostrand, Princeton, N.J., pp. 1–3, 78–99.
Sorimachi, K., and Ui, N., 1974. An improved chromatographic method for the analysis of iodoamino acids in tyroglobulin, *J. Biochem. (Tokyo)* 76:39–45.
Spangenberg, D. B., 1974. Thyroxine in early strobilation in *Aurelia aurita*, *Am. Zool.* 14:825–831.
Stanbury, J. B., Ermans, A. M., Hetzel, B. S., Pretell, E. A., and Querido, A., 1974. Endemic goitre and cretinism: Public health significance and prevention, *WHO Chron.* 28:220–228.
Sterling, K., 1979. Thyroid hormone action at the cell level, *N. Engl. J. Med.* 300:117–123; 173–177.
Taurog, A., Howells, E. M., and Nachimson, H. I., 1966. Conversion of iodate to iodide in vivo and in vitro, *J. Biol. Chem.* 241:4686–4693.
Taurog, A., 1978. Thyroid hormone synthesis and release, in *The Thyroid*, S. C. Werner and S. H. Ingbar (eds.), Harper & Row, New York, pp. 31–61.
Tong, W., 1971. Thyroid hormone synthesis and release, in *The Thyroid*, S. C. Werner and S. H. Ingbar (eds.), Harper & Row, New York, pp. 24–40.
Underwood, E. J., 1977. Trace elements in human and animal nutrition, Academic Press, 4th ed., pp. 271–301.
Von Hippel, F., Wolfe, S., and La Cheen, C., 1982. Potassium iodide policy (Letters to the Editor), *Science* 218:295; 1983. 221:906.
Yalow, R. S., 1982. Potassium iodide distribution (Letters to the Editor), *Science* 217:295–296; 218:742.

Fluorine

4

Harold H. Messer

4.1. Introduction

Fluorine has been known to occur in biological tissues for almost 200 years, with demonstrations by Berzelius and others early in the nineteenth century of its occurrence in teeth. Fluorine first attracted widespread attention in biology because of its chronic toxic effects on both teeth and bone. The decade of the 1930s was remarkable for the association of both mottled enamel and bone changes with excessive fluorine ingestion. Mottled enamel had been recognized as a distinct entity in the late nineteenth century, and the defect was subsequently ascribed first to the drinking water and then, in 1931, to fluoride in the drinking water. The association of a low caries rate with mottled enamel and the demonstration that caries reduction could be achieved with fluoride intakes too low to cause mottled enamel led to the adoption in 1945 of artificial water fluoridation as a public health measure.

Effects of fluoride generated by industrial sources on surrounding vegetation had been thoroughly documented by the early 1900s. Skeletal abnormalities in humans were attributed to occupational exposure to fluoride in 1932 and to high dietary intakes in the same year. Both were demonstrated in several countries by the end of the decade. As in the case of dental caries, observations of toxic effects of fluoride on the skeleton have led to recognition of potential therapeutic benefits in the prevention and treatment of osteoporosis.

Interest in the biology of fluoride today transcends its toxic and beneficial effects in humans, although these have undoubtedly provided a major impetus for the study of the element. The question of essentiality has been addressed

---

Harold H. Messer • Department of Oral Biology, University of Minnesota, Minneapolis, Minnesota 55455.

repeatedly but has invariably led to an equivocal answer. Fluoride ion has been used as an important tool in biochemistry as an enzyme inhibitor and more recently as an activator of adenylate cyclase. The ubiquitous distribution of fluoride throughout biological fluids and tissues has led to the investigation of potential effects beyond those on the teeth and skeleton. Reliable analytical methods for fluoride, particularly with regard to low levels encountered in non-mineralized tissues, have proved a major challenge. Only within the past 10–15 years have reliable methods for plasma ionic fluoride determinations been developed. A good understanding of many areas of fluoride biology is thus very recent or still emerging.

By virtue of close similarities in the hydrated ions, fluoride substitutes for hydroxyl ion with high efficiency. This substitution accounts for many of the major effects of fluoride in biology, not only in mineralized tissues but seemingly also in its effects on enzymes. In contrast, fluoride does not displace other halides. Unlike chloride, fluoride readily enters cells; unlike fluoride, chloride is not incorporated into the lattice structure of apatite, the mineral component of bones and teeth. Fluoride is not accumulated in the thyroid or salivary glands and does not compete with iodide in the thyroid even at toxic intakes of fluoride.

A brief note on terminology is in order. The widespread use of the term *fluoride* rather than *fluorine* is followed in this chapter, reflecting both common usage and the predominant chemical form in which the element is found in nature. *Fluorine* is used whenever nonionic occurrence or simple reference to the element is implied. Fluoride concentration has traditionally been reported in parts per million (ppm). This is still the most useful means of expressing tissue levels on a dry or ash weight basis, and tradition makes it the most readily comprehended expression of fluoride levels in drinking water. In biological fluids, culture media, and buffers, molarity is used throughout this chapter. For cross-reference, 1 mM fluoride is equal to 19 ppm; 1 ppm equals approximately 0.05 mM.

4.2 Fluoride in Cells and Tissues

4.2.1 Mineralized Tissues

Mineralized tissues contain approximately 99% of total body fluoride, with the vast majority found in bone. The mineral component of vertebrate hard tissues is principally apatite, a basic calcium phosphate with a theoretical formula of $Ca_{10}(PO_4)_6(OH)_2$. The mineral phase is predominantly crystalline, occurring as small crystallites embedded in an organic matrix. While fluoride is only one of many ions occurring as major or minor contaminants of apatite, it is unusual in its efficiency of incorporation into crystal structure, by substitution for OH^-

ions. F^- ion is well able to substitute for OH^-; the ionic radius of F^- (1.29 Å) is very similar to that of OH^- (1.33 Å), and both ions share the same charge and primary hydration number (5). By contrast, Cl^- has a much larger ionic radius (1.81 Å) and a primary hydration number of 2, so that halide substitution in biological apatites is largely confined to fluoride.

Fluoride may be incorporated into apatite in two ways: direct incorporation during initial crystal formation, or displacement of OH^- ions from previously deposited mineral. Both occur *in vivo*, with the latter probably predominating, at least in adult life. The displacement reaction occurs according to the equation

$$Ca_{10}(PO_4)_6(OH)_2 + 2F^- \rightarrow Ca_{10}(PO_4)_6F_2 + 2\ OH^-$$

The extent of F^- substitution for OH^- is generally low, so that a mixed apatite results, $Ca_{10}(PO_4)_6(OH)_{2-x}F_x$. Fluoride concentrations in mineralized tissues are generally expressed as parts per million (ppm). Pure fluorapatite has a theoretical fluoride content of approximately 38,000 ppm, although this value is rarely approached in biological apatites. A fluoride content of 1000 ppm, as might be encountered in bone or dental enamel, represents a substitution of F^- for OH^- of only approximately 1 : 40. With the exception of fish enameloid, most mineralized tissues contain less than 10,000 ppm F. The extent of fluoride deposition is dependent on age, fluoride exposure, and rate of turnover of the tissues. In descending order, the highest fluoride concentrations are found in cementum, bone, dentin, and enamel (Singer and Armstrong, 1962). Of these, cementum (covering the roots of teeth) has received the least attention because of the difficulty in obtaining material for analysis.

Bone. Attempting to define a normal range for bone fluoride content is largely meaningless. In humans, values as low as 50 ppm in a newborn rib and as high as 15,000 ppm in adult rib (from a patient with severe fluorosis) have been reported (Singer *et al.*, 1974), and even higher levels have been recorded in experimental animals. Values for adult human bone in association with moderate fluoride intakes are commonly in the range 1000–5000 ppm F. The level varies with location, remodeling activity, and vascularity; cancellous bone has a higher fluoride content than cortical bone, and the deeper portions of cortical bone have less fluoride than the periosteal surface layers.

For a given site, the major variables affecting bone fluoride concentrations are fluoride intake and age. Bone plays a major role in the regulation of extracellular fluid fluoride concentration, by rapidly sequestering a large proportion of an absorbed fluoride dose (see under Section 4.4.1 "Transport, Distribution, and Regulation"). As a consequence, bone accumulates fluoride throughout life, although the rate of accumulation may decline in the elderly (National Academy of Sciences, 1971). Similarly, bone fluoride content increases with increasing fluoride intake; an approximately linear relationship exists at moderate fluoride intakes, but the proportion declines at very high intakes. There is little evidence to support the concept of a steady-state condition in which bone fluoride levels remain constant in the presence of a continuing high fluoride intake. Nonetheless, bone fluoride may decline if fluoride intake is reduced substantially. Since mineral dissolution occurs as an integral part of normal bone turnover and remodeling, fluoride may be mobilized from the skeleton if the extracellular fluid fluoride concentration is reduced as a result of a reduced fluoride intake.

Enamel. Despite its significance in the prevention of dental caries, the fluoride content of enamel is lower than that of other mineralized tissues. A dramatic exception is the enameloid of certain fishes, which may consist of almost pure fluorapatite (LeGeros and Suga, 1980). In one species of Selaciens that may inhabit either fresh water (0.1 ppm F) or sea water (1.3 ppm F), the fluoride content of enameloid was unaffected by environmental fluoride and averaged approximately 38,000 ppm in both cases. Other species from salt or fresh water showed similar but low fluoride concentrations of 2–3000 ppm F,

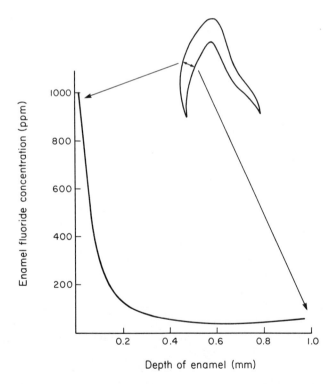

Figure 4-1. Idealized fluoride "gradient" in enamel, demonstrating a rapid decrease in concentration from surface layers to deeper portions of enamel.

and dentin from all species contain 2–4000 ppm F. The mechanism by which some species are able to concentrate fluoride and deposit virtually pure fluorapatite is unknown, but this represents the only known example of complete substitution of fluoride for hydroxyl ions during the formation of a biological apatite.

In mammalian enamel, with much lower fluoride concentrations, the distribution of fluoride is highly asymmetric. The highest fluoride concentration occurs at the enamel surface and declines rapidly with increasing depth (Figure 4-1). The fluoride content of deeper layers of enamel is exceedingly low (30–50 ppm) while the most superficial enamel (surface layer of 1–5 μm thickness) may contain 1000 ppm F or more (Weidmann and Weatherell, 1970). The fluoride "gradient" is established during initial mineral deposition and is enhanced by further fluoride uptake by the enamel surface both before tooth eruption (from extracellular fluid) and after eruption (from saliva, drinking water, etc.). There is not a large difference in the fluoride concentration of surface enamel between subjects from fluoridated (1 ppm F) and nonfluoridated (0.1 ppm F) communities, despite the difference in caries susceptibility. While there are difficulties in assigning precise values to surface enamel fluoride levels, based on sampling problems and the steepness of the fluoride "gradient," the difference appears to be only approximately two-fold, and both are low (500–1000 ppm) relative to other mineralized tissues (Brudevold *et al.*, 1960).

4.2.2 Cells and Soft Tissues

The fluoride content of soft tissues is generally low unless the tissue is prone to direct environmental contamination (e.g., lung) or to pathological calcification (e.g., aorta) (Smith *et al.*, 1960). Values of 2–5 ppm on a dry weight basis (i.e., approximately 0.5–2 ppm fresh weight) have been reported for human soft tissues, and these values were not substantially influenced by fluoride intake (except for kidney) or by age (Smith *et al.*, 1960). Analytical problems may in fact make these values and others obtained 20 or more years ago a significant overestimate.

In contrast to chloride, fluoride is found intracellularly, although the intracellular fluoride concentration is lower than that of extracellular fluid. The intracellular–extracellular fluoride ratio varies among tissues. Armstrong and Singer (1980) reported ratios of approximately 0.8 for liver and 0.4 for muscle, obtained *in vivo* (rats). Ratios determined *in vitro*, using much higher extracellular fluoride concentrations (3–3.5 mM versus μM levels *in vivo*), tend to be lower: 0.27 for rat liver cells in primary cell suspension and 0.3–0.4 for L cells (mouse fibroblast line) (Repaske and Suttie, 1979; Hongslo *et al.*, 1980b).

The basis for the lower intracellular fluoride concentration relative to extracellular fluid levels appears to be the existence of a transmembrane pH gradient. In both mammalian and bacterial cells, the membrane is considered to be largely impermeable to fluoride ion F^-, and the diffusible species is the undissociated acid HF (Whitford et al., 1977). The intracellular pH of microorganisms is generally alkaline relative to that of the culture medium, so that dissociation of HF occurs following passage across the cell membrane, leading to fluoride accumulation inside the cell. The intracellular pH of mammalian cells is generally lower than that of extracellular fluid (e.g., liver intracellular pH 7.2–7.4; muscle, 6.9–7.0, versus plasma, 7.4 (Armstrong and Singer, 1980). Thus the pH gradient is in the reverse direction and intracellular fluoride concentrations are lower than those of extracellular fluid. Hence muscle, with the lower intracellular pH, has a lower fluoride concentration than liver with the higher intracellular pH. Similarly, the lowering of extracellular pH from 7.4 to 6.7 increased the intracellular–extracellular fluoride ratio of human epithelial cells *in vitro* from 0.4 to 0.8 (Helgeland and Leirskar, 1976).

4.2.3 Extracellular Fluid

The analytical problems confronting fluoride analyses in blood have been largely overcome within the past decade (Guy et al., 1976), and the total picture of the nature, concentration, and regulation of blood fluorides has been clarified enormously as a result. Fluoride is present in both plasma and cellular elements, with plasma fluoride predominant (approximately 75%) and physiologically more important. Fluoride occurs in plasma in both ionic and nonionic forms, and ionic fluoride is the form subject to physiological regulation.

Nonionic or organically bound fluoride is conveniently considered first. It is present in variable concentrations in the plasma or serum of humans and animals, and its concentration is independent of total fluoride intake and plasma ionic fluoride concentrations. The concentration averaged approximately 1–2 μM in subjects from both industrial and rural communities in the United States (Guy et al., 1976) and approximately 0.5 μM in subjects from rural China (Belisle, 1981). Lower values, ranging down to undetectable levels, have been reported in animals. Perfluoro-octanoic acid has been proposed as the predominant organic species, based on analyses of blood from subjects living in the United States (Guy et al., 1976). The widespread industrial use of perfluorocarbons has led to the suggestion of environmental contamination, although this has been questioned in the case of rural Chinese and animals (Belisle, 1981). Presumably, a variety of naturally occurring and industrially generated fluorocarbons contribute to varying extents. The health implications of these organically bound fluorine compounds are seemingly minimal. In addition to covalently bound fluorine in plasma, the binding of fluoride by amides (Emsley et al.,

1981) raises the possibility of the complexing of fluoride by plasma proteins. Armstrong (1982) has not been able to confirm this.

Ionic fluoride concentration in plasma or serum is dependent on long-term fluoride intake as well as a number of other variables, but is in the low micromolar range for normally encountered fluoride intakes (Guy *et al.*, 1976). In one study (Guy *et al.*, 1976), plasma ionic fluoride concentration showed a remarkably linear relationship with fluoride intake as measured by fluoride content of the drinking water. Mean values increased from 0.4 μM for subjects consuming nonfluoridated water to 4.3 μM for subjects consuming water containing 5.6 ppm F, although a 5- to 10-fold range in individual values was encountered at any given level of fluoride intake. Other studies have confirmed that plasma ionic fluoride levels increase with increasing fluoride intake, but the mean values for a particular fluoride intake vary considerably from one study to another. Serum ionic fluoride values show a positive correlation with age (Hanhijärvi, 1975; Cowell and Taylor, 1981), although the increase is not great. Taves and Guy (1979) attributed the age-related increase to an increase in bone fluoride levels, since a strong relationship between fasting serum ionic fluoride concentrations and bone fluoride levels has been described for both humans and experimental animals. Variations in bone turnover rate result in corresponding changes in serum fluoride levels, with an increase in response to increased bone resorption (e.g., hyperparathyroidism or parathyroid hormone administration) and a decrease in response to inhibition of bone resorption (e.g., calcitonin administration) (Waterhouse *et al.*, 1979). Renal function and acid–base balance also influence plasma ionic fluoride concentrations. Even in the absence of any complicating factors, plasma fluoride values show a small diurnal variation and a larger but transient increase in response to oral doses of soluble fluorides (Ekstrand, 1977; Cowell and Taylor, 1981). Regulation of plasma ionic fluoride concentrations, and the roles of the skeleton and kidney in that regulation, will be considered in more detail under Section 4.4.1, "Transport, Distribution, and Regulation."

4.3 Fluorine Deficiency and Function

4.3.1 Deficiency

The absence of unequivocal evidence for the essentiality of fluorine makes it difficult to describe specific manifestations of deficiency. Nonetheless, fluorine is increasingly regarded as an essential trace element. Defects associated with low fluoride intakes tend to be so nonspecific that they do not provide useful insights into possible underlying biochemical changes. Defects attributed to fluorine deficiency fall into several categories, as described later. For the most part, they are confined to experimental animals for which diets extremely low in fluorine have been developed. I do not consider that dental caries or osteoporosis warrants the designation of a fluorine deficiency disease.

Growth and Reproduction. Of the many studies using low fluoride diets (as low as 0.005 ppm), only two have shown impaired growth relative to animals fed identical diets supplemented with fluoride. Schroeder et al. (1968) reported a reduced growth rate and longevity in female mice. Schwartz and Milne (1972) used a trace element-controlled isolator system to demonstrate a growth-promoting effect in rats of fluoride supplements (2.5–7.5 ppm) to an otherwise low-fluoride diet (0.04–0.46 ppm). In both instances, the effect was small. Similar responses to cadmium and lead, as well as other trace elements that are unquestionably essential, have also been reported for rats in the isolator system.

Studies of reproduction have often been hampered by poor fertility resulting from the highly refined diets, and most have failed to demonstrate any effect of fluoride intake (reviewed in Messer et al., 1974). We reported a progressive loss of fertility in mice fed diets containing 0.1–0.3 ppm F, which was prevented or cured by fluoride supplementation (Messer et al., 1972, 1973). In contrast, Tao and Suttie (1976), using an identical diet, did not observe any loss of reproductive capacity, and we have since been unable to confirm our earlier observations (Ophaug et al., 1980). The diet used in these studies contains approximately 60% whole wheat flour, with variable (and generally low) levels of trace metals; the diet prepared by Tao and Suttie (1976) had three times as much copper and five times as much iron as the diet used in our earlier study. The best explanation for the reduced fertility is thus a marginal intake of trace metals, and that fluoride supplements enhanced the absorption of trace metals.

Anemia. We also reported a greater severity of anemia of pregnancy and infancy in mice fed the same low-fluoride diet as the preceding, compared with mice fed the same diet but supplemented with fluoride in the drinking water (Messer et al., 1972). Since these anemias are related to a low iron intake at times of increased physiological demand and since the diet was known to be marginal in iron, a similar pharmacological action of fluoride in promoting iron absorption is the most logical explanation for the finding. Wegner et al. (1976) showed that the anemia is of the microcytic, hypochromic type typical of iron deficiency, and that iron supplementation abolished the response to fluoride. An effect of fluoride is thus confined to marginal iron intakes and cannot be construed as evidence of fluorine deficiency. The effect will be considered in more detail in Section 4.3.2.

Mineralized Tissues. Fluoride shows a strong predilection for mineralized tissues and is associated with incipient mineralization. While it has been proposed on theoretical grounds that fluoride might be required to initiate mineral deposition (Newesely, 1961), defective mineralization has not been demonstrated *in vivo* despite stringent attempts to minimize exposure to fluoride. Even after several generations on extremely low intakes, the bones of experimental animals contain easily measurable quantities of fluoride, well in excess of the minimum calculated to be necessary for mineral precipitation (10^{-8}–10^{-7} M). Crenshaw and Bawden (1981) have presented evidence for the high affinity of an enamel matrix protein for fluoride, and they speculated that the fluoride incorporated into enamel during early enamel formation could be bound to the matrix rather than associated with the mineral phase. If this is the case, it could participate in the nucleation of crystal formation.

At the opposite extreme, both osteoporosis and dental caries have occasionally been considered manifestations of fluorine deficiency, and the Food and Nutrition Board (1980) classifies fluorine as an essential trace element on the basis of its beneficial effects on dental health. Even though bone or enamel mineral with a low fluoride content may be more soluble than mineral with a high fluoride content under selected conditions, osteoporosis and dental caries are much more complex diseases than simple dissolution phenomena, and the role of fluoride in preventing the diseases is much more complex than a reduction in mineral solubility. Both diseases can be prevented despite a low fluoride intake and can occur despite a high fluoride intake. Beneficial effects of fluoride in preventing dental caries and osteoporosis are best regarded as pharmacological effects (National Academy of Sciences, 1971).

Biochemical Defects. Small differences (10–20%) in enzyme activities between high- and low-fluoride animals have been occasionally reported, and in most instances the activities have been lower in low-fluoride animals. Such enzymes include bone acid and alkaline phosphatase and liver isocitrate dehydrogenase (reviewed in Messer *et al.*, 1974). The differences in enzyme activity cannot be readily correlated with clinical manifestations of deficiency, and they do not qualify as pertinent, specific biochemical changes accompanying a deficiency state, as would be expected for an essential trace element.

In summary, sufficient evidence to qualify fluorine as an essential trace element is lacking. The majority of effects ascribed to fluorine supplementation are probably pharmacological, including the prevention of dental caries and osteoporosis as well as the amelioration of anemia and infertility in experimental animals. A requirement for fluorine in biological mineralization involving apatites may exist, but is not experimentally verifiable. The consistent failure of often heroic efforts to demonstrate an unquestioned fluorine deficiency has condemned the element to the indefinite status of "possibly essential."

4.3.2 Functions of Fluoride

The major functions of fluoride are related to effects on the mineral component of bones and teeth. In this context, the conferring of protection against pathological demineralization has received by far the greatest attention. Effects on initial mineralization may be equally important, although the low levels of fluoride sufficient to promote mineral deposition have hampered investigation of this role *in vivo*. The interaction of fluoride with other nutrients, with possible implications for human health, has periodically attracted some small degree of interest.

Formation of Mineralized Tissues

The initial deposition of apatitic mineral in bones and teeth involves transitional forms of calcium phosphates that are converted to the crystalline apatite. In simple aqueous systems *in vitro*, fluoride promotes the formation of apatite from metastable solutions of calcium and phosphate, and an absolute requirement for fluoride for this process to occur *in vivo* has been suggested (Newesely, 1961). Brown (1966) calculated that exceedingly low fluoride concentrations (10^{-8}–10^{-7} M) are sufficient to initiate mineral precipitation, and fluoride concentrations of this magnitude are invariably exceeded in biological systems. The observation of Crenshaw and Bawden (1981) alluded to earlier (Section 4.3.1) of a fluoride-binding protein in early enamel matrix, may shed new light on the mineralization of enamel, which has received little attention relative to other mineralized tissues. The binding of fluoride to this matrix protein preceding the initiation of mineralization would provide more direct evidence for a role of fluoride in mineral deposition, at least in this tissue.

Inhibition of Mineral Dissolution

A major role of fluoride in preventing dental caries and osteoporosis is related to its effects on mineral solubility. In both diseases, the mineral component of the tissue undergoes acid dissolution; in dental caries, the acid is produced by plaque bacteria on the surface of enamel, while in osteoporosis the bone mineral is resorbed by osteoclasts or other bone cells. As was pointed out earlier (Section 4.2.1), the extent of substitution of fluoride for hydroxyl in biological apatites is generally low, and the difference between "low" and "high" fluoride apatites is small. In examining effects of fluoride on mineral solubility, then, we are concerned not with a comparison of pure fluorapatite versus pure hydroxyapatite, but with the effects of partial substitution over a relatively narrow range. We will consider two aspects of mineral solubility: (1) simple dissolution phenomena and (2) effects on crystallinity that may influence dissolution rates.

If apatite is prepared *in vitro* by precipitation from solutions of calcium and phosphate in the presence of varying amounts of fluoride, mixed apatites of known fluoride substitutions are obtained, and the fluoride is distributed evenly throughout the apatite. Apatites ranging from almost pure hydroxyapatite to almost pure fluorapatite have been prepared in this manner (Moreno *et al.*, 1973). Solubility at a given pH can then be assessed by measuring the calcium concentration in supernatant fluid in equilibrium with an excess of solid apatite. Under these conditions, the incorporation of fluoride into hydroxyapatite results in a marked reduction in solubility, with a measurable effect beginning at 10–20% substitution (approximately 4000–8000 ppm F) and a maximum effect achieved at approximately 50–75% substitution (20,000–30,000 ppm F). The bone fluoride concentration in subjects consuming water containing 5–8 ppm F for 50 years or more, and hence protected against osteoporosis (Bernstein et al., 1966) will be in excess of 5000 ppm, compared with only approximately 1000 ppm in subjects from nonfluoridated communities (National Academy of Sciences, 1971). These differences in bone fluoride concentration are consistent with measurable (if small) differences in mineral solubility, based on the data of Moreno et al. (1973). The picture is much less clear-cut for the protection of enamel against caries. In one study, based on enamel biopsies averaging 5 μm in depth, the surface enamel fluoride concentration of subjects in a nonfluoridated community was approximately 500 ppm, compared with 900 ppm for subjects consuming optimally fluoridated water (Brudevold et al., 1960). In addition to the small differences between the two groups, both are so low that they fall below the degree of substitution necessary for effects on solubility. Shallower sampling techniques (1 μm or less, compared with 5 μm in the preceding study) yield higher estimates of surface enamel fluoride concentrations and magnify the difference between high- and low-fluoride subjects. Nonetheless, differences in enamel solubility appear too small to account fully for the dramatic caries preventive effect of fluoride, and it has been necessary to invoke alternative explanations.

The first of these is related to effects on enamel crystals. In the hydroxyapatite unit cell, each hydroxyl ion is surrounded by a triangle of three calcium atoms that are coplanar with one axis of the unit cell. Hydroxyl ions are slightly displaced from the plane of the calcium atoms, while fluoride is coplanar with them. The ionic radius of fluoride is only slightly smaller than that of the hydroxyl ion, so that its substitution does not result in distortion of the unit cell structure; rather, the greater electronegativity and the coplanar position result in increased bond strength and a small reduction in unit cell dimensions. Stabilization of the crystal structure promotes increased crystal size and crystal perfection, with a consequent reduction in chemical reactivity (summarized in Myers, 1975). Thus the rate of dissolution may be decreased.

A second possibility involves nonuniform distribution of fluoride in individual enamel crystals. The incorporation of fluoride into enamel occurs largely after mineralization is complete and continues even after tooth eruption. This form of incorporation occurs by the displacement by fluoride of hydroxyl ions from already formed crystallites, so that some preference for the more readily accessible surface unit cells might be predicted (Brown et al., 1977). If fluoride is incorporated into a majority of the surface unit cells of each apatite crystal, solubility characteristics of pure fluorapatite might be achieved despite a relatively low overall fluoride content. Other, more complex, thermodynamic explanations have also been offered (Moreno et al., 1977), as well as the promotion of remineralization (the redeposition of mineral in partially demineralized enamel). Additional effects unrelated to mineral include antibacterial activity and morphological effects during tooth formation. These are largely beyond the scope of this chapter.

The use of fluoride in the treatment (as opposed to the prevention) of osteoporosis deserves special mention. Large oral doses of fluoride (50–75 mg NaF/day) are given, generally in conjunction with calcium supplements; vitamin D supplements have also been commonly used as part of the treatment regimen, but their replacement by estrogen has recently been advocated (Riggs et al., 1982). The intent is not simply to inhibit bone resorption as discussed earlier, but to promote new bone formation (Jowsey et al., 1979). Fluoride is considered to act in this instance by stimulating osteoblastic activity, but in the absence of calcium supplements defective mineralization is observed. The effect of fluoride in the treatment of osteoporosis thus mimics the pattern of osteosclerosis seen in chronic fluoride toxicity (see Section 4.4.4), and the boundary between pharmacologic and toxic effects of fluoride is blurred considerably.

Unexplained Roles

Two effects of fluoride at pharmacological levels, apparently unrelated to the interaction of fluoride with mineral, have possible implications in human nutrition. These are the effects on lipid and iron metabolism, neither of which has been studied intensively or explained at the biochemical level.

As described earlier (Section 4.3.1), a high fluoride intake ameliorated iron deficiency anemia in mice (Messer et al., 1972; Wegner et al., 1976). Wegner et al. (1976) described the phenomenon in some detail but did not investigate the mechanism by which fluoride influenced iron status at marginal iron intakes. The hematocrits of pups reared by dams fed a low-fluoride, marginal iron (29 ppm) diet declined from birth to 15 days of age, and the decline was less severe when the dams drank water containing 50 ppm F. (This level of fluoride intake is considered nontoxic in rodents.) The total body iron content of pups of fluoride-supplemented dams was 27% higher than that of pups of low-fluoride mothers. The differences in total body iron reflected differences in the iron content of mother's milk, although the difference in total body iron content of the pups (27%) was smaller than the difference in the iron content of the milk (51%). A higher percentage of a dose of ^{59}Fe given to 15-day-old pups of the two groups was retained in the high fluoride group. The best explanation for the effect is that at marginal iron intakes, fluoride enhances the absorption and/or retention of iron, although the mechanism by which this might occur has not been explored. All effects of fluoride were abolished by a high iron intake.

Effects of fluoride on lipid metabolism have also been described. Suttie and Phillips (1960) reported that very high fluoride intakes decreased the absorption of dietary lipid in rats, and this was confirmed by perfusion experiments (Schnitzer-Polokoff and Suttie, 1980). A 30% inhibition occurred at an estimated duodenal fluoride concentration of 3 mM, and the effect was specific to lipid, in that concomitant glucose absorption was not affected. Even though the fluoride concentration in the perfusate was high, Schnitzer-Polokoff and Suttie (1980) did not consider it sufficient to result in gross toxicity. On the basis of comparative effects of fluoride on the absorption of triglycerides and free fatty acids and on lymph lipid profiles, they concluded that fluoride probably inhibited lipid absorption by inhibiting reesterification of free fatty acids during passage through mucosal epithelial cells. Townsend and Singer (1977) described effects of fluoride on lipid metabolism at lower fluoride intakes. Using a low-fluoride diet containing 6.8% fat and 0.25% cholesterol, they observed lower serum cholesterol, phospholipid and triglyceride levels in guinea pigs consuming 25 or 50 ppm F in the drinking water than in animals consuming deionized water, whereas serum free fatty acids and liver lipid profiles were unaffected by fluoride intake. Again, these observations are consistent with a reduced intestinal absorption of lipid, although the mechanism of action of fluoride was not investigated. By contrast, Vatassery *et al.* (1980) reported that a high fluoride intake (25 ppm F in the drinking water) led to elevated serum and liver lipids in guinea pigs fed a diet containing 18% lipid for prolonged periods.

The human implications of the effects of fluoride on lipid metabolism are uncertain. Epidemiological studies have suggested that atherosclerosis may be less severe in subjects consuming high fluoride waters (4–6 ppm F), and Taves (1978) attempted to relate a reduced mortality from cardiovascular disease to consumption of fluoridated water.

4.4 Metabolism and Toxicity of Fluoride

4.4.1 Metabolism

Transfer of Fluoride Across Cell Membranes

Understanding of the mechanisms involved in fluoride transfer across cell membranes and through cell layers (e.g., epithelia of the gastrointestinal tract) has increased enormously during the past decade. Because of its application to numerous areas of fluoride metabolism, it is worth considering general principles in some detail before examining its role in absorption, distribution, and excretion. Mention has already been made of the phenomenon in considering intracellular fluoride concentrations in Section 4.2.2.

Active transport mechanisms such as have been demonstrated for chloride do not appear to exist for fluoride, except in one instance of a fluoride-resistant cell line (see under Section 4.4.2, "Effects on Cells"). The dissimilarity between fluoride and chloride ions in ionic radius and hydration do not allow for efficient competition between fluoride and chloride for the chloride transport system. Consequently, fluoride is found intracellularly as well as extracellularly, although there is commonly a concentration gradient across the cell membrane. This may occur in either direction: Bacterial cells tend to accumulate fluoride, while mammalian cells generally have a lower intracellular fluoride concentration than is found in extracellular fluid.

Early observations of an effect of pH on the severity of fluoride toxicity to cells led Borei (1945) to propose that fluoride crosses cell membranes only as the undissociated acid HF, and that the membrane is essentially impermeable to fluoride ion F^-. Assuming passive diffusion of HF across the cell membrane, and no intracellular binding of fluoride, the distribution between intracellular and extracellular compartments will thus be a function of the pH difference across the cell membrane (Figure 4-2). Whitford and coworkers, in an extensive series of papers, have investigated the phenomenon in bacteria, epithelia, and urinary fluoride excretion and provided convincing evidence of its validity. In fact, the distribution of fluoride between intracellular and extracellular waters can be used as a means of determining intracellular pH.

Hydrogen fluoride is a weak acid with a pKa of 3.45, and dissociates according to the equation:

$$HF \rightleftharpoons H^+ + F^-$$

The proportions of undissociated HF to fluoride ion F^- at a given pH can be

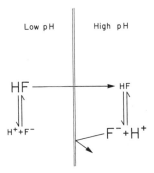

Figure 4-2. Proposed mechanism for fluoride transfer across biological membranes (adapted from Whitford and Pashley, 1979). The membrane is permeable to undissociated hydrogen fluoride HF but not to fluoride ion F^-. A pH gradient across a cell membrane promotes the transfer (by passive diffusion) of HF from the region of low pH to the region of high pH, and fluoride is then trapped by dissociation to fluoride ion F^-.

estimated from the Henderson–Hasselbalch equation. It is readily apparent that at physiological pH (e.g., extracellular fluid pH = 7.4), F^- ion is the predominant species by a ratio of approximately 10^4 : 1, while at the pH of the stomach (approximately pH 1) more than 99% is present as undissociated HF. In most instances, at least for mammalian cells, the pH will be near neutrality and hence the concentration of HF will be extremely low; if ionic fluoride concentration is 1 μM, the concentration of HF will be approximately 0.1 nM. In addition, the difference in pH across a cell membrane or a cell layer will generally be no more than a few tenths of a pH unit. Even so, differences of this magnitude will result in intracellular–extracellular HF ratios of approximately 0.5, which is the same order of magnitude as differences in ionic fluoride concentrations (Whitford *et al.*, 1977; Armstrong and Singer, 1980).

The phenomenon can be demonstrated most readily in bacteria, where extracellular pH can be varied systematically over a wide range and fluoride distribution can be readily correlated with independent measurements of intracellular pH (Whitford *et al.*, 1977). Similar techniques have been applied to mammalian cells *in vitro* (Helgeland and Leirskar, 1976), and Whitford and coworkers have also studied effects in whole animals by manipulation of acid-base balance and by use of the urinary bladder as a model for epithelial fluoride transfer (reviewed in Whitford and Pashley, 1979).

Absorption

The absorption of ingested fluoride occurs throughout the stomach and small intestine, with the rate and extent of absorption influenced by the size and solubility of the ingested dose and by the presence of other dietary factors. In addition, absorption of inhaled fluoride in the respiratory tract is highly efficient and may constitute a significant source of fluoride in occupationally exposed individuals. This will not be considered further here.

Site of Absorption. The rapidity of absorption of fluoride following oral ingestion implies that absorption occurs in the stomach, and this has been confirmed by radiotracer studies of ligated gastrointestinal segments and everted sac techniques (Carlson *et al.*, 1960; Stookey *et al.*, 1964). Fluoride is thus almost unique as a nutrient in being absorbed at this site. Absorption also occurs throughout the small intestine, with similar rates of absorption in different segments. The relative contribution of the different sites to the absorption of an oral dose has not been systematically investigated. While Stookey *et al.* (1964) reported that absorption from the stomach was only half that from the small intestine *in vitro*, their study did not take into account normal stomach pH or the relative surface areas of the different sites in comparing rates of absorption. At least in man, the stomach is probably the major site of absorption.

Rate and Extent of Absorption. Fluoride absorption is generally considered to be rapid, although in the small intestine it is much slower than that of chloride or water (Armstrong et al., 1961). On average, approximately 50% of a moderate dose of a soluble fluoride will be absorbed in 30 minutes in an intact animal, with complete absorption after 90 minutes. Wagner (1962) reported that 50% of a small dose of fluoride was absorbed from the ligated stomach of the rat in the first hour. While intestinal absorption has been described as more rapid than that from the stomach using everted sac techniques (Stookey et al., 1964), the question of pH effects on absorption rate makes that claim uncertain.

The extent of fluoride absorption is very variable, depending on the solubility of the fluoride compound ingested and on the presence of modifying factors in the diet. Under normal conditions, approximately 80% of total dietary fluoride is absorbed by man. Soluble fluorides, such as sodium fluoride and sodium fluorosilicate, as are commonly used in water fluoridation, are almost completely absorbed (86–97%) (Largent, 1961; Cremer and Büttner, 1970). The availability of fluoride in foods is generally lower than that from water, and less soluble sources of fluoride, such as bone meal, are poorly absorbed (less than 50%). Despite evidence in experimental animals that mineral components of the diet (calcium, magnesium, phosphate) may influence absorption of fluoride, these constituents do not adversely affect human fluoride absorption at practical levels in the diet (Spencer et al., 1980). Aluminum hydroxide, widely used as an antacid in clinical medicine, markedly inhibits fluoride absorption under conditions of standard therapeutic use (greater than 50% inhibition) (Spencer et al., 1980). Of other dietary factors investigated, trace metals do not appear to influence fluoride absorption to a significant extent, while a high-fat diet enhances fluoride absorption in rats via a delay in gastric emptying (this occurred at high fat and toxic fluoride intakes) (McGown and Suttie, 1974).

Mechanism. Fluoride absorption is considered to occur by passive diffusion. In the intestine, fluoride transfer across the intestinal wall required a concentration gradient, was not influenced by metabolic inhibitors, and was not temperature dependent (Stookey et al., 1964). Absorption across the stomach wall has not been investigated, but Whitford et al. (1976) proposed that the low pH of the stomach, leading to a high concentration of the undissociated acid HF forms the basis for the rapid absorption of fluoride from this site. Whether diffusion as undissociated HF also occurs in the small intestine, which has a much higher luminal pH than the stomach, is less clear. At the high pH of the small intestine, the concentration of HF will be extremely low, and the HF gradient small. By contrast, the concentration of F^- will be high, and the F^- gradient also high.

Transport, Distribution, and Regulation

Fluoride occurs in plasma as the inorganic fluoride ion F^- and as organically bound fluorine, as discussed in Section 4.2.3. By analogy with the binding of many ions and small molecules to transport proteins in plasma, the possibility of a similar binding of fluoride has been raised repeatedly. The question is not yet fully resolved, but the demonstration of covalently bound fluorine in compounds such as perfluoro-octanoic acid (Guy et al., 1976) has cast doubt on the concept. These low-molecular-weight compounds are considered to be environmentally derived. The fluorine, since it is covalently bound, is not displaceable by a large excess of inorganic fluoride ion, and the concentration of organically bound fluorine bears no apparent relationship to total fluoride intake or to plasma ionic fluoride concentration. The reported binding of fluoride ion by amides (Emsley et al., 1981) again raised the possibility of an association of fluoride with plasma proteins, but such binding would tend to be nonspecific and in any case has not been confirmed by Armstrong (1982). At present, it must be concluded that there is no evidence of a fluoride transport protein in plasma.

The rapid absorption of fluoride following oral ingestion leads to a prompt rise in plasma ionic fluoride concentration, which subsequently declines as fluoride is removed from the circulation. Soft tissue fluoride levels are preserved within narrow limits and appear little affected by long-term fluoride intake or by short-term fluctuations in plasma levels. Nonetheless, soft tissue fluoride is readily exchangeable with extracellular fluid fluoride, as demonstrated by the rapid distribution of ^{18}F following intravenous injection.

Short-term fluctuations in plasma ionic fluoride concentration following the ingestion of fluoride are relatively small and transient. A single dose of 1.5 mg fluoride (as NaF) to young men, representing the equivalent of a total daily fluoride intake encountered in a fluoridated community, resulted in a five-fold increase in plasma fluoride, with the peak occurring 1 hour after ingestion and a decline to fasting levels in 3–5 hours (Ekstrand, 1977). Larger doses (up to 10 mg F) resulted in a greater elevation of longer duration, and it is apparent that the chronic ingestion of water with a high fluoride content or of large fluoride supplements for the treatment of osteoporosis will result in a persistent elevation of plasma fluoride.

Obviously, plasma ionic fluoride concentration is not precisely controlled by homeostatic mechanisms comparable to those for calcium, sodium, or chloride. Nevertheless, regulatory mechanisms exist to minimize fluctuations in plasma fluoride, particularly by damping the extent and duration of increases in response to large intakes, but also by maintaining relatively constant levels during periods of low fluoride intake. The mechanism of regulation involves the dual action of bone and kidney (Smith, 1966). The generally low extent of tubular reabsorption of fluoride results in rapid excretion. Deposition in bone, however, appears to serve as the principal mechanism in removing fluoride from the circulation, at least in the short term. Fluoride is sequestered by incorporation into apatite, as discussed earlier. In young adult rats, more than 50% of an oral dose of ^{18}F was absorbed and in turn removed from the circulation in the first hour (Wallace, 1953). Uptake by bone accounted for 44% of the administered dose and urinary excretion accounted for an additional 13%. In a series of studies, the fraction of an administered dose deposited in bone was 1.5–4.5 times greater than the fraction excreted in the urine. In humans consuming moderate fluoride intakes, approximately 50% of ingested fluoride is deposited in the skeleton and 50% excreted in the urine (Largent, 1961).

The proportions of an ingested dose that are deposited in the skeleton or excreted in the urine vary more substantially than the preceding numbers indicate, depending on age, growth rate, degree of previous exposure (skeletal "saturation"), and presence of disease (Smith, 1966; Hodge et al., 1970). Subjects with a limited previous exposure to fluoride tend to deposit more fluoride in bone than those consuming a relatively constant quantity over a prolonged period. Growing individuals also incorporate relatively more fluoride into the skeleton than adults. Human infants were reported to excrete only approximately 40% of ingested fluoride, compared with 50% for relatively unexposed adults and 90% in "steady state" adults with a prolonged elevated fluoride intake (Hodge et al., 1970). Largent (1959), in a study of his own fluoride balance, described only small variations from the 50% value over a range of intakes from 1–8 mg fluoride per day. Subjects with very high fluoride intakes for 30 years were almost in balance, excreting approximately 90% of ingested fluoride in the urine. At times of reduced fluoride intake, fluoride may also be mobilized from the skeleton to maintain extracellular fluid fluoride concentrations.

Excretion

Excretion of fluoride occurs primarily via the urine, which accounts for approximately 90% of total excretion. Fecal excretion accounts for the bulk of the remainder, with variable but generally minor losses in sweat (reviewed in Hodge et al., 1970).

Fluoride is excreted very rapidly in the urine. In man, ^{18}F appears in the urine within a few minutes of oral ingestion, and 20–30% of an ingested dose appears in the urine within 3–4 hours (Carlson et al., 1960; Hodge et al., 1970). The extent of urinary excretion varies in relation to fluoride intake, age, previous skeletal exposure, and so on, as was considered earlier, but always acounts for a major proportion of fluoride intake and may exceed 90% of total intake (Largent, 1961). Fluoride is removed from the circulation via glomerular filtration, and the rapidity of fluoride excretion is related to high renal blood flow and to the low efficiency of tubular reabsorption. In two humans, tubular reabsorption of fluoride was 51% and 63%, compared with 99.5% for chloride (Carlson et al., 1960). Much larger fluctuations have been reported in experimental animals (20–70%), ranging as high as 95% in diuresis experiments (reviewed in Whitford and Pashley, 1979).

The mechanism of tubular reabsorption appears to involve diffusion of undissociated HF. Whitford and coworkers have demonstrated in rats, dogs, and humans, via disturbances of acid–base balance, that tubular reabsorption of fluoride was inversely related to urinary pH (Whitford and Pashley, 1979). The pH of renal collecting duct fluid varies according to acid–base balance, but may be as low as 4.5. At this pH, approximately 10% of the total fluoride is in the form of undissociated HF. The higher pH of the peritubular fluid leads to an extremely steep HF gradient across the tubular epithelium, favoring reabsorption. The difference in total fluoride concentrations between tubular fluid and interstitial fluid also favors reabsorption. Following diffusion, HF is "trapped" as the dissociated fluoride ion and returned to the circulation. Even at high urinary pH, exceeding that of extracellular fluid, proximal intratubular pH is lower than that of blood, with an increase in the distal tubule. Thus some tubular reabsorption of fluoride will still occur. Alkalosis favors enhanced fluoride excretion, while acidosis promotes fluoride retention, with elevations of plasma and bone fluoride levels.

Whitford has argued on the basis of observations summarized earlier that acid–base balance is the major determinant of total body fluoride balance, via its effects on urinary excretion. While urinary fluoride excretion is clearly correlated with urinary pH, it is also dependent on urinary flow rate, as demonstrated in studies involving diuresis, partial nephrectomy, and renal failure. Fractional fluoride excretion was closely related to urinary flow rate (Schiffl et al., 1981), and these workers concluded that flow rate is a major factor in fluoride excretion under normal physiological conditions. Massmann (1981), in a study of almost 400 healthy adults with low fluoride intakes, did not observe a significant correlation between urinary pH and fluoride excretion. Acid–base balance may be the predominant factor influencing urinary fluoride excretion at physiological extremes, but appears to be one of only several determinants in normal, healthy individuals.

Fecal excretion also accounts for some loss of fluoride, although fecal fluoride presumably represents a combination of unabsorbed dietary fluoride and fluoride excreted directly into the large intestine (Hodge et al., 1970). Fecal fluoride increases after inhalation of gaseous hydrogen fluoride, and intravenously injected ^{18}F appears in the intestinal tract, indicating actual loss into the intestine. Diffusion from blood vessels rather than secretion via saliva, bile, or pancreatic juice appears to constitute the major source of fecal fluoride.

Maternal–Fetal Exchange

The placental transfer of fluoride from the maternal to the fetal circulation has been demonstrated on numerous occasions in both man and experimental animals (reviewed in Gedalia, 1970). In humans, ^{18}F administered intravenously to the mother was detected in fetal blood within 4 minutes, the earliest time at which a fetal blood sample could be obtained (Ericsson and Malmnas, 1962). In sheep the intravenous injection of a small dose of fluoride (2 mg F as a 100 ppm solution of NaF) led to rapid increases in both maternal and fetal serum fluoride concentrations. The peak concentration in both maternal and fetal blood occurred 1 minute after injection, the earliest sampling time (Maduska et al., 1980). Despite the evidence for a very rapid placental transfer of fluoride, the concept of a "placental barrier" restricting passage of fluoride has been repeatedly postulated (see Gedalia, 1970). The ionic fluoride concentration in human cord blood taken at delivery was only approximately 75% of maternal blood fluoride levels (Shen and Taves, 1974), although Armstrong et al. (1970) did not find a significant difference between maternal and cord blood obtained during cesarean section. Fetal bone fluoride levels tend to be very low (but increase during fetal growth), and the deposition of ^{18}F in the fetal skeleton following administration to the mother is much lower than that in the maternal skeleton. The question of a "barrier" is complicated by the development of calcific deposits in the placenta near term, which may sequester fluoride rather than permitting its transfer to the fetus. While the evidence is not conclusive, limitations on the supply of fluoride to the fetus probably reflect the ability of the maternal skeleton and kidneys to remove the bulk of an administered fluoride dose very rapidly, rather than suggesting the presence of an active placental barrier to fluoride transfer.

Fluoride in Secretions

Both milk and saliva have fluoride concentrations similar to plasma ionic fluoride levels, and variations in plasma levels will be reflected in these secretions. The analysis of milk has been hampered by analytical difficulties, but even the limited number of studies conducted indicate milk fluoride concentrations of 1 μM or less, representing a meager dietary source of the element to infants. Salivary fluoride concentrations and their fluctuations in response to blood levels have been investigated more thoroughly because of the implications for saliva as a source for posteruptive fluoride uptake by tooth enamel. Unstimulated ductal saliva, which to a large extent represents an ultrafiltrate of serum, has an ionic fluoride concentration of 1–2 μM, with small variations related to long-term fluoride intake. Following the oral ingestion by adult humans of soluble fluorides in the range 1–10 mg F, salivary fluoride concentrations parallel closely those of plasma, with a rapid elevation of up to 10 times the resting level (Ekstrand, 1977). The peak fluoride concentration is reached in approximately 30 minutes and is generally lower than that of plasma; it is followed by a decline to resting values over a period of hours. The total quantity of fluoride secreted in saliva per 24 hours is only approximately 20–50 μg. Nonetheless, this may constitute a significant source of fluoride for uptake by surface enamel, and in combination with salivary calcium and phosphate may contribute to remineralization of incipient enamel caries.

The variability of electrolyte concentrations in sweat with the rate of sweating makes the quantitation of sweat fluoride very difficult. Sweat fluoride concentration tends to be lower than that of plasma and is not markedly influenced by fluctuations in plasma fluoride concentrations (Henschler *et al.*, 1975). Its importance as a route for fluoride loss seems to be minor.

4.4.2 Toxicity

The toxicity of fluoride has received considerable attention over many years. Chronic toxicity via excessive fluoride in water supplies or from industrial exposure has been reported in many parts of the world, manifested as dental fluorosis (mottled enamel) and, at higher fluoride intakes, skeletal fluorosis. Acute fluoride toxicity has resulted from accidental or intentional exposure to a variety of fluorine compounds, and deaths from acute fluoride poisoning are still reported episodically. Mechanisms of fluoride toxicity are still only partially understood, and the broad spectrum of effects of fluoride makes it difficult to assign clinical manifestations of toxicity to specific metabolic actions. Although extrapolation to the clinical picture is largely speculation, toxic effects of fluoride at the subcellular and cellular levels are logically considered first.

Effects on Enzymes

Effects of fluorine on enzymes *in vitro* occur for the most part at such high concentrations that they are unlikely to have any physiological significance. The 10^{-3}–10^{-2} M F$^-$ concentrations commonly encountered in biochemical studies as necessary to inhibit enzyme activity are far removed from the typical intracellular fluorine concentrations in the micromolar range. Nonetheless, intracellular fluoride ion concentrations of sufficient magnitude to affect enzyme activity may be achieved in acute fluorine toxicity. Fluoride ion effects on enzymes may be either inhibitory or, more rarely, stimulatory.

Inhibition of Enzyme Activity. Inorganic fluorides inhibit a very large number of enzymes, and fluoride has been commonly used as a nonspecific inhibitor during characterization studies of enzymes. The F$^-$ concentration necessary for inhibition ranges from 10^{-7} to 10^{-1} M, with the majority of enzymes requiring 10^{-3} to 10^{-2} M F$^-$ (Wiseman, 1970). Liver microsomal esterases are unique in showing inhibition by fluoride ion in the range of 0.5–20 μM (Haugen and Suttie, 1974). Enzymes are generally considered sensitive to fluoride if they are inhibited in the range of 10^{-5}–10^{-4} M F$^-$. At high concentrations, the risk of artifactual inhibition is enhanced, particularly by the formation of insoluble F-metal ion–substrate complexes, as in the case of reverse transcriptase (Srivastava *et al.*, 1981).

The enzymes inhibited by fluoride and the characteristics of the inhibition vary enormously, and few generalizations can be made. Inhibition may be competitive, noncompetitive, or mixed (Cimasoni, 1971). Fluoride has been reported to inhibit metalloenzymes containing copper, manganese, zinc, nickel, heme iron, and nonheme iron by binding to the metal in the active site (reviewed by Wiseman, 1970; Nowak and Maurer, 1981). Enzymes requiring divalent cations (particularly Mg^{2+}) as cofactors are frequently inhibited, and in some instances the inhibition by fluoride ion is enhanced by inorganic phosphate (e.g., enolase, succinic dehydrogenase). Other enzymes requiring divalent cations as cofactors, however, are unaffected by fluoride. Lactate dehydrogenase, a zinc metalloenzyme, is inhibited by fluoride, but the inhibition appears to involve binding of fluoride to histidine or arginine in the pyruvate binding site rather than to the metal (Anderson, 1981). Wiseman (1970) has pointed out that the potent inhibitory effect of fluoride on enzymes relative to other anions would not have been predicted based on its position in the Hofmeister lyotropic series. The similarity of fluoride ion to hydroxyl ion in terms of ionic radius and primary hydration number (see Section 4.2.1) suggests that F^- may act as an analog of OH^-, which is theoretically a much stronger inhibitor.

Enolase merits further consideration; the inhibition of glycolysis via an inhibition of enolase is frequently invoked both as part of the acute toxic effect of fluorides in mammals and in the inhibition of acid production by oral bacteria following topical fluoride applications. It is probably the most intensively studied of all enzymes affected by fluoride with investigations spanning 40 years since the first detailed report by Warburg and Christian in 1942. Enolase is not particularly sensitive to fluoride inhibition, requiring concentrations of 10^{-4}–10^{-2} M for substantial effects *in vitro* (Wiseman, 1970).

Enolase catalyzes the conversion of 2-phosphoglycerate to phosphoenolpyruvate and water. The active site contains a divalent cation binding site (Mg^{2+} or Mn^{2+}) and a phosphoryl binding site. Inhibition by fluoride is competitive and is enhanced by inorganic phosphate. Inhibition occurs by the formation of a tightly bound enzyme–metal ion–F–Pi complex in which the F^- interacts directly with the metal ion in the active site and indirectly with the phosphate bound at the phosphoryl binding site (Nowak and Maurer, 1981). These authors conclude that F^- serves as a possible analog of the OH^- group involved in the gain or loss of water as part of the normal reaction mechanism. A similar substitution of F^- for OH^- has been suggested in the inhibition of inorganic pyrophosphatase, leading to a more stable enzyme–metal ion–substrate–F complex. The parallel with the substitution of OH^- by F^- in mineralized tissues, with an increased chemical stability, is readily apparent.

Acute fluoride toxicity clearly involves the inhibition of enzymes. The relationship between *in vitro* and *in vivo* concentrations necessary for enzyme inhibition is simply not known, so that the specific enzymes most severely affected have not been identified. Intracellular fluoride concentrations achieved during acute fluoride toxicity (see the following discussion) appear capable of inhibiting a range of enzymes, and attempts to relate clinical manifestations of toxicity to a few highly sensitive enzymes may prove fruitless.

Stimulation of Enzyme Activity. Fluoride is known to stimulate only three enzymes, the best known of which is adenylate cyclase (the others are a cyclic GMP phosphodiesterase from vertebrate photoreceptors and a bacterial enzyme system catalyzing the conversion of mevalonate to phytoene) (Rasenick and Bitensky, 1980; Suzue, 1962). Adenylate cyclase is the only enzyme with important implications for biological effects of fluorine. A four- to ten-fold activation of adenylate cyclase by fluoride ion *in vitro* was first reported by Rall and Sutherland in 1958, and this has since been confirmed for the enzyme from a wide variety of mammalian tissues. Activation occurs in the range 1–25 mM F^-.

The mechanism of activation is slowly being elucidated. Adenylate cyclases contain two components (excluding the hormone receptor unit): a catalytic unit that binds the substrate ATP and a regulatory unit (N protein) that binds guanine nucleotides. The guanine nucleotide regulatory protein has now been shown to be necessary for fluoride activation of adenylate cyclase, and fluoride is considered to bind to the regulatory protein rather than the catalytic unit, as was previously assumed (Downs *et al.*, 1980). While details are not yet known, it appears that fluoride interacts with magnesium ion and/or guanine nucleotides in association with their binding to the regulatory protein (Iyengar and Birnbaumer, 1981). Beyond that the mechanism by which the regulatory protein controls the activity of the catalytic unit and the effect of fluoride in enhancing activity are not well understood.

The biological significance of the activation of adenylate cyclase by fluoride is uncertain. The effect has been generally considered an *in vitro* phenomenon limited to broken cell preparations (Perkins, 1973) and occurring at such high fluoride concentrations that *in vivo* effects would not be predicted. Several recent reports, however, have indicated increased cyclic AMP production by whole cells *in vitro,* in response to high fluoride ion concentrations in the culture medium (Allmann and Shahed, 1977). This has been true of some cell types (e.g., hepatocytes) but not others (e.g., fibroblasts) (Holland *et al.*, 1980). Increases in both tissue adenylate cyclase activity and tissue cyclic AMP have been described in experimental animals given fluorine intakes ranging from toxic levels to as little as 1 ppm in the drinking water (Susheela and Singh, 1982; Kleiner and Allmann, 1982). Elevated urinary cyclic AMP levels in humans and experimental animals have also been reported by some authors (Kleiner and Allmann, 1982) but not others (Ophaug *et al.*, 1979). The increased tissue adenylate cyclase activity in rabbits receiving toxic doses of fluoride for 6 months (Susheela and Singh, 1982) presumably reflects an increased synthesis of the enzyme rather than activation and occurred despite a small increase in tissue fluorine levels, which remained in the micromolar range. It appears to be secondary to other effects of the fluorine toxicity.

Effects of an elevated fluoride intake on tissue and urinary cyclic AMP levels must await confirmation, particularly in view of the huge discrepancy between tissue fluoride levels and those necessary for *in vitro* activation of adenylate cyclase. The fundamental role of cyclic AMP in regulation of cellular activity suggests that changes in tissue cyclic AMP levels should be accompanied by measurable metabolic changes. To date, convincing evidence of such changes is lacking, but the possibility of a role for adenylate cyclase in mediating some of the toxic effects of fluoride cannot be ignored.

Effects on Cells

Probably most cell culture studies involving fluorides have used fluoride ion simply as a metabolic inhibitor. An example would be an investigation attempting to establish a relationship between a particular cellular activity (e.g., phagocytosis, active transport) and specific metabolic pathways serving as potential sources of energy for that activity. Fluoride concentrations of 20–200 mM, which commonly have been employed in such studies, are so much higher than those achieved in acute, severe fluoride toxicity that they yield little insight into mechanisms of toxicity. In any event, such high fluoride concentrations in the presence of millimolar concentrations of calcium and phosphate as are found in must culture media involve the real risk of precipitation of calcium salts or at least the complexing of ionic calcium; interpretation of effects as the result of fluoride toxicity then becomes extremely difficult. Matthews (1970) has reviewed the literature thoroughly on the use of fluoride as a metabolic inhibitor in cell culture studies and has emphasized that fluoride inhibits not only glycolysis but oxidative phosphorylation in general.

Several other aspects of the effects of fluoride on cells are of much greater import in the toxicology of fluoride. These include (1) the concentrations of fluoride necessary to inhibit cellular activity, and their relationship to levels achieved *in vivo* in acute and chronic fluorine toxicity, (2) intracellular fluoride concentrations in relation to extracellular levels (including fluoride resistance), and (3) mechanisms by which cells regulate intracellular fluoride concentrations.

Isolated reports of growth inhibition by micromolar concentrations of fluoride can probably be ignored, since they could not be confirmed. The fluoride levels necessary to inhibit cell growth *in vitro* are somewhat dependent on cell type and culture conditions, but are generally of the order 0.1–10 mM F$^-$, and most frequently are in the middle of that range (Matthews, 1970; Repaske and Suttie, 1979; Hongslo *et al.*, 1980a). Established cell lines appear more sensitive than primary cell cultures, and kidney cells are more sensitive than liver cells (Hongslo *et al.*, 1980a.) Fluoride-resistant cell lines have been isolated (Repaske and Suttie, 1979) and fluoride-resistant liver cells but not kidney cells were obtained by previously exposing animals to high fluoride intakes for prolonged periods before cell isolation (Hongslo et al., 1980b). The extent of fluoride resistance is not great: inhibition of cell growth requires approximately double the fluoride concentration that is toxic to sensitive cells.

Hongslo et al. (1980a) consider protein synthesis the most sensitive cellular activity, while Drescher (1974) reported a simultaneous inhibition of DNA, RNA, and protein synthesis. Drescher (1974) considered this as evidence for a primary effect of fluoride on some other aspect of cellular activity, with energy production a logical target. Glycolytic activity, however, was reduced by only 25%, and cellular ATP levels remained unaffected despite a 50% reduction in cell growth. The primary cellular effect of fluoride thus remains uncertain. Activation of adenyl cyclase does not appear to be a mechanism for growth inhibition by fluoride (Holland et al., 1980).

As discussed earlier, intracellular fluoride levels are generally lower than extracellular fluid levels, with intracellular–extracellular ratios of 0.4–0.8 for tissues in vivo (Armstrong and Singer, 1980) and ratios of 0.25–0.4 for cells in culture (Repaske and Suttie, 1979; Hongslo et al., 1980b). These ratios do not vary substantially over a large range of fluoride concentrations, including severely toxic levels. Fluoride resistance in cells able to tolerate high extracellular fluoride levels appears related to a low intracellular level rather than changes in sensitivity of enzymes to fluoride inhibition. In at least one cell line, this is achieved by an active extrusion mechanism by the cell, which is temperature and ouabain sensitive and saturable (Repaske and Suttie, 1979). This was not found for primary liver cell cultures obtained from animals previously exposed to high fluorine intakes, and some alternative adaptation mechanism must be invoked to explain resistance induced in vivo (Hongslo et al., 1980b).

Within the limits of extrapolating from in vitro to in vivo conditions, fluoride ion concentrations in the millimolar range (approximately 0.1–10 mM) appear necessary for the inhibition of enzymes and of metabolic activities of whole cells. The question thus becomes how closely these levels are approached in vivo in acute and chronic fluorine toxicity. Plasma (and hence extracellular fluid) ionic fluoride concentrations, even in subjects consuming water with sufficient fluoride to produce manifestations of chronic toxicity (greater than 5 ppm), are still less than 5 μM (Guy et al., 1976), two to three orders of magnitude below in vitro inhibitory levels. Acute fluoride toxicity is accompanied by vastly higher plasma ionic fluoride concentrations. Lethal doses of fluorides result in plasma levels of 1.5–2 mM before death in rats and rabbits, while the fluoride concentrations of blood and soft tissues obtained at autopsy from fluoride-poisoned humans averaged approximately 0.5 mM (Hodge and Smith, 1981). Clearly, these values are well within the range of concentrations sufficient for the inhibition of many enzymes and cellular activities. The explanation of chronic toxicity, particularly the seemingly selective effects on ameloblasts and osteoblasts, remains a major challenge.

Acute Toxicity

Acute fluoride toxicity follows the ingestion of excessive quantities of soluble fluoride compounds. Solubility is an important factor, since rapid extensive absorption is necessary for the onset of toxicity. Sodium fluoride is the most commonly encountered compound in acute fluoride poisoning; calcium fluoride, which is much less soluble, does not appear to lead to acute illness. Intakes in the range 1–5 mg F/kg body weight (approximately 150–750 mg NaF for a 70-kg man) produce transient signs of toxicity, primarily gastrointestinal distress lasting for a few hours (Spoerke et al., 1980). A lethal dose is approximately 2–5 g soluble fluoride (i.e., 5–10 g NaF), with death occurring in 2–4 hours in the absence of treatment (Hodge and Smith, 1981).

Hodge and Smith (1981) classify the effects of acute fluoride toxicity into four major categories: (1) enzyme inhibition, (2) calcium complex formation, (3) shock, and (4) specific organ injury. To these should be added the gastrointestinal disturbances that are presumably caused by preabsorptive effects of ingested fluoride. Vomiting, diarrhea, and abdominal pain are the most frequently described signs of acute fluoride poisoning, and more severe gastrointestinal tract damage may result from the formation of hydrofluoric acid in the stomach (Largent, 1961). Corrosive changes in the stomach are commonly observed at autopsy, and vomitus burns may also occur in the throat, in the mouth, and on contacted skin.

Following absorption of large quantities of fluoride, blood and tissue fluoride concentrations increase from micromolar to millimolar levels, as considered previously. Despite the potential effects on many enzymes at these levels, little effort has been directed toward identifying specific enzymes most severely affected. The acute symptomatology is not readily related to the inhibition of particular enzymes.

A prominent feature of acute fluoride poisoning is a profound hypocalcemia, which is often ascribed to the complexing of calcium by fluoride. There are problems with this interpretation, related to the presence of sufficient fluoride in extracellular fluid to complex the calcium and remove it from the circulation. Alternative explanations are possible. The maintenance of extracellular fluid calcium levels requires the active extrusion of calcium from bone, and in the absence of a traditional "calcium pump," the production of lactate has been implicated in this calcium extrusion (bone is a major site of lactate production in the body) (reviewed in Messer, 1982). In view of the inhibition of glycolysis by fluoride and the predilection of fluoride for bone, the inhibition of lactate production could become a major contributor to the hypocalcemia. In rats the hypocalcemic action of fluoride requires the presence of the thyroid gland, implying a role for calcitonin in the hypocalcemia (Baker, 1974). This phenomenon has not been thoroughly investigated.

Many features of shock accompany acute fluoride toxicity. A drop in blood pressure and an increased rate and depth of respiration, followed by respiratory arrest and refractory ventricular fibrillation have been described in humans and dogs (National Academy of Sciences, 1971; Baltazar et al., 1980). Baltazar et al. 1980) reported electrocardiographic changes reminiscent of hyperkalemia in a patient who subsequently died of acute fluoride toxicity, and they confirmed the existence of hyperkalemia accompanying similar ECG patterns in two dogs experimentally exposed to lethal doses of fluoride. They speculated that the elevated serum potassium levels resulted from reduced Na^+, K^+-ATPase activity in erythrocytes, secondary to inhibition of glycolysis and thus ATP production.

Chronic Toxicity

Chronic toxicity is much more difficult to explain than acute toxicity in terms of specific metabolic effects of fluoride. The extracellular fluoride levels remain in the low micromolar range, far below the threshold for *in vitro* effects on enzymes or whole cells. Except for possible effects on kidney and thyroid at high fluoride intakes for long periods, chronic fluoride toxicity is largely confined to mineralized tissues.

The earliest detectable effect is on tooth enamel, resulting in "mottling" or the presence of patchy white or brown opacities in enamel. This can be produced only during the period of enamel formation, although it is not observed until many years later, when the affected teeth erupt. Even at optimal fluoride intakes for caries reduction (i.e., 1 ppm in the drinking water in temperate climates), approximately 10% of subjects show detectable mottling; at double the optimal intake enamel fluorosis is a public health problem, and at four times the optimal intake virtually all subjects are affected (Largent, 1961). The epidemiology and clinical features of enamel fluorosis have been thoroughly documented (Bhussry, 1970) and need not be repeated here.

The major question related to enamel fluorosis is the underlying defect produced at such low extracellular fluid fluoride concentrations. Enamel fluorosis in humans is rare when plasma ionic fluoride is approximately 1 μM (on a population basis), but is present in a substantial percentage of subjects with plasma ionic fluoride levels of 2 μM and 100% of subjects with 3–4 μM fluoride. In rats, enamel fluorosis occurs when plasma ionic fluoride levels are maintained at 3–5 μM or when they surge intermittently to 10 μM (Angmar-Mansson et al., 1976). Whether these low fluoride concentrations are capable of affecting ameloblasts, enamel matrix, or mineralization is not known. The ameloblast would have to be an exquisitely sensitive cell to be affected by such low fluoride levels. Recent studies of the fluoride-binding characteristics of enamel proteins, however, suggest that changes in the binding of fluoride to early enamel matrix, with possible implications for mineral deposition or maturation, could be responsible. The predominant protein of early enamel matrix has a high affinity for fluoride, and the fluoride initially deposited in enamel is protein-bound rather than bound in mineral (Crenshaw and Bawden, 1981). At physiological concentrations of fluoride, the protein is approximately half-saturated, whereas two- to four-fold increases in fluoride would substantially increase fluoride binding. This excessive binding during the initial phase of enamel deposition could conceivably alter the characteristics of mineral deposition and/or the subsequent removal of the enamel matrix protein that normally occurs during enamel maturation.

At higher fluoride intakes, skeletal fluorosis (osteosclerosis) occurs, but intakes of 20–80 mg F per day for 10–20 years are considered necessary for the fluorosis to become crippling (National Academy of Sciences, 1971). Lower fluoride intakes lead to radiographic evidence of increased bone density, which is often considered beneficial. Clinical and radiographic features of osteosclerosis have been adequately described elsewhere (Singh and Jolly, 1970). Plasma ionic fluoride concentrations at intakes consistent with the development of osteosclerosis are likely to be approximately 5–20 μM (by extrapolation from data of Guy et al., 1976). These concentrations of fluoride are very low relative to enzymic and cellular effects *in vitro,* although the predilection of fluoride for bone may lead to a localized elevation of fluoride concentration in the vicinity of bone cells.

Bone fluoride levels may become extremely high in severe fluorosis (up to 15,000 ppm), and there can be little doubt that the incorporation of fluoride into bone mineral, with effects on physical and chemical properties, contributes to the clinical picture of osteosclerosis. Fluoride inhibits bone resorption in model systems, primarily via an effect on bone mineral. In addition, effects on bone cells are noted *in vivo.* Both osteoblastic and osteoclastic activity are stimulated, with increased bone turnover (Brearley and Storey, 1970). Osteoblastic activity predominates at higher intakes of fluoride (Jowsey et al., 1972), and a progressive increase in bone deposition occurs, including the aberrant bone (exostoses) associated with crippling fluorosis.

The mechanism of action of fluoride on bone cells is not understood. Given the low extracellular fluid fluoride concentrations and the stimulatory effect on osteoblastic activity, a direct action of fluoride either on enzyme activity or on the proliferation and differentiation of bone cells cannot be readily explained. A secondary hyperparathyroidism, occurring in response to a reduced solubility of bone mineral and hence a reduced tendency to bone resorption, has been frequently invoked (Rich and Feist, 1971). The data in support of such a mechanism are equivocal.

4.5 Summary

Fluorine is a possibly essential trace element that is best known for its beneficial effects in man and its acute and chronic toxicity. It is rapidly absorbed and enters cells readily but is preferentially taken up by mineralized tissues and exerts its major effects there.

The passage of fluoride across biological membranes appears to involve the passive diffusion of undissociated hydrogen fluoride HF, rather than as fluoride ion F^-. This has been demonstrated in bacterial cells, in mammalian cells *in vitro*, and in tubular reabsorption in the kidney. It has not been shown to occur in absorption but could explain the rapid absorption that occurs in the stomach.

Many of the major biological actions of fluoride can be explained in terms of its substitution for hydroxyl ion, which fluoride ion closely resembles in ionic radius, hydration number, and charge. This is true for its role in mineralized tissues, where the substitution results in effects on mineral dissolution. It may also explain the efficiency of enzyme inhibition. In the case of enolase and pyrophosphatase, which have been examined in considerable detail, the substitution of fluoride ion for the hydroxyl ion that normally participates in the enzymic reaction results in the formation of stable, less reactive complexes.

The acute toxic effects of fluoride can be explained in terms of elevation of extracellular and intracellular fluoride concentrations to levels consistent with enzyme inhibition (millimolar range). Chronic toxicity occurs at extracellular fluid fluoride concentrations in the low micromolar range and includes both inhibitory and stimulatory effects on cells that cannot be readily explained in terms of known biochemical effects.

References

Allmann, D. W., and Shahed, A. R., 1977. Effect of NaF on cyclic AMP and glucose metabolism in isolated hepatocytes, *J. Cell Biol.* 75:A178.

Anderson, S. R., 1981. Effects of halides on reduced NAD binding properties and catalytic activity of beef heart lactic dehydrogenase, *Biochemistry* 20:464–467.

Angmar-Mansson, B., Ericsson, Y., and Eckberg, O., 1976. Plasma fluoride and enamel fluorosis, *Calcif. Tiss. Res.* 22:77–84.

Armstrong, W. D., 1982. Is fluoride hydrogen bonded to amides? *J. Dent. Res.* 61:292.

Armstrong, W. D., and Singer, L., 1980. Fluoride tissues distribution: intracellular fluoride concentrations, *Proc. Soc. Exp. Biol. Med.* 164:500–506.

Armstrong, W. D., Singer, L., and Markowski, E. L., 1970. Placental transfer of fluoride and calcium, *Am. J. Obstet. Gynecol.* 107:432–434.

Armstrong, W. D., Venkateswarlu, P., and Singer, L., 1961. Intestinal transport of labeled fluoride, chloride and water, *J. Dent. Res.* 40:727.

Baker, K. L., 1974. Fluoride and tetracycline-induced changes in rat serum calcium and phosphate levels, *Arch. Oral Biol.* 19:717–723.

Baltazar, R. F., Mower, M. M., Reider, R., Funk, M., and Salomon, J., 1980. Acute fluoride poisoning leading to fatal hyperkalemia, *Chest* 78:660–663.
Belisle, J., 1981. Organic fluorine in human serum: natural versus industrial sources, *Science* 212:1509–1510.
Bernstein, D. S., Sadowsky, N., Hegsted, D. M., Gurl, C. D., and Stare, F. J., 1966. Prevalence of osteoporosis in high and low fluoride areas in North Dakota, *J. Am. Med. Assoc.* 198:499–504.
Bhussry, B. R., 1970. Chronic toxic effects on enamel organ, in *Fluorides and Human Health*, WHO Monograph Series No. 59, WHO, Geneva, pp. 230–238.
Borei, H., 1945. Inhibition of cellular oxidation by fluoride, *Ark. Kim., Mineral. Geol.* 20A:1–125.
Brearley, L. J., and Storey, E., 1970. Osteofluorosis in the rabbit: macroscopic and radiographic changes, *Pathology* 2:231–247.
Brown, W. E., 1966. Crystal growth of bone mineral, *Clin. Orthop.* 44:205–217.
Brown, W. E., Gregory, T. M., and Chow, L. C., 1977. Effects of fluoride on enamel solubility and cariostasis, *Caries Res.* 11 (suppl. 1):118–141.
Brudevold, F., Steadman, L. T., and Smith, F. A., 1960. Inorganic and organic components of tooth structure, *Ann. N.Y. Acad. Sci.* 85:110–132.
Carlson, C. H., Armstrong, W. D., and Singer, L., 1960. Distribution and excretion of radiofluoride in the human, *Proc. Soc. Exp. Biol. Med.* 104:235–239.
Cimasoni, G., 1971. Fluoride and enzymes, in *Fluoride in Medicine*, T. L. Vischer (ed.), Hans Huber Publishers, Bern, pp. 14–26.
Cowell, D. C., and Taylor, W. H., 1981. Ionic fluoride: A study of its physiological variation in man, *Ann. Clin. Biochem.* 18:76–83.
Cremer, H. D., and Büttner, W., 1970. Absorption of fluorides, in *Fluorides and Human Health*, WHO Monograph Series, No. 59, WHO, Geneva, pp. 75–91.
Crenshaw, M. A., and Bawden, J. W., 1981. Fluoride-binding by organic matrix from early and late developing bovine fetal enamel determined by flow rate dialysis, *Arch. Oral Biol.* 26:473–476.
Downs, R. W., Spiegel, A. M., Singer, M., Reen, S., and Aurbach, G. D., 1980. Fluoride stimulation of adenylate cyclase is dependent on the guanine nucleotide regulatory protein, *J. Biol. Chem.* 255:949–954.
Drescher, M. J., 1974. Effects of sodium fluoride on growth and metabolism of L cells, Thesis, University of Wisconsin, Madison.
Ekstrand, J., 1977. A micro method for the determination of fluoride in blood plasma and saliva, *Calcif. Tiss. Res.* 23:225–228.
Emsley, J., Jones, D. J., Miller, J. M., Overill, R. E., and Waddilove, R. A., 1981. An unexpectedly strong hydrogen bond: ab initio calculations and spectroscopic studies of amide-fluoride systems, *J. Am. Chem. Soc.* 103:9–16.
Ericsson, Y., and Malmnas, C., 1962. Placental transfer of fluoride investigated with ^{18}F in man and rabbit, *Acta Obstet. Gynecol. Scand.* 41:144–158.
Food and Nutrition Board, 1980. *Recommended Dietary Allowances*, National Academy of Sciences, Washington, D. C.
Gedalia, I., 1970. Distribution of fluorides. 4. Distribution in placenta and fetus, in *Fluorides and Human Health*, WHO Monograph Series, No. 59, WHO, Geneva, pp. 128–134.
Guy, W. S., Taves, D. R., and Brey, W. S., 1976. Organic fluorocompounds in human plasma: Prevalence and characterization, in *Biochemistry Involving Carbon–Fluorine Bonds*, R. Filler (ed.), ACS Symposium Series, American Chemical Society, Washington, D.C., pp. 117–134.
Hanhijärvi, H., 1975. Inorganic plasma fluoride concentrations and its renal excretion in certain physiological and pathological conditions in man, *Fluoride* 8:198–207.
Haugen, D. A., and Suttie, J. W., 1974. Fluoride inhibition of rat liver microsomal esterases, *J. Biol. Chem.* 249:2723–2731.

Helgeland, K., and Leirskar, J., 1976. pH and the cytotoxicity of fluoride in an animal cell culture system, *Scand. J. Dent. Res.* 84:37–45.

Henschler, D., Büttner, W., and Patz, J., 1975. Absorption, distribution in body fluids, and bioavailability of fluoride, in *Calcium Metabolism, Bone and Metabolic Diseases*, F. Kuhlencordt and H. P. Kruse (eds.), Springer-Verlag, Berlin, pp. 111–121.

Hodge, H. C., and Smith, F. A., 1981. Fluoride, in *Disorders of Mineral Metabolism*, Vol. 1, F. Bronner and J. W. Coburn (eds.), Academic Press, New York, pp. 439–483.

Hodge, H. C., Smith, F. A., and Gedalia, I., 1970. Excretion of fluorides, in *Fluorides and Human Health*, WHO Monograph Series, No. 59, WHO, Geneva, pp. 141–161.

Holland, R. I., Hongslo, J. K. and Christoffersen, T., 1980. On the role of cyclic AMP in the cytotoxic effect of fluoride, *Acta Pharmacol. Toxicol.* 46:66–72.

Hongslo, C. F., Hongslo, J. K., and Holland, R. I., 1980a. Fluoride sensitivity of cells from different organs, *Acta Pharmacol. Toxicol.* 46:73–77.

Hongslo, J. K., Hongslo, C. F., Hasvold, O., and Holland, R. I., 1980b. Reduced fluoride sensitivity of liver cells from rats chronically exposed to fluoride, *Acta Pharmacol. Toxicol.* 47:355–358.

Iyengar, R., and Birnbaumer, L., 1981. Hysteretic activation of adenylyl cyclases. I. Effect of Mg ion on the rate of activation by guanine nucleotides and fluoride, *J. Biol. Chem.* 256:11036–11041.

Jowsey, J., Riggs, B. L., Kelly, P. J., and Hoffman, D. L., 1972. Effect of combined therapy with sodium fluoride, vitamin D and calcium in osteoporosis, *Am. J. Med.* 53:43–49.

Jowsey, J., Riggs, B. L., and Kelly, P. J., 1979. Fluoride in the treatment of osteoporosis, in *Continuing Evaluation of the Use of Fluorides*, E. Johansen, D. R. Taves, and T. O. Olsen (eds.), AAS Selected Symposium 11, Westview Press, Boulder, Colo., pp. 111–123.

Kleiner, H. S., and Allmann, D. W., 1982. The effects of fluoridated water on rat urine and tissue cAMP levels, *Arch. Oral Biol.* 27:107–112.

Largent, E. J., 1959. Excretion of fluoride, in *Fluorine and Dental Health*, J. C. Muhler and M. K. Hine (eds.), Indiana University Press, Bloomington, pp. 128–156.

Largent, E. J., 1961. *Fluorosis*, Ohio State University Press, Columbia, Ohio.

LeGeros, R. Z., and Suga, S., 1980. Crystallographic nature of fluoride in enameloids of fish, *Calcif. Tiss. Int.* 32:169–174.

Maduska, A. L., Ahokas, R. A., Anderson, G. D., Lipshitz, J., and Morrison, J. C., 1980. Placental transfer of intravenous fluoride in the pregnant ewe, *Am. J. Ostet. Gynecol.* 136:84–86.

Massmann, W., 1981. Reference values for renal excretion of fluoride, *J. Clin. Chem. Clin. Biochem.* 19:1039–1041.

Matthews, J., 1970. Changes in cell function due to inorganic fluoride, in *Pharmacology of Fluorides, Part 2, Handbook of Experimental Pharmacology*, Vol. XX, F. A. Smith (ed.), Springer-Verlag, Berlin, pp. 98–143.

McGown, E. L., and Suttie, J. W., 1974. Influence of fat and fluoride on gastric emptying of rats, *J. Nutr.* 104:909–915.

Messer, H. H., 1982. Bone cell membranes, *Cell. Orthop.* 166:256–276.

Messer, H. H., Wong, K., Wegner, M., Singer, L., and Armstrong, W. D., 1972. Effect of reduced fluoride intake by mice on haematocrit values, *Nature (New Biol.)* 240:218–219.

Messer, H. H., Armstrong, W. D., and Singer, L., 1973. Fertility impairment in mice on a low fluoride intake, *Science* 177:893–894.

Messer, H. H., Armstrong, W. D., and Singer, L., 1974. Essentiality and function of fluoride, in *Trace Element Metabolism in Animals*, Vol. 2, W. G. Hoekstra, J. W. Suttie, H. E. Ganther, and W. Mertz (eds.), University Park Press, Baltimore, pp. 425–437.

Moreno, E. C., Kresak, M., and Zahradnik, R. T., 1974. Fluoridated hydroxyapatite solubility and caries formation, *Nature* 247:64–65.

Moreno, E. C., Kresak, M., and Zahradnik, R. T., 1977. Physicochemical aspects of fluoride-apatite system relevant to the study of dental caries, *Caries Res.* 11 (suppl. 1):142–171.

Myers, H. M., 1975. The mechanism of the anticaries action of fluoride ion, in *Fluorides and Dental Caries*, E. Newbrun (ed.), Charles C. Thomas, Springfield, Ill., pp. 99–112.

National Academy of Sciences, 1971. *Fluorides*, National Academy of Sciences, Washington, D.C.

Newesely, H., 1961. Changes in crystal types of low solubility calcium phosphates in the presence of accompanying ions, *Arch. Oral Biol.* 6 (suppl.):174–180.

Nowak, T., and Maurer, P. J., 1981. Inhibition of yeast enolase. 2. Structural and kinetic properties and complexes determined by nuclear relaxation rate studies, *Biochemistry* 20:6901–6911.

Ophaug, R. H., Wong, K. M., and Singer, L., 1979. Lack of effect of fluoride on urinary cAMP excretion in rats, *J. Dent. Res.* 58:2036–2039.

Ophaug, R. H., Singer, S., Wong, K. M., and Messer, H. H., 1980. Effects of protracted fluoride intake on reproduction in mice, *J. Dent. Res.* 59:311.

Perkins, J. P., 1973. Adenyl cyclase, *Adv. Cyclic Nuc. Res.* 3:1–64.

Rasenick, M. M., and Bitensky, M. W., 1980. Partial purification and characterization of a macromolecule which enhances fluoride activation of adenylate cyclase, *Proc. Nat. Acad. Sci.* 77:4628–4632.

Repaske, M. G., and Suttie, J. W., 1979. Fluoride resistance in cell cultures, in *Continuing Evaluation of the Use of Fluorides*, E. Johansen, D. R. Taves and T. O. Olsen (eds.), AAAS Selected Symposium 11, Westview Press, Boulder, Colo., pp. 223–240.

Rich, C., and Feist, E., 1971. The action of fluoride on bone, in *Fluoride in Medicine*, T. L. Vischer (ed.), Hans Huber Publishers, Bern, pp. 70–87.

Riggs, B. L., Seeman, E., Hodgson, S. F., Taves, D. R., and O'Fallon, W. M., 1982. Effect of the fluoride/calcium regimen on vertebral fracture occurrence in postmenopausal osteoporosis, *New Engl. J. Med.* 306:446–450.

Schiffl, H., Hofmann, U., Huggler, M., and Binswanger, U., 1981. Renal fluoride excretion: Experimental evaluation of the role of extracellular volume status during intact and impaired kidney function, *Nephron* 29:245–249.

Schnitzer-Polokoff, R., and Suttie, J. W., 1981. Effect of fluoride on the absorption of dietary fat in rats, *J. Nutr.* 111:537–544.

Schroeder, H. A., Mitchener, M., Balassa, J. J., Kanisawa, M., and Nason, A. P., 1968. Zirconium, niobium, antimony and fluorine in mice: Effects on growth, survival and tissue levels, *J. Nutr.* 95:95–101.

Schwartz, K., and Milne, D. B., 1972. Fluorine requirement for growth in the rat, *Bioinorg. Chem.* 1:331–338.

Shen, Y. W., and Taves, D. R., 1974. Fluoride concentration in the human placenta and maternal and cord blood, *Am. J. Obstet. Gynecol.* 119:205–207.

Singer, L., and Armstrong, W. D., 1962. Comparison of fluoride contents of human dental and skeletal tissues, *J. Dent. Res.* 41:154–157.

Singer, L., Armstrong, W. D., Zipkin, I., and Frazier, P. D., 1974. Chemical composition and structure of fluorotic bone, *Clin. Orthop.* 99:303–312.

Singh, A., and Jolly, S. S., 1970. Chronic toxic effects on the skeletal system, in, *Fluorides and Human Health*, WHO Monograph Series, No. 59, WHO, Geneva, pp. 238–249.

Smith, F. A., 1966. Metabolism of inorganic fluoride, in *Pharmacology of Fluorides, Part 1, Handbook of Experimental Pharmacology*, Vol. XX, F. A. Smith (ed.), Springer-Verlag, Berlin, pp. 53–140.

Smith, F. A., Gardner, D. E., Leone, N. C., and Hodge, H. C., 1960. The effects of the absorption of fluoride. V. The chemical determination of fluoride in human soft tissues following prolonged ingestion of fluoride at various levels, *Arch. Indust. Health* 21:330–332.

Spencer, H., Kramer, L., Osis, D., Wiatrowski, E., Norris, C., and Lender, M., 1980. Effect of calcium, phosphorus, magnesium and aluminum on fluoride metabolism in man, *Ann. N.Y. Acad. Sci.* 355:181–193.

Spoerke, D. G., Bennett, D. L., and Gullekson, D. J. K., 1980. Toxicity related to acute low dose sodium fluoride ingestions, *J. Fam. Pract.* 10:139–140.

Srivastava, S. K., Gillerman, E., and Modak, M. J., 1981. The artifactual nature of fluoride inhibition of reverse transcriptase and associated ribonuclease H, *Biochem. Biophys. Res. Comm.* 101:183–188.

Stookey, G. K., Dellinger, E. L., and Muhler, J. C., 1964. *In vitro* studies concerning fluoride absorption, *Proc. Soc. Exp. Biol. Med.* 115:298–301.

Susheela, A. K., and Singh, M., 1982. Adenyl cyclase activity following fluoride ingestion, *Toxicol. Lett.* 10:209–212.

Suttie, J. W., and Phillips, P. H., 1960. Fat utilization in the fluoride fed rat, *J. Nutr.* 72:429–434.

Suzue, G., 1962. The enzymic synthesis of bacterial phytoene from mevalonic acid-2-^{14}C, *J. Biochem.* 51:246–252.

Tao, S., and Suttie, J. W., 1976. Evidence for a lack of an effect of dietary fluoride level on reproduction in mice, *J. Nutr.* 106:1115–1122.

Taves, D. R., 1978. Fluoridation and mortality due to heart disease, *Nature* 272:361–362.

Taves, D. R., and Guy, W. S., 1979. Distribution of fluoride among body compartments, in *Continuing Evaluation of the Use of Fluorides*, E. Johansen, D. R. Taves, and T. O. Olsen (eds.), AAAS Selected Symposium 11, Westview Press, Boulder, Colo., pp. 159–185.

Townsend, D., and Singer, L., 1977. Effect of fluoride on the serum lipids of guinea pigs, *J. Nutr.* 107:97–103.

Vatassery, G. T., Ophaug, R. H., and Singer, L., 1980. The effect of fluoride intake on the total lipid, cholesterol and vitamin E levels in sera and liver of guinea pigs on high fat diet, *Life Sci.* 27:1961–1966.

Wagner, M. J., 1962. Absorption of fluoride by the gastric mucosa in the rat, *J. Dent. Res.* 41:667–671.

Wallace, P. C., 1953. The metabolism of ^{18}F in normal and chronically fluorosed rats, University of California Report UCRL-2196, Berkeley, California.

Waterhouse, C., Taves, D., and Munzer, A., 1980. Serum inorganic fluoride: Changes related to previous fluoride intake, renal function and bone resorption, *Clin. Sci.* 58:145–152.

Wegner, M. E., Singer, L., Ophaug, R. H., and Magil, S. G., 1976. The interrelation of fluoride and iron in anemia, *Proc. Soc. Exp. Biol. Med.* 153:414–418.

Weidmann, S. M., and Weatherell, J. A., 1970. Distribution of fluorides. 3. Distribution in hard tissues, in *Fluorides and Human Health*, WHO Monograph Series, No. 59, WHO, Geneva, pp. 104–128.

Whitford, G. M., and Pashley, D. H., 1979. The effect of body fluid pH on fluoride distribution, toxicity and renal clearance, in *Continuing Evaluation of the Use of Fluorides*, E. Johansen, D. R. Taves, and T. O. Olsen (eds.), AAAS Selected Symposium 11, Westview Press, Boulder, Colo., pp. 187–221.

Whitford, G. M., Pashley, D. H., and Stringer, G. I., 1976. Fluoride renal clearance: A pH-dependent event, *Am. J. Physiol.* 230:527–532.

Whitford, G. M., Schuster, G. S., Pashley, D. H., and Venkateswarlu, P., 1977. Fluoride uptake by *Streptococcus mutans* 6715, *Infect. Immun.* 18:680–687.

Wiseman, A., 1970. Effect of inorganic fluoride on enzymes, in *Pharmacology of Fluorides, Part 2, Handbook of Experimental Pharmacology*, Vol. XX, F. A. Smith (ed.), Springer-Verlag, Berlin, pp. 48–97.

Manganese 5

Carl L. Keen, Bo Lönnerdal, and Lucille S. Hurley

5.1 Introduction

Manganese, although widely distributed in the biosphere, occurs in only trace amounts in animal tissues, in which concentrations of 2–4 µg/g (<0.06 µM) are considered high. Despite these low concentrations, manganese is essential for several biological functions, but its precise biochemical roles have not been delineated. Although manganese has long been known to be a constituent of animal tissues (Bertrand and Medigreceanu, 1913), it was first shown to be required by animals when Kemmerer and coworkers (1931) and Orent and McCollum (1931) demonstrated poor growth in mice and abnormal reproduction in rats fed diets devoid of the element. Today it is known that manganese deficiency results in a wide variety of structural and physiological defects. Alternatively, an excess of the element can also result in severe pathologies, particularly of the central nervous system. In this chapter, current knowledge of the biochemistry of manganese is reviewed; where possible, studies are emphasized in which the biochemistry of manganese has been studied in relation to the intact animal.

5.2 Manganese Concentration in Animal Tissues

The concentration of manganese in animal tissues is very low. The total amount of manganese in an adult human has been estimated to be 10–20 mg (Schroeder *et al.*, 1966). These low concentrations necessitate the use of very sensitive analytical methods. Neutron activation analysis and X-ray fluorescence spectroscopy have been used extensively for the measurement of manganese in

---

Carl L. Keen, Bo Lönnerdal, and Lucille S. Hurley • Department of Nutrition, University of California, Davis, California 95616.

tissues and fluids. Using these methods, concentrations of manganese as low as 1 ng/ml have been reported for spinal fluid, whole blood, and urine (Cotzias *et al.*, 1968). While neutron activation analysis and X-ray fluorescence spectroscopy have the necessary sensitivity for measuring tissue manganese, these methods are not readily available and are very expensive. An alternative method that is readily available and relatively inexpensive is atomic absorption spectrophotometry. Flame atomic absorption spectrophotometry can be used for measuring manganese when it is present in the parts per million range (μg/g), while graphite furnace atomic absorption spectroscopy is used to measure manganese in the parts per billion (ng/g) range with a high degree of accuracy (Clegg *et al.*, 1982).

Manganese tends to be highest in tissues rich in mitochondria, as it is more concentrated in mitochondria than in cytoplasm or other cell organelles. Hair can accumulate high concentrations of manganese, and it has been suggested that hair manganese levels may reflect body manganese stores (Hidiroglou, 1979). High concentrations of manganese are usually found in pigmented structures, such as the retina, dark skin, and melanin granules (Van Woert *et al.*, 1967). The concentration of manganese in the skeleton is relatively high (2–8 μg/g) and can account for up to 25% of total body manganese; however, it is apparently not readily mobilized (Underwood, 1977). Most of the bone manganese is believed to be deposited in the inorganic portion of bone, but a small proportion is associated with the organic matrix (Fore and Morton, 1952).

Typical concentrations of manganese in several tissues of the rat are shown in Table 5-1. Unlike some other trace elements, the concentration of manganese

Table 5-1. Manganese in Rat Tissues

Tissue	Mn[a] (mg/kg)
Liver	1.95 ± 0.15
Pancreas	1.35 ± 0.16
Testis	1.33 ± 0.10
Intestine	0.74 ± 0.07
Kidney	0.72 ± 0.05
Brain	0.47 ± 0.02
Heart	0.41 ± 0.04
Spleen	0.24 ± 0.09
Thymus	0.24 ± 0.04
Milk	0.21 ± 0.03
Lung	0.11 ± 0.02
Muscle (soleus)	0.07 ± 0.01
Blood	0.012 ± 0.002

[a] Mean ± SEM: $N = 6$.

in various tissues is fairly similar in most species and is relatively constant with age. In striking contrast to several of the other essential trace elements, the fetus does not normally accumulate liver manganese before birth, and values are actually lower in fetal liver than in adult liver (Meinel *et al.*, 1979). However, the feeding of a diet high in manganese to the mother during pregnancy increases the manganese concentration of the fetal liver. The observation of low fetal liver manganese is particularly surprising when considering the very low concentrations of this element in milk (5–20 µg/L Lönnerdal *et al.*, 1981). It may be hypothesized that the lack of high fetal storage of manganese is related to the toxicity of high concentrations of the element and the inability of the young animal to excrete manganese from the body (see the following discussion).

The partition between free and bound manganese in tissues has recently received attention. Using electron paramagnetic resonance (EPR), Mn^{2+} concentration of rat hepatocyte cells has been reported to be approximately 35 nmol/ml cell water (Ash and Schram, 1982). The percentage of free to bound Mn^{2+} in hepatocytes was shown to be dependent on the physiological state; the free Mn^{2+} concentration of whole cell was 0.71 nmol/ml cell water in the fed rats and 0.25 nmol/ml cell water in fasted rats. Thus there was considerably more intracellular Mn^{2+} bound to molecules in hepatocytes from fasted rats than in those from fed rats. It has been suggested that this change in free Mn^{2+} concentration of fasting could reflect an increased amount or activity of gluconeogenic enzymes that are dependent on manganese activation such as phosphoenolpyruvate carboxykinase. The kinetic constant for activation of phosphoenolpyruvate carboxykinase by Mn^{2+} is 1.7 µM, when Mg guanosine 5′-triphosphate and free Mg^{2+} are in excess. Manganese can activate the enzyme up to 10-fold, primarily by affecting Vmax; thus, since the intracellular free Mn^{2+} concentration is near the activation constant for phosphoenolpyruvate carboxykinase, fluctuations in its level would be predicted to affect the activity of the enzyme *in vivo* (Ash and Schramm, 1982). In a broader context, it has been suggested that changes in intracellular free Mn^{2+} may be an important mechanism of cellular metabolic control, in a manner analogous to that of free Mg^{2+} and Ca^{2+} (Williams, 1982). The mechanism(s) underlying the changes in cellular free Mn^{2+} levels are not known but could be related to mitochondrial uptake or release of Mn^{2+}, possibly through Ca^{2+} channels (Jeng and Shamoo, 1980) or the induction of molecules that bind Mn^{2+}. Supporting the former idea is the observation that isoforms of phosphoenolpyruvate carboxykinase responded differently to manganese activation; both Mn^{2+} sensitive and insensitive forms were found (Schramm *et al.*, 1981; Brinkworth *et al.*, 1981).

The possibility that the mitochondrial Ca^{2+} transport system also transports Mn^{2+} is intriguing, as it suggests that metabolic signals that cause release of Ca^{2+} from mitochondria, an event necessary for maximizing phosphorylase activity for gluconeogenesis, would also stimulate release of Mn^{2+}, which would then activate phosphoenolpyruvate carboxykinase. The release of Ca^{2+} and Mn^{2+} from mitochondria through a common mechanism would allow them to act synergistically to increase gluconeogenesis.

Treatment of hepatocytes with digitonin, which makes plasma membranes permeable to ions, results in the release of about 13 nmol Mn^{2+}/ml cell water, indicating that about one-third of total cell Mn^{2+} is extramitochondrial and is not tightly bound to proteins. Since the cellular concentration of the three manganese metalloenzymes—pyruvate carboxylase, arginase, and superoxide dismutase—is approximately 20 μM, the majority of bound manganese in liver can be accounted for by these three enzymes.

5.3 Metabolism of Manganese

5.3.1 Absorption

Studies on the absorption and excretion of manganese have been carried out primarily with the isotope ^{54}Mn, a gamma emitter that has a half-life of 312 days. The mechanisms of manganese absorption are not well understood; this is an area requiring considerably more research. Absorption of the element apparently occurs equally well throughout the length of the small intestine (Thomson *et al.*, 1971). The absorption of dietary manganese is relatively poor and is not thought to be normally under homeostatic control. For the adult human, it has been reported that approximately 3–4% of dietary manganese is absorbed, a figure that is similar to that reported for the adult rat. The percentage of manganese absorbed from a meal is, to a large extent, independent of the amount of manganese in the diet, or the body burden of the element (Mena, 1981).

The percentage of manganese absorbed from a meal can be influenced by other dietary factors. Ethanol has been reported to increase the intestinal transport of manganese *in vitro* and *in vivo* over a period of 4 hours (Schafer *et al.*, 1974). Consistent with this observation, prolonged ethanol feeding can result in increased hepatic manganese levels that are in part reflected by increased manganese superoxide dismutase activity (Dreosti *et al.*, 1982; Keen *et al.*, 1983b). High levels of dietary calcium, phosphorus, and phytate have been shown to increase the requirements for manganese of several species, possibly by adsorption of manganese in the intestinal tract, resulting in a reduction of soluble element that can be absorbed (Davies and Nightingale, 1975).

Considerable work has been done on the interaction of iron and manganese. Matrone and coworkers (1959) demonstrated an interaction between these elements affecting hemoglobin production. They found that high concentrations of dietary manganese depressed the formation of hemoglobin in piglets, presumably by decreasing iron absorption. A similar mechanism probably explains the anemia observed in sheep grazing on manganese-rich pastures (Underwood, 1977). Conversely, in chickens, high levels of dietary iron accentuate the severity of perosis, presumably by decreasing manganese absorption (Wilgus and Patton, 1939).

The interaction of manganese and iron has also been studied in iron-deficient animals. Humans with iron deficiency anemia have manganese absorption values as high as 7%, twice normal levels (Mena, 1981). Thomson and Valberg (1972), using iron-deficient rats, found that the passage of iron, cobalt, and manganese from the gastrointestinal lumen to the body of the animal had characteristics of carrier-mediated transport, with the order of transport being cobalt > iron > manganese. These investigators also reported that the mucosal and serosal binding sites for the metals were different, with manganese, cobalt, and iron competing for binding sites at both steps.

The absorption and retention of manganese from foods low in iron, such as milk, are relatively high. If milk is supplemented with iron, the percentage of manganese absorbed is reduced (Gruden, 1977). Using weanling mice, it has been shown that the lower absorption of manganese from iron-supplemented milk than from nonsupplemented milk was reflected by a significantly lower concentration of manganese in the liver (Keen et al., 1984a). Although the functional significance of this finding remains to be determined, the level of iron supplementation was similar to that of iron-fortified infant formulas. Therefore this study has important implications for infant nutrition, as an unwanted side effect of iron supplementation may be induction of suboptimal manganese nutrition.

Increased absorption of manganese has been reported during pregnancy in sows (Kirchgessner et al., 1981) and with coccidiosis infection in chickens (Southern and Baker, 1983). The mechanisms underlying these changes in manganese absorption are not known.

In addition to the influence of other dietary factors on manganese absorption, the molecular localization of the element may have a considerable impact on its bioavailability. An example of this point can be taken from studies on the localization of manganese in milk and its absorption from this food. The concentration of manganese in human milk is quite low, about 10 times lower than that of other species investigated (Lonnerdal et al., 1981). The concentration varies during the lactation period in a pattern dissimilar to that of other trace elements; concentrations are high in early lactation (6 µg/L), then decline to a level around 4 µg/L, and subsequently, in late lactation, increase to about 6–8 µg/L (Vuori, 1979). Cow's milk contains 20–50 µg manganese per liter and infant formulas contain widely varying amounts (0–7800 µg/L) (Lonnerdal et al., 1983a). The concentration of manganese in both human and cow's milk has been reported to be influenced by maternal manganese intake (Lonnerdal et al., 1981). However, it is not known if the increased milk manganese concentrations seen with dietary supplementation are due to correction of a previous suboptimal intake, or if manganese is different from other trace elements, for which a strong effect of dietary intake on the concentration in milk is not seen (Vuori *et al.*, 1980).

Little is known about the biochemistry of manganese in milk. An extrinsic label of manganese added to skimmed human milk has been found to bind to two protein fractions (Chan *et al.*, 1982). The major fraction was estimated to have a molecular weight of 407,000 daltons and the minor fraction, a molecular weight of 128,000 daltons. The high-molecular-weight fraction appeared homogeneous with respect to molecular weight and charge. The bound manganese was in a metal-to-protein ratio of 1 : 1. Cow's milk was found to bind manganese in three fractions, all different from those in human milk. The molecular weights of the fractions in cow's milk were 234,000, 83,000, and less than 1000. Lönnerdal *et al.* (1984) found that in human milk the major part of manganese (71%) was located in the decaseinated whey fraction, while 11% of the manganese in human milk was bound to casein, and 18% was bound to the lipid fraction. In contrast, manganese in cow's milk was predominantly bound to casein (67%); 32% was in the whey, and very little (1%) was in the lipids. Within the whey fraction, 75% of the manganese in human milk was associated with a protein with a molecular weight of about 80,000; 19% to high molecular weight proteins; and 6% to low molecular weight ligands. The 80,000 MW protein was identified as lactoferrin by immunoaffinity chromatography. Supporting this result obtained by extrinsic labeling, lactoferrin was isolated from human milk whey and by means of atomic absorption spectrophotometry was found to contain the majority of native manganese. However, the amount of manganese bound to lactoferrin was considerably lower than the amount of iron bound to this protein.

There is very little information regarding the bioavailability of manganese from milk. Lonnerdal *et al.* (1983b) found that manganese added to human milk resulted in significantly higher liver manganese levels in weanling mice than in those fed cow's milk with added manganese. In contrast, Bates *et al.* (1983), using the everted gut sac method in rats, found that the transport and uptake of manganese was less in the presence of human milk and its isolated manganese ligands than it was from bovine milk or infant formula. Further studies are clearly needed on the bioavailability of manganese from various milk sources.

5.3.2 Transport and Tissue Distribution

Manganese absorbed from the gastrointestinal tract enters into the portal blood, where most of it becomes bound to plasma α_2-macroglobulin. One manganese atom apparently binds to one globulin molecule (Cotzias *et al.*, 1960). While manganese is found in erythrocytes, this manganese is incorporated during erythropoiesis. Free manganese (Mn^{2+}) and α_2-macroglobulin bound Mn^{2+} in the portal blood are rapidly taken up by the liver. Like iron, a fraction of the manganese is oxidized (to Mn^{3+}), possibly by ferroxidase I, and is then bound to the plasma transport protein transferrin. Transferrin-bound manganese can be taken up by extrahepatic tissue (Gibbons et al., 1976). The association constant of manganese for transferrin is lower than that of iron (Worwood, 1974). Foradori and Dinamarca (1972) have reported that transferrin molecules bind manganese and iron simultaneously. Manganese flux in plasma is extremely rapid; for humans it has been reported that only 1% of intravenously injected ^{54}Mn is still present in blood after 10 minutes (Cotzias et al., 1968).

Within the cells, manganese is found predominantly in mitochondria (Maynard and Cotzias, 1955), and thus tissues rich in mitochondria, such as liver, kidney, and pancreas have relatively high manganese concentrations and take up much of an absorbed or injected ^{54}Mn dose. The uptake and release of manganese by mitochondria may be via a Ca^{2+} carrier (Jeng and Shamoo, 1980). It has been shown by Hughes et al. (1966) that the amount of manganese leaving the liver for extrahepatic tissues can be increased by glucocorticoid hormones or ACTH injection. However, the physiological significance of this observation is not clear as adrenalectomy does not alter tissue manganese levels, except to raise concentrations slightly in animals fed high manganese diets (Hughes et al., 1966). Clearly, hormonal regulation of manganese metabolism is an area needing more research.

The retention of ingested ^{54}Mn by the adult human has been estimated at 10% 14 days after feeding an oral dose. For premature and term infants, the 10-day retention of ^{54}Mn has been cited as 16 and 8%, respectively (Mena, 1981). The high retention of manganese by premature infants may be the result of increased manganese absorption by the immature gastrointestinal tract, immaturity of mechanisms involved in manganese excretion, and/or a higher requirement for manganese because of tissue synthesis in rapid growth. Kirchgessner et al. (1981), using everted intestinal sacs from rats of varying ages, showed a decrease in manganese absorption with increasing age that was not a function of changing diet. Miller et al. (1975) have shown that for several animal species there is a virtual absence of manganese excretion in the neonatal period because of low bile fluid output. The observation of high manganese retention by infants suggests that they may be particularly susceptible to manganese toxicity.

In adult men, manganese absorbed by the oral route has a half-life in the body of about 10 days. In striking contrast, manganese given by injection into the systemic circulation has a half-life of about 40 days. This difference in half-life reflects the longer turnover time of manganese in extrahepatic tissues than in the liver, as a higher proportion of manganese injected into the systemic system will go into these tissues (Mena, 1981). Mahoney and Small (1968) have reported that the clearance of an intravenous dose of ^{54}Mn can be described by a curve characterized by two exponential components. The major part (~70%) of an injected dose is eliminated by a "slow" pathway with a half-time of 39 days. The minor part of the dose is eliminated by the "fast" component and has a half-life of 4 days. In a subject who had a low manganese intake for 6 months, the proportion of ^{54}Mn eliminated by the "slow" pathway increased to 84% (rather than 70%). The half-time of this pathway increased to 90 days while that of the "fast" component remained the same. A large dose of oral manganese subsequently given to the same subject markedly increased the rate of excretion of ^{54}Mn. Preloading with manganese for 35 days caused a dramatic decrease in the fraction of ^{54}Mn eliminated by the "slow" pathway with less effect on half-time. Thus both long-term manganese status and dietary manganese intake appear to affect the elimination rate.

An interaction between iron status and manganese excretion was also shown by these investigators in that an iron-deficient subject showed a decreased percentage of manganese eliminated by the "slow" pathway followed by a decrease in half-time. In contrast, oral iron therapy caused a decrease in elimination rate.

5.3.3 Excretion

Absorbed manganese which is not retained by the body is almost totally excreted with the feces, with only trace amounts found in the urine. If the rectum is ligated, manganese excretion is effectively abolished (Papavasiliou *et al.*, 1966). It is mainly through excretion of manganese into the intestinal tract that homeostatic regulation of manganese levels in tissues occurs, although evidence has been provided that absorption can play a role in manganese homeostasis in some circumstances (Britton and Cotzias, 1966; Abrams *et al.*, 1976; King *et al.*, 1980). In contrast to the absorptive process, the excretion of manganese is affected only by the body burden of the element and is not influenced by other dietary metal ions (Cotzias and Greenough, 1958). Manganese that is taken up by the liver and not shunted to extrahepatic tissues is rapidly incorporated into the mitochondrial and lysosomal compartments (Suzuki and Wada, 1981). While it has been reported that manganese is also rapidly incorporated into the nuclear fraction, these findings may be the result of contamination into the nuclear fraction by the mitochondrial fraction.

Suzuki and Wada (1981) have suggested that low-molecular-weight manganese complexes from the lysosomes are subsequently transferred to bile for excretion. The excretion of manganese through bile is very rapid and apparently occurs in two "waves." The first wave is due to the clearance of initially absorbed manganese; the second is due to a combination of initially absorbed manganese and manganese from the enterohepatic circulation (Bertinchamps *et al.*, 1966). In rats, intravenously injected ^{54}Mn is cleared rapidly from the blood and can be shown to concentrate in bile within an hour after injection. Within 48 hours most of the injected dose will be eliminated through the biliary route. This rapid transfer of manganese into bile is in contrast to the relatively slow kinetics observed for several other metals, including iron, copper, zinc, cadmium, lead, and mercury, suggesting that an active transport system with high specificity exists for manganese transfer into bile. Several substances, including L-Dopa, dopamine, glucagon, and cyclic AMP, have been reported to depress biliary excretion of manganese (Papavasiliou et al., 1968). The mechanism(s) by which these compounds reduce the transfer of manganese into bile is not known, but it is thought to involve cyclic AMP changes in liver.

Small amounts of manganese are also excreted through the pancreatic juice; these amounts are increased if the bile flow is interrupted (Papavasiliou et al., 1966). In addition to bile and pancreatic juice, a small amount of manganese is excreted through the intestinal wall in the duodenum, jejunum, and to a lesser extent, the ileum (Bertinchamps et al., 1966). It should be pointed out that much of the work investigating biliary excretion of manganese has been done with acutely, rather than chronically, cannulated animals. For biliary excretion of copper and zinc, it has been shown that the cannulation process perturbs normal secretion for up to 5 days (Schneeman et al., 1983). It is not known if there is a similar effect of cannulation on biliary excretion of manganese.

5.4 Biochemistry of Manganese

5.4 Manganese Chemistry

The characteristic oxidative state of manganese in solution, in metal enzyme complexes, and in metalloenzymes is Mn^{2+}. Complexes of Mn^{2+} have a coordination number of 6 and thus form an octahedral coordination complex (O'Dell and Campbell, 1971; Ash and Schramm, 1982). Divalent manganese is usually high spin d^5, so there is no crystal field stabilization energy. As a result, Mn^{2+} complexes can distort from regular octahedral geometry with little increase in energy. Divalent manganese, like Fe^{3+}, has a high affinity for imidazole, in contrast to other divalent cations, such as Zn^{2+}, Cu^{2+}, and Cd^{2+}, which have higher affinities for thiol. Overall the binding affinities of ligands for Mn^{2+} are much lower than those for Zn^{2+} and Cu^{2+}. For these reasons, manganese would not be predicted to compete for protein binding sites with Zn^{2+} or Cu^{2+} (Luckney and Venugopal, 1977).

Mn^{3+} is important *in vivo*. For example, Mn^{3+} is the oxidative state of manganese in manganese superoxide dismutase (see the following discussion) and is the form in which transferrin binds manganese, and probably the form of manganese that interacts with Fe^{3+}. In addition, Borg and Cotzias (1962) have reported that the interaction between Mn^{2+} and phenothiazine drug derivatives in brain tissue may depend on a change in valence state ($Mn^{2+} - Mn^{3+}$). Phenothiazine derivatives have a broad spectrum of pharmacological properties, including tranquillization, antiemesis, antipruritic and antibiotic action, antagonism of serotonin, and hypocholesterolemic effects. Borg and Cotzias (1962) proposed that a common denominater in these drug effects may be an interaction with manganese that involves free radical generation. The chelation of Mn^{3+} in biologic systems should be more avid than that of Mn^{2+} because of its smaller ionic radius. Ingested Mn^{2+} is thought to be converted into Mn^{3+} in the alkaline medium of the duodenum (O'Dell and Campbell, 1971). The concentration of Mn^{4+} is very low because of the low solubility of its dioxides.

Most manganese is paramagnetic, and the presence of the unpaired electron endows the metal ion with magnetic properties that can be utilized in nuclear magnetic and electron spin resonance (EPR) studies. Free Mn^{2+} shows a sextet hyperfine spectrum in EPR, while bound Mn^{2+} does not. This property of manganese has been exploited for studying the interactions of dissociable manganese substrate and enzyme. It can also be utilized for studying enzyme-metal-ion-substrate complexes, where the metal normally occurring *in vivo* is not paramagnetic, but can exchange with Mn^{2+}. For example, Mn^{2+} has been used as a probe to study the Zn^{2+} containing enzymes leucine aminopeptidase and carboxypeptidase (Taylor *et al.*, 1982; Niccolai *et al.*, 1982).

Magnetic resonance techniques provide a method of investigation in which the structure and formation of Mn^{2+} containing compounds of low and high molecular weights, their metabolism, and transport can be studied. A comprehensive review of the use of Mn^{2+} as a magnetic relaxation probe in studies of enzymatic mechanisms and of structures of metal-containing biomacromolecules has recently been published (Niccolai *et al.*, 1982).

5.4.2 Manganese as a Cofactor and in Metalloenzymes

Metal-Activated Reactions

Like other essential metals, manganese can function both as an enzyme activator, where the element acts as a dissociable cofactor, and as a constituent of a metalloenzyme, where the element has a structural or functional role. For manganese-activated reactions, the enzyme–metal interaction can involve either the chelation of the metal ion with a phosphate-containing substrate (particularly ATP) or a direct interaction of the metal ion with the protein. The chemistry of the Mn^{2+} ion is similar to that of the Mg^{2+} ion. Therefore, for most enzymatic reactions that are activated by Mn^{2+}, the activation is nonspecific, with magnesium also able to function as the metal activator. Examples of nonspecific

Mn^{2+} activated enzymes include kinases, decarboxylases, hydrolases, and transferases (Vallee and Coleman, 1964).

It is, of course, important to emphasize that the action of Mn^{2+} and Mg^{2+} can be quite different. For example, both complex with ATP; however, Mn^{2+} interacts with all three phosphate groups and with either the pyridine ring or a water molecule, while Mg^{2+} interacts with the β and γ-phosphates only, resulting in different interactions with ATP-activated proteins (Cohn and Hughes, 1962). It has been shown that, for at least one enzyme, β-galactosidase from Escherichia coli, binding of Mn^{2+} or Mg^{2+} affects the substrate acted upon. This enzyme has four subunits, each of which binds one Me^{2+}. Interestingly, the binding of Mn^{2+} is cooperative, while Mg^{2+} is noncooperative. When Mn^{2+} is bound by the enzyme, allolactose is the substrate, whereas when Mg^{2+} is bound, the preferred substrate is lactose (Woulfe-Flanagan and Huber, 1978).

It must be emphasized that the demonstration of manganese activation *in vitro* does not prove a role for the element in enzyme activation *in vivo*. In addition, a problem with those enzymes not specifically activated by manganese is the difficulty of making direct correlations between pathological defects seen with manganese deficiency and specific biochemical lesions. There are some enzymes, however, such as glycosyl transferase, that are highly specific for Mn^{2+} activation and that can be studied with regard to tissue pathology. In general, the proteins that are specifically activated by Mn^{2+} bind manganese more tightly than those that can also be activated by Mg^{2+}; as a result, the addition of Mn^{2+} to a crude preparation of the enzyme will have little effect on its activity. In contrast, proteins that can be activated by either Mn^{2+} or Mg^{2+} tend to have a higher dissociation constant and the activity of the enzyme in crude preparations increases upon the addition of added Mn^{2+}.

It has been proposed by McEuen (1981) that proteins that can be activated by either Mn^{2+} or Mg^{2+} can be grouped into three classes: those in which the optimal activity induced by Mn^{2+} is greater than the optimal activity induced by Mg^{2+}; those that have similar optimal activities with Mn^{2+} or Mg^{2+} addition; and those in which Mg^{2+} can induce an activity greater than that of Mn^{2+}. The use of classification is limited at present by lack of detailed information on the cellular concentration of free Mn^{2+} and Mg^{2+} and of association and dissociation constants for Mn^{2+} and Mg^{2+} for the different enzymes under cellular conditions. For example, *in vitro,* glutamine synthetase can use either Mn^{2+} or Mg^{2+} for its stability and activity, but the affinity of this enzyme for Mn^{2+} ions is 400 times that of Mg^{2+} ions; however, it is not known which metal activates the enzyme *in vivo* (O'Dell and Campbell, 1971). When more information of this type becomes available, the proposed classification of McEuen could provide a useful way of identifying those proteins for which manganese activation has an important *in vivo* function and that can be affected by fluxes in intracellular manganese concentrations.

An interesting exception to the Mn^{2+}/Mg^{2+} activated enzymes is phosphoenolpyruvate carboxykinase (EC 4.1.1.32). For this manganese-activated enzyme Fe^{2+}, not Mg^{2+}, is the principal competing cation (MacDonald et al., 1978). Activation by Fe^{2+}, however, required the presence of a protein called "ferroactivator." This protein has a MW of 10^5 and is distinct from ferroxidase. Ferroactivator is found only in the cytoplasm and has been reported to increase in tissues undergoing active gluconeogenesis, such as occurs with starvation and diabetes, indicating that part of the control of phosphoenolpyruvate carboxykinase is through Fe^{2+} (MacDonald et al., 1978). Activation of isolated phosphoenolpyruvate carboxykinase by manganese requires the presence of sulfate (MacDonald et al., 1978). Some examples of manganese-specific proteins and manganese-activated enzymes that may be influenced by nutritional status follow. For a more detailed discussion of manganese-activated enzymes and manganese metalloproteins, the excellent review by McEuen is recommended (1981). For a general examination of the roles of metals in enzyme catalysis, the review by Mildvan (1970) is helpful.

As stated previously, one exception to the nonspecific activation of enzymes by manganese is the manganese-specific activation of glycosyl transferases. That the activating role of manganese for this enzyme class cannot be fulfilled *in vivo* by other divalent cations is demonstrated by the abnormal glycosaminoglycan metabolism associated with manganese deficiency (Leach, 1971). Evidence that the *in vivo* activity of the glycosyl transferase enzymes is suboptimal in manganese-deficient animals has been provided by the *in vitro* studies of Leach et al. (1969). These investigators showed that when enzymes necessary for chondroitin sulfate synthesis in the 105,000-g particulate fraction from control and manganese deficient chicken tissues were compared, the preparations from deficient tissues incorporated more radioactive substrate than did those of controls, demonstrating a higher number of receptor sites. Effects of manganese deficiency that have been linked to abnormal glycosaminoglycan and glycoprotein metabolism include skeletal abnormalities, congenital ataxia, and impaired egg shell formation (Leach, 1971; Hurley, 1981). The general reaction may be depicted as follows:

UDP-sugar-nucleotide + acceptor → product (acceptor-sugar) + UDP

The reason that manganese alone activates the glycosyl transferase enzymes *in vivo* is not clear, as Co^{2+} will activate these enzymes *in vitro,* but feeding high levels of Co^{2+} does not alleviate the signs of manganese deficiency (Leach et al., 1969). A specific example of a glycosyl transferase that has been shown to require manganese is galactosyl transferase (EC 2.4.1.38) isolated from milk. This enzyme has two metal ion binding sites, one of which is specific for Mn^{2+}

(Tsopanakis and Herries, 1978). Galactosyl transferase has been studied by Morrison and Ebner (1971) with regard to the action of manganese on the protein. They found that the reactants added in the order of Mn^{2+}, UDP-sugar, and acceptor. The manganese and the enzyme formed a complex that did not dissociate after each catalytic cycle. A similar essential activating role for manganese has also been proposed for xylosyltransferase (McNatt et al., 1976). This enzyme, isolated from chick cartilage, catalyzes the transfer of xylose in the formation of the linkage between proteins and polysaccharides. Cartilage isolated from manganese-deficient chickens was found to have a higher rate of xylose incorporation than control cartilage, again suggesting a higher concentration of xylose acceptor sites in the manganese-deficient tissue *in vivo*.

Manganese Metalloenzymes

In contrast to other essential transition elements, relatively few manganese metalloenzymes have been identified in higher organisms. Manganese-containing metalloenzymes include arginase, pyruvate carboxylase, and superoxide dismutase. Arginase (EC 3.5.3.1; MW 120,000), the cytosolic enzyme responsible for urea formation, contains 4 g-atoms of Mn^{2+} per mole of enzyme, two of which cannot be removed without irreversible loss of activity. The other two can be removed by exhaustive dialysis with a resultant loss of 50% of activity. Addition of Mn^{2+} to the system restores activity to 100% (Hirsch-Kolb et al., 1971). The activity of arginase is affected by dietary status. Thus rats fed diets deficient in manganese (0.7 ppm Mn) had levels of liver arginase activity that were only 30% of those found in rats fed diets adequate in manganese (50 ppm), while liver arginase activity in rats fed diets containing 6 ppm manganese were 40% of control values (Kirchgessner and Heiseke, 1978). The functional significance of lower arginase activity in manganese-deficient animals is not known. However, it seems apparent that in birds, the pathological lesions characteristic of manganese deficiency cannot be ascribed to an effect on this enzyme, since birds do not make urea. In contrast to manganese-deficient animals, rats rendered diabetic by streptozotocin injection showed liver arginase activity two to three times higher than normal. This increased activity occurred in parallel with a higher concentration of liver manganese and was not due to a larger amount of arginase protein. Thus the activity of arginase can apparently be modulated by the intracellular concentration of manganese (Bond et al., 1983).

Pyruvate carboxylase (EC 6.4.1.1; MW 500,000) is found in the mitochondria and has 4 moles of Mn^{2+} per mole of enzyme (Scrutton et al., 1972). The enzymatic reaction of pyruvate carboxylase follows:

$$\text{pyruvate} + HCO_3^- + ATP \xrightarrow{\text{Acetyl-CoA } Mg^{2+}} \text{Oxalacetate} + ADP + Pi$$

Pyruvate carboxylase is essential for the synthesis of carbohydrate from pyruvate. Manganese has been shown to be on the outer surface of the enzyme and it has been suggested that the Mn^{2+} in pyruvate carboxylase activates the methyl group of pyruvate by its electron-withdrawing properties. The methyl group of pyruvate then serves as the acceptor for carbon dioxide in the formation of oxalacetate (Fung and Mildvan, 1973). While pyruvate carboxylase normally contains manganese, in chickens fed a manganese-deficient diet, magnesium substitued for manganese in the enzyme with no apparent loss of activity (Scrutton et al., 1972). In agreement with the findings for avians, we have observed that pyruvate carboxylase activity is not reduced in livers from adult manganese-deficient rats unless they are stressed by fasting (Baly et al., 1984b). This enzyme is discussed further later (see carbohydrate metabolism section).

A third metalloenzyme that contains manganese is manganese superoxide dismutase (EC 1.15.1..1; MW 80,000). Superoxide dismutases catalyze the reaction of superoxide anion to hydrogen peroxide (Fridovich, 1975).

$$2O_2 + 2H^+ \rightarrow O_2 + H_2O_2$$

In mammals and avian species superoxide dismutases exist in two forms, one containing manganese and the other containing copper and zinc (MW 32,500). Intracellular localization of the two forms of superoxide dismutases appears to differ; most copper zinc superoxide dismutase activity is located in the cytosol, while manganese superoxide dismutase occurs primarily in the mitochondrial matrix. Manganese superoxide dismutase isolated from chicken liver has a molecular weight of 80,000 and contains four subunits of equal size, each containing one atom of manganese. In contrast to pyruvate carboxylase and arginase, the manganese in resting superoxide dismutase is in the trivalent state. The catalytic cycle of this enzyme involves reduction and then reoxidation of the metal center during successive encounters with O_2. While isoenzymes of manganese superoxide dismutases have been found, it is not known if they differ in physiological function (Lonnerdal et al., 1979).

The function of manganese superoxide dismutase, while not yet completely elucidated, is thought to be related to protection of the cell from free radical damage. If the activity of the enzyme is sufficiently depressed, superoxide anions could react with hydrogen peroxide to produce hydroxyl radicals, which could initiate lipid peroxidation potentially resulting in deleterious effects on membranes. Consistent with a controlling effect of MnSOD on lipid peroxidation rates are data showing a correlation between manganese superoxide dismutase activity and lipid peroxidation during postnatal development of the rat; as hepatic superoxide dismutase activity increased, lipid peroxidation decreased (Yoshioka et al., 1977).

The activity of manganese superoxide dismutase is affected by dietary status.

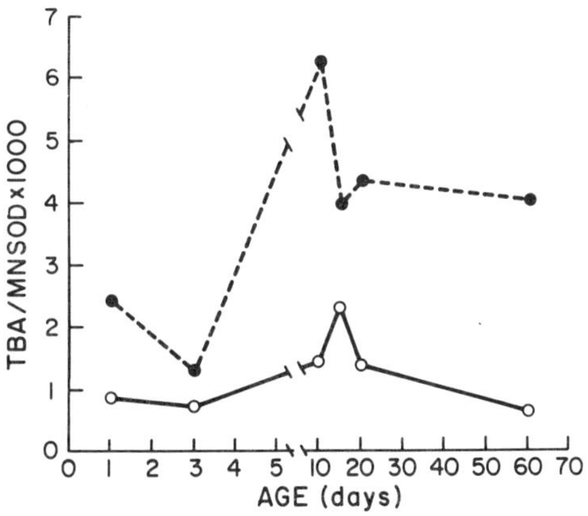

Figure 5-1. Lipid peroxidaton as measured by TBA reacting products (absorbance at 532 nm) in liver mitochondria for control (—) and Mn-deficient (- - - -) rats. Isolated mitochondria were incubated in Tris-HCl buffer, pH 7.4, with the addition of oxygen initiators. (From Zidenberg-Cherr et al., 1983.)

In mice, the activity of this enzyme was significantly lower in liver, brain, heart, and lung of deficient animals than in controls (de Rosa et al., 1980). In chickens, there was a depressed activity of manganese superoxide dismutase in liver after only 7 days of feeding a manganese-deficient diet to hatchlings. The activity of the enzyme was quickly elevated to normal by feeding manganese. Concomitant with the decline in activity of manganese superoxide dismutase, the activity of copper zinc superoxide dismutase was increased, suggesting a compensatory response. Similarly, Paynter (1980) has shown that the heart tissue of rats is especially sensitive to the effect of dietary manganese on manganese superoxide dismutase activity, and correlated the level of manganese in the diet to the activity of the enzyme. Similarly, offspring from rats fed a manganese-deficient diet had lower than normal liver manganese superoxide dismutase activity. Furthermore, they showed higher than normal lipid peroxidation in isolated liver mitochondria (Figure 5-1) (Zidenberg-Cherr et al., 1983). An increased rate of lipid peroxidation in manganese-deficient animals could be one of the underlying biochemical lesions that lead to the membrane damage seen with manganese deficiency (see the following discussion).

Work with *Lactobacillus plantarum* strongly supports the idea that adequate levels of superoxide dismutase activity are essential for normal cell viability. This and some related species of lactic acid bacteria are unusual in that they lack manganese superoxide dismutase. However, these bacteria accumulate intracellular manganese concentrations greater than those normally found, and much of this increased manganese is found in the divalent state, complexed to organic acids. The Mn^{2+} organic acid complexes have been shown to function as superoxide scavengers, although their catalytic rate is low compared to the enzyme (Archibald and Fridovich, 1982). If the bacteria are grown on a manganese-deficient medium, the intracellular manganese concentration is reduced, the superoxide anion scavenging ability of the cell is reduced and the cells are more susceptible to the lethality of the antihemorrhagic drug plumbagin, whose metabolism yields superoxide anion (Archibald and Fridovich, 1981).

The activity of manganese superoxide dismutase can be induced under conditions that result in an increased production of superoxide radicals, such as chronic exposure to hyperbaric oxygen (Stevens and Autor, 1977), ozone exposure (Dubick and Keen, 1983) and prolonged ethanol ingestion (Dreosti *et al.*, 1982; Keen *et al.*, 1983b), again supporting the hypothesis that manganese superoxide dismutase functions as an enzyme involved in cellular protection.

5.4.3 Manganese and Carbohydrate Metabolism

A relationship between manganese and abnormal carbohydrate metabolism was first suggested by Rubenstein and coworkers (1962). They reported the case of a diabetic patient resistant to insulin therapy who responded to oral doses of manganese chloride with decreasing blood glucose levels. Interestingly, these investigators tried manganese supplementation partly because of the patient's statement that his diabetic condition could be controlled to some extent by an extract of lucerne (alfalfa, *Medicago sativa*), an old South African folk medicine treatment for diabetes. Upon analysis it was found that the alfalfa contained a high concentration of manganese. In contrast to manganese, oral supplements of zinc, magnesium, cobalt, or iron had no effect on the patient's blood sugar levels. The authors hypothesized that manganese acted by inhibiting the release or function of glucagon. Consistent with the observation of Rubenstein *et al.*, patients suffering from chronic manganism show a prolonged reactionary hypoglycemia following intravenous glucose tolerance testing (Hassenein *et al.*, 1966). A hypoglycemic effect of manganese has also been reported in rabbits (Pignatari, 1932) and dogs (Bellotti, 1956). Large-scale studies evaluating the therapeutic effect of manganese supplementation in treatment of diabetes have not been reported.

In experimental animals, an essential role of manganese in carbohydrate metabolism was demonstrated by Everson and Shrader (1968). These investigators found that guinea pigs born to manganese-deficient dams and fed manganese-deficient diets to 60 days of age had abnormal glucose tolerance curves and decreased ultilization of glucose. Tissue pathology in these animals included hypertrophied pancreatic islet tissue with degranulated β-cells and an increased number of α-cells (Shrader and Everson, 1968). All of these signs of manganese deficiency were reversed following dietary manganese supplementation for 2 months. Interestingly, the pancreatic lesions seen in the manganese-deficient guinea pigs appear to be similar to those of human infants born to diabetic mothers. In an intriguing study by Shani *et al.* (1972), the sand rat *(Psammonys obesus)*, whose natural diet is high in manganese, was found to develop an insulin-resistant diabetes when fed a commercial rat feed. The diabetic condition cleared up upon reintroduction of the manganese-rich natural diet.

Manganese deficiency could affect carbohydrate metabolism through an effect on insulin metabolism. Based on the work of Everson and Shrader (1968), it could be hypothesized that manganese deficiency might result in impaired synthesis and/or secretion of insulin from the pancreas. Such abnormalities could explain, at least partially, the altered glucose metabolism observed in manganese-deficient animals. Recent data from this laboratory supports the idea that one of the biochemical lesions of manganese deficiency is abnormal insulin metabolism (Baly *et al.*, 1984a). Second-generation manganese-deficient rats fed diets containing 1 ppm manganese were compared to control rats fed diets containing 45 ppm manganese. Fasting blood glucose and insulin levels were similar in the two groups, but the deficient rats responded to a glucose load with a diabetic type of glucose tolerance curve. Using an isolated perfused pancreas preparation, it was found that the first phase of insulin release, reflecting insulin output due to release of stored hormone, was lower in deficient than in control rats, suggesting decreased insulin stores in the deficient rat. However, upon analysis,

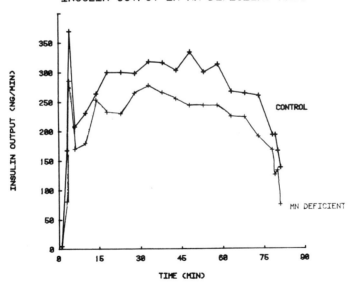

Figure 5-2. Data illustrating the time course of insulin output in control and Mn-deficient rats in response to a continuous 78-minute glucose stimulus beginning at minute zero. A perfused pancreas system is used. (From Baly *et al.*, 1984a.)

pancreatic insulin levels were not remarkably different between the two groups. The second phase of insulin release, reflecting insulin biosynthesis, was also lower in the deficient rats (Figure 5-2). Thus dietary manganese deficiency appears to affect carbohydrate metabolism, at least in part, through synthesis and release of insulin. An additional way in which manganese deficiency could produce abnormal carbohydrate metabolism would be through reduced insulin sensitivity resulting from abnormal peripheral insulin metabolism. No information on this possibility is available.

Finally, the effect of manganese deficiency on insulin production could be through the destruction of pancreatic β-cells. Diabetogenic agents such as alloxan have been postulated to function via the production of superoxide anion radicals (Cohen and Heikkila, 1974). The activity of manganese superoxide dismutase in pancreatic islet cells is low relative to other tissues (Grankvist *et al.*, 1981); thus a reduction in its activity with manganese deficiency may make the pancreas particularly susceptible to free radical damage.

A direct effect of manganese on gluconeogenesis has been shown using a rat liver perfusion method (Mangnall *et al.*, 1976). The addition of alanine to livers producing glucose from lactate caused a shift to the utilization of alanine as the main gluconeogenic substrate, but when manganese was added to the perfusion system, both substrates were used for gluconeogenesis, resulting in a net increase in glucose output. However, with isolated hepatocytes from fasted rats, Rognstad (1981) found that the addition of Mn^{2+} (0.5 mM) to the culture system in the absence of gluconeogenic hormones had no effect on increasing gluconeogenesis from lactate and pyruvate. When glucagon and epinephrine were added to the hepatocytes, an additive increase in their effects on gluconeogenesis was seen, but only in the presence of Mn^{2+}. The effect of Mn^{2+} on hormonal additivity was not duplicated by Fe^{2+}, Co^{2+}, Cd^{2+}, or Zn^{2+}. The mechanisms underlying the effect of Mn^{2+} in inducing hormonal additivity are not known. Friedmann and Rasmussen (1970) reported that the effect of Mn^{2+} on gluconeogenesis is not mediated via an effect on c-AMP. An additional observation by Rognstad (1981) was that the addition of Mn^{2+} increased the specific activity of glucose formed from lactate and $NaH^{14}CO_3$. The author speculated that the additional specific activity could be due to an increased exchange reaction of either pyruvate carboxylase or phosphoenolpyruvate carboxykinase, suggesting that one of these enzymes was stimulated by Mn^{2+}.

Keen *et al.* (1984b) have recently observed a rapid rise in blood glucose following a single dose of intraperitoneal manganese in rats (40 mg Mn/kg BW). Thus the effect of manganese on increasing glucose output from the liver can be also demonstrated *in vivo*. The increase was observed at two doses (10 and 40 mg Mn/kg BW) and was dose dependent. For both doses, maximal response was seen at about 1 hour postinjection, with blood glucose levels decreasing 4 hours after injection. In some cases blood glucose decreased to values below normal by 8 hr after injection of the high dose. The effect of injected manganese occurred in both fed and fasted rats, suggesting that while manganese could have stimulated glycogen breakdown, gluconeogenesis must also have been increased.

The mechanism(s) by which manganese deficiency affects carbohydrate metabolism are not known. One possibility is that it affects gluconeogenesis directly. The first steps in gluconeogenesis involve the carboxylation of pyruvate to oxaloacetate followed by the phosphorylative decarboxylation of oxaloacetate to phosphoenolypyruvate. The enzymes involved in this process are pyruvate carboxylase, a manganese metalloenzyme, and phosphoenolpyruvate carboxykinase, a manganese-activated enzyme (Bentle and Lardy, 1976). Both of these enzymes are important in early postnatal life when the newborn depletes liver glycogen and begins to utilize gluconeogenesis to maintain glucose levels. In the rat, pyruvate carboxylase emerges during the late fetal period and increases during the first week of life, while phosphoenolpyruvate carboxykinase activity appears after birth, increasing rapidly the first day, and then more slowly until it reaches its maximum at the fourth postnatal day (Ballard and Hanson, 1967). In our studies on the effects of dietary manganese deficiency, it has been observed repeatedly that mortality of offspring of manganese-deficient animals is highest within the first few days postpartum. It is possible that these animals that die are not able to maintain their blood glucose levels because of an inadequate level of activity of either pyruvate carboxylase and phosphoenolpyruvate carboxykinase or both. Pyruvate carboxylase is considered a regulatory enzyme and thus could be one metabolic point at which manganese could affect carbohydrate metabolism. However, we recently found in rats that the activity of liver pyruvate carboxylase appears to be unaffected by manganese deficiency; it was similar in offspring of control and deficient animals from birth to 30 days of age. Similarly, pyruvate carboxylase activity did not differ between adult control and manganese-deficient rats in the fed state. However, in adult rats fasted for 48 hours, the liver pyruvate carboxylase activity was lower in manganese deficient rats than in controls. Thus under stress the deficient rat may not have sufficient manganese to allow synthesis of optimal levels of the enzyme (Baly *et al.*, 1984b). While it is not known if the apparently normal activity of pyruvate carboxylase in the fed manganese-deficient rat is due to substitution of magnesium for manganese in the metalloenzyme, as reported in chickens (Scrutton *et al.*, 1972), it is evident that the effect of manganese deficiency on carbohydrate metabolism does not occur through changes in pyruvate carboxylase under normal conditions.

Phosphoenolpyruvate carboxykinase is not generally considered a rate-limiting enzyme in the adult animal. During the neonatal period, however, when its activity normally increases rapidly, a depression of this increase may result in diminished gluconeogenic capacity. This might occur if the intracellular pool of free Mn^{2+} were affected by dietary manganese deficiency. Such a mechanism would be consistent with the suggestion of Schramm (1982) that Mn^{2+} may function as a regulatory signal in cellular metabolism, and specifically, that the activities of phosphoenolpyruvate carboxykinase and arginase may be modulated by shifts in intracellular Mn^{2+} concentrations. It is known that total cell manganese is reduced in manganese deficiency.

In addition to its activation of phosphoenolpyruvate carboxykinase, manganese has been shown to have an activating effect on the glucose-6-phosphate-dependent form of glycogen synthase isolated from human placenta when sulfate is present in the incubation mixture, by increasing the Vmax of the enzyme (Huang and Robinson, 1977). The authors speculated that the effect of insulin on increasing glycogen synthesis may be mediated in part through increasing cytosolic Mn^{2+} concentrations, and thus activating glycogen synthase. A reduction in glycogen levels with manganese deficiency would exacerbate the effects of a reduced rate of gluconeogenesis.

Manganese may also be involved in carbohydrate metabolism as it relates to lectin chemistry. Lectins (phytohemagglutinins) are known to contain metal ions, and in some cases evidence has been presented for a requirement of Mn^{2+} for activity (Lis and Sharon, 1981). However, in many cases, other transition metal ions can substitute for Mn^{2+} without loss of biological activity, as demonstrated for concanavalin A (Agrawal and Goldstein, 1968) and *Dolichos biflorus* lectin (Borrebaeck et al., 1981). Most lectins appear to contain Mn^{2+} in combination with Ca^{2+}; removal of these cations abolishes carbohydrate binding properties for some lectins, while others seem more independent of cations for their binding to carbohydrates (Lonnerdal et al., 1983c). In general, the physiological role of manganese-containing lectins in mammalian systems is not well characterized.

5.4.4 Manganese and Lipid Metabolism

The relationship of manganese to lipid metabolism has not been defined, but it appears to have a lipotropic action, and possibly an interaction with choline. In early studies by Amdur and coworkers (1946), it was observed that manganese as well as choline prevented the deposition of excess fat in the rat liver, with the two substances acting synergistically. Consistent with the idea of an interaction between manganese and choline is the work of Barak and colleagues (1971), who found reduced liver manganese concentrations in choline-deficient rats. These investigators suggested that manganese may be essential for normal lipoprotein structure.

Manganese-deficient mice have also been reported to have enlarged deposits of abdominal fat and fatty livers. Abnormal fat metabolism could be one of the underlying biochemical lesion(s) resulting in the ultrastructural abnormalities seen in the tissues of the manganese-deficient mouse. Reported changes include alterations in the integrity of cell membranes, swollen and irregular endoplasmic reticulum, and elongated mitochondria with stacked cristae (Bell and Hurley, 1973). Detailed studies on changes in the lipid composition of cell membranes in manganese-deficient animals are needed. The effect of manganese deficiency on cell membrane integrity could also be due to an increased rate of lipid peroxidation of membranes, as the activity of manganese superoxide dismutase is lower in manganese-deficient animals than in controls (see preceding discussion). Increased body fat has also been reported in pigs fed a manganese-deficient diet for a prolonged period of time. Manganese deficiency also affects lipid synthesis in *Aspergillus niger;* however, in contrast to mice and pigs, manganese-deficient fungi show lower than normal lipid content, mainly because of a reduction in lipids containing a glycerol moiety, such as triglycerides and phospholipids (Othofer et al., 1979).

High levels of dietary manganese have also been reported to influence lipid metabolism when weanling rats were fed high-fat diets (Baquer et al., 1982). Supplementation of these diets with manganese resulted in increased activity of several glycolytic enzymes, including hexokinase, glyceraldehyde-3-phosphate dehydrogenase, enolase, lactate dehydrogenase, and glycerol-3-phosphate dehydrogenase. Increases were also found for enzymes of the pentose phosphate pathway and of lipogenesis. Thus the supplementation of manganese to the high-fat diet resulted in an increased potential for glucose oxidation and for lipogenesis.

The mechanism(s) by which excess manganese elicited these enzyme changes is not known, but it was suggested that an effect on insulin metabolism could be involved. Another possibility was that manganese might change levels of cyclic nucleotides, which then act as second messengers (Baquer et al., 1982). That manganese could affect cyclic nucleotide metabolism is supported by the observations that Mn^{2+} can be an optimal activator of guanylate cyclase (Goldberg and Haddox, 1977) and that it may be an activator for phosphodiesterase (Robinson et al., 1971).

Another step at which manganese may be critical for lipid metabolism is as a cofactor in steroid biosynthesis. Enhancement by manganese of cholesterol synthesis has been demonstrated in rat liver preparations (Curran and Azarnoff, 1961); manganese added to the medium increased the incorporation of C^{14} acetate into cholesterol. The metabolic site in cholesterol synthesis at which manganese is required is apparently in the formation of farnesyl pyrophosphate from the condensation of geranyl and isopentenyl pyrophosphate. The enzyme catalyzing this reaction, farnesyl pyrophosphate synthase, is manganese activated, with magnesium having little effect (Benedict et al., 1965). The influence of manganese deficiency on cholesterol synthesis *in vivo* is not clear. Hypocholesterolemia was observed in one case reported to be human manganese deficiency (Doisy, 1972). It was suggested that low cholesterol synthesis could affect sex hormone synthesis and thus explain, in part, the abnormal reproduction seen in manganese-deficient animals.

Recently, Roby and coworkers (1982) have reported that male weanling Sprague–Dawley rats fed a manganese-deficient diet had lower plasma and hepatic cholesterol levels than controls. Even when 2% cholesterol and cholic acid were added to the diet, plasma, but not hepatic, cholesterol was lower in manganese-deficient than in manganese-sufficient rats. In our laboratory, we have observed that plasma cholesterol concentration was significantly lower in second-generation manganese-deficient Sprague–Dawley rats than in controls. However, not all manganese-deficient rats had low plasma cholesterol, despite liver manganese levels similar to those of affected animals. Thus the effect of manganese deficiency on plasma cholesterol is not always an early event (Keen et al., unpublished data). Klimis-Tavantzis and coworkers (1983) did not find a pronounced effect of manganese deficiency on cholesterol metabolism in estrogen-treated chicks, laying hens, or rats of the Wistar and genetically hypercholesterolemic RICO strains. They did find that manganese deficiency significantly decreased plasma LDL cholesterol concentration in hypercholesterolemic RICO rats and tended to decrease it in Wistar rats, but plasma VLDL and HDL cholesterol were unaffected. They suggested that the finding of low LDL cholesterol could be due to an alteration in VLDL synthesis, composition, or metabolism or a change in LDL synthesis or metabolism such that there was a net reduction in its production or an increase in its catabolism. In this study, manganese deficiency had no effect on hepatic cholesterol synthesis or lipid concentration in either Wistar or RICO rats. An intriguing difference between the two strains was that manganese deficiency had no effect on fatty acid synthesis in hypercholesterolemic and normocholesterolemic RICO rats, but it resulted in a marked reduction in fatty acid synthesis in Wistar rats. The mechanism(s) underlying this strain difference are not known. Strain variation in response to manganese deficiency might explain the finding of hypocholesterolemia in manganese-deficient Sprague–Dawley rats but not in Wistar and RICO rats. The reduced hepatic fatty acid synthesis in the Wistar strain could be due to a reduction in acetyl CoA carboxylase activity, as Fletcher and Myant (1961) have shown, in cell-free fractions of rat liver that manganese is a cofactor for this enzyme.

5.4.5 Manganese and Brain Function

Manganese is essential for normal function of the brain. Manganese-deficient rats are more susceptible to convulsions than are normal rats and show electroencephalographic recordings similar to those of epileptics (Hurley et al., 1963). The importance of these observations is underscored by the recent report by Papavasiliou and coworkers (1979), who found that whole blood manganese was significantly lower in epileptics than in controls. Furthermore, blood manganese concentration correlated with the frequency of seizures. Patients with only a few seizures a year had blood manganese levels similar to those of controls, while those with frequent seizures (one to three per month) had low concentrations. Blood manganese levels were not correlated with blood levels of the anticonvulsive drugs taken by these patients, indicating that the low blood manganese was not attributable to drug therapy. Tanaka (1977) has also reported that one-third of children with convulsive disorders of unknown origin had whole-blood managanese concentrations significantly lower than normal. The biochemical lesions occurring in the brain due to manganese deficiency have not been well defined. Manganese toxicity can also severely affect the brain (see the following discussion).

In the human the highest brain manganese concentrations are in the pineal gland, olfactory bulb, median eminence of the hypothalamus, and basal ganglia (Bonilla et al., 1982; Barbeau et al., 1976). In the rat the highest concentrations are in the median eminence, hippocampus, midbrain, and cerebellum (Donaldson et al., 1973). Manganese in the brain accumulates mainly in pigmented structures, such as the melanocytes of the substantia nigra. Although the function of manganese and melanin in the brain have not been defined, their interaction may be important with regard to metabolism of biogenic amines. It has been suggested that if excessive amounts of manganese accumulate in melanocytes, the concentration of semiquinone radicals in these cells may increase, particularly if compounds such as phenothiazines are present. The increased production of semiquinone radicals may then block normal dopamine metabolism (Papavasiliou, 1981). Abnormal dopamine metabolism is a characteristic of chronic manganese toxicity (see the following discussion).

The biological functions of manganese in the brain are not well understood. The element may be needed for normal membrane structure and stability, as discussed earlier. In addition, Yip and Dain (1970) have shown that maximal activity of UDP-galactose : GN_2 ganglioside galactosyltransferase is obtained with Mn^{2+}; thus normal ganglioside synthesis may be dependent on adequate Mn^{2+}. Manganese is apparently necessary for normal metabolism of biogenic amines. Mn^{2+} has also been suggested to be an important activator for catechol O-methyl transferase (COMT), the enzyme that transfers the methyl group of 5-adenosylmethionine to the 3-hydroxyl group of catechol. It has been suggested that some drugs that affect COMT activity may do so in part by chelating manganese (Tagliamonte et al., 1970). There is strong evidence that part of the effect of manganese on catecholamine metabolism may be through an interaction with c-AMP. Katz and Tenenhouse (1973) have shown that in cell-free systems, Mn^{2+} can increase c-AMP accumulation, in part by reducing ATP and ADP turnover. The activity of adenylate cyclase in vitro can also be modified by Mn^{2+}. This may occur through two mechanisms: first, there may be a direct stimulation of the enzyme by Mn^{2+}; second, the dopamine stimulated increase in adenylate cyclase activity can be reduced by Mn^{2+}, suggesting that the amounts of dopamine and Mn^{2+} and their ratio may be important cell regulators of adenylate cyclase activity and hence c-AMP levels (Walton and Baldessarini, 1976).

Functional interaction between manganese and biogenic amines has been demonstrated. Injections of dopamine, epinephrine, and levodopa increased the intracellular concentration of simultaneously injected $^{54}Mn^{2+}$ in mice. Since glucagon or c-AMP injections produced similar effects, these results suggest that the biogenic amines apparently function by increasing c-AMP levels in the cell (Papavasiliou *et al.*, 1968). Based on work done with other trace elements, Papavasiliou (1981) has suggested that catechols and ethanolamines can form a complex with Mn^{2+} that may be important for their binding, transport, and storage.

5.5 Manganese Nutrition

5.5.1 Manganese Deficiency

In the previous section some of the biochemical lesions associated with manganese deficiency were considered. Some of the gross manifestations of these lesions are now discussed. For a more extensive review of the teratogenic effects of manganese deficiency, the recent review by Hurley (1981) should be consulted.

Experimental Animals

The essentiality of adequate amounts of manganese in the diet was established in 1931 by Kemmerer *et al.* in mice and by Orent and McCollum in rats. The latter workers observed that the offspring of manganese-deficient rats had high neonatal mortality. That manganese deficiency could occur outside of a laboratory environment was demonstrated by Wilgus *et al.* in 1936, who established that the economically significant disease of chickens "slipped tendon disease" or perosis was due to dietary manganese deficiency. Manganese deficiency has subsequently been produced in every species studied (Underwood, 1977).

One of the major effects of manganese deficiency is on development of the skeleton. Chick embryos from manganese-deficient hens showed chondrodystrophy, characterized by a shortening of the legs, wings, and lower mandible and by a globular contour of the head (Lyons and Insko, 1937). The effect of prenatal manganese deficiency on skeletal development in rats has been studied extensively (Hurley, 1981). The offspring of females fed a manganese-deficient diet from weaning show disproportionate growth of the skeleton at birth, which is characterized by severe shortening of the radius, ulna, tibia, and fibula in proportion to body length, as well as disproportionate growth of the skull, which is shorter, wider, and higher than normal in relation to skull length. Additional skeletal defects seen in rats are curvature of the spine, a localized dysplasia of the tibial epiphysis resulting in an abnormal knee joint, and anomalous development of ossification of the inner ear. The skeletal lesions found with manganese deficiency are apparently due to abnormal cartilage and bone matrix formation because of a defect in the synthesis of mucopolysaccharides, which, as discussed previously, is dependent on manganese for activation of glycosyl transferases. Calcium metabolism is not altered in manganese deficiency (Leach et al., 1969; Hurley, 1981). Lesions seen in the ruminant skeleton with manganese deficiency have been reviewed recently (Hidiroglou, 1979, 1980).

One of the more dramatic effects of prenatal manganese deficiency is congenital irreversible ataxia, which is characterized by lack of equilibrium and retraction of the head. This ataxia is due to abnormal development of the otoliths, calcified structures in the vestibular portion of the inner ear that are required for normal body-righting reflexes (Erway et al., 1966, 1970). Total or partial absence of otoliths has been reported for manganese-deficient mice, rats, guinea pigs, and chicks. In manganese-deficient mouse fetuses, mucopolysaccharide synthesis in the otolithic matrix is depressed; thus the abnormal development of the inner ear that leads to ataxia is probably the result of low glycosyl transferase activity (Hurley, 1981).

In addition to the skeletal and inner ear lesions caused by manganese deficiency, there are also ultrastructural abnormalities. In liver, kidney, heart, and pancreas of aged manganese-deficient mice, disorganization and dilation of the rough endoplasmic reticulum were observed. The vascular portion of the liver cell reticulum was increased, and there was enlargement of the Golgi apparatus. The mitochondria were elongated and stacked on each other, with cristae parallel to the outer membrane, in contrast to the normal perpendicular arrangement. Outer cell membrane damage was also a finding in these animals (Bell and Hurley, 1973). The low levels of manganese superoxide dismutase that are found in these tissues could be related to the morphological changes observed. The observation of an increased rate of mitochondrial lipid peroxidation in manganese-deficient animals supports this idea (Zidenberg-Cherr et al., 1983). The membrane abnormalities may also be due in part to a direct interaction of Mn^{2+} ions with membranes. Metal ions are well known to affect the stability of the lipid bilayer structure, thus influencing the fluidity of the membrane. While it is known that cellular Mg^{2+} fluctuations can affect membrane fluidity, it is not known if Mn^{2+} has a similar role.

Reproductive dysfunction has been observed with manganese deficiency.

Female rats severely deficient in manganese show absent or irregular estrous cycles and will not mate. As with some other nutrient deficiencies, there may be a delay in the opening of the vaginal orifice. Severely manganese-deficient male rats exhibit sterility associated with degeneration of the seminal tubules, lack of spermatozoa, and accumulation of degenerating cells in the epididymis. Reproduction problems associated with manganese deficiency have also been reported for cattle, goats, rabbits, and guinea pigs, and it has been suggested that manganese supplementation can improve the reproductive capacity of many ruminants (Underwood, 1977; Hidiroglou, 1979). The biochemical lesion(s) underlying the reproductive dysfunctions have not been delineated. Doisy (1972) has suggested that they may be the result of defective steroid synthesis. This hypothesis has yet to be extensively tested. High concentrations of manganese have been reported for the corpus luteum (Hidiroglou and Shearer, 1976).

Human Manganese Deficiency

There has been one case reported of dietary manganese deficiency in a human subject (Doisy, 1972). This volunteer developed manganese deficiency following its accidental omission from a purified diet that was being used to study the effects of vitamin K deficiency. Signs that were thought to be associated with manganese deficiency included weight loss, reddening of his black hair, reduced growth of hair and nails, dermatitis, hypercholesterolemia, and inability to elevate depressed clotting proteins in response to vitamin K administration. A problem with this case is that it is difficult to ascertain which of these pathologies were due solely to the manganese deficiency. While the case by Doisy is the only report of frank manganese deficiency in man, several other investigations suggest that suboptimal manganese status may not be unusual. Poor manganese status, as defined by low hair and plasma manganese levels, has been reported in children suffering from the inborn errors of metabolism maple syrup urine disease and phenylketonuria, even though dietary intake of manganese was similar to that of control children. It is not known if these low levels are functionally significant (Hurry and Gibson, 1982). Liver and heart manganese levels have been reported to be lower than normal in children suffering from protein-calorie malnutrition (Lehmann et al., 1971).

Finally, it has been suggested that manganese deficiency may be related to hydralazine disease, a syndrome that simulates systemic lupus erythematosis or rheumatoid arthritis. Hydralazine disease occurs in some people taking the antihypertensive drug hydralazine, and it has been reported to be responsive to manganese therapy (Comens, 1956). Manganese supplementation also reduces some aspects of the disease in chickens, dogs, and rats given hydralazine (Comens, 1956; Hurley et al., 1963).

While, at the present time, manganese deficiency has only been implicated in a few human disorders, this may be due in part to lack of a practical method for assessing an individual's manganese status. In our laboratory, we have found recently that whole blood manganese can be a useful indicator of whole body manganese status (Keen *et al.*, 1983a). Manganese was measured in whole blood, as well as in tissues of experimental animals (rats) fed purified diets containing adequate (45 μg Mn/g diet) or deficient (1 μg Mn/g diet) levels of the element for 60 days. Blood manganese levels were measured by flameless atomic absorption spectrophotometry (AAS). Whole blood manganese concentration was 40% lower in deficient than in control animals, with average values of 8 and 14 μg/L, respectively. Using flame AAS, after wet ashing (Clegg *et al.*, 1982), liver samples from deficient and control animals had manganese concentrations of 0.91 and 2.53 μg/g, respectively. Thus the low blood levels of manganese in the deficient rats reflected low levels of the element in soft tissue. When the diets of the two groups were switched, whole blood levels changed in parallel with soft tissue changes. For the deficient animals fed control diet, blood and tissue manganese levels reached control values within 5 days. This rapid effect on tissue manganese is consistent with the data of Hurley and Everson (1963), showing that manganese-deficient females supplemented with manganese for 24 hours on day 14 of pregnancy gave birth to young whose survival to day 28 approached that of controls; none of the offspring were ataxic. Control animals given the deficient diet maintained their initial blood and tissue manganese levels for several weeks. These observations support the idea that blood manganese levels are a useful indicator of soft tissue manganese levels, and not just a reflection of recent diet history.

5.5.2 Genetic Interaction and Manganese Metabolism

The importance of manganese during perinatal development is demonstrated by several genetic mutants that possess errors of manganese metabolism. Interactions affecting development can be classified into two groups. The first type of interaction involves a single mutant gene, whose phenotypic expression can be reduced or prevented by pre-, post-, or perinatal nutritional manipulation. The second type involves strain differences that produce differential responses to a dietary deficiency of the element (Hurley, 1981).

Examples of the first category of gene-nutrient interactions are mutant genes in mice and mink that produce alterations in manganese metabolism. Most extensively studied has been the relationship between the *pallid* gene in mice and manganese metabolism (Erway *et al.*, 1966, 1970, 1971). The mutant gene *pallid* in mice is characterized in part by pale coat color and ataxia that is caused by missing or absent otoliths (Lyon, 1953). Supplementation of the diet with very high levels of manganese (1500–2000 µg Mn/g diet) during pregnancy prevents the effect of the gene on otolith development and on occurrence of congenital ataxia. The mutant gene itself and the effect of the gene on pigmentation are unaltered by the manganese supplementation. Cotzias et al. (1972) showed that mice with the *pallid* gene differed from normal mice in the metabolism of radiomanganese, in that there was prolonged retention of an injected dose. The concentration of manganese in bone and brain but not liver or kidney was lower than normal in the mutant mice. In addition, it was shown that the metabolism of biogenic amines is abnormal in the pallid mouse. Compared to controls, there was a reduced synthesis of dopamine from injected levodopa, and there was a lower ability to synthesis brain serotonin from L-tryptophan. The biochemical lesions underlying these problems are not known.

A gene analogous to *pallid*, *screwneck,* has been identified in mink. Again, this mutant is characterized by a pale coat color, abnormal or missing otoliths in the inner ear, and ataxia. This economically important genetic lesion has also been shown to be alleviated by dietary manganese supplementation (Erway and Mitchell, 1973). The precise biochemical lesions caused by the *pallid* and *screwneck* genes have not been identified, but they provide excellent models with which to study manganese metabolism.

Representative of the second category of gene-nutrient interactions is the observation by Gallup and Norris (1939) that there are breed and strain differences in the amount of manganese required to prevent perosis in the chick. A second example of strain influence on manganese metabolism is from this laboratory (Hurley and Bell, 1974). When pregnant mice of several different strains were fed diets containing either a normal (45 µg Mn/g diet) or a low level of manganese (3 µg Mn/g diet), the effects on otolith development of their fetuses were different, depending on their genetic background. All the strains showed normal otolith development when fed the normal level of manganese; however, with the lower manganese diet, some of the strains showed 30% of normal otolith development in their fetuses, while the others showed only 5% of normal otolith development. The biochemical differences underlying these strain variations are not known, but possible sites include differences in manganese absorption by the mother, differences in maternal blood manganese transport, differences in placental manganese transport, and different fetal requirements with regard to glycosyl transferase enzymes.

A final example of strain variation in response to manganese deficiency is provided by the work of Klimis-Tavantzis et al. (1983) discussed earlier, which showed that the Wistar strain of rats differed from the RICO rat strain in the effect of manganese deficiency on hepatic fatty acid synthesis. Again, the explanation for this strain difference is not known.

Manganese

5.5.3 Human Requirements

As with most nutrients, estimation of the human requirement for manganese is based primarily on balance studies. Engel *et al.* (1967) have reported that girls 6–10 years of age need about 1.25 mg Mn/day. Greger *et al.* (1978) reported that young teenage girls need an excess of 3 mg/d to maintain balance, and McLeod and Robinson (1972) found that an intake of 2.5–3.2 mg/d was required for women 19–22 years of age. More recently, Rao and Rao (1982) reported that an intake of 3.72–4.15 mg/day, was required to maintain balance in adult Indian males. Taken together, a manganese intake of 35 µg/kg/day should be adequate to maintain manganese balance for most individuals; however, Guthrie and Robinson (1977) have reported that subjects with an intake of less than 2 mg/d do not show any signs of manganese deficiency. In a recent survey of 100 premenopausal women, self-selected diets were found to provide an average of 3.1 mg Mn/day. Most of the manganese in the diet was from breads and cereals (Gibson and Scythes, 1982). Manganese intake by exclusively breast-fed infants at 1, 2, and 3 months of age has been reported to be 0.19, 0.6, and 0.5 µg/kg/day, considerably lower than the figure of 35 µg/kg/day given earlier (Vuori, 1979). It has been reported that young infants have a negative manganese balance (Widdowson, 1969).

The dietary manganese requirements for several laboratory species and domestic animals are discussed in detail in Underwood (1977). The manganese requirments for different species vary considerably. In terms of dietary concentration (ppm), approximate requirements can be estimated as follows: rat, 45; mouse, 45; rabbit, 10; cattle, 20; birds, 50; and pigs, 40.

5.5.4 Manganese Content of Foods

The manganese content of foods varies greatly. The highest concentrations are found in nuts, grains, and cereals (2–14 mg/kg), intermediate levels in vegetables (0.5–2 mg/kg) and low levels in dairy products, red meat, poultry and fish (0.1–1. mg/kg) (Wenlock *et al.*, 1979). Relatively high concentrations are found in coffee and tea, and these sources can account for as much as 10% of daily manganese intake for some individuals (McLeod and Robinson, 1972).

5.6 Manganese Toxicity

Although manganese can produce toxic effects if taken into the body in excessive amounts, it is not normally considered a nutritional problem. Incidents of oral manganese poisoning in man have been reported by Kawamura *et al.*, (1941), who described a series of cases of manganese toxicity in individuals who consumed well water that had been contaminated by high amounts of manganese released from discarded batteries. Such incidents have also been reported by Banta and Markesbery (1977), who reported one case of manganese toxicity in a man who had consumed large doses of minerals for several years. In contrast to the few reports of oral manganese poisoning in man, there is an extensive literature on manganese toxicity in humans who chronically inhale high concentrations of airborne manganese, a situation that can occur in manganese mines, steel mills, and some chemical industries. Although cases of manganese toxicity in man occurring because of environmental pollution have been reported since the nineteenth century, the biochemical lesions underlying the pathology are still not totally understood.

The principal organ affected by manganese toxicty is the brain, probably because of its slower turnover of the element relative to other tissues. Recently, the neurotoxicity resulting from manganese has attracted increased attention because of the use of manganese-containing compounds (methylcyclopentadienyl manganese tricarbonyl) as replacements for lead in gasoline (Ter Harr *et al.*, 1975). In addition, it has been reported that children living in urban areas may have elevated blood levels of both manganese and lead (Joselow *et al.*, 1978).

Cotzias and coworkers (1968) reported several cases of manganese toxicity in Chilean manganese miners. Signs of toxicity in these individuals occurred only after several months or several years of exposure. Manganese toxicity is often first manifested by severe psychiatric symptoms, including hyperirritability, violent acts, and hallucinations, all of which have been referred to as manganic madness. If the individual is removed from the toxic environment, some improvement in psychiatric manifestations can occur. With progression of the disease, there is a permanent crippling neurological disorder of the extrapyramidal system, the morphologic lesions of which are very similar to those of Parkinson's disease. Based on autopsy findings of low dopamine levels in the basal ganglia and substantia nigra of individuals who had suffered from manganese toxicity, treatment with levodopa (L-β-3,4-dihydroxyphenylalanine), a precursor of dopamine that can cross the blood–brain barrier, was initiated in several patients with manganism. Treatment with levodopa has resulted in considerable improvement of the symptoms of the disease, but the pathological lesions are apparently permanent, and this drug treatment must be continued throughout the life of the individual (Mena *et al.*, 1970).

Chelation therapy has been tried in patients with manganism, but it has not been efficacious. The reason for this is that following removal of the individual from the high manganese environment, there is a rapid loss of the excess manganese from the body, including the brain, but the tissue lesions that occurred during exposure to the excess manganese do not heal. Thus chelation therapy has little effect in treating patients with advanced manganese poisoning (Papavasiliou, 1981). In cases of acute manganese poisoning, on the other hand, chelation therapy may be useful. Polyaminocarboxylic acids have been shown to be superior to thiol chelators in treatment of acute manganese toxicity (Tandon and Khandelwal, 1982).

The biochemical abnormalities that result in manganism have not been well delineated. Experimental manganism has been produced in nonhuman primates. Neff et al. (1969) reported that in monkeys with early manganism, there is a marked reduction in dopamine and serotonin concentrations in the caudate nucleus, in some cases prior to histopathological lesions. Thus the early signs of manganism are apparently due to biochemical and not anatomical defects. This suggests that the initial lesion may be a disturbance in a regulatory function of manganese on neurotransmitter metabolism. Mena (1981) has suggested that one component of the early biochemical lesion may be the displacement by Mn^{2+} of catecholamines from the ATP complex in the storage vesicles of chromaffin granules in the adrenal medulla.

Gianutsos and Murray (1982) have reported alterations in brain dopamine and γ-amino-*n*-butyric acid (GABA) following prolonged administration of inorganic or organic manganese compounds to mice. The inorganic manganese ($MnCl_2$) was fed in the diet at a concentration of 4% for up to 6 months, while the organic manganese (methylcyclopentadienyl manganese tricarbonyl) was given by intraperitoneal injection (5 mg Mn/kg BW) on alternating days for up to 3 weeks. The effects of inorganic and organic manganese poisoning were very similar. In both cases, there was a reduction in striatal and limbic dopamine concentrations and an increase in striatal GABA content. There was no effect of manganese on choline acetyltransferase activity, supporting the findings of Lai et al. (1981) that manganese toxicity does not have an early effect on cholinergic neurons. The biochemical lesions underlying the changes in dopamine and GABA levels are not known.

Bonilla (1978) has reported that glutamic acid decarboxylase activity is not affected by manganese toxicity, despite changes in GABA content, while Lai et al. (1981) reported that the enzyme could be affected. The observations by Gianutsos and Murray (1982) that GABA content is affected only in areas where dopamine is altered suggests that the changes in GABA could be secondary to dopamine neuronal damage. Bonilla (1980) has shown that manganese toxicity exerts a biphasic effect on tyrosine hydroxylase activity, with levels increased during the first month of exposure to high levels of oral manganese and decreased with long-term exposure. Since manganese has been reported not to inhibit tyrosine hydroxylase *in vitro* (Deskin et al., 1980), the decreased dopamine content seen with time would appear to be the result of damaged cells.

Donaldson et al. (1980) have suggested that cellular damage could occur through the effects of neurotoxic metabolites generated from a manganese-enhanced autooxidation of dopamine. An additional mechanism by which the manganese-induced tissue pathology could occur is through an effect on lipid peroxidation. Shukla and Chandra (1981) found a reduction in brain lipid peroxidation *in vitro* with increasing brain manganese concentrations. Since it has been proposed that nonenzymic lipid peroxidation has a significant role in the maintenance of the redox potential of brain and perhaps in membrane remodeling, a manganese-induced inhibition in lipid peroxidation could result in tissue injury.

An increased superoxide scavenging ability by the manganese-loaded cell could also affect dopamine β-hydroxylase, as this enzyme may be regulated by free radicals such as O_2^- generated during ascorbate reduction. Donaldson and coworkers (1982) have suggested that alterations in cholinergic neurotransmitter receptor binding can occur with manganese toxicity as the synthesis of cyclic GMP, a regulatory hormone in controlling cholinergic function, is dependent on guanylate cyclase, an enzyme activated by free radicals. Decreases in brain dopamine content with prolonged manganese toxicity have been reported for the rat (Morgan and Huffman, 1976). Cotzias *et al.* (1974) have also shown that brain neurotransmitter metabolism can be affected in animals receiving excess manganese by the oral route. However, their results were quite different from those of Morgan and Huffman (1976). In the study by Cotzias *et al.* (1974) mice fed milk diets containing 1000 ppm manganese showed significantly higher brain manganese and dopamine concentrations than did mice fed normal levels of manganese. It is not known if the finding of increased brain dopamine in the manganese-poisoned mouse—compared to decreased brain dopamine in the manganese-poisoned monkey, rat, adult mouse, or man—is due to a difference in the age of the animals or if it reflects different stages of the pathology. It is interesting to note that the level of manganese in the milk diet (1000 ppm) would not be considered a toxic level in a solid diet, demonstrating the importance of bioavailablility from the diet.

For more detailed discussions of manganese toxicity and the central nervous system, the recent reviews by Mena (1981) and Papavasiliou (1981) may be examined.

In addition to central nervous system pathology in experimental and domestic animals, a frequent finding with oral manganese toxicity is an iron-responsive microcytic anemia (National Academy of Sciences, 1980). As discussed earlier, this is due mainly to an inhibition of gastrointestinal iron absorption. Reproductive dysfunction has also been reported in manganese-poisoned cattle and rats. Both males and females have been reported to show reduced reproductive capabilities with very high levels of dietary manganese (3500 ppm) (Laskey *et al.*, 1982).

Finally, in rats, excess manganese can induce intrahepatic cholestasis, which can result in both morphological and functional alterations in the liver. The observation that the hepatic lesions are similar to those found in humans with drug-induced cholestasis has led to use of the manganese-poisoned rat as a model for studying the mechanisms underlying cholestasis. It has been reported that one component of the insult is an interaction between manganese and bilirubin, and it has been proposed that this complex may alter the canalicular membranes, or the pericanalicular microfilament network (Lamirande and Plaa, 1979).

5.7 Manganese in Relation to Immunocompetence and Cancer

In recent years interest has been generated regarding the influence of trace elements on host immunocompetence and oncogenesis. A comprehensive review of this subject has been recently published (Beach et al., 1982). Data collected to date have indicated that adequate manganese nutrition is necessary for normal antibody production. Experimental animals fed manganese-deficient diets have shown deficient antibody synthesis and/or secretion. When manganese was added to the diets of experimental animals, there was an improved antibody production. However, if manganese is added in excessive amounts, inhibition in antibody production may occur. The mechanism(s) by which manganese affects antibody synthesis or release have not been worked out, although the negative effect of excess manganese is thought to involve the plasma membrane (McCoy et al., 1979; Hart, 1978). Manganese has also been reported to influence immunocompetence by affecting neutrophil and macrophage function, with excess amounts of the element acting to inhibit normal chemotaxis (Rabinowitch and Dertefano, 1973).

In addition to its direct effects on the immune system, manganese at high concentrations has been shown by several groups to be a potent mutagenic agent, with one mechanism being the induction of frame shift and point mutations (Beach et al., 1982). Both nuclear and mitochondrial genes have been shown to be affected by high manganese (Fukunaga and Mizuguchi, 1982). The mutagenicity of Mn^{2+} is probably due to its ability to affect the activity of many polymerases and nucleases. Although Mn^{2+} may be needed for normal activity of some nucleases and polymerases (McEuen, 1981), it has been shown that it can alter the specificity of others, if it is present at high concentrations. Thus excess Mn^{2+} will decrease the selectivity of DNA polymerase I isolated from E. coli for ribonucleotides and deoxyribonucleotide (Barnes, 1978) and can affect the specificity of isolated RNA polymerase from E. coli (Paddock et al., 1974).

These types of findings suggest that the mutations induced by manganese result, in part, from a decrease in the fidelity of DNA synthesis, because of an effect on DNA polymerase. That an effect of excess Mn^{2+} in the intact cell can result in altered selectivity of polymerases has been strongly supported by the work of Falchuck and coworkers (1978). These investigators reported that cells of Euglena gracilis grown on a zinc-deficient medium had manganese concentrations 35 times that of control cells. The manganese-loaded cells contained a unique isoenzyme of RNA polymerase and synthesized mRNA, which had an unusually high (G + C)/(A + U) ratio. That some of the effects were due to the high cellular Mn^{2+} concentrations, as well as to the zinc deficiency, was indicated by the observation that increasing the Mn^{2+} concentration of the assay medium from 1 to 10 mmol resulted in an approximate 50% reduction of UMP : CMP incorporation into RNA by RNA polymerases I and II from the control cells and a 90% reduction for the novel polymerase isolated from the zinc-deficient cells. Thus these results show that the base composition of RNA can be affected by the concentration of manganese.

It is not known if manganese deficiency is mutagenic. Manganese is known to bind directly to DNA. Based on its dissociation constant Wilburg and Neuman (1957) have suggested that DNA binds manganese more strongly than other metals; thus a reduction in DNA-bound manganese could result in conformational changes leading to decreased fidelity in its transcription. Additionally, RNA polymerase II is significantly more activated by Mn^{2+} than by Mg^{2+} (Vaisius and Horgen, 1980). DNA polymerases are also activated by Mn^{2+}. Thus manganese has a role in initiating protein synthesis by stimulating DNA and RNA polymerases. The effect of a reduction in cellular Mn^{2+} on RNA polymerase activity and DNA polymerase activity is not known.

Excess manganese has also been shown to be carcinogenic; it can act as a direct carcinogen or as a cocarcinogen. For asbestos, manganese can be a cocarcinogen by inhibiting the catabolism of benz(a)pyrene, the agent underlying asbestos carcinogenicity (Beach *et al.*, 1982).

The extent to which manganese may be involved in abnormal immune function or oncogenesis in man is unknown, although it is intriguing to point out that a characteristic of several malignancies is the pronounced reduction of manganese superoxide dismutase in the transformed cells (Oberley and Buettner, 1979). Research on the role of manganese in immunocompetence and oncogenesis is needed.

5.8 Summary

Despite the essentiality of manganese, its metabolism and biochemistry are still poorly understood. Although a deficiency or toxicity of this element has pathological consequences, the underlying biochemical lesions have not been well defined.

Absorption of manganese is not well regulated and is influenced by several dietary factors. Homeostatic regulation of manganese is mainly through excretion of the element into the intestinal tract via bile. Absorbed manganese is transported on α_2-macroglobulin and transferrin; retained manganese concentrates in mitochondria-rich tissues. Within a cell, much of the element is localized in the mitochondria. Fluctuations in the intracellular concentration of free Mn^{2+}, like that of Mg^{2+} and Ca^{2+}, may be a mechanism of cellular metabolic control. Manganese functions both as an enzyme activator and as a constituent of metalloenzymes. It is the preferred metal cofactor for a number of glycosyltransferases, and much of the connective tissue defects seen with manganese deficiency may be explained by depression in activity of these enzymes. Manganese metalloenzymes that are affected by manganese status include arginase and superoxide dismutase. Some of the membrane abnormalities seen with manganese deficiency may be the result of depressed superoxide dismutase activity, with subsequent increased lipid peroxidation.

Manganese is involved in carbohydrate metabolism, as a deficiency of the element may produce abnormal insulin metabolism and abnormal glucose tolerance. Acute manganese toxicity also perturbs carbohydrate metabolism. The relationship of manganese to lipid metabolism has not been defined, but it appears to have a lipotropic action and is involved in cholesterol and fatty acid synthesis.

Manganese is essential for normal brain function, at least in part through its role in the metabolism of biogenic amines. Manganese toxicity is a serious health hazard in some industries, resulting in a permanently crippling neurological disorder of the extrapyramidal system. The role of manganese in immunocompetence and oncogenesis is poorly defined.

The adequacy of manganese nutrition in man has not been well characterized, in part because of lack of methods for the detection of suboptimal manganese status. Future studies defining manganese status of various population groups at risk, coupled wtih a greater understanding of the element's function in metabolism, will further delineate the role of this element in biological systems.

References

Abrams, E., Lassiter, J. W., Miller, W. J., Neathery, M. W., Gentry, R. P., and Scarth, R. D., 1976. Absorption as a factor in manganese homeostasis, *J. Anim. Sci.* 42:630–636.

Agrawal, B. B., and Goldstein, I. J., 1968. Protein–carbohydrate interaction. VII. Physical and chemical studies on concanavalin A, the hemagglutinin of Jack Bean, *Arch. Biochem. Biophys.* 124:218–229.

Amdur, M. O., Norris, L. C., and Heuser, G. F., 1946. The lipotrophic action of manganese, *J. Biol. Chem.* 164:783–784.

Archibald, F. S., and Fridovich, I., 1982. The scavenging of superoxide radical by manganous complexes: In vitro, *Arch. Biochem. Biophys. 214:452–463*.

Ash, D. E., and Schramm, V. L., 1982. Determination of free and bound manganese (II) in hepatocytes from fed and fasted rats, *J. Biol. Chem.* 257:9261–9264.

Ash, D. E., and Schramm, V. L., 1982. Determination of free and bound manganese (II) in hepatocytes from fed and fasted rats, *J. Biol. Chem.* 257:9261–9264.

Ballard, F. J., and Hanson, R. W., 1967. Phosphoenolypyruvate carboxykinase and pyruvate carboxylase in developing rat liver, *Biochem. J.* 104:866–871.

Baly, D. L., Currey, D. L., Keen, C. L. and Hurley, L. S., 1984a. Effect of manganese deficiency on insulin secretion and carbohydrate homeostasis, *J. Nutr.*: in press.

Baly, D. L., Keen, C. L., Curry, D. L., and Hurley, L. J., 1984b. Effects of manganese deficiency on carbohydrate metabolism, in Proceedings of the Fifth Symposium on Trace Element Metabolism in Man and Animals. Aberdeen, Scotland, June 29-July 4, 1984, in press.

Banta, G., and Markesbery, W. R., 1977. Elevated manganese levels associated with dementia and extrapyramidal signs, *Neurology* 27:213–216.

Baquer, N. Z., Hothersall, J. S., Sochor, M., and McLean, P., 1982. Bio-inorganic regulation of pathways of carbohydrate and lipid metabolism. 1. Effect of iron and manganese on the enzyme profile of pathways of carbohydrate metabolism in adipose tissue during development, *Enzyme* 27:61–68.

Barbeau, A., Inoue, N., and Cloutier, T., 1975. Role of manganese in dystonia, in *Advances in Neurology*, Vol. 14, R. Elridge and S. Fahn (eds.), Raven Press, New York, pp. 339–352.

Barak, A., Keefer, R., and Tuma, D., 1971. The possible role of manganese in hepatic lipid transport, *Nutr. Rep. Int.* 3:243–246.

Barnes, W. M., 1978. DNA sequencing by partial ribosubstitution, *J. Molec. Biol.* 119:83–99.

Bates, J., Chan, W., Mahood, A., and Rennert, O. M., 1983. Human milk, bovine milk and formula ligand-bound manganese transport in rat, *Fed. Proc.* 42:817.

Beach, R. S., Gershwin, M. E., and Hurley, L. S., 1982. Zinc, copper, and manganese in immune function and experimental oncogenesis, *Nutr. Cancer* 3:172–191.

Bell, L., and Hurley, L. S., 1973. Ultrastructural effects of manganese deficiency in liver, heart, kidney and pancreas of mice, *Lab. Invest.* 29:723–736.

Bellotti, R. M., Ravera, M. and Abbona, C., 1956. Studio in vivo dell' attiveta di alcuna sali de manganese sur metabolism intermedio degii idrati di carbonio, *Arch. Maragliano Patol. Clin.* 12:683–690.

Benedict, C., Kett, J., and Porter, J., 1965. Properties of farnesyl pyrophosphate synthetase of pig liver, *Arch. Biochem. Biophys.* 110:611–621.

Bentle, L. A., and Lardy, H. A., 1976. Interactions of anions and divalent metal ions with phosphoenolpyruvate carboxykinase, *J. Biol. Chem.* 251:2916–2921.

Bertinchamps, A. J., Miller, S. T., and Cotzias, G. C., 1966. Interdependence of routes excreting manganese, *Am. J. Physiol.* 211:217–224.

Bertrand, G., and Medigreceanu, M. F., 1913. Recherches sur la presence du manganese dans la serie animale, *Ann. de L'inst. Pasteur* 27:282–288.

Bond, J. S., Failla, M. L., and Unger, D. F., 1983. Elevated manganese concentration and arginase activity in livers of streptozotocin-induced diabetic rats, *J. Biol. Chem.* 258:8004–8009.

Bonilla, E., 1978. Increased GABA content in caudate nucleus of rats after chronic manganese chloride administration, *J. Neurochem.* 22:551–552.

Bonilla, E., 1980. L-tyrosine hydroxylase activity in the rat brain after chronic oral administration of manganese chloride, *Neurobehav. Toxicol.* 2:37–41.

Bonilla, E., Salazar, E., Villasmil, J., and Villalobos, R., 1982. The regional distribution of manganese in the normal human brain, *Neurochem. Res.* 7:221–227.

Borg, D. C., and Cotzias, G. C., 1962. Interaction of trace metals with phenothiazine drug derivatives. I–III. *Proc. Natl. Acad. Sci.* 48:617–652.

Borrebaeck, C. A. K., Lonnerdal, B., and Etzler, M. E., 1981. Metal ion content of *Dolichos biflorus* lectin and effect of divalent cations on lectin activity, *Biochemistry* 20:4119–4122.

Brinkworth, R. J., Hanson, R. W., Fullin, F. A., and Schramm, V. L., 1981. Mn^{2+}-sensitive and insensitive forms of phosphoenolpyruvate carboxykinase (GTP), *J. Biol. Chem.* 256:10795–10802.

Britton, A. A., and Cotzias, G. C., 1966. Dependence of manganese turnover on intake, *Am. J. Physiol.* 203:203–206.

Chan, W.-Y., Bates, J. M., Jr., and Rennert, O. M., 1982. Comparative studies of manganese binding in human breast milk, bovine milk and infant formula, *J. Nutr.* 112:642–651.

Clegg, M. S., Keen, C. L., Lonnerdal, B., and Hurley, L. S., 1982. Analysis of trace elements in animal tissues. III. Determination of manganese by graphite furnace atomic absorption spectrophotometry, *Biol. Trace Element Res.* 4:145–156.

Cohen, G., and Heikkila, R. E., 1974. The generation of hydrogen peroxide, superoxide radical, and hydroxyl radical by 6-hydroxydopamine, dialuric acid and related cytotoxic agents, *J. Biol. Chem.* 249:2447–2452.

Cohn, M., and Hughes, T. R., 1962. Nuclear magnetic resonance spectra of adenosine di and triphosphate. II. Effect of complexing with divalent ions, *J. Biol. Chem.* 237:176–181.

Comens, P., 1956. Manganese depletion as an etiologic factor in hydralazine, *Am. J. Med.* 20:944–945.

Cotzias, G. C., and Bertinchamps, J., 1960. Transmanganin, the specific manganese-carrying protein of human plasma, *J. Clin. Invest.* 39:979.

Cotzias, G. C., and Greenough, J. J., 1958. The high specificity of the manganese pathway through the body, *J. Clin. Invest.* 37:1298–1305.

Cotzias, G. C., Horiuchi, K., Fuenzalida, S., and Mena, I., 1968. Chronic manganese poisoning: Clearance of tissue manganese concentrations with persistence of the neurological picture, *Neurology* 18:376–382.

Cotzias, G. C., Papavasiliou, P. S., Mena, I., Tang, L. C., and Miller, S. T., 1974. Manganese and catecholamines, in *Advances in Neurology*, Vol. 5, F. H. McDowell and A. Barbeau (eds.), Raven Press, New York, pp. 235–243.

Cotzias, G. C., Tang, L. C., Miller, S. T., Sladic-Simic, D., and Hurley, L. S., 1972. A mutation influencing the transportation of manganese, L-dopa, and L-tryptophan, *Science* 176: 410–412.

Curran, G. L., and Azarnoff, D. L., 1961. Effect of certain transition elements on cholesterol biosynthesis, *Fed. Proc.* 20:Suppl. 10, 109–111.

Davies, W. T., and Nightingale, R., 1975. The effects of phytate on intestinal absorption and secretion of zinc, and whole body retention of zinc, copper, iron and manganese in rats, *Br. J. Nutr.* 34:243–258.

de Rosa, G., Keen, C. L., Leach, R. M., and Hurley, L. S., 1980. Regulation of superoxide dismutase activity by dietary manganese, *J. Nutr.* 110:795–804.

Deskin, R., Bursian, S. J., and Edens, F. W., 1980. An investigation into the effects of manganese and other divalent cations on tryrosine hydroxylase activity, *Neurotoxicology* 2:75–81.

Doisy, E., Jr., 1972. Micronutrient controls of biosynthesis of clotting proteins and cholesterol, in *Trace Substances in Environmental Health*, Vol. VI, D. Hemphill (ed.), University of Missouri, Columbia, pp. 193–199.

Donaldson, J., LaBella, F. S., and Gesser, D., 1980. Enhanced autooxidation of dopamine as a possible basis of manganese neurotoxicity, *Neurotoxicology* 2:53–64.

Donaldson, J., McGregor, D., and Labella, F., 1982. Manganese neurotoxicity: a model for free radical mediated neurodegeneration? *Can. J. Physiol. Pharmacol.* 60:1398–1405.

Donaldson, J., St. Pierre, T., Minnich, J. L., and Barbeau, A., 1973. Determination of Na^+, K^+, Mg^{2+}, Zn^{2+} and Mn^{2+} in rat brain regions, *Can. J. Biochem.* 51:87–92.

Dubick, M. A., and Keen, C. L., 1983. Tissue trace elements and lung superoxide dismutase activity in mice exposed to ozone, *Toxicol. Lett.* 17:355–360.

Dreosti, I. E., Manuel, S. J., and Buckley, R. A., 1982. Superoxide dismutase (EC 1.15.1.1) manganese and the effect of ethanol in adult and fetal rats, *Br. J. Nutr.* 48:205–210.

Engel, R. W., Price, N. O., and Miller, R. F., 1967. Copper, manganese, cobalt, and molybdenum balance in pre-adolescent girls, *J. Nutr.* 92:197–204.

Erway, L., Hurley, L. S., and Fraser, A., 1966. Neurological defect: Manganese in phenocopy and prevention of a genetic abnormality of inner ear, *Science* 152:1766–1768.

Erway, L., Hurley, L. S., and Fraser, A., 1970. Congenital ataxia and otolith defects due to manganese deficiency in mice, *J. Nutr.* 100:643–654.

Erway, L., Fraser, A., and Hurley, L. S., 1971. Prevention of congenital otolith defect in *Pallid* mutant mice by manganese supplementation. *Genetics* 67:97–108.

Erway, L. C., and Mitchell, S. E., 1973. Prevention of otolith defect in pastel mink by manganese supplementation, *J. Hered.* 64:111–119.

Everson, G. J., and Shrader, R. E., 1968. Abnormal glucose tolerance in manganese-deficient guinea pigs, *J. Nutr.* 94:89–94.

Falchuk, K. H., Hardy, C., Ulpino, L., and Vallee, B. L., 1978. RNA metabolism, manganese, and RNA polymerases of zinc-sufficient and zinc-deficient *Euglena gracilis*, *Proc. Natl. Acad. Sci.* 75:4175–4179.

Flanagan, H. W., and Huber, R. E., 1978. Cooperative binding of Mn^{2+} and non-cooperative binding of Mg^{2+} to β-galactosidase *(E. coli)*, *Biochem. Biophys. Research Comm.* 82:1079–1083.

Fletcher, K., and Myant, N., 1961. Effect of some cofactors on the synthesis of fatty acids and cholesterol in cell-free preparations of rat liver, *J. Physiol. (Lond.)* 155:498–505.

Foradori, A., and Dinamarca, M. G., 1972. Transferrina o transmanganina definicion de la proteina de transporte de manganeso en el plasma humano, *Rev. Med. Chile* 100:148–153.

Fore, H., and Morton, R. A., 1952. Microdetermination of manganese in biological material by a modified catalytic method, *Biochem. J.* 51:594–598.

Fridovich, I., 1975. Superoxide dismutases, *Ann. Rev. Biochem.* 44:147–159.

Friedmann, N., and Rasmussen, H., 1970. Calcium, manganese and hepatic gluconeogenesis, *Biochem. Biophys. Acta* 222:41–52.

Fukunaga, M., and Mizuguchi, Y., 1982. The effects of propidium on nuclear and mitochondrial mutation induced in yeast by manganese, *Chem. Pharm. Bull.* 30:2889–2893.

Fung, C. H., and Mildvan, A. J., 1973. Interaction of pyruvate with pyruvate carboxylase and pyruvate kinase as studied by paramagnetic effects on ^{13}C relaxation rate, *Biochemistry* 12:620–629.

Gallup, W. D., and Norris, L. C., 1939. The amount of manganese required to prevent perosis in the chick, *Poultry Sci.* 18:76–82.

Gianutsos, G., and Murray, M. I., 1982. Alterations in brain dopamine and GABA following inorganic or organic manganese administration, *Neurotoxicology* 3:75–82.

Gibbons, R. A., Dixon, S. N., Hallis, K., Russell, A. M., Sanson, B. F., and Symonds, H. W., 1976. Manganese metabolism in cows and goats, *Biochem. Biophys. Acta.* 444:1–10.

Gibson, R. S., and Scythes, C. A., 1982. Trace element intakes of women, *Br. J. Nutr.* 48:241–248.

Goldberg, N. D., and Haddox, M. K., 1977. Cyclic GMP metabolism and involvement in biological regulation, *Ann. Rev. Biochem.* 46:823–896.

Grankvist, K., Marklund, S. L., and Taljedal, I-B., 1981. CuZn-superoxide dismutase, Mn-superoxide dismutase, catalase and glutathione peroxidase in pancreatic islets and other tissues in the mouse, *Biochem. J.* 199:393–398.

Greger, J. L., Balinger, P., Abernathy, R. P., Bennett, O. A., and Peterson, T., 1978. Calcium, magnesium, phosphorus, copper and manganese balance in adolescent females, *Am. J. Clin. Nutr.* 31:117–121.

Gruden, N., 1977. Suppression of transduodenal manganese transport by milk diet supplemented with iron, *Nutr. Metab.* 21:305–309.

Guthrie, B. E., and Robinson, M. F., 1977. Daily intakes of manganese, copper, zinc and cadmium by New Zealand women, *Br. J. Nutr.* 38:55–63.

Hart, D. A., 1978. Evidence that manganese inhibits an early event during stimulation of lymphocytes by mitogens, *Exp. Cell Res.* 113:139–150.

Hassanein, M., Ghaleb, H. A., Haroun, E. A., Hegazy, M. R., and Khayyal, M. A. H., 1966. Chronic manganism: Preliminary observations on glucose tolerance and serum proteins, Br. Ind. Med. 23:67–70.

Hidiroglou, M., 1979. Manganese in ruminant nutrition: A review, *Can. J. Anim. Sci.* 59:217–236.

Hidiroglou, M., 1980. Zinc, copper and manganese deficiencies and the ruminant skeleton: A review, *Can. J. Anim. Sci.* 60:579–590.

Hidiroglou, M., and Shearer, D. A., 1976. Concentration of manganese in the tissues of cycling and anestrous ewes, *Can. J. Comp. Med.* 40:306–309.

Hirsch-Kolb, H., Kolb, H. J., and Greenberg, D. M., 1971. Nuclear magnetic resonance studies of manganese binding of rat liver arginase, *J. Biol. Chem.* 246:395–401.

Huang, K-P., Chen, C. H-J., and Robinson, J. C., 1978. Glycogen synthesis by choriocarcinoma cells in vitro, *J. Biol. Chem.* 253:2596–2603.

Hughes, E. R., Miller, S. T., and Cotzias, G. C., 1966. Tissue concentrations of manganese and adrenal function, *Am. J. Physiol.* 211:207–210.

Hurley, L. S., 1981. Teratogenic aspects of manganese, zinc, and copper nutrition, *Physiol. Rev.* 61:249–295.

Hurley, L. S., and Bell, L. T., 1974. Genetic influence on response to dietary manganese deficiency in mice, *J. Nutr.* 104:133–137.

Hurley, L. S., and Everson, G. J., 1963. Influence on timing of short-term supplementation during gestation on congenital abnormalities of manganese-deficient rats, *J. Nutr.* 79:23–27.

Hurley, L. S., Wooley, D. E., Rosenthal, F., and Timiras, P. S., 1963. Influence of manganese on susceptibility of rats to convulsions. *Am. J. Physiol.* 204:493–496.

Hurry, V. J., and Gibson, R. S., 1982. The zinc, copper and manganese status of children with malabsorption syndromes and inborn errors of metabolism, *Biol. Trace Element Res.* 4:157–174.

Huang, K., and Robinson, J. C., 1977. Effect of manganese(ous) and sulfate on activity of human placental glucose-6-phosphate dependent form of glycogen synthase, *J. Biol. Chem.* 252:3240–3244.

Hysell, D. K., Moore, W., Stara, J. F., Miller, R., and Campbell, K. I., 1974. Oral toxicity of methylcyclopentadienyl manganese tricarbonyl (MMT) in rats, *Environ. Res.* 7:158–168.

Jeng, A. Y., and Shamoo, A. E., 1980. Isolation of a Ca^{2+}-carrier from calf heart inner mitochondrial membrane, *J. Biol. Chem.* 255:6897–6903.

Joselow, M. M., Tobias, E., Koehler, R., Coleman, S., Bogden, J., and Gause, D., 1978. Manganese pollution in the city environment and its relationship to traffic density, *Am. J. Pub. Health* 68:557–560.

Katz, S., and Tenenhouse, A., 1973. The relation of adenyl cyclase to the activity of other ATP utilizing enzymes and phosphodieterase in preparations of rat brain; mechanism of stimulation of cyclic AMP accumulation of adrenalin, ovabain and Mn^{++}, *Br. J. Pharmacol.* 48:516–526.

Kawamura, R., Ikuta, H., Fukuzumi, S., Yamada, R., Tsubaki, S., Kodama, T., and Kurata, S., 1941. Intoxication by manganese in well water, *Kisasato Arch. Exp. Med.* 18:145–169.

Keen, C. L., Clegg, M. S., Lonnerdal, B., and Hurley, L. S., 1983a. Whole blood manganese as an indicator of body manganese status. *New Engl. J. Med.* 308:1230.

Keen, C. L., Fransson, G. B., and Lönnerdal, B., 1984a. Supplementation of milk with iron bound to lactoferrin using weanling mice. II Effects on tissue manganese, zinc and copper, *J. Pediat. Gastroenterol. and Nutr.* 3:256–261.

Keen, C. L., Baly, D. L., and Lonnerdal, B. 1984b. Metabolic effects of high doses of manganese in rats, *Biol. Trace Element Res.* In press.

Keen, C. L., Tamura, T., Lonnerdal, B., Hurley, L. S., and Halsted, C. H., 1983b. Effect of chronic ethanol feeding on the activity of superoxide dismutase in monkeys, *Am. J. Clin. Nutr.* 35:836.

Kemmerer, A. R., Elvehjem, C. A., and Hart, E. B., 1931. Studies on the relation of manganese to the nutrition of the mouse, *J. Biol. Chem.* 92:623–630.

King, B. D., Lassiter, J. W., Neathery, M. N., Miller, W. J., and Gentry, R. P., 1980. Effect of lactose, copper and iron on manganese retention and tissue distribution in rats fed dextrose–casein diets, *J. Anim. Sci.* 50:452–458.

Kirchgessner, M., and Heiseke, D., 1978. Arginase-Aktiuitat in der leber washsender ratten bei Mn-mangel, *Int. Z. Vit. Ern. Forschung.* 48:75–78.

Kirchgessner, M., Schwarz, F. J., and Roth-Maier, D. A., 1981. Changes in the metabolism (retention, absorption, excretion) of copper, zinc, and manganese in gravidity and lactation, in: *Trace Element Metabolism in Man and Animals* (TEMA-4), J. McC Howell, J. M. Gawthorne, and C. L. White (eds.), Australian Academy of Sciences, Canberra, pp. 85–88.

Klimis-Tavantzis, D. J., Leach, R. M., and Kris-Etherton, P. M., 1983. The effect of dietary manganese deficiency on cholesterol and lipid metabolism in the Wistar rat and in the genetically hypercholesterolemic RICO rat, *J. Nutr.* 113:328–338.

Lai, J. C. K., Leung, T. K. C., and Lim, L., 1981. Brain regional distribution of glutamic acid decarboxylase, choline acetyltransferase and acetylcholinesterase in the rat: Effect of chronic manganese chloride administration after two years, *J. Neurochem.* 3:1443–1448.

Lamirande, E. D., and Plaa, G. L., 1979. Dose and time relationships in manganese-bilirubin cholestasis, *Toxicol. App. Pharmacol.* 49:257–263.

Laskey, J. W., Rehnberg, G. L., Hein, J. F., and Carter, S. D., 1982. Effects of chronic manganese (Mn_3O_4) exposure on selected reproductive parameters in rats, *J. Toxicol. Environ. Health* 9:677–687.

Leach, R. M., 1971. Role of manganese in mucopolysaccharide metabolism, *Fed. Proc. Fed. Am. Soc. Exp. Biol.* 30:991–994.

Leach, R. M., Muenster, A. M., and Wein, E. M., 1969. Studies on the role of manganese in bone formation. II. Effect upon chrondroitin sulfate synthesis in chick epiphyseal cartilage, *Arch. Biochem. Biophys.* 133:22–28.

Lehmann, B. H., Hansen, J. D. L., and Warren, P. J., 1971. The distribution of copper, zinc and manganese in various regions of the brain and other tissues of children with protein-calorie malnutrition, *Br. J. Nutr.* 26:197–202.

Lis, H., and Sharon, N., 1981. Lectins in higher plants, in *The Biochemistry of Plants*, A. Marcus (ed.), Vol. 6, Academic Press, New York, pp. 371–447.

Lönnerdal, B., Borrebaeck, C. A. K., Etzler, M. E., and Errson, B., 1983c. Dependence on cations for the binding activity of lectins as determined by affinity electrophoresis, *Biochem. and Biophys. Res. Commun.*, 115:1069–1074.

Lönnerdal, B., Keen, C. L., Ohtake, M., and Tamura, T., 1983a. Iron, zinc, copper, and manganese in infant formulas, *Am. J. Dis. Child.*, 137:433–437.

Lönnerdal, B., Keen, C. L., and Hurley, L. J., 1984. Manganese binding proteins in human and cow's milk, *Am. J. Clin. Nutr.*, in press.

Liinerdal, B., Keen, C. L., Ontake, M., and Tamura, T., 1983a. Iron, zinc, copper, and manganese in infant formulas, *Am. J. Dis. Child.*, 137:433–437.

Lönnerdal, B., Keen, C. L., and Hurley, L. J., 1984. Manganese binding proteins in human and cow's milk, *Am. J. Clin. Nutr.*, in press.

Lönnerdal, B., Keen, C. L., and Hurley, L. S., 1983b. Manganese binding in human milk and cow's milk—an effect on bioavailability, *Fed. Proc.* 42:926.

Luckney, T. D., and Venugopal, B., 1977. Protein-metal interactions, in *Metal Toxicity in Mammals*, Vol. I, Plenum, New York, pp. 120–123.

Lyon, M. F., 1953. Absence of otoliths in the mouse: An effect of the pallid mutant, *J. Genet.* 51:638–650.

Lyons, M., and Insko, W. M., 1937. Chondrodystrophy in the chick embryo produced by manganese deficiency in the diet of the hen. *Ky. Agric. Exp. Station Bull.* No. 371.

MacDonald, M. J., Bentle, L. A., and Lardy, H. A., 1978. P-enolpyruvate carboxykinase ferroactivator, distribution, and the influence of diabetes and starvation, *J. Biol. Chem.* 253:116–124.

Mahoney, J. P., and Small, W. J., 1968. Studies on manganese. III. The biological half-life of radiomanganese in man and factors which effect this half life, *J. Clin. Invest.* 47:643–653.

Mangnall, D., Giddings, A. E. B., and Clark, R. G., 1976. Studies of gluconeogenesis in rat liver using a once-through perfusion technique. Effects of manganese ions, *Int. J. Biochem.* 7:293–299.

Matrone, G., Hartman, R. H., and Clawson, A. J., 1959. Manganese–iron antagonism in the nutrition of rabbits and baby pigs, *J. Nutr.* 67:309–317.

Maynard, L. S., and Cotzias, G. C., 1955. The partition of manganese among organs and intracellular organelles of the rat, *J. Biol. Chem.* 214:489–495.

McCoy, J. H., Kenney, M. A., and Gillham, B., 1979. Immune response in rats fed marginal, adequate and high intakes of manganese, *Nutr. Rep. Int.* 19:165–172.

McEuen, A. R., 1981. Manganese metalloproteins and manganese-activated enzymes, in *Inorganic Biochemistry*, H. A. O. Hill (ed.), Royal Society of Chemistry, Burlington House, London, pp. 249–282.

McLeod, B. E., and Robinson, M. F., 1972. Metabolic balance of manganese in young women, *Br. J. Nutr.* 27:221–227.

McNatt, M. L., Fisher, F. M., Elders, M. J., Kilgore, B. S., Smith, W. G., and Hughes, E. R., 1976. Uridine diphosphate xylosyltransferase activity in cartilage from manganese-deficient chicks, *Biochem. J.* 160:211–216.

Meinel, B., Bode, J. C., Koenig, W., and Richter, F. W., 1979. Contents of trace elements in the human liver before birth, *Biol. Neonate* 36:225–232.

Mena, I., 1981. Manganese, in *Disorders of Mineral Metabolism*, F. Bronner and J. W. Coburn (eds.), Academic Press, New York, pp. 233–270.

Mena, I., Court, J., Fuenzalida, S., Papavasiliou, P. J., and Cotzias, G. C., 1970. Modification of chronic manganese poisoning: Treatment with L-Dopa or 5-OH tryptophane, *New Engl. J. Med.* 282:5–10.

Mildvan, A. S., 1970. Metals in enzyme catalysis, in *The Enzymes*, Vol. II, P. D. Boyer (ed.), Academic Press, New York, pp. 445–536.

Miller, S. T., Cotzias, G. C., and Evert, H. A., 1975. Control of tissue manganese: Initial absence and sudden emergence of excretion in the neonatal mouse, *Am. J. Physiol.* 229:1080–1084.

Morgan, W. W., and Huffman, R. D., 1976. The effect of chronic manganese intoxication on the content and turnover of rat brain catecholamines, *Anat. Rec.* 184:484.

Morrison, J. F., and Ebner, K. E., 1971. Studies on galactosyltransferase: Kinetic investigations with glucose as the galactosyl group acceptor, *J. Biol. Chem.* 246:3985–3998.

National Academy of Sciences, 1980. Manganese, in *Mineral Tolerance of Domestic Animals*, National Academy Press, Washington, pp. 290–303.

Neff, N. H., Barrett, R. E., and Costa, E., 1969. Selective depletion of caudate nucleus dopamine and serotonin during chronic manganese dioxide administration to squirrel monkeys, *Experientia* 15:1140–1141.

Niccolai, N., Tiezzi, E., and Valensin, G., 1982. Manganese (II) as magnetic relaxation probe in the study of biomechanisms and of biomacromolecules, *Chem. Rev.* 82:359–384.

Oberley, L. W., and Buettner, G. R., 1979. Role of superoxide dismutase in cancer: A review, *Cancer Res.* 39:1141–1149.

O'Dell, B. L., and Campbell, B. J., 1971. Trace elements: Metabolism and metabolic function, in *Comprehensive Biochemistry*, Vol. 21, M. Florkin and E. H. Stotz (eds.), American Elsevier, New York, pp. 179–266.

Orent, E. R., and McCollum, E. V., 1931. Effects of deprivation of manganese in the rat, *J. Biol. Chem.* 92:651–678.

Orthofer, R., Kubicek, C. P., and Rohr, M., 1979. Lipid levels and manganese deficiency in citric acid producing strains or *Aspergillus niger*. *FEMS Microbiol. Lett.* 5:403–406.

Paddock, G. V., Heindell, H. C., and Salser, W., 1974. Deoxysubstitution in RNA by RNA polymerase in vitro: A new approach to nucleotide sequence determinations, *Proc. Nat. Acad. Sci. USA* 71:5017–5021.

Papavasiliou, P. S., 1978. Manganese and the extrapyramidal system, in *Electrolytes and Neuropsychiatric Disorders*, P. A. Alexander (ed.), SP Medical and Scientific, New York, pp. 187–225.

Papavasiliou, P. S., Kutt, H., Miller, S. T., Rosal, V., Wang, Y. Y., and Aronson, R. B., 1979. Seizure disorders and trace metals: Manganese tissue levels in treated epileptics, *Neurology* 29:1466–1473.

Papavasiliou, P. S., Miller, S. T., and Cotzias, G. C., 1966. Role of liver in regulating distribution and excretion of manganese, *Am. J. Physiol.* 211:211–216.

Papavasiliou, P. S., Miller, S. T., and Cotzias, G. C., 1968. Functional interactions between biogenic amines, 3',5', cyclic AMP and manganese, *Nature* 220:74–75.

Paynter, D. J., 1980. Changes in activity of the manganese superoxide dismutase enzyme in tissues of the rat with changes in dietary manganese, *J. Nutr.* 110:437–447.

Pignatari, F. J., 1932. Glicimia et lipemia nella intossicazione de manganese, *Folia Med.* 18:484–500.

Rabinovitch, M., and Destefano, M. J., 1973. Macrophage spreading in vitro. II. Manganese and other metals as inducers or as cofactors for induced spreading, *Exp. Cell Res.* 79:423–430.

Rao, C. N., and Rao, B. S. N., 1982. Copper, manganese and cobalt balances in Indian adult men and estimation of daily requirement of copper and manganese, *Nutr. Rep. Inter.* 26:1113–1121.

Robinson, G. A., Butcher, R. W., and Sutherland, E. W., 1971. in *Cyclic AMP* Academic Press, New York, p. 72.

Roby, M. J., Vann, K. L., Freeland-Graves, J. H., and Shorey, R. L., 1982. Plasma and liver cholesterol in the manganese-deficient rat, *Fed. Proc.* 41:786.

Rognstad, R., 1981. Manganese effects on gluconeogenesis, *J. Biol. Chem.* 256:1608–1610.

Rorsman, P. and Hellman, B., 1983. The interaction between manganese and calcium fluxs in pancreatic β-cells, *Biochem. J.* 210:307–314.

Rubenstein, A. H., Levin, N. W., and Elliott, G. A., 1962. Manganese-induced hypoglycemia, *Lancet* ii:1348–1351.

Schafer, D. F., Stephenson, D. V., Barak, A. J., and Sorrell, M. F., 1974. Effects of ethanol on the transport of manganese by small intestine of the rat, *J. Nutr.* 104:101–104.

Schneeman, B. O., Lonnerdal, B., Keen, C. L., and Hurley, L. S., 1983. Zinc and copper in rat bile and pancreatic fluid: Effects of surgery, *J. Nutr.* 113:1165–1168.

Schramm, V. L., 1982. Metabolic regulation: Could Mn^{2+} be involved? *Trends Biochem. Sci.* 7:369–371.

Schramm, V. L., Fullin, F. A., and Zimmerman, M. D., 1981. Kinetic studies of the interaction of substrates, Mn^{2+}, and Mg^{2+} with the Mn^{2+}-sensitive and -insensitive forms of phosphoenolpyruvate carboxykinase, *J. Biol. Chem.* 256:10803–10808.

Schroeder, H. A., Balassa, J. J., and Tipton, I. H., 1966. Essential trace metals in man: Manganese, a study in homeostasis, *J. Chronic Dis.* 19:545–571.

Scrutton, M. C., Griminger, P., and Wallace, J. C., 1972. Pyruvate carboxylase: Bound metal content of the vertebrate liver enzyme as a function of diet, *J. Biol. Chem.* 247:3305–3313.

Shani, J., Ahronson, Z., Sulman, F. G., Mertz, W., Frenkel, A., and Kraicer, P. F., 1972. Insulin-potentiating effect of salt bush *Atriplex halimus*) ashes, *Isr. J. Med. Sci.* 8:757–758.

Shrader, R. E., and Everson, G. J., 1968. Pancreatic pathology in manganese-deficient guinea pigs, *J. Nutr.* 94:269–281.

Shukla, G. S., and Chandra, S. V., 1981. Manganese toxicity: Lipid peroxidation in rat brain, *Acta Pharmacol. Toxicol.* 48:95–100.

Shukla, G. S., and Chandra, S. V., 1982. Effects of manganese on carbohydrate metabolism and mitochondrial enzymes in rats, *Acta Pharmacol. Toxicol.* 51:209–216.

Southern, L. L., and Baker, D. H., 1983. *Eimeria acervulina* infection in chicks fed deficient or excess levels of manganese, *J. Nutr.* 113:172–177.

Stevens, J. B., and Autor, A. P., 1977. Induction of superoxide dismutase by oxygen in neonatal rat lung, *J. Biol. Chem.* 252:3509–3514.
Suzuki, H., and Wada, O., 1981. Role of liver lysosomes in uptake and biliary excretion of manganese in mice, *Environ. Res.* 26:521–528.
Tagliamonte, A., Tagliamonte, P., and Gessa, G. L., 1970. Reserpine-like action of chloropromazine on rabbit basal ganglia, *J. Neurochem.* 17:733–738.
Tanaka, Y., 1977. Manganese: Its neurological and teratological significance in man, *Natl. Mtg. Chem. Soc.*, Chicago.
Tandon, S. K., and Khandelwal, S., 1982. Chelation in metal intoxication. XII. Antidotal efficacy of chelating agents on acute toxicity of manganese, *Arch. Toxicol.* 50:19–25.
Taylor, A., Sawan, S. and James, T. L., 1982. Structural aspects of the inhibition complex formed by N-(leucyl)-o-aminobenzenesulfonate and manganese with Zn^{2+}–Mn^{2+} leucine aminopeptidase (EC 3.4.11.1), *J. Biol. Chem.* 257:11571–11576.
Ter Harr, G. L., Griffing, M. E., Brandt, M., Oberding, D. G., and Kapron, M., 1975. Methylcyclopent-adienyl manganese tricarbonyl as an antiknock: Composition and fate of manganese exhaust products, *J. Air Pollution Cont. Assoc.* 25:858–860.
Thomson, A. B. R., Olatunbosum, D., and Valberg, L. S., 1971. Interrelation of intestinal transport system of manganese and iron, *J. Lab. Clin. Med.* 78:643–655.
Thomson, A. B. R., and Valberg, L. S., 1972. Intestinal uptake of iron, cobalt and manganese in the iron-deficient rat, *Am. J. Physiol.* 223:1327–1329.
Tsopanakis, A. D., and Herries, D. G. 1978. Bovine galactosyl transferase. Substrate; manganese complexes and the role of manganese ions in the mechanism, *Eur. J. Biochem.* 83:179–188.
Underwood, E. J., 1977. *Trace Elements in Human and Animal Nutrition*, 4th ed., Academic Press, New York, pp. 170–195.
Vaisius, A. C., and Horgen, P. A., 1980. The effects of several divalent cations on the activation or inhibition of RNA polymerase II, *Arch. Biochem. Biophys.* 203:553–564.
Vallee, B. L., and Coleman, J. E., 1964. Metal coordination and enzyme action, in *Comprehensive Biochemistry*, Vol. 12, M. Florkin and E. Stotz (eds.), Elsevier, New York, pp. 165–235.
Van Woert, M. H., Nicholson, A., and Cotzias, G. C., 1967. Mitochondrial functions of polymelanosomes, *Comp. Biochem. Physiol.* 22:477–485.
Versieck, J., Barbier, F., Speecke, A., and Hoste, J., 1974. Manganese, copper, and zinc concentrations in serum and packed blood cells during acute hepatitis, chronic hepatitis and post hepatic cirrhosis, *Clin. Chem.* 20:1141–1145.
Vuori, E., 1979. Intake of copper, iron, manganese and zinc by healthy, exclusively breast-fed infants during the first 3 months of life, *Br. J. Nutr.* 42:407–411.
Vuori, E., Makinen, S. M., Kara, R., and Kuitunen, P., 1980. The effects of dietary intakes of copper, iron, manganese and zinc on the trace element content of human milk, *Am. J. Clin. Nutr.* 33:227–231.
Walton, K. G., and Baldessarini, R. J., 1976. Effects of Mn^{2+} and other divalent cations on adenylate cyclase activity in rat brain, *J. Neurochem.* 27:557–564.
Wenlock, R. W., Buss, D. H., and Dixon, E. J., 1979. Trace nutrients. 2. Manganese in British food, *Br. J. Nutr.* 41:253–261.
Widdowson, E. M., 1969. Trace elements in human development, in *Mineral Metabolism in Paediatrics*, B. D. Burland (ed.), Blackwell Scientific Publications, Oxford, England, pp. 85–98.
Wilberg, J. S., and Neuman, W. F., 1957. The binding of bivalent metals by deoxyribonucleic and ribonucleic acids, *Arch. Biochem. Biophys.* 72:66–76.
Wilgus, H. S., Norris, L. C., and Heuser, G. F., 1936. The role of certain inorganic elements in the cause and prevention of perosis, *Science* 84:252–253.

Wilgus, H. S. and Patton, A. R., 1939. Factors affecting manganese utilization in the chicken, *J. Nutr.* 18:35–45.

Williams, R. J. P., 1982. Free manganese(II) and iron(II) cations can act as intracellular cell controls, *FEBS Lett.* 140:3–10.

Worwood, M., 1974. Iron and the tracer metals, in *Iron Metabolism in Biochemistry and Medicine*, A. Jacobs and M. Worwood (eds.), Academic Press, New York, p. 336.

Woulfe-Flanagan, H., and Huber, R. E., 1978. Cooperative binding of Mn^{2+} and non-cooperative binding of Mg^{2+} to β-galactosidase (*E. coli*), *Biochem. Biophys. Res. Comm.* 82:1079–1083.

Yip, G. B., and Dain, J. A., 1970. The enzymic synthesis of ganglioside. II. UDP-galactose:N-acetylgalactosaminyl-(CN-acetylneuraminyl) galactosyl-glucosyl-ceramide galactosyltransferase in rat brain. *Biochem. Biophys. Acta* 206:252–260.

Yoshioka, T., Utsunio, K., and Sekiba, K., 1977. Superoxide dismutase activity and lipid peroxidation of the rat liver during development, *Biol. Neonate* 32:147–153.

Zidenberg-Cherr, S., Keen, C. L., Lonnerdal, B., and Hurley, L. S., 1983. Superoxide dismutase activity and lipid peroxidation: Developmental correlations affected by manganese deficiency, *J. Nutr.* 113:2498–2504.

Cobalt

6

Roland S. Young

6.1 Introduction and History

Studies on the biochemistry of cobalt can be said to have commenced around 1934, when the cause of serious disorders of cattle and sheep, in various parts of the world, was finally traced to a deficiency of this element in their diet. Farmers and ranchers had realized for a very long time that if cattle and sheep were grazed continuously on these areas, the animals would lose appetite and weight, become anaemic and weak, and finally die. Agricultural scientists endeavored for many years to find an explanation by seeking toxic elements in the soil and vegetation, parasitic infestation, or deficiencies of the major elements essential for animal nutrition. When investigations turned to minor elements such as iron, and conflicting results were obtained, it was eventually discovered that the small quantity of cobalt present in some iron compounds was actually the curative agent.

It was soon proved that cattle and sheep could be kept healthy on the formerly "unhealthy" pastures by adding small quantities of cobalt compounds to feed, salt licks, water, or top dressings for herbage (Underwood, 1977; Young, 1979; Davies, 1980). The correction of cobalt deficiency in cattle and sheep has sometimes been effected by the administration of cobalt-containing pellets or granules, often termed *bullets*. These are usually an oxide of cobalt mixed with china clay, and are given directly to the animal as a pill.

The importance of cobalt in ruminant nutrition has led to an enormous amount of work throughout the world on the determination of cobalt in soils, plants, feedstuffs, herbage, waters, and fertilizers. Investigations have extended to the biochemistry of cobalt in animals, humans, microorganisms, and enzymes.

Roland S. Young • Consulting Chemical Engineer, 605, 1178 Beach Drive, Victoria, B.C. V8S 2M9, Canada.

A steady flow of results from many parts of the world for over 40 years has provided substantial data on total and available cobalt in soils. Total cobalt is the amount found in the sample after complete decomposition by either acid treatment or fusion. Available cobalt is that portion which is extracted from the soil by treatment with dilute acids or salt solutions; the most popular extractant is 2.5% by volume acetic acid. Available cobalt as a percentage of total cobalt shows a very wide range, 1–95%, but is usually between 5 and 20%.

Total cobalt may range from about 0.3 mg kg^{-1} of soil in markedly cobalt-deficient areas to 1000 mg kg^{-1} soil over mineralized regions; most soils fall within 2–40 mg kg^{-1}. Available cobalt varies from about 0.01 to 6.8 mg kg^{-1} of soil, but is usually in the range 0.1–2 mg kg^{-1}.

From the middle of the nineteenth century, cobalt was detected in plant material by the occasional investigator, but the first systematic studies were published in 1930 (Bertrand and Mokragnatz, 1930). When it became evident, a few years later, that cobalt deficiencies in livestock existed in many localities throughout the world, investigations into the cobalt content of plants increased rapidly. Data have been compiled in recent years on the cobalt content of a number of plant materials, hays and pastures, and trees and shrubs (Young, 1979).

For over 50 years it has been known that whole liver was effective in the dietary treatment of pernicious anemia. Many years later, the pure antipernicious anemia factor, now known as vitamin B_{12}, in which cobalt plays a key role, was isolated and its structure described (Smith, 1965; Hodgkin, 1969). The discovery of vitamin B_{12} extended the studies on cobalt to human foods and to investigations into the effects of cobalt on microorganisms and enzymes.

These measurements were greatly facilitated by the introduction about 40 years ago of photoelectric colorimeters, and the development of improved analytical methods for the determination of small quantities of cobalt. Around 1960, the rapid extension of the technique of atomic absorption spectroscopy provided the analytical chemist with another valuable procedure for rapidly determining low concentrations of cobalt.

6.2 Cobalt and Its Compounds in Cells and Tissues

In practically all of its compounds, cobalt has a valence of 2 or 3; the divalent is the more stable form in the simple salts and the trivalent in the complexes. The simple cobaltic ion, standing between the stable ferric and the nonexistent trivalent nickel ion, is so unstable that it nearly always oxidizes its surroundings. A very large number of stable trivalent cobalt complexes exist, particularly in the form of amines.

Another important fact to bear in mind is that cobalt, like the other elements of Group VIII, has marked catalytic power.

Once inside an organism, cobalt, like all metal ions, is handled much as is oxygen. It is stored, transported, and even inserted into chosen environments.

6.2.1 Cobalt in Soils

The amount of cobalt sorbed by soil appears to depend chiefly on the organic matter and pH. Cobalt is preferentially bound at noncationic sites in ammoniacal marine sediment humus (Aggarwal and Desai, 1980), and organic compounds of a peat compost increased the fixation of cobalt in soil (Kovsh and Tolchel'-nikov, 1980). It was found that humic acids had the highest number of carboxyl groups interacting with cobalt, and the highest cobalt-complexing capacity (Tuev et al., 1980).

In a study of the degree of retention of metals by sand, it was reported that the retention of copper, lead, nickel, and zinc was higher than that of cobalt, whereas cadmium, manganese, and mercury were lower (Penchev and Velikov, 1982).

It has been shown in a study of transport mechanisms of elements in the ocean that cobalt is directly incorporated into pelagic clays at the sediment–water interface (Li, 1981).

6.2 Cobalt in Plants

From the extensive determinations of cobalt in soils and plants, cobalt deficiency studies have led to a very interesting development in plant physiology. Over many years, a large number of investigators in various countries have reported a beneficial effect on plant growth from small additions of cobalt. In experiments designed to determine the effect of cobalt on crop yield, addition of this element is made in one of four ways:

1. Adding solid, water-soluble cobalt compounds to the soil, sand, or water culture.
2. Soaking the seed or tuber, prior to planting, in a dilute solution of a water-soluble cobalt compound.
3. Spraying the plant foliage at an early stage with a water-soluble cobalt compound.
4. Dusting the seeds, before sowing, with a finely powdered water-soluble cobalt compound.

The great majority of workers have reported an improvement in yield and quality from cobalt additions. Many plant physiologists now believe that cobalt must be included among the elements essential for plant growth, albeit in very small quantities. It is difficult to carry out rigid experiments to test this hypothesis, because traces of cobalt are invariably present in the reagents used to make nutrient solutions for such studies.

Numerous studies have been made on the uptake of cobalt by plants. Most workers found cobalt higher in legumes than in grasses or cereals. It has been reported that winter rye, timothy, and potatoes took up the same amount of cobalt from the soil.

In most plants, cobalt has been found to accumulate mainly in the roots. It was observed that the cobalt content of seeds of a number of vegetables and fruits was two to three times higher than in their edible parts. Investigators have reported that tree barks are five to seven times richer in cobalt than are woody tissues, and that birch leaves contained two to three times more cobalt than the trunk.

As a general rule, the cobalt concentration of plants increases during the growth period, then decreases after flowering. Usually the second crop of pasture grass has a lower cobalt content than the first cutting.

More cobalt was picked up by plants in light soils with a low organic matter content than in soils rich in organic constituents. In a study on the solution of metal oxides by decomposing plant materials, it was found that dissolved cobalt is in a complex form in true solution and does not undergo ion exchange to any extent. In sand cultures the highest uptake of cobalt was observed from organic complexes obtained from aerobic fermentation, whereas in water cultures a cobalt–anaerobic solution or a cobalt–mineral solution were more effective.

Several workers have reported lower cobalt levels in plants grown in soils over calcareous rocks or on limed soils. Plants on alkali soils showed a lower cobalt content than the same plants grown on other soils. Higher cobalt was found in plants from acid soils, in the pH range 4.9–6.2 (Gille and Graham, 1971). Large increases in the uptake of cobalt by Sudan grass were reported when aluminum chloride, calcium chloride, and ferrous chloride were added to the soil.

Some plants are hyperaccumulators of cobalt. Tissues of higher plants generally contain less than 1 ppm cobalt on a dry basis, unless this element was added to the substrate. The wheat grasses *Agropyron desertorum* and *Agropyron intermedium* over a wide area in Nevada were found to contain over 20 ppm cobalt (Lambert and Blincoe, 1971). *Astragulus* sp. and *Artemesia* sp., common in various parts of the world where sheep-raising is practiced, are strong accumulators of cobalt, containing up to 100 ppm in dried plant material. On mineralized hillocks in Zaire, a number of plant species have been found to contain well in excess of 1000 ppm cobalt in their dried mass.

6.1.3 Cobalt in Animals

A number of investigations have been reported on cobalt in meat, organs, and blood of ruminants. The liver from normal cattle has been found to contain 0.076–0.201 mg kg^{-1} cobalt, and liver from animals suffering from cobalt deficiency contained 0.056 mg kg^{-1}. The cobalt content of liver is higher than in other organs. In cows, cobalt is present in the blood at reported values, in mg kg^{-1} of blood 0.025, 0.0315 for fetal blood, and 0.0213 for maternal blood, 0.1 for sterile cows, 0.041–0.084 for dry cows, and 0.128–0.233 for pregnant cows. The cobalt content of the myocardium of cattle was found to be 0.033 mg kg^{-1}, and that of calves 0.022 mg kg^{-1}.

The cobalt content of sheep liver has been given, in mg kg^{-1}, as 0.04–1.0 for normal animals and 0.001–0.002 for unhealthy sheep. Cobalt in the blood of sheep ranged from 0.016 to 0.088 mg kg^{-1}. The blood of milking sheep contained three times the cobalt of nonmilking animals.

The cobalt content of cattle hair varied from 0.179 to 0.234 mg kg^{-1}. It has been frequently demonstrated that cobalt additions to the feed of sheep resulted not only in weight gains, but also in improved wool yields.

Papers have reported cobalt in milk, in mg kg^{-1}, from 0.0008 to 0.208, the wide range obviously reflecting differences in soil, feed, season, breed of dairy cattle, and other factors (Young, 1979).

The addition of cobalt to poultry feed increased the iron and cobalt content of blood and bone marrow and gave a significant increase of total nitrogen in blood serum.

The average cobalt content in blood of chickens has been placed at 0.045 mg kg^{-1}, and that of liver at 0.12 mg kg^{-1}. The average cobalt in shelled eggs has been reported as 0.03 mg kg^{-1}, and in uncooked chicken meat 0.0057 mg kg^{-1}.

The cobalt contents of a number of plant and animal products and fish used in human nutrition have been compiled (Young, 1979). Cobalt in the liver of cattle, chicken, cod, pig, and sheep is considerably higher than in other parts of the animal, bird, or fish.

The cobalt content of human blood and body tissues is reported in a number of papers. In erythrocytes, cobalt ranged from 0.059 to 0.13 mg kg^{-1}, in blood serum from 0.0055 to 0.40 mg kg^{-1}, and in whole blood it averaged 0.238 mg kg^{-1}. Blood cobalt is somewhat higher in men than in women at all ages. Cobalt in adult tissues ranged from a mean value of 0.0003 mg kg^{-1} in serum to 0.07 mg kg^{-1} wet weight in liver. The highest level of cobalt is found in the nonpregnant woman, and the lowest at the time of childbirth. Female hair contained significantly more cobalt than male hair. About 12–23% of the intake of cobalt is excreted in urine. The amount of cobalt in human milk decreases continuously with the period of lactation. Cobalt and iron appear to share a similar absorptive mechanism in the proximal intestine, but not in the distal.

An extensive compilation of trace elements in fertilizers, issued a few years ago, has a number of references to cobalt (Swaine, 1962). The values ranged from 0.001 mg kg^{-1} to 100 mg kg^{-1}, but the majority were about 1 mg kg^{-1}. A more recent table on cobalt contents of fertilizers gives a range from 0 to 3000 mg kg^{-1}, with the majority containing 1–6 mg kg^{-1} (Young, 1979).

A study of the aerial deposition of cobalt in Britain has indicated that this amounts to 0.15–0.41 mg/m^2 per year. This is sufficient to satisfy most of the crop requirement; in sparsely settled regions of the world, deposition might be only one-tenth of this value.

A compilation of cobalt contents of waters has been published (Young, 1979); the content of cobalt in water, like that of other microelements, is usually expressed as μg L^{-1} or parts per billion. Sea waters vary appreciably from about 0.01 to 4.6 μg L^{-1}. Fresh waters show an even greater range in cobalt, with most values falling within 0.1 to 10 μg L^{-1}, but some spring, mineral, and mine waters contain more cobalt. Canadian drinking water supplies, for example, range in cobalt from <2 to 6 μg L^{-1}, with a median value of <2 (Meranger et al., 1981). In natural waters, cationic forms of cobalt predominate, but appreciable amounts of neutral and small quantities of organic forms are present (Kulmatov et al., 1982).

Domestic sewage sludge in Japan contained 4 mg kg^{-1} cobalt, and cobalt in food accounted for about 50% of this element in domestic sewage sludge (Tai et al., 1980). A study in Holland showed that after a yearly application of 18 tons of sewage sludge for five years, the treated plots had a cobalt content of 12.4 mg kg^{-1} compared with 2 mg kg^{-1} on the untreated plots (Hemkes et al., 1980).

A compilation of the cobalt content of human foods is available (Young, 1979). Foods containing a high cobalt content include beet greens, bread, buckwheat, cabbage, figs, green onions, molasses, mushrooms, pears, radishes, spinach, tomatoes, and turnip greens; all these have a cobalt content of about 0.2 mg kg^{-1}. Low-cobalt foods, containing less than about 0.05 mg kg^{-1}, include apples, apricots, bananas, carrots, cassava, cherries, coffee, corn, eggplant, oats, pepper, potatoes, rice, salt, sweet potatoes, wheat, and yams. It is worthy of note that cassava, corn, potatoes, rice, and wheat, the staple food of so many humans, are usually low in cobalt.

6.3 Cobalt Deficiency and Function

Some workers have found that 2.5 mg total cobalt per kilogram of soil is the minimum for avoiding cobalt deficiency in ruminants; other investigators have reported values of 2.8–10 mg total cobalt per kilogram of soil. A number of workers have accepted 0.3 mg available cobalt per kilogram of soil as the minimum for healthy animals, whereas others have favored values of 1.35–<2 mg available cobalt per kilogram of soil (Young, 1979).

Many workers have given a value of about 0.07–0.10 mg cobalt per kilogram of dry feed as the minimum quantity required for the maintenance of health in cattle and sheep. A few investigators have reported figures ranging from 0.1 to 2 mg cobalt per kilogram of dry feed as the minimum for preventing cobalt deficiency.

6.3.1 Cobalt in Animal Nutrition

A number of investigators have reported weight gains and an improvement in the general development of pigs from cobalt additions. Increased weight gains have been demonstrated for the following additions, in milligrams of cobalt per kilogram of feed, 0.075–2.1. Cobalt supplements increase the hemoglobin content and concentration of cobalt in the liver and kidneys, and improve nitrogen metabolism and the reproductive capacity of pigs.

Additions of cobalt to rations for rabbits, in quantities ranging from 0.07 to 0.1 mg kg^{-1} of feed, resulted in increased weight gains. When rabbits were fed diets containing vegetables rich in cobalt, gains were also recorded.

The importance of rats and mice in laboratory investigations in biochemistry and other life sciences is reflected in an extensive literature on the effect of cobalt on these animals. Following the injection of cobalt chloride, the main organs that accumulated cobalt were the liver and kidneys; a large part of the injected dose was excreted rapidly. About 27% of cobalt-60 administered to rats was absorbed by the tissues; after two months only about 4% remained in the organism. When supplies of iodine were low, 0.04–0.4 mg cobalt per rat per day gave higher weights of thyroid, but over 0.4 mg tended to suppress the thyroid function. Cobalt is without effect against liver necrosis in the rat. Cobalt chloride in aerosol form caused acute lung edema. When cobalt was administered in milk, absorption from the gastrointestinal tract was 40%.

Additions of cobalt to the feed of bees, buffaloes, carp, foxes, geese, mink, and silkworms have proved beneficial (Young, 1979).

6.3.2 Cobalt in Human Nutrition

The average uptake of cobalt by humans is 0.03–0.3 mg per day, and a positive balance has been reported with a daily diet of 0.03 mg of cobalt.

Vitamin B_{12} is a red crystalline solid, fairly soluble in water and lower alcohols, but not in most other organic solvents. Its empirical formula is $C_{63}H_{88}O_{16}N_{14}PCo$, giving it a cobalt content of 4.35%. The structure of vitamin

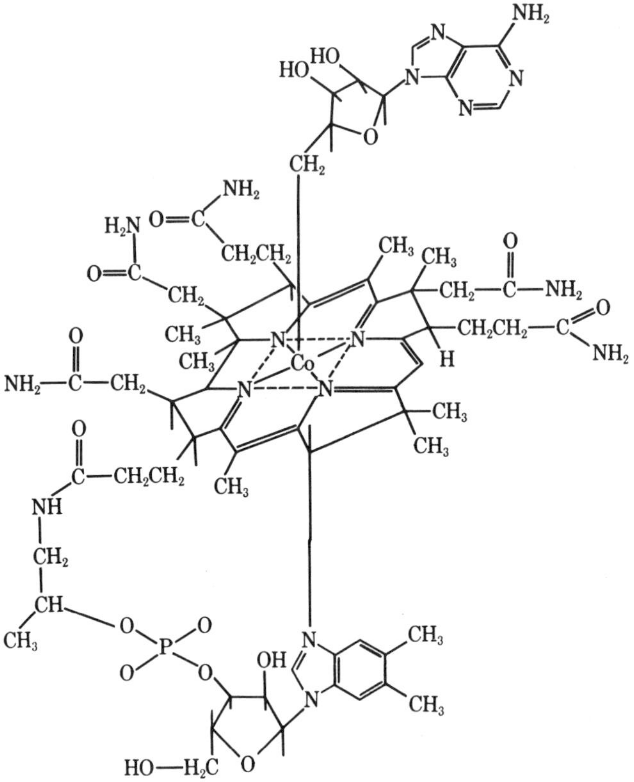

Figure 6-1. The structure of vitamin B_{12}.

B_{12} is shown in Figure 6-1. The central cobalt atom is in the trivalent state, but is easily reduced to the divalent condition. Vitamin B_{12} contains four pyrrole rings, as does heme, but these rings are partially reduced whereas heme rings are fully conjugated. Also, in heme the groups binding the pyrrole rings together are all alike, but in vitamin B_{12} one bridging-CH-group is missing, so there is a five-membered ring in place of a six-membered one. There is also a direct cobalt–carbon bond. Vitamin B_{12} brings about molecular rearrangements, moving an organic group from one carbon atom in the substrate to another.

Vitamin B_{12} is the first vitamin found to contain a metal, the only cobalt-containing compound in the human body, and the most complex nonpolymer yet found in nature. It is also one of the most physiologically potent compounds; only about 1 μg per day is required in human nutrition. It is essential, in conjunction with an enzyme, in at least 10 reactions, although only one of these is present in man, where it affects growth and red blood cell formation.

Vitamin B_{12} is the only vitamin that is synthesized exclusively by microorganisms. Some bacteria and protozoa cannot synthesize the vitamin; other bacteria and actinomycetes make far more than they need and are actually employed for the manufacture of the vitamin. The most concentrated natural synthesis occurs in the forestomach of ruminants. Bacterial synthesis of the vitamin also takes place in the gut of other species, including humans. In some animals and in man, however, this synthesis occurs too low in the gastrointestinal tract for the vitamin to be absorbed; an external dietary source is required.

Most agricultural scientists now believe that cobalt deficiency in ruminants is essentially a deficiency of vitamin B_{12}. The pathological consequences of vitamin B_{12} deficiency form a syndrome, notable features of which are neurological and muscular lesions, in which the metabolic consequences of hepatic damage may play a significant role (Fell, 1981). The following daily additions have been shown to raise the level of vitamin B_{12} to satisfactory ranges and overcome cobalt deficiencies: for sheep, 2–5 mg cobalt chloride, and for cows, 20–30 mg cobalt chloride.

6.4 Metabolism and Toxicity of Cobalt

6.4.1 Effect of Cobalt on Plants

Yields of barley, buckwheat, and oats were increased by soil additions of cobalt. Cobalt fertilization by presoaking seeds has increased the yields of barley, buckwheat, oats, and wheat; foliar fertilization improved the yields of millet and wheat.

Corn yields have invariably been increased by addition of cobalt to the soil; soaking seeds in cobalt solutions and the use of foliar sprays also proved beneficial to this crop.

The yield and quality of cotton and flax have been improved by cobalt additions to the soil and by presoaking seeds in a cobalt solution.

Cobalt additions to soil and foliar spraying with dilute cobalt solutions have increased the yield and improved the quality of grapes.

Most investigators report beneficial effects on the yield and quality of legumes when cobalt is added to the soil; all workers agree on an improvement from soaking legume seeds in a cobalt solution.

Some scientists have indicated that cobalt is an essential element for growth of soybean plants when they depend on the symbiotic relation with *Rhizobium japonicum* for the fixation of atmospheric nitrogen. Other investigators have reported that the legume itself, as distinct from the symbiont, requires cobalt.

Most studies have shown that the yield and quality of sugar beets are improved by cobalt additions to the soil, by soaking seeds, foliar applications, or dusting seeds with cobalt compounds.

Additions of cobalt to soil have been consistently beneficial to the yield of vegetables (Young, 1979).

The addition of cobalt to the soil, or the presoaking of seeds in a dilute cobalt solution, have been found to increase the amount of chlorophyll in barley, buckwheat, corn, grapes, legumes, oats, sugar beets, vegetables, trees, and miscellaneous plants.

Cobalt additions to soil, or seed treatment with a cobalt solution, increased the protein content of buckwheat, corn, oats, and wheat. Spraying grapevines with a dilute cobalt solution increased the protein content of the leaves three- to fivefold. Most workers report the addition of cobalt increased the protein content of clover, legume hay, lupine, and peas. The protein content of potatoes has been agumented by soaking the tubers, before planting, in 0.05% cobalt nitrate solution. The protein content of peanuts has been improved by treating the seed and foliage with cobalt. Spraying the foliage of sunflowers with a cobalt solution increased the protein content of the leaves.

Numerous investigations have been carried out on the influence of cobalt additions on the quantity of sugars or other carbohydrates in plants. Cobalt increased the starch content of corn and wheat and the sugar content of corn. The accumulation of sugar in grapes is favored by the addition of cobalt. Starch and sugars in lupines have been increased by cobalt applications to soil, seed, and foliage. All investigators have reported an increased sugar content in sugar beets by additions of cobalt to soil, seed-soaking solutions, foliar sprays, or seed-dusting powders. The starch content of potatoes and the sugar content of tomatoes have been improved by cobalt.

The effects of cobalt on enzyme activity in plants have been reported in a number of publications. Cobalt additions increased the activity of catalase in cotton and of catalase, ascorbic oxidase, polyphenoloxidase, and peroxidase in grapevines. Cobalt enhanced the activity of catalase and dehydrogenase in potatoes and of catalase and peroxidase in tomatoes. In legumes, cobalt was found to exert a positive effect on the enzyme activity of nodules.

The transpiration rates of many plants were improved by cobalt additions.

An investigation into the translocation of cobalt in the tomato plant indicated that this element was transported predominantly as an inorganic cation in the cell exudate.

Cobalt has been reported to improve the drought-resistance of barley, corn, oats, and soybeans; its effect has been related to its capacity to increase the bound water content, maintain a high protein content, and increase the rates of synthesis and migration of carbohydrates from the leaves to the fruit-bearing organs.

6.4.2 Effect of Cobalt on Animals

The literature furnishes abundant evidence of the beneficial effect of cobalt supplements on milk production. From 10 to 40 mg cobalt chloride per head per day increased milk yield. It has been observed that the cobalt content of milk drops gradually at the beginning of the lactation period and increases toward the end. It has been reported that the addition of 0.1 mg cobalt per kilogram of feed increased the sugar and total nitrogen of milk, but decreased the content of acetoacetic and β-hydroxybutyric acids, ammonia, and urea.

The addition of cobalt to feedstuffs of cattle and sheep has improved the digestibility of nutrients and utilization of food. Cobalt additions have increased the concentration of hemoglobin and erythrocytes in the blood of ruminants, increased the protein of blood and serum, and produced favorable effects on nitrogen metabolism. Additions of cobalt chloride to cattle increased the carotene content of their blood, but did not alter carotene levels in sheep. Augmenting feedstuffs with cobalt has been reported to improve the reproductive capacity of cattle and sheep and to stimulate the lactation of ewes.

Enzyme activity in ruminants has been increased by administration of cobalt compounds. Amylase, diastase, lipase, nuclease, and trypsin in sheep and arginase activity in cattle were improved by cobalt supplements.

Many papers have shown that the addition of small quantities of cobalt compounds to poultry rations has improved the growth and development of chickens. Cobalt stimulated the bacterial synthesis of vitamin B_{12} in the intestines of chicks, and the element is nutritionally important for chicks that are on diets lacking choline and vitamin B_{12}. Additions of cobalt to the feed, in milligrams per kilogram of dry feed, have varied appreciably, from about 0.005 to 1.2.

It was observed many years ago that food products from places known for endemic goiter usually had low cobalt contents. The application of cobalt, even in the absence of iodine, seemed to have a beneficial effect on the incidence of goiter. Investigations have found that cobalt activates thyroid gland activity in the case of an iodine insufficiency, but best results are obtained by the combined use of cobalt and iodine.

6.4.3 Effect of Cobalt on Microorganisms

Many interesting observations have been made on the relation of cobalt to microbial biochemistry. Cultures of actinomycetes from cobalt-containing muds in stagnant reservoirs have been found to produce vitamin B_{12}. Addition of 0.05–5 parts 10^{-6} cobalt as cobalt chloride to a culture medium increased vitamin B_{12} synthesis from actinomycetes.

Cobalt stimulated growth in three species of algae at concentrations of 0.002–0.2 parts 10^{-6}, depending on the species; the addition of 2 parts 10^{-6} or more inhibited growth. Another study found that a concentration of cobalt exceeding 0.04 parts 10^{-6} retarded the growth of three algal species. It has been reported that 0.4 parts 10^{-6} cobalt is necessary for optimum growth of four algal species. Marine algae have a relatively high cobalt content, the ratio Co : Ni approaching 1; in seawater it is about 0.3.

The toxicity of cobalt toward bacteria varies with the species, and in general occupies an intermediate position among cations. Growth of *Streptomyces griseus* and *B. subtilis* was not affected by 2 parts 10^{-6} cobalt, but exposure to dilute solutions of cobalt sulfate decreased the growth and virulence of both typhoid and dysentery baccili (Young, 1979).

Nearly all investigators report a beneficial effect on the activities of nitrogen-fixing bacteria on the roots of legumes from the addition of small quantities of cobalt. A variety of nitrogen-fixing organisms, such as *Azotobacter, Clostridium,* and *Rhizobium,* require cobalt for growth and for the synthesis of vitamin B_{12} compounds (Evans *et al.*, 1965).

Several scientists found that cobalt increased the bacteriostatic action of some antibiotics.

The action of cobalt on fungi has interested a number of workers. The yield of *Aspergillus niger* decreased consistently with increasing additions of 0.1–50 mg cobalt nitrate per liter of nutrient solution. Another study showed that increments of cobalt sulfate up to 0.002% increased the growth and weight of *Aspergillus niger* and *Penicillium glaucum,* but a concentration of 0.033% reduced growth to below that of the check (Beeson, 1950). The toxic limits of cobalt for fungi such as *Aspergillus niger, Penicillium oxalium,* and *P. expansum* were 1500–1600 parts 10^{-6}; cobalt salts were less injurious than were mercury or silver, but more detrimental than cadmium, lead, or nickel.

A recent study of 87 fungal species has indicated that fungi contain more cobalt than higher plants.

Treatment of yeasts with a small quantity of cobalt increases cell numbers, cell sizes, and proteins in the resulting yeast-containing products.

Cobalt additions have increased the enzyme activity of alkaline β-glycerophosphatase, dehydrogenase, hydrogenase, nitrate reductase, oxidase, pancreatic amylase and lipase, phosphatase, proteinase, and zymosan.

6.4.4 Toxicity of Cobalt

Like other micronutrients, cobalt, when added in very small quantities to soil, plants, animal feed, and human foods, is generally beneficial; larger amounts, however, inhibit growth and development and can induce severe toxicity. The inherent variability of biological systems is responsible for wide differences between beneficial and toxic concentrations of cobalt in organisms and enzymes.

A few investigators have studied the toxic efects of cobalt in soils and plants. It has been observed that 0.04–2 ppm cobalt inhibited growth in various algal species. Above 5–20 ppm cobalt depressed growth of various strains of actinomycetes. The growth of some bacteria is decreased by 1–2 ppm cobalt.

The toxic limits of cobalt for *Aspergillus niger, Penicillium oxalicum,* and *P. expansum* have been given as 1500–1600 ppm. A solution of 0.1% cobalt retarded the growth of the fungus *Sclerotinia*, responsible for storage rot of sunflowers.

Dehydrogenase activity of *Rhizobia* is inhibited by 0.1–50 mg of cobalt per liter. In a liquid culture of *Actinomycetes cinerosus*, 0.25–40 parts 10^{-6} cobalt partially or completely inhibited the formation of protease.

In Italy, cobalt was toxic to oats at concentrations greater than 10–40 mg kg^{-1} of soil, in all types of the latter. Toxicity to an oat crop was induced by the presence in the soil solution of cobalt at a concentration of 0.3–1.4 mg liter.$^{-1}$. The seedling tops of corn were injured by a cobalt content in the soil of 25 mg kg^{-1} of soil.

High cobalt levels of 2200 ppm occur in the soils of some areas of Zaire, but there is no evidence that this is an important factor controlling plant growth and distribution (Shewry *et al.,* 1979; Malaisse *et al.,* 1979; Morrison *et al.,* 1979).

Growing cattle can consume up to 50 mg cobalt per 45 kg body weight without ill effects, and sheep can tolerate up to 160 mg daily for at least eight weeks without harmful results, but higher dosages are injurious (Young, 1979).

On soil heavily contaminated with industrial waste, cabbages had a content of 0.69 mg cobalt per kilogram of plant material, and carrots contained 0.88 mg kg^{-1}, but these had no detrimental effect when fed to rabbits.

The toxicity of dusts from the production and use of cemented carbides has been investigated. These carbides, usually tungsten carbide, are commonly bonded with cobalt powder. The machining operations in which these carbides are employed give rise to fine dusts or aerosols that have been of occasional concern to those in the field of industrial hygiene. A British study of hard metal, or tungsten carbide, disease found the incidence to be 1 in 255 workers and stated that pulmonary response may be due to sensitivity. The investigators also reported that cobalt may be the toxic agent, though this metal has not been found in post mortem analyses of lung tissue, perhaps because of its high solubility in plasma.

The effect of cobalt on the human body is of a temporary nature. Many workers with occupational exposure to grinding of hard metal show signs compatible with an allergic alveolitis; symptoms disappear when the patient is absent from work for more than a month. It appears to be related to exposure to cobalt dissolved in coolants necessary for grinding hard metal (Sjoegren *et al.,* 1980).

Cobalt-containing cutting fluid after machining cobalt-containing alloys can give rise to a concentration of this element in the air of the working area of about 0.02 mg per cubic metre causing eczema (Einarsson et al., 1979).

It is of interest to note that a nonferrous alloy, containing about 65% cobalt, has been used for over 40 years in osteosynthesis and in dental prosthesis. Its high strength and compatibility with body tissues and fluids make it very suitable for these purposes.

The radioisotope cobalt-60, prepared by exposing cobalt to the radiations of an atomic pile, has been extensively used as a radioactive tracer in studies on the behavior of this element in soils, plants, and animals. Tracer techniques are very useful in nutritional studies. An element that occurs in only a few parts per hundred million parts of feed can be traced through the digestive tract to its location in the tissues of a 1000-lb animal without disturbing its normal physiology.

6.5 Conclusion

The decorative qualities of cobalt compounds have been used for 4000 years, but it was well into this century before the remarkable chemical and physical properties of cobalt were put to the service of mankind in magnets, high-temperature alloys, cemented carbides, wear-resistant alloys, catalysts, and other industrial applications. Less than 50 years have elapsed since serious disorders of cattle and sheep in many parts of the world were traced to a deficiency of cobalt; their correction by the addition of cobalt compounds to the soil or feedstuffs has been of enormous aid to world agriculture. It is only in the last few decades that the structure of vitamin B_{12} has been established, demonstrating the key role of cobalt in this essential vitamin for human nutrition. In the same time span, the use of the radioisotope cobalt-60 as a tracer in studies on soils, plants, and animals has provided many new insights in biology and biochemistry. It is not unrealistic to anticipate that the coming years will see advances in our knowledge of the biochemical behavior of cobalt and in the development of further valuable uses for this versatile element.

References

Aggarwal, P. N., and Desai, M. V. M., 1980. Selective interaction of metal ions with humus, *Bull. Radiat. Prot.* 3:23–25.

Beeson, K. C., 1950. Cobalt occurrence in soils and forages in relation to a nutritional disorder in ruminants, *Inf. Bull.* No. 7, U.S. Department of Agriculture, Washington, D.C.

Bertrand, G., and Mokragnatz, M., 1930. Sur la répartition du nickel et du cobalt dans les plantes, *Bull. Soc. Chim.* 47:326–331.

Davies, B. E., 1980. *Applied Soil Trace Elements*, Wiley, New York.

Einarsson, O., Erikson, E., Lindstedt, G., and Wahlberg, J. E., 1979. Dissolution of cobalt from hard metal alloys by cutting fluids, *Contact Dermatitis* 5:129–132.

Evans, H. J., Russell, S. A., and Johnson, G. V., 1965. Role of cobalt in organisms that fix atmospheric nitrogen, Non-heme iron proteins, role energy conversion symp., Yellow Springs, Ohio, pp. 303–313.

Fell, B. F., 1981. Pathological consequences of copper deficiency and cobalt deficiency, *Phil. Trans. Roy. Soc. Lond. Ser. B.* 294(1071):153–169.

Gille, G. L., and Graham, E. R., 1971. Isotopically exchangeable cobalt. Effect of soil pH and ionic saturation of a soil, *Soil Sci. Soc. Am. Proc.* 35:414–416.

Hemkes, O. J., Kemp, A., and Van Broekhoven, L. W., 1980. Accumulation of heavy metals in the soil due to annual dressings with sewage sludge, *Neth. J. Agric. Sci.* 28:228–237.

Hodgkin, D. C., 1969. Vitamin B_{12}, *Proc. Roy. Inst. Gt. Br.* 42(199), Pt. 6:377–396.

Kovsh, N. V., and Tolchel'nikov, Yu. S., 1980. Effect of peat composts on trace element mobility in sod podzol soils of various mechanical compositions, *Vestn. Leningr. Univ. Biol.* (3):104–109; *Chem. Abstr.* 1981, 94:14362.

Kulmatov, R. A., Rakhmatov, U., and Kist, A. A., 1982. Migration forms of mercury, zinc, and cobalt in natural waters, *Zh. Anal. Khim.* 37:393–398; *Chem. Abstr.* 1982, 96:168271.

Lambert, T. L., and Blincoe, C., 1971. High concentrations of cobalt in wheat grasses, *J. Sci. Food Agr.* 22(1):8–9.

Li, Y. H., 1981. Geochemical cycles of elements and human perturbation, *Geochim. Cosmochim. Acta* 45:2073–2084.

Malaisse, F., Gregoire, J., Morrison, R. S., Brooks, R. R., and Reeves, R. D., 1979. Copper and cobalt in vegetation of Fungurume, Shaba Province, Zaire, *Oikos* 33:472–478.

Meranger, J. C., Subramanian, K. S., and Chalifoux, C., 1981. Survey for cadmium, cobalt, chromium, copper, nickel, lead, zinc, calcium, and magnesium in Canadian drinking water supplies, *J. Assoc. Off. Anal. Chem.* 64(1):44–53.

Morrison, R. S., Brooks, R. R., Reeves, R. D., and Malaisse, F., 1979. Copper and cobalt uptake by metallophytes from Zaire, *Plant Soil* 53:535–539.

Penchev, P., and Velikov, B., 1982. Experimental studies of the simultaneous migration of heavy metals in sands of average coarseness, *Izv. Vyssh. Ucheln. Zaved., Geol. Razved.* 25(2):110–116; *Chem. Abstr.* 1982, 96:168302.

Shewry, P. R., Woolhouse, H. W., and Thompson, K., 1979. Relationships of vegetation to copper and cobalt in the copper clearings of Haut Shaba, Zaire, *Bot. J. Linn. Soc.* 79:1–35.

Sjoegren, I., Hillerdal, G., Andersson, A., and Zetterstrohm, O., 1980. Hard metal lung disease: Importance of cobalt in coolants, *Thorax* 35:653–659.

Smith, E. L., 1965. *Vitamin B_{12}*, 3rd ed., Wiley, New York.

Swaine, D. J., 1962. The trace-element content of fertilizers, *Commonw. Bur. Soil Sci. Tech. Commun.* No. 52, Farnham Royal, England.

Tai, S., Okada, M., and Sudo, R., 1980. Origin of heavy metal contained in domestic sewage sludge, *Kokuritsu Kogai Kenkyusho Kenkyu Hokoku* 14:203–211; *Chem. Abstr.* 1981, 94:144755.

Tuev, N. A., Chebaevskii, A. I., and Shelikh, A. F., 1980. Reaction of cobalt with various fractions of humus compounds in sodpodzolic soils, *Vestn. Leningr. Univ. Biol.* (2):92–95; *Chem. Abstr.* 1980, 93:131301.

Underwood, E. J., 1977. *Trace Elements in Human and Animal Nutrition*, 4th ed., Academic Press, London.

Young, R. S., 1979. *Cobalt in Biology and Biochemistry*, Academic Press, London.

Molybdenum 7

K. V. Rajagopalan

7.1 Introduction and History

Among the members of the second transition series, molybdenum is the only element definitely known to have specific biological functions. With its ability to exist in oxidation states from -2 to $+6$ and coordination ranging from 4 to 8, the metal has an extraordinarily complex chemistry. No doubt the biological functions of molybdenum stem from the versatility of its chemistry, which allows it to exist in multiple valence states and to participate in facile ligand exchange. Thus it is known that in most of the molybdoenzymes so far examined, the metal attains valence states of $+4$, $+5$, and $+6$ during the catalytic cycle and that the molybdenum center is capable of binding exogenous ligands such as substrates and inhibitors. In all cases, the metal is tightly bound to the host enzymes, which invariably are complex proteins containing two or more different prosthetic groups. Molybdenum-dependent enzymes have been identified in all species of living systems and facilitate such processes as conversion of N_2 and NO_3^- to ammonia in plants and lower organisms, growth of microorganisms on compounds such as NO_3^-, purines and pyridines as carbon and nitrogen sources, utilization of NO_3^- as electron sink, and conversion of sulfite to sulfate in animals. This repertoire of catalytic activities makes molybdenum an essential element for plants and animals, and conditionally so for microorganisms.

Even though the presence of molybdenum in the ashes of plants was discovered in 1900, it was not until 1930 that its role in nitrogen metabolism was realized, with the discovery that the metal was essential for the growth of the nitrogen-fixing organism *Azotobacter* (Bortels, 1930). This was confirmed for

---

K. V. Rajagopalan • Department of Biochemistry, Duke University Medical Center, Durham, North Carolina 27710.

other nitrogen-fixing systems, including leguminous bacteria, and ultimately led to the characterization of nitrogenase as a molybdenum-containing enzyme (Bulen and LeComte, 1966). In parallel with this development, it was also discovered that molybdenum was essential for the utilization of nitrate by fungi and plants and that molybdenum supplementation of soils in various endemic regions of the world resulted in increased plant growth. These studies ultimately led to the finding that nitrate reductase from *Neurospora crassa* is a molybdenum-containing enzyme (Nicholas and Nason, 1954). The essentiality of molybdenum for growth of microorganisms and plants on N_2 or NO_3^- was thus demonstrated to be due to the presence of the metal in nitrogenase and nitrate reductase.

The discovery that the molybdenum content of the diet had a marked influence on the xanthine oxidase activity of rat tissues (De Renzo *et al.*, 1953; Richert and Westerfeld, 1953) was the first indication that molybdenum could have a role in animal metabolism. The presence of molybdenum in purified xanthine oxidase in amounts similar to the flavin content of the enzyme (Richert and Westerfeld, 1953) and subsequent identification of the metal in rabbit liver aldehyde oxidase (Mahler *et al.*, 1954) established molybdenum as a functional trace element in animals. But the essentiality of the metal for humans was not proved until the discovery of sulfite oxidase as a molybdoenzyme (Cohen *et al.*, 1971) shortly after the report of genetic deficiency of the enzyme in a child exhibiting severe neurological abnormalities (Mudd *et al.*, 1967). More recently, the marked pathological sequelae caused by genetic deficiency of the molybdenum cofactor (Johnson *et al.*, 1980), an essential vehicle for almost all the biological functions of the metal, has further underlined the importance of molybdenum for normal human development.

This chapter summarizes the information available to date on the biochemical functions of molybdenum, the distribution and biological availability of the metal, the nature of the molybdenum cofactor, and the toxicological aspects of exposure to high levels of the metal. Owing to the rather broad scope of this chapter, wherever possible the reader is referred to the most recent and comprehensive review articles rather than to the original papers. Apologies are therefore offered to those authors of papers to which direct citation is not made.

7.2 Molybdenum and Its Compounds in Cells and Tissues

As mentioned earlier, all known biological functions of molybdenum are attributable to molybdoenzymes, in which the metal is present in very tight association. Ionic molybdenum, either as the cation or as molybdate, has not been shown to exhibit or promote any catalytic or regulatory activity at the levels at which it is found in cells. Indeed, it is not known whether inorganic molybdenum is present in cells to any extent. In one study Johnson *et al.* (1977) found that the hepatic Mo of rats maintained on standard laboratory chow was entirely present in macromolecular association, partly as known molybdoenzymes and the remainder as a pool of "molybdenum cofactor." The latter entity highlights the fact that in all molybdoenzymes the metal is present not in direct association with the protein but as a complex structure containing other components. Nitrogenase is unique among molybdoenzymes in that its molybdenum moiety is part of a complex metal cluster containing iron and sulfide also (Shah and Brill, 1977). In all the other known molybdoenzymes, whatever the source, the metal is present as part of the universal "molybdenum cofactor," the nature of which will be discussed.

The technique that has been most useful, and indeed essential, for probing the molybdenum sites of diverse enzymes is electron paramagnetic resonance (EPR). Since there are numerous reviews dealing with EPR studies on molybdoenzymes, only the salient conclusions are presented here. Similarly, extended X-ray absorbance fine structure (EXAFS) analysis has recently been applied to molybdoenzymes and has provided information, albeit controversial, regarding the number and chemical nature of the ligand atoms in the coordination field of Mo in several enzymes. Some of these data will be cited.

7.2.1 Molybdenum-Containing Enzymes

A list of the known molybdenum enzymes and the reactions catalyzed by them is presented in Table 7-1. Of these, xanthine dehydrogenase, aldehyde oxidase, purine hydroxylase, and pyridoxal oxidase are generally referred to as molybdenum hydroxylases (Coughlan, 1980), since they catalyze the hydroxylation of substrates using the elements of water. These enzymes act on a wide range of substrates, including purines, pteridines, pyridines, and aldehydes. In principle, the same type of reaction is catalyzed by sulfite oxidase, CO oxidase, and formate dehydrogenase, but each of these enzymes displays rigorous substrate specificity. The reaction catalyzed by the nitrate reductases can be considered as the reverse of the oxidative hydroxylation catalyzed by the other enzymes. Finally, the reaction catalyzed by nitrogenase is mechanistically quite different, in conformity with the uniqueness of this enzyme as a molybdoprotein. Because of this, nitrogenase is discussed separately, while the others are treated as an interrelated group of enzymes.

Nitrogenase

Although all forms of life depend on the availability of nitrogen at its most reduced level, no higher plant or animal is capable of reducing N_2 to ammonia, the process known as nitrogen fixation. Yet this process is a vital cog in the

Table 7-1. Molybdenum-Containing Enzymes from Diverse Sources

Enzyme	Source	Reaction catalyzed	Other prosthetic groups
Xanthine dehydrogenase	Animals	$RH + H_2O \rightarrow ROH + 2e + 2H^+$	FAD, Fe/S, MPT[a]
	Plants		
	Microorganisms		
Aldehyde oxidase	Animals	$RH + H_2O \rightarrow ROH + 2e + 2H^+$	FAD, Fe/S, MPT
Pyridoxal oxidase	D. melanogaster	$RH + H_2O \rightarrow ROH + 2e + 2H^+$	
Nicotinic acid hydroxylase	Bacteria	$RH + H_2O \rightarrow ROH + 2e + 2H^+$	FAD, Fe/S, MPT, Se
Purine hydroxylase	A. nidulans	$RH + H_2O \rightarrow ROH + 2e + 2H^+$	FAD, Fe/S, MPT
Sulfite oxidase	Animals	$SO_3^{2-} + H_2O \rightarrow SO_4^{2-} + 2e + 2H^+$	heme, MPT
	Plants		
	Bacteria		
CO dehydrogenase	Bacteria	$CO + H_2O \rightarrow CO_2 + 2e + 2H^+$	FAD, Fe/S, MPT
Nitrate reductase, assimilatory	Plants	$NO_3^- + 2e + 2H^+ \rightarrow NO_2^- + H_2O$	FAD, heme, MPT
	Microorganisms		
Nitrate reductase, respiratory	Bacteria	$NO_3^- + 2e + 2H^+ \rightarrow NO_2^- + H_2O$	Fe/S, MPT
Formate dehydrogenase	Bacteria	$HCOOH + 2e + 2H^+ \rightleftharpoons CO_2 + 2H_2O$	Se, Fe/S, MPT
Nitrogenase	Microorganisms	$N_2 + 6e + 8H^+ \rightarrow 2NH_4^+$	Fe/S

[a] MPT is molybdopterin.

nitrogen cycle that links various forms of life. Those species that cannot perform this function are therefore dependent on those that can. The latter group is relatively small and includes symbiotic microorganisms that are associated with some higher plants, some autotrophic photosynthetic bacteria and algae, and a few nonphotosynthetic bacteria. Despite the limited number of species of nitrogen-fixing organisms, it has been estimated that the annual global biological N_2 fixation is about 120×10^6 metric tons, a figure to be compared to 40×10^6 metric tons of chemically fixed fertilizer nitrogen (Burris, 1980). Because of the dependence of world food production on the availability of nitrogen fertilizer, it is no surprise that a great deal of interest has been generated in the genetic engineering of the nitrogenase system.

Purified nitrogenases from all sources display a great deal of similarity of structure and function (Hardy and Burns, 1974). The enzyme is a complex of two proteins, one containing Mo and Fe and the other, Fe only. The Mo–Fe protein (also called component 1 or dinitrogenase), with a molecular weight of 220,000–245,000, contains 2 Mo and 30–32 each of Fe and S^{2-} and has an $\alpha_2\beta_2$ structure with the two types of subunits differing slightly in size. The Fe protein (also called component 2 or dinitrogen reductase) has a molecular weight of about 60,000 and contains one Fe_4–S_4 cluster. Nitrogenase is capable of reducing a large number of nitrogen-containing compounds as well as transferring electrons to protons. ATP is an absolute requirement for the reaction and is concomitantly hydrolyzed to ADP and P_i. The reactions catalyzed by nitrogenase are shown in Figure 7-1.

Shah and Brill (1977) demonstrated that the molybdenum component of nitrogenase could be separated from the protein in an active form. This has been confirmed by several other groups, but the exact chemical structure of the material, called Fe–Mo cofactor, is still not known. The cofactor contains Fe and S^{2-} and the overall stoichiometry appears to be 1 Mo : 8 Fe : 6 S^{2-}.

EPR spectroscopy, which has been a powerful tool for probing the molybdenum centers of other molybdenum enzymes, has shown that nitrogenase does not exhibit any signals due to an isolated Mo atom. The observed EPR signals are attributable to the cluster comprising the cofactor (Rawlings et al., 1978). In line with this, EXAFS analysis of nitrogenase indicates that the coordination sphere of the Mo includes three or four Mo–S bonds of 2.35 Å length and two or three Mo–Fe neighbors at a distance of 2.72 Å (Cramer et al., 1978). Several cluster structures have been proposed, but none have been agreed upon.

Xanthine Oxidase–Dehydrogenase (Coughlan, 1980)

Enzymes catalyzing the hydroxylation of xanthine to uric acid are widely distributed in nature. In lower organisms the enzyme enables the utilization of purines as nitrogen sources. The role of the enzyme in higher organisms is not

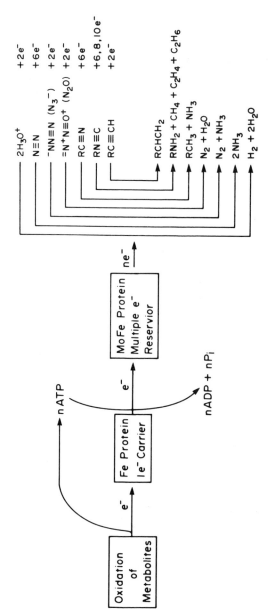

Figure 7-1. The reactions catalyzed by nitrogenase. (Modified from Hardy and Burns, 1973.)

clear. Humans genetically deficient in the enzyme show no symptoms of illness and are usually not identified unless they show the effects of xanthine stones (Wyngaarden, 1978).

Xanthine oxidase from bovine milk is the most thoroughly studied of all molybdoenzymes. The enzyme is a dimer of 155,000 MW subunits and contains one Mo, one FAD, and two Fe_2/S_2 clusters per subunit. Xanthine-oxidizing enzymes from all sources have similar structure and composition (Coughlan, 1980). The enzyme catalyzes a variety of reactions, the most studied of which is

$$\text{Xanthine} + H_2O + O_2 \rightarrow \text{uric acid} + H_2O_2$$

It is now well established that the oxidative hydroxylation of xanthine occurs at the Mo site and that the reduction of O_2 to H_2O_2 occurs at the flavin site (Coughlan, 1980). A considerable fraction of the O_2 reduction is by univalent electron transfer yielding O_2^-, the superoxide anion (Fridovich, 1970). The oxidase activity is a characteristic of mammalian enzymes, since the analogous enzymes from avian, fungal, and some bacterial sources are dehydrogenases using NAD^+ or other electron carriers as physiological electron acceptor. The mammalian enzymes undergo a conversion from dehydrogenase to oxidase during isolation (Waud and Rajagopalan, 1976). Thus the O_2^- flux of the mammalian enzymes is probably highly attenuated *in vivo*.

The molybdenum center of xanthine oxidase has been extensively studied by EPR (Bray, 1980). It is known that the metal exists in valence states of $+6$, $+5$, and $+4$ during catalysis, of which the $+5$ state is paramagnetic. The mechanism of catalysis appears to involve abstraction of a hydride ion from the substrate by the Mo^{6+} to yield Mo^{4+}, with concomitant hydroxylation of the substrate to yield product. The Mo^{4+} is reoxidized to Mo^{6+} by two rapid one-electron transfers, presumably via the Fe–S centers, to the FAD and thence to the exogenous electron acceptor. Such a mechanism implies that a substrate-derived hydrogen and a water molecule would be in the coordination sphere of the Mo during catalysis. Analysis of EPR nuclear hyperfine interactions has provided strong physical evidence that this in fact occurs (Bray, 1975).

The xanthine-oxidizing enzymes are inactivated by reagents such as cyanide, arsenite, and methanol. These inhibitors have pronounced effects on the Mo^{5+} EPR spectrum (Coughlan *et al.*, 1969). In the case of arsenite and methanol the effects are due to interaction with the As and proton nuclei, respectively. EPR has thus provided conclusive proof that these inhibitors interact directly with the Mo. Inhibition by cyanide is due to the removal of a unique sulfur moiety of the enzyme as SCN^- (Massey and Edmonson, 1970), and there is strong evidence that the sulfur is a terminal S^{2-} ligand of the molybdenum (Wahl and Rajagopalan, 1982). EXAFS studies with xanthine dehydrogenase (Cramer *et al.*, 1981) and

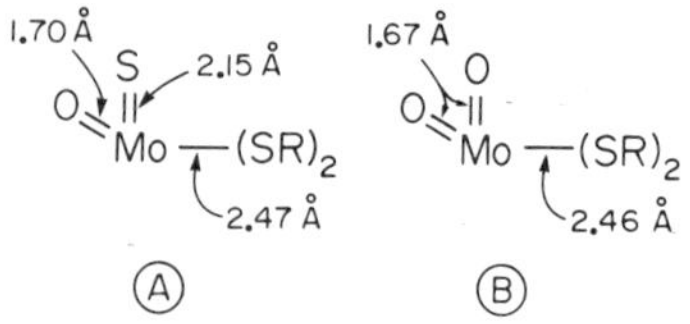

Figure 7-2. Proposed structures of the Mo sites of (A) native and (B) cyanide-treated (desulfo)xanthine dehydrogenase of chicken liver based on EXAFS data.

xanthine oxidase (Bodras et al., 1980) have shown the sulfur ligand of the Mo, at a bond distance of 2.15 Å, is replaced by an oxo group at a bond distance of 1.67 Å in cyanide-treated enzymes, in keeping with the chemical and EPR data. The proposed ligation of the Mo in native and cyanolyzed xanthine dehydrogenase, on the basis of the EXAFS data, is shown in Figure 7-2.

Xanthine oxidase is capable of oxidizing a wide variety of compounds, including purines, pteridines, pyrimidines, pyridines, and aldehydes (Rajagopalan, 1980). However, the physiological significance of some of these reactions is not known. Allopurinol, a pyrazalopyrimidine that is clinically used for the treatment of gout, is oxidized by the enzyme to oxypurinol, which is a tight-binding active-site directed inhibitor of the enzyme and thus blocks urate production. It has been shown that oxypurinol inhibition results from the binding of the molecule to the Mo, which is thereby trapped in the +4 state (Coughlan, 1980). Inactive enzyme generated by this procedure can be reactivated completely by treatment with suitable oxidizing agents.

The literature on xanthine oxidase and the related group of enzymes is quite immense. The topics covered in recent reviews include, to mention a few, a comparative discussion of the chemistry and enzymology of the proteins (Coughlan, 1980), EPR properties (Bray, 1975), relationship to animal and human health (Hainline and Rajagopalan, 1983), and relevance to mechanisms of biological toxicity as well as detoxification (Rajagopalan, 1980).

Aldehyde Oxidase (Rajagopalan and Handler, 1968)

Studies on purified rabbit liver aldehyde oxidase have shown that the enzyme is structurally and chemically quite similar to xanthine-oxidizing enzymes, with which it coexists in many organisms. The rabbit liver enzyme exhibits a broad range of substrate specificity quite similar to that of xanthine oxidase with differences in detail (Rajagopalan, 1980). For example, while both enzymes oxidize hypoxanthine to xanthine, further conversion of xanthine to uric acid is effected by xanthine oxidase but not by aldehyde oxidase. Another notable difference is that rabbit liver aldehyde oxidase converts N^1-methylnicotinamide to the 6-pyridone at physiological pH, but xanthine oxidase does so only at much higher pH. The *in vivo* formation of the pyridone, which is a urinary excretory product, is probably due solely to aldehyde oxidase. Withal, the normal physiological role of aldehyde oxidase is unknown. Genetic deficiency of the enzyme in humans has not been discovered. Significantly, aldehyde oxidase is not inactivated by allopurinol even though it oxidizes it to oxypurinol, which is a potent inhibitor of xanthine oxidase. The marked difference in the effect of allopurinol on the two enzymes can be used to differentiate between the reactions catalyzed *in vivo* by the two enzymes. A remarkable property of rabbit liver aldehyde oxidase is its extreme sensitivity to detergents (Rajagopalan and Handler, 1964), a property not exhibited by xanthine oxidase. Enzyme activity inhibitable by detergents has also been identified in hog liver and human liver. It remains to be seen whether aldehyde oxidases from other sources such as *Drosophila* and potato also exhibit sensitivity to detergents.

The chemical and EPR properties of the Mo center of aldehyde oxidase are similar to those of xanthine-oxidizing enzymes from animal sources (Barber *et al.*, 1982). The same review articles mentioned with respect to xanthine oxidase also contain similar information about aldehyde oxidase.

Sulfite Oxidase (Rajagopalan, 1980a)

Sulfite oxidase is the third and latest molybdenum-containing enzyme shown to be present in animal tissues. The substrate for the enzyme is sulfite, which is the penultimate product in the pathway of metabolism of the sulfur moiety of methionine and cysteine. The product, inorganic sulfate, is the major sulfur compound in the urine, with sulfate esters representing a small fraction of organically linked sulfate. No other substrate for the enzyme has been identified, in contrast to the promiscuity exhibited by xanthine oxidase and aldehyde oxidase. Sulfite oxidase is primarily a hepatic enzyme and is located in the intermembrane space of mitochondria. It has been shown that mitochondrial cytochrome c is in fact the physiological electron acceptor for the enzyme. The proposed catalytic mechanism of the enzyme is shown in Figure 7-3.

Sulfite oxidase has been purified from bovine, chicken, rat, and human livers. The purified enzymes are structurally and chemically similar to one another and are dimers of 55,000–60,000 MW, identical subunits each of which contains one Mo and one cytochrome b_5-type heme (Rajagopalan, 1980a). Oxidation of sulfite is catalyzed by the Mo center, resulting in the reduction of Mo^{6+} to Mo^{4+}. The reoxidation of the Mo^{4+} occurs by univalent steps, generating Mo^{5+} in the process. The electrons are transferred to the heme, which is the site of reduction of cytochrome c.

The two prosthetic groups of sulfite oxidase are located on distinct domains. Mild proteolysis has been successfully employed to isolate a catalytically active molybdenum domain of rat liver sulfite oxidase (Johnson and Rajagopalan, 1977). This domain is the only known case of a protein with only Mo as the prosthetic group. The absorption spectrum of the Mo domain of rat liver sulfite oxidase is

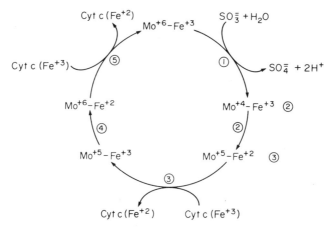

Figure 7-3. Suggested reaction mechanism of sulfite oxidase. The steps are (1) oxidation of sulfite by Mo^{6+} and formation of Mo^{4+}; (2) 1 e transfer from Mo^{4+} to the heme; (3) reoxidation of reduced heme by cytochrome c; (4) electron transfer from Mo^{5+} to the heme; (5) formation of oxidized enzyme by 1 e transfer to cytochrome c.

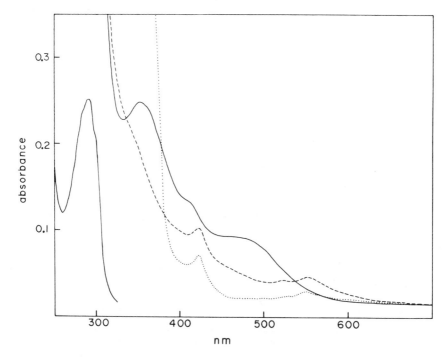

Figure 7-4. Absorption spectrum of the Mo domain of rat liver sulfite oxidase. ———, oxidized; ---- reduced with SO_3^{2-}; and . . . . reduced with dithionite.

shown in Figure 7-4. The spectrum is sensitive to changes in pH and to the presence of phosphate, both of which have been shown to perturb the EPR spectrum of Mo^{5+} in the enzyme.

Administration of tungsten to rats was used to create molybdenum deficiency (Johnson et al., 1974). Inactive, Mo-free sulfite oxidase isolated from the livers of the tungsten-fed rats was immunologically identical to native enzyme. Chemical and EPR studies showed that about 30–40% of the molecules contained W in place of Mo and that the substituted W had the same ligand field as the Mo in native enzyme. This was the first instance of a W-containing protein, although in an inactive state. It was concluded that the more negative oxidation-reduction potential of W as compared to Mo was the reason for the lack of catalytic activity in the W-containing molecules. The EPR properties of sulfite oxidase have been reviewed earlier (Rajagopalan, 1980a).

Oxidized sulfite oxidase is insensitive to cyanide, in contrast to the purine-oxidizing enzymes. This observation would suggest that the unusual "cyanolyzable" sulfur, the terminal ligand of Mo in the latter enzymes, is not present in sulfite oxidase. EXAFS studies by Cramer et al. (1981) have shown, in fact, that the sulfite oxidase Mo center contains two oxo ligands, and no Mo–S bond at 2.15 Å. The ligand field of Mo in sulfite oxidase is thus similar to that of cyanide-treated xanthine dehydrogenase shown in Figure 2.

Genetic deficiency of sulfite oxidase in humans is characterized by severe brain damage, mental retardation, and dislocation of ocular lenses and results in increased urinary output of sulfite, S-sulfocysteine, and thiosulfate and a marked decrease in sulfate output. The molecular basis of the pathology of the disease is not known. The two main possibilities are (1) accumulated toxicity from the higher levels of sulfite in some critical organ and (2) the absence of sulfate required for the formation of sulfated lipids, proteins, and small molecules. The severe pathophysiology of the disease attests to the essentiality of molybdenum for humans.

Nitrate Reductase (Hewitt and Notton, 1980)

Reduction of nitrate to nitrite is a reaction that occurs in virtually all plants and in a broad spectrum of microorganisms. In plants and in some of the microorganisms assimilatory nitrate reduction initiates the process of ammoniagenesis. In other organisms respiratory nitrate reductase enables the anaerobic utilization of nitrate as the terminal electron acceptor for energy-generating electron transport systems. Both the soluble assimilatory reductase and the membrane-bound respiratory or dissimilatory reductase are totally dependent on molybdenum for their activity. Respiratory nitrate reductase has been identified in numerous procaryotic organisms, while the assimilatory enzyme has been described in eucaryote organisms and a few procaryotes (Hewitt and Notton, 1980).

Assimilatory nitrate reductases have been purified from plant, fungal, and algal sources to varying degrees of purity. Those that have been characterized contain Mo and a b_5-type heme, and either have tightly bound FAD or are totally dependent on added FAD for their activity (Hewitt and Notton, 1980). The molecular weights and subunit structures also show wide variations. The electron transport sequence shown in Figure 7-5 has been proposed for the assimilatory enzyme. The physiological electron donor is either NADPH or NADH, depending on the source of the enzyme. The oxidation of reduced pyridine nucleotide with any electron acceptor requires the enzymic FAD, and reduction of NO_3^- definitely involves the Mo center. The role of the heme group is yet to be defined. While EPR or EXAFS information on nitrate reductase is not available, insensitivity of oxidized enzyme to inactivation by cyanide indicates that the Mo center does not contain a cyanolyzable sulfur. Reversible inactivation of reduced enzyme in the presence of cyanide has been reported for the *Chlorella* nitrate reductase (Lorimer *et al.*, 1974) and has been attributed to trapping of the Mo in a reduced state. This behavior of nitrate reductase is emulated by xanthine dehydrogenase and sulfite oxidase, the reduced forms of which are also reversibly inactivated by cyanide (Coughlan *et al.*, 1980). In the latter case valency of the trapped Mo has been suggested to be $+3$ or $+2$.

The best characterized of the respiratory nitrate reductases is the enzyme from *Escherichia coli*, which is repressed by O_2 and induced by NO_3^-. The membrane-bound enzyme has been solubilized as a complex of about 800,000 MW, containing two types of subunits, of 142,000 MW (subunit A) and 60,000 MW (subunit B), with a Mo content of about 4 g atoms per mole (MacGregor *et al.*, 1974). A more recent procedure for large-scale isolation of the enzyme yielded a preparation containing 0.8 g atom of Mo, 16 g atoms of Fe, and 14 g atoms of acid-labile S/200,000 g of protein (Adams and Mortenson, 1982).

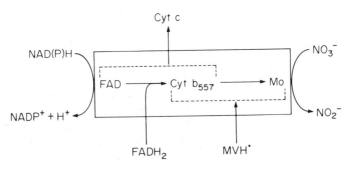

Figure 7-5. Proposed electron transfer reactions of assimilatory nitrate reductase. Reduction of cytochrome c is at the FAD or at the heme; oxidation of reduced methylviologen is either at the heme or at the Mo site.

Chaudhry and MacGregor (1983) have shown that the prosthetic groups and the catalytic activity reside in subunit A.

Vincent and Bray (1978) have examined the EPR properties of *E. coli* nitrate reductase. They found that the resting oxidizing enzyme gave a Mo^{5+} EPR spectrum quite similar in properties to that of substrate-reduced hepatic sulfite oxidase. The signal was sensitive to pH and was also influenced by NO_3^- and NO_2^-. The effects of these anions support the proposal that the Mo center of the enzyme is the site of reduction of NO_3^-.

Formate Dehydrogenase (Ljungdahl, 1980)

Enzymes catalyzing the oxidation of formate to CO_2, known as formate dehydrogenases, have been identified in bacteria, yeasts, and plants. Only the bacterial enzymes are dependent on Mo for their activity. A wide variety of natural electron acceptors, including O_2, NAD(P), quinones, nitrate and H^+ have been reported for the enzyme from different sources. In aerobically grown *E. coli*, oxidation of formate is coupled to O_2 reduction, whereas in anaerobically grown cells either H^+ or NO_3^- is used as the electron acceptor. The primary protein moiety involved in formate oxidation has not been characterized but is assumed to be the same under all conditions of growth. Expression of formate dehydrogenase activity in *E. coli* is completely dependent on the presence of both Mo and Se in the medium (Ljungdahl, 1980). The nutritional dependence on these elements is reflected in the presence of stoichiometric amounts of Mo and Se in purified preparations of formate dehydrogenase. No spectroscopic studies have been reported on any of the purified enzymes.

The clostridial formate dehydrogenases function as CO_2 reductases rather than as formate dehydrogenases (Ljungdahl, 1980). The most significant development in this area is the recent finding that formate dehydrogenase from *Clostridium thermoaceticum* contains tungsten rather than molybdenum (Yamamoto *et al.*, 1983). The purified enzyme had a molecular weight of 340,000 with an $\alpha_2\beta_2$ type of subunit structure and contained 2 W, 2 Se, 36 Fe, and 50 S per molecule. The presence of W, rather than Mo, in this enzyme is in marked contrast to observations made with *E. coli* that growth on W abolishes formate dehydrogenase activity.

Nicotinic Acid Hydroxylase

Nicotinic acid hydroxylase is a molybdoenzyme isolated from the anaerobic organism *Clostridium barkeri* (Dilworth, 1983). The presence of Mo in the enzyme is in keeping with the fact that the reaction catalyzed by the enzyme (i.e., formation of 6-hydroxynicotinic acid from nicotinic acid) is analogous to those of aldehyde oxidase and xanthine dehydrogenase. The enzyme has a molecular weight of 300,000 and contains FAD and Fe–S centers, thus showing structural similarity to the two enzymes as well. The substrate specificity of the enzyme has not been explored. An enzyme catalyzing the hydroxylation of nicotinic acid and a wide range of other substrates, including purines, pteridines, and aldehydes, has been purified from *Aspergillus nidulans* and shown to be a molybdoprotein (M. P. Coughlan, personal communication). The *Aspergillus* enzyme is quite resistant to inactivation by cyanide, the half-time for the reaction being several hours, rather than minutes. The enzyme is distinct from the xanthine dehydrogenase present in the organisms but is structurally similar to it. The clostridial enzyme is a selenoprotein, but it is not known whether the *Aspergillus* enzyme contains selenium.

Carbon Monoxide Oxidase (Meyer, 1983)

A group of aerobic bacteria belonging to species of different genera, including *Pseudomonas, Azomonas, Alcaligenes, Bacillus, Arthrobacter,* and *Azotobacter,* are able to grow autotrophically in air, using carbon monoxide as the sole carbon and energy source (Meyer, 1980). This unusual pathway of CO assimilation is initiated by the enzyme CO oxidase. Meyer (1982) has purified the enzyme from several strains of *Pseudomonas* and has reported that the properties of the different oxidase preparations are quite similar and possibly identical. The physiological electron acceptor for the enzyme has not been identified. The absorption spectrum, molecular weight, and cofactor composition of CO oxidase are strikingly similar to those of xanthine dehydrogenases from animal sources, with the Mo : FAD to Fe–S ratio being 1 : 1 : 4. CO oxidase resembles *Aspergillus* nicotinic acid hydroxylase in its refractoriness to inactivation by cyanide. A fascinating behavior of the enzyme is the four-fold activation elicited by treatment with selenite. The rate of activation parallels the rate of incorporation of 2 Se per enzyme molecule (Meyer, 1983). Under similar conditions selenite has no effect on the animal xanthine-oxidizing enzymes.

7.2.2 The Molybdenum Cofactor

For much of the 30 years since the characterization of xanthine oxidase as a molybdoenzyme little was known about the forces of interaction between the metal and the several proteins that are dependent on it for their activities. That the affinity of the proteins for the metal was very high was apparent from the retention of stoichiometric amounts of the metal in purified preparations. But in the absence of a system for resolution and reconstitution of the metal, information regarding the form of the metal required for binding to the proteins could not be obtained. One of the most exciting developments in recent years has been

Molybdenum

the unraveling of the nature of the molybdenum cofactor, a hypothetical link between two molybdoenzymes in *Aspergillus nidulans* at the beginning, but established now to be a universal essential component of all but one of the known molybdenum-containing enzymes.

The concept of the molybdenum cofactor had its origins in the genetic studies of Pateman *et al.* (1964), who identified a class of pleiotropic mutants of *Aspergillus nidulans* which lacked both nitrate reductase and xanthine dehydrogenase. These cnx mutants belonged to five distinct genetic loci, the coordinate expression of which was necessary for the formation of the postulated cofactor. A chemical basis for the cofactor was provided by the studies of Nason and coworkers with the analogous mutant *nit-1* of *Neurospora crassa*. These investigators showed that the inactive nitrate reductase present in extracts of induced cultures of the mutant could be activated by treatment with extracts of wild-type *N. crassa* as well as several *Neurospora* mutants lacking nitrate reductase only (Nason *et al.*, 1970), but more importantly, also with acid-treated preparations of purified molybdoenzymes from a wide variety of sources (Nason *et al.*, 1971). No activation was observed with molybdate or with several sulfur-containing complexes of the metal. These workers, and others subsequently, pointed out the extreme lability of the reconstituting activity of the cofactor.

Insight into the chemical nature of the cofactor has been derived by using

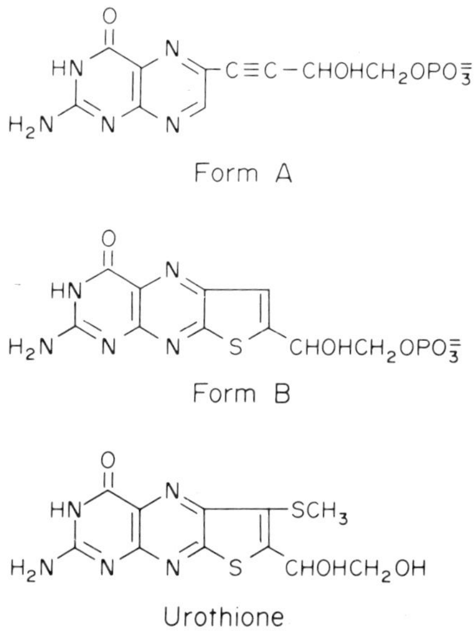

Form A

Form B

Urothione

Figure 7-6. The structures of the Form A and Form B derivatives of the Mo cofactor and urothione.

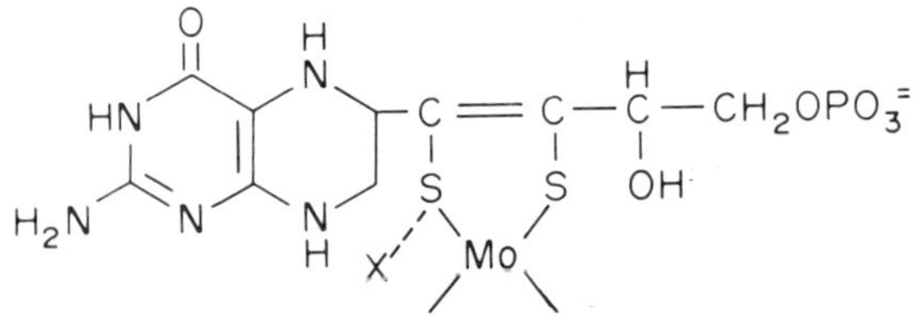

molybdenum cofactor
(proposed structure)

Figure 7-7. Tentative structure of the Mo cofactor.

the strategy of deliberately inactivating the cofactor and elucidating the structures of the inactive forms. Rajagopalan and coworkers (Rajagopalan *et al.*, 1982; Johnson and Rajagopalan, 1982) have shown that the organic moiety of the cofactor contains a pterin ring with a 6-alkyl side chain containing a phosphate ester. Two major oxidized fluorescent derivatives of the pterin have been isolated, and tentative structures have been assigned to them (Figure 7-6).

Form A is a normal pterin whereas Form B contains a substituted thieno (3,2-g) pterin ring system, which has previously been observed in urothione, a compound isolated from urine (also shown in Figure 7-6). The striking similarity in structure between Form B and urothione has led to the finding that the two molecules are metabolically related. Using the technique of high-pressure liquid chromatography, Johnson and Rajagopalan (1982) showed that urothione was present in all control urines but could not be detected in the urine of any of several patients who have recently been characterized as being deficient in molybdenum cofactor. The severe illness associated with cofactor deficiency shows that the cofactor, like Mo, is essential for normal human development.

On the basis of the presence of two sulfurs in urothione and the evidence for the cofactor being a reduced compound, the following tentative structure has been proposed for active molybdenum cofactor (Figure 7-7).

The molybdenum cofactor is postulated to be a complex of the metal with the pterin, which has been termed *molybdopterin*. The postulated ligation of the metal to the two sulfur atoms of molybdopterin is in accordance with the results of EXAFS studies that have provided evidence for two to three Mo–S distances of this type in each of the molybdenum enzymes so far examined. It must be mentioned that the Fe–Mo cofactor of nitrogenase does not appear to contain molybdopterin, in keeping with the uniqueness of the cofactor. The phosphate group of the cofactor is essential for activity, since treatment of cofactor preparation with alkaline phosphatase abolishes the ability to reconstitute *nit-1* nitrate reductase. The pterin is shown to be in the tetrahydro form, but this is yet to be proved.

Table 7-2. Molybdoenzymes Containing Molybdopterin or Similar Materials

Enzyme	Source
Xanthine oxidase	Bovine milk
Xanthine dehydrogenase	Chicken liver
Aldehyde oxidase	Rabbit liver
Sulfite oxidase	Human, rat, and chicken livers; *Thiobacillus novellus*
Nitrate reductase	*Neurospora crassa*
	Chlorella vulgaris
	Squash
Purine hydroxylase	*Aspergillus nidulans*
Formate dehydrogenase	*Methanobacterium formicicum*
	Clostridium thermoaceticum
Nicotinic acid hydroxylase	*Clostridium barkeri*

By testing for the ability to generate Form A under standardized conditions it has been found that *E. coli* mutants lacking molybdenum cofactor activity, the *nit-1* mutant of *Neurospora crassa* and liver samples of cofactor-deficient patients are deficient in molybdopterin. This is very strong evidence for molybdopterin being the sole organic moiety of the molybdenum cofactor.

It was stated earlier that nitrogenase is unique among the molybdoenzymes in having a different cofactor. A list of all the purified enzymes that have been examined and shown to contain molybdopterin, by their ability to give rise to Form A under defined conditions, is shown in Table 7-2. The universality of molybdopterin as the organic component of the molybdenum cofactor is evident from the diversity of the nature and the distribution of these enzymes.

7.2.3 General Aspects of Molybdenum Biochemistry

It is now apparent that the tight association of Mo with enzyme proteins is mediated by molybdopterin. Substrates for molybdoenzymes include small molecules such as N_2, NO_3^-, SO_3^{2-} (or HSO_3^-), CO, and CO_2 (or HCO_3^-) and complex heterocyclic molecules such as purines and pteridines. It is clear that, even though all these enzymes except nitrogenase appear to contain the same molybdenum cofactor, different specificities have been conferred on the molybdenum center in each protein through interaction with the protein itself. In all instances the Mo center actively participates in the catalytic reaction leading to substrate modification, either by oxidation or by reduction. There is no known system in which Mo acts as an internal electron carrier in a multicomponent electron transport chain.

EPR studies on xanthine oxidase, xanthine dehydrogenase, aldehyde oxidase, and sulfite oxidase have shown that the Mo atom exists in valence states of $+6$, $+5$, and $+4$ during catalysis. The EPR signal shape of Mo^{5+} is characteristic of an isolated Mo^{5+} nucleus interacting with its ligand field. The lineshapes of the spectra seen with different enzymes are remarkably alike, indicating the similarities of the coordination field of the metal in different enzymes. In xanthine oxidase, xanthine dehydrogenase, and aldehyde oxidase, a terminal sulfur is liganded to the Mo and is replaced by an oxo group upon treatment with cyanide. The terminal sulfur is not present in sulfite oxidase and nitrate reductase. The presence or absence of a terminal sulfur is yet to be established in the case of CO oxidase, formate dehydrogenase, purine hydroxylase, and nicotinic acid hydroxylase. All of these enzymes appear to be more resistant to cyanolysis than are the enzymes that are known to contain the cyanolyzable sulfur. Recently, Dilworth (1983) has reported that the selenium component of nicotinic acid hydroxylase is released from the protein upon denaturation and that its chromatographic elution behavior coincides with those of Mo and pterin. This finding raises the intriguing possibility that in some enzymes Se rather than S is a terminal ligand of Mo. This selectivity could be explained if the incorporation of the terminal ligand is enzyme catalyzed. There is, in fact, evidence for this, as demonstrated with the *ma-1* mutant of *Drosophila melanogaster*. It has long been known that this mutant is pleiotropic for Mo enzymes and is deficient in xanthine dehydrogenase, aldehyde oxidase, and pyridoxal oxidase (Glassman, 1965). Wahl *et al.* (1982) have recently shown that treatment of *ma-1* fly extracts with sulfide and dithionite, a procedure that has been shown to regenerate the terminal sulfur in cyanolyzed chicken liver xanthine dehydrogenase, leads to reconstitution of xanthine dehydrogenase activity to the same levels as present in wild-type fly extracts. It is apparent that the *ma-1* mutation leads to the accumulation of desulfo forms of the three hydroxylases. In conformity with this, *ma-1* flies contain normal levels of sulfite oxidase, an enzyme that is not dependent on the terminal sulfur.

Despite the similarities of line shapes of Mo^{+5} EPR signals, the oxidation-reduction potentials of the Mo centers of various enzymes cover a surprisingly large span (Table 7-3). The corresponding $Mo^{6+} \rightarrow Mo^{5+}$ and $Mo^{5+} \rightarrow Mo^{4+}$ potentials are quite similar for milk xanthine oxidase, chicken liver xanthine dehydrogenase, and rabbit liver aldehyde oxidase. Removal of the terminal sulfur leads to more negative potentials in all cases. The oxidation reduction potentials of the Mo in *E. coli* nitrate reductase have the most positive values, whereas those of hepatic sulfite oxidase and *Chlorella* nitrate reductase are similar to each other and intermediate in the overall span of potentials. The wide range in potential is interesting in the context of the evidence for the presence of the same molybdenum cofactor in all the enzymes. The observed differences in potentials could be either because of the effect of the individual proteins on the

Table 7-3. Oxidation-Reduction Potentials of Molybdenum Enzymes

Enzyme	E_1 (mV) $Mo^{6+} \rightarrow Mo^{5+}$	E_2 (mV) $Mo^{5+} \rightarrow Mo^{4+}$
Aldehyde oxidase		
(Rabbit liver, active)	−359	−351
(Rabbit liver, desulfo)	−439	−401
Xanthine dehydrogenase		
(Chicken liver, active)	−357	−337
(Chicken liver, desulfo)	−397	−433
Xanthine oxidase		
(Bovine milk, active)	−355	−355
(Bovine milk, desulfo)	−354	−386
Sulfite oxidase		
(Chicken liver)	38	−163
Nitrate reductase		
(*Chlorella vulgaris*)	−34	−54
(*Escherichia coli*)	180	220

Mo or because of the differences in the oxidation states of the pterin ring, or both.

7.3 Nutritional Aspects of Molybdenum

With any dietary constituent, especially a trace element like Mo, it is important to have information on the relationship between dietary levels and states of deficiency, adequacy, and toxicity. In the case of molybdenum there are no documented instances of deficiency or toxicity in normal human populations. Since the subject of the nutritional aspects of Mo in animals has been covered in recent reviews (Mills and Bremner, 1980; Hainline and Rajagopalan, 1983), only a brief summary is presented here.

7.3.1 Molybdenum in the Diet

Schroeder *et al.* (1970) and Tsongas *et al.* (1980) have presented data on the Mo content of various foodstuffs, including packaged finished products. The metal has widespread occurrence, with legumes, meats, and milk being some of the richer sources. The average daily intake of the metal by U.S. adults was calculated to be about 350 μg by Schroeder *et al.* and 120–240 μg by Tsongas *et al.* Since there are no documented cases of Mo deficiency in human populations, it would seem that the widespread occurrence of the metal ensures an adequate dietary supply in all types of regimen.

Schroeder *et al.* (1970) also measured the Mo content of various human tissues and found that liver, kidney, and the adrenal had the highest concentrations. It may be pointed out that liver and kidney are also the tissues with high activities of xanthine dehydrogenase, aldehyde oxidase, and sulfite oxidase. Bone and skeletal muscle contained a large fraction of the total body content of Mo, but the function of the metal in those tissues is unknown.

Intestinal absorption of molybdenum appears to be a saturable process, through a carrier that is also used for sulfate uptake (Underwood, 1977). Dietary SO_4^{2-} has been shown to decrease tissue Mo content in several species. Since SO_4^{2-} is the product of the reaction catalyzed by the molybdoenzyme sulfite oxidase, it is conceivable that competition between sulfate and molybdate is a mechanism for controlling sulfate concentrations in the body. The induction of Mo deficiency by high levels of WO_4^{2-} presumably reflects competition at the uptake level.

7.3.2 Molybdenum Deficiency

The only natural occurrence of Mo deficiency in humans has been that due to a genetic deficiency of the molybdenum cofactor in several severely ill children (Duran *et al.*, 1979; Johnson *et al.*, 1980) displaying marked neurological abnormalities, dislocation of the lens, and mental retardation, resulting in fatalities in the more severe cases. Biochemically, the disease is detectable by the increased urinary excretion of sulfite, thiosulfate, *S*-sulfocysteine, hypoxanthine, and xanthine. In the two cases examined, hepatic content of xanthine dehydrogenase, sulfite oxidase, molybdenum cofactor activity, and Mo itself were below the levels of detection. Supplementation of the patients' diets with molybdenum had no effect on the pattern of abnormal urinary metabolites, showing that a simple dietary deficiency of the metal is not the causative factor.

The pathological symptoms of molybdenum cofactor deficiency are quite like those of simple sulfite oxidase deficiency. Since individuals with xanthine dehydrogenase deficiency are known to be normal, it would seem that the essentiality of Mo and the cofactor for human health stems from its function in sulfite oxidase. Since sulfite oxidase deficiency leads to accumulation of SO_3^{2-} and absence of SO_4^{2-} in the tissues, the pathology of the disease could be due either to accumulated toxicity from constant low levels of sulfite or to the abnormal development of the neurological system in the absence of SO_4^{2-}, a component of sulfolipids and sulfoproteins in several tissues, especially the brain, in the early stages of development.

While natural dietary deficiency of Mo is difficult to achieve even experimentally, deficiency of the metal in experimental animals can be produced by administration of large amounts of tungstate in the diet for several weeks (Johnson *et al.*, 1974). The hepatic activities of sulfite oxidase and xanthine dehydrogenase can be decreased to less than 1% of control levels without any apparent effects on the physical well-being of the animals. Gunnison *et al.* (1981) have used this procedure to correlate tissue sulfite oxidase activity with the urinary levels of thiosulfate and R–S–SO$_3^-$ in tungsten-treated rats. At residual sulfite oxidase levels of 1% of control, the W-treated rats were considered to be metabolically equivalent to humans retaining 10% of normal human hepatic sulfite oxidase content. Gunnison *et al.* found that the urinary thiosulfate and *S*-sulfocysteine of the deficient animals were 10–20% of those seen with sulfite-oxidase-deficient patients. Thus even at hepatic sulfite oxidase levels of 1% of control, the W-treated rats were not nearly as metabolically affected as the patients with genetic sulfite oxidase deficiency. No attempt was made by these investigators to exclude SO$_4^{2-}$ from the diet. Thus the experimental model requires additional refinements before it can answer the question about the origins of the pathology of the human disease. In the meantime the technique of Gunnison and co-workers could be the basis for a noninvasive appraisal of the sulfite oxidase status of even humans.

Lack of knowledge concerning the primary physiological role of aldehyde oxidase makes it impossible to determine whether absence of this enzyme compounds the effects of sulfite oxidase deficiency in the cofactor-deficient patients. Similarly, the absence of other, so far unidentified molybdoenzyme(s) could also contribute to the pathology. Apropos of this it is of interest that the Mo content of normal human liver is considerably greater than the Mo present as xanthine dehydrogenase, aldehyde oxidase, sulfite oxidase, and the storage form of molybdenum cofactor.

7.3.3 Molybdenum Toxicity

Oddly enough, the earliest implication of molybdenum in animal health or disease was the identification of the role of the metal in the etiology of "tearts," a copper deficiency disease, in ruminant animals from several different geographical areas (Mills and Bremner, 1980). Administration of copper supplements invariably reversed the effects of molybdenum.

In attempts at setting up a rat model for this molybdenum–copper interrelationship, it was found that very much higher levels of molybdate were needed to produce a similar effect. It thus became evident that there is a marked difference between ruminant and nonruminant animals in the levels of Mo needed to produce the alteration in Cu metabolism leading to the deficiency state. In other studies on ruminants it was also found that increases in the sulfur content of the diet, either as SO_4^{2-} or as organic sulfur compounds, potentiated the Mo–Cu antagonism (Mills and Bremner, 1980). Since rumen bacteria generate sulfide from sulfate, the possibility was considered that thiomolybdates, generated by the interaction of molybdate with the sulfide generated by the rumen bacteria, were the compounds responsible for producing Cu deficiency. Strong evidence for this was provided by the finding that administration of low levels of tetrathiomolybdate readily caused Cu deficiency in rats.

Analysis of the Cu present in the plasma and kidney of rats treated with tetrathiomolybdate has shown that all of the Cu is precipitable with trichloroacetic acid and exists as a 2 : 1 complex with Mo. Trapping of the body Cu in this nonutilizable complex is apparently responsible for the symptoms of Cu deficiency. Of course in nonruminants there is no S^{2-} in the gut to produce the thiomolybdates. The exact structure of the complex containing Cu, Mo, and $S^=$ has not been determined. It is also not known whether conditions that produce Cu deficiency also interfere with normal metabolism of molybdenum.

7.4 Conclusion

The identification of molybdopterin as the organic component of the molybdenum cofactor common to all molybdoenzymes except nitrogenase has several ramifications. Whether the pterin ring itself participates in some manner in the catalytic cycle, and whether it has the same state of reduction in all enzymes remain to be investigated. The pterin ring is also present in folic acid and in biopterin, two compounds with established biochemical functions. It has been shown that guanosine triphosphate is the precursor of both of these pterins (Brown, 1971). Whether molybodopterin synthesis shares some of the early steps with those of folic acid and biopterin and whether animal tissues can synthesize molybdopterin *de novo* are questions that have to be answered. The proposed structure of the molybdenum cofactor is as yet tentative and has to be proved conclusively. Since the genetic disease of molybdenum cofactor deficiency should in theory be treatable with some form of the pterin that can circumvent the genetic lesion, studies on the biosynthesis and assimilation of molybdopterin could have clinical usefulness.

7.5 Summary

The trace element molybdenum is essential for the activities of several enzymes in diverse organisms. In nitrogenase the metal is present in a unique cluster containing Fe, Mo, and S^{2-}, whereas in all other molybdoenzymes the metal exists as the Mo cofactor, a complex of the metal with a novel organic molecule, termed *molybdopterin*. The Mo cofactor-containing enzymes can be further subdivided into two groups, one in which the metal is additionally coordinated to two oxo ligands and the other in which one of the oxo ligands is replaced by a terminal sulfide. In the latter enzymes, removal of the sulfido ligand by cyanolysis leads to loss of catalytic activity.

Nutritionally based molybdenum deficiency due to inadequate amounts of the metal in the diet does not occur under normal circumstances. Deficiency can be induced in experimental animals by feeding high levels of tungsten. Tungsten treatment thus provides a potentially useful means of studying experimental molybdenum deficiency, especially in the context of the recently discovered genetic disorder in humans, associated with molybdenum cofactor deficiency. Tissues of the cofactor deficient patients are devoid of molybdopterin, the sulfur-containing organic moiety of the cofactor. Molybdopterin is also absent in pleiotropic mutants of *Echerichia coli, Neurospora crassa,* and *Drosophila melanogaster,* which have combined deficiencies of all molybdoenzyme activities. The severe neuropathology seen in cofactor-deficient patients attests to the essentiality of Mo and of molybdopterin for normal human development.

References

Adams, M. W. W., and Mortenson, L. E., 1982. The effect of cyanide and ferricyanide on the activity of the dissimilatory nitrate reductase of *Escherichia coli, J. Biol. Chem.* 257:1791–1799.

Barber, M. J., Siegel, L. M., Coughlan, M. P., and Rajagopalan, K. V., 1982. Spectral properties of rabbit liver adlehyde oxidase—a complex metalloprotein, *Biochemistry* 21:3561–3568.

Bodras, J., Bray, R. C., Garner, C. D., Gutteridge, S., and Hasnain, S. S., 1980. X-ray absorption spectroscopy of xanthine oxidase. The molybdenum centers of the functional and the desulfo forms, *Biochem. J.* 191:499–508.

Bortels, H., 1930. Molybdan als Katalysator bei biologischen Strickstoffbindung, *Arch. Microbiol.* 1:333–342.

Bray, R. C., 1975. Molybdenum iron-sulfur flavin hydroxylases and related enzymes, *Enzymes* 12:299–419.

Bray, R. C., 1980. The reactions and the structures of molybdenum centers in enzymes, *Adv. Enzymol. Relat. Areas Mol. Biol.* 51:107–167.

Brown, G. M., 1971. The biosynthesis of pteridines, *Adv. Enzymol.* 35:35–77.

Bulen, W. A., and LeComte, J. R., 1966. The nitrogenase system from *Azotobacter*: two enzyme requirement for N_2 reduction, ATP-dependent H_2 evolution, and ATP hydrolysis, *Proc. Natl. Acad. Sci.* 56:979–986.

Burris, R. H., 1980. The global nitrogen budget—science or seance? In W. E. Newton and W. H. Orme-Johnson (eds.), *Nitrogen Fixation,* Vol. I, University Park Press, Baltimore, pp. 3–16.

Chaudhry, G. R., and MacGregor, C. H., 1983. *Escherichia coli* nitrate reductase subunit A: Its role as the catalytic site and evidence for its modification, *J. Bacteriol.* 154:387–394.

Cohen, H. J., Fridovich, E., and Rajagopalan, K. V., 1971. Hepatic sulfite oxidase. A functional role for molybdenum, *J. Biol. Chem.* 246:374–382.

Coughlan, M. P., 1980. Aldehyde oxidase, xanthine oxidase and xanthine dehydrogenase: Hydroxylases containing molybdenum, iron-sulfur and flavin, in *Molybdenum and Molybdenum-Containing Enzymes,* M. P. Coughlan (ed.), Pergamon Press, New York, pp. 119–185.

Coughlan, M. P., Rajagopalan, K. V., and Handler, P., 1969. The role of molybdenum in xanthine oxidase and related enzymes, *J. Biol. Chem.* 244:2658–2663.

Coughlan, M. P., Johnson, J. L., and Rajagopalan, K. V., 1980. Mechanisms of inactivation of molybdoenzymes by cyanide, *J. Biol. Chem.* 255:2694–2699.

Cramer, S. P., Hodgson, K. O., Gillum, W. O., and Mortenson, L. E., 1978. The molybdenum site of nitrogenase. Preliminary structural evidence from x-ray absorption spectroscopy, *J. Am. Chem. Soc.* 100:3398–3407.

Cramer, S. P., Rajagopalan, K. V., and Wahl, R. C., 1981. The molybdenum sites of sulfite oxidase and xanthine dehydrogenase. A comparison by EXAFS, *J. Am. Chem. Soc.* 103:7721–7727.

De Renzo, E. C., Kaleita, E., Heytler, P., Oleson, J. J., Hutchings, B. L., and Williams, J. H., 1953. THe nature of the xanthine oxidase factor, *J. Am. Chem. Soc.* 75:753.

Dilworth, G. L., 1983. Occurrence of molybdenum in the nicotinic acid hydroxylase from *Clostridium barkeri,* *Arch. Biochem. Biophys.* 221:565–569.

Duran, M., Korteland, J., Beemer, C., v.d. Heiden, C., de Bree, P. K., Brink, M., and Wadman, S. K., 1979. Variability of sulfituria: combined deficiency of sulfite oxidase and xanthine oxidase, in *Models for the Study of Inborn Errors of Metabolism,* F. A. Hommes (ed.), Elsevier/North Holland, Amsterdam, pp. 103–107.

Fridovich, I., 1970. Quantitative aspects of superoxide production by milk xanthine oxidase, *J. Biol. Chem.* 245:4053–4057.

Glassman, E., 1965. Genetic regulation of xanthine dehydrogenase in *Drosophila malanogaster,* *Fed. Proc.* 24:1243–1251.

Gunnison, A. F., Farruggella, T. J., Chiang, G., Dulak, L., Zaccardi, J., and Birkner, J., 1981. A sulfite-oxidase-deficient rat model: metabolic characterization, *Fd. Cosmet. Toxicol.* 10:209–220.

Hainline, B. E., and Rajagopalan, K. V., 1983. Molybdenum in animal and human health, in *Trace Elements in Health,* J. Rose (ed.), Butterworths, London, pp. 150–166.

Hardy, R. W. F., and Burns, R. C., 1973. Comparative biochemistry of iron–sulfur proteins and dinitrogen fixation, in *Iron–Sulfur Proteins,* Vol. I, W. Lovenberg (ed.), Academic Press, New York, pp. 65–110.

Hewitt, E. J., and Notton, B. A., 1980. Nitrate reductase systems in eukaryotic and prokaryotic organisms, in *Molybdenum and Molybdenum-Containing Enzymes,* M. P. Coughlan (ed.), Pergamon Press, New York, pp. 273–325.

Johnson, J. L., 1980. The molybdenum cofactor common to nitrate reductase, xanthine dehydrogenase and sulfite oxidase, in *Molybdenum and Molybdenum-Containing Enzymes,* M. P. Coughlan (ed.), Pergamon Press, New York, pp. 345–383.

Johnson, J. L., and Rajagopalan, K. V., 1977. Tryptic cleavage of rat liver sulfite oxidase. Isolation of molybdenum and heme domains, *J. Biol. Chem.* 252:2017–2025.

Johnson, J. L., and Rajagopalan, K. V., 1982. Structural and metabolic relationship between the molybdenum cofactor and urothione, *Proc. Natl. Acad. Sci.* 79:6856–6860.

Johnson, J. L., Jones, H. P., and Rajagopalan, K. V., 1977. *In vitro* reconstitution of demolybdosulfite oxidase by a molybdenum cofactor from rat liver and other sources, *J. Biol. Chem.* 252:4994–5003.

Johnson, J. L., Rajagopalan, K. V., and Cohen, H. J., 1974. Molecular basis of the biological function of molybdenum. Effect of tungsten on xanthine oxidase and sulfite oxidase in the rat, *J. Biol. Chem.* 249:859–866.

Johnson, J. L., Waud, W. R., Rajagopalan, K. V., Duran, M., Beemer, F. A., and Wadman, S. K., Inborn errors of molybdenum metabolism. Combined deficiencies of sulfite oxidase and xanthine oxidase in a patient lacking the molybdenum cofactor, *Proc. Natl. Acad. Sci.* 77:3715–3719.

Ljungdahl, L. G., 1980. Formate dehydrogenases: Role of molybdenum, tungsten and selenium, in *Molybdenum and Molybdenum-Containing Enzymes*, M. P. Coughlan (ed.), Pergamon Press, New York, pp. 463–486.

Lorimer, G. H., Gewitz, H-S., Völker, W., Solomonson, L. P., and Vennesland, B., 1974. The presence of bound cyanide in the naturally inactivated form of nitrate reductase of *Chlorella vulgaris*, *J. Biol. Chem.* 249:6074–6079.

MacGregor, C. H., Schnaitman, C. A., Normansell, D. E., and Hodgins, M. G., 1974. Purification and properties of nitrate reductase from *Escherichia coli* K12, *J. Biol. Chem.* 249:5321–5327.

Mahler, H. R., Mackler, B., Green, D. E., and Bock, R. M., 1954. Studies on metalloproteins. III. Aldehyde oxidase: a molybdoprotein, *J. Biol. Chem.* 210:465–480.

Massey, V., and Edmondson, D., 1970. On the mechanism of inactivation of xanthine oxidase by cyanide, *J. Biol. Chem.* 245:6595–6598.

Meyer, O., 1980. Using a carbon monoxide to produce single-cell protein, *BioScience* 30:405–407.

Meyer, O., 1982. Chemical and spectral properties of carbon monoxide: methylene blue oxidoreductase, *J. Biol. Chem.* 257:1333–1341.

Meyer, O., 1983. Biology of aerobic carbon monoxide-oxidizing bacteria, *Ann. Rev. Microbiol.* 37:277–310.

Mills, C. F., and Bremner, I., 1980. Nutritional aspects of molybdenum in animals, in *Molybdenum and Molybdenum-Containing Enzymes*, M. P. Coughlan (ed.), Pergamon Press, New York, pp. 517–542.

Mudd, H. S., Irreverre, F., and Laster, L., 1967. Sulfite oxidase deficiency in man: Demonstration of the enzymatic defect, *Science* 156:1599–1602.

Nason, A., Antoine, A. D., Ketchum, P. A., Frazier, W. A. III, and Lee, D. K., 1970. Formation of assimilatory nitrate reductase by inter-cistronic complementation in *Neurospora crassa. Proc. Natl. Acad. Sci.* 65:137–144.

Nason, A., Lee, K-Y., Pan, S-S., Ketchum, P. A., Lamberti, A., and Devries, J., 1971. *In vitro* formation of assimilatory NADPH : nitrate reductase from a *Neurospora* mutant and a component of molybdoenzymes, *Proc. Natl. Acad. Sci.* 68:3242–3246.

Nicholas, D. J. D., and Nason, A., 1954. Molybdenum and nitrate reductase. II. Molybdenum as a constituent of nitrate reductase, *J. Biol. Chem.* 207:353–360.

Pateman, J. A., Cove, D. J., Rever, B. M., and Roberts, D. B., 1964. A common cofactor for nitrate reductase and xanthine dehydrogenase which also regulates the synthesis of nitrate reductase, *Nature* 201:58–60.

Rajagopalan, K. V., 1980. Xanthine oxidase and aldehyde oxidase, in *Enzymatic Basis for Detoxification*, W. B. Jacoby (ed.), Academic Press, pp. 295–309.

Rajagopalan, K. V., 1980a. Sulfite oxidase (sulfite : ferricytochrome c oxidoreductase), in *Molybdenum and Molybdenum-Containing Enzymes*, M. P. Coughlan (ed.), Pergamon Press, New York, pp. 241–272.

Rajagopalan, K. V., and Handler, P., 1968. Metalloflavoproteins, in *Biological Oxidations*, T. P. Singer (ed.), Interscience, New York, pp. 301–337.

Rajagopalan, K. V., Johnson, J. L., and Hainline, B. E., 1982. The pterin of the molybdenum cofactor, *Fed. Proc.* 41:2608–2612.

Rawlings, J., Shah, V.K., Chisnell, J. R., Brill, W. J., Zimmerman, R., Munck, E., Orme-Johnson, W. H., 1978. Novel metal cluster in the Fe–Mo cofactor from nitrogenase: Spectroscopic evidence, *J. Biol. Chem.* 253:1001–1004.

Richert, D. A., and Westerfeld, W. W., 1953, Isolation and identification of the xanthine oxidase factor as molybdenum, *J. Biol. Chem.* 203:915–923.

Schroeder, H. A., Balassa, J. J., and Tipton, I. H., 1970. Essential trace metals in man: Molybdenum, *J. Chron. Dis.* 23:481–499.

Shah, V. K., and Brill, W. J., 1977. Isolation of an iron–molybdenum cofactor from nitrogenase, 1977, *Proc. Natl. Acad. Sci.* 74:3249–3253.

Tsongas, T. A., Meglen, R. R., Walravens, P. A., and Chappell, W. R., 1980. Molybdenum in the diet: An estimate of average daily intake in the United States, *Am. J. Clin. Nutr.* 33:1103–1107.

Underwood, F. J., 1977. Molybdenum, in *Trace Element Metabolism in Animals*, F. J. Underwood (ed.), Academic Press, New York, pp. 116–140.

Vincent, S. P., and Bray, R. C., 1978. Electron-paramagnetic-resonance studies on nitrate reductase from *Escherichia coli* K12, *Biochem. J.* 171:639–647.

Wahl, R. C., and Rajagopalan, K. V., 1982. Evidence for the inorganic nature of the cyanolyzable sulfur of molybdenum hydroxylases, *J. Biol. Chem.* 257:1354–1359.

Wahl, R. C., Warner, C. K., Finnerty, V., and Rajagopalan, K. V., 1982. *Drosophila melanogaster ma-1* mutants are defective in the sulfuration of desulfo Mo hydroxylases, *J. Biol. Chem.* 257:3958–3962.

Waud, W. R., and Rajagopalan, K. V., 1976. The mechanism of conversion of rat liver xanthine dehydrogenase from an NAD-dependent form (Type D) to an O_2-dependent form (Type O), *Arch Biochem. Biophys.* 172:365–379.

Wyngaarden, J. B., 1978. Hereditory xanthinuria, in *The Metabolic Basis of Inherited Disease*, J. B. Stanbury, J. B. Wyngaarden, and D. S. Fredrickson (eds.), McGraw-Hill, New York, pp. 1037–1044.

Yamamoto, I., Saiki, T., Liu, S-M., and Ljungdahl, L. G., 1983. Purification and properties of NADP-dependent formate dehydrogenase from *Clostridium thermoaceticum*, a tungsten-selenium-iron protein, *J. Biol. Chem.* 255:1826–1832.

Chromium

8

Janet S. Borel and Richard A. Anderson

8.1 Introduction

Chromium is an essential element for animals and humans. Insufficient dietary chromium leads to signs and symptoms similar to those associated with diabetes and/or cardiovascular diseases. The dietary chromium intake of normal individuals is often less than the suggested mimimum intake. Children with protein-calorie malnutrition, diabetics, and elderly and middle-aged subjects have all been shown to respond to supplemental Cr. Dietary trends such as consuming more highly processed foods that are often not only low in Cr but also stimulate increased losses of body Cr stores may exacerbate existing problems associated with marginal dietary intake of chromium by humans.

8.2 Chromium: Physical and Chemical Properties

Chromium is a white, hard, brittle metal of the first transition series, m.p. 1903° ± 10°, atomic number 24 and atomic weight 51.996 g/mol. Four stable isotopes of chromium exist with mass numbers 50 (4.31%), 52 (83.76%), 53 (9.55%), and 54 (2.38%) (Anderson, 1981a). Five radioactive isotopes can be produced but only ^{51}Cr, with a half-life of 27.8 days, is commercially available and is used for radiotracer studies. The other radioisotopes of chromium have half-lives of less than 1 day.

Chromium can occur in any oxidation state from −2 to +6, but the most common oxidation states are 0, +2, +3, and +6. Divalent chromium com-

---

Janet S. Borel and Richard A. Anderson • Beltsville Human Nutrition Research Center, U.S. Department of Agriculture, Beltsville, Maryland 20705.

pounds are readily oxidized to the trivalent state on exposure to air and thus Cr(II) would not be expected to occur in biological systems. Hexavalent chromium, predominantly linked with oxygen as either chromate (CrO_4^{2-}) or dichromate ($Cr_2O_7^{2-}$), is a strong oxidizing agent. Those ions are easily reduced to Cr(III) in acidic solutions. Trivalent chromium is the most stable oxidation state and the one most likey to exist in biological systems. However, chromium forms very stable "sandwich complexes" with a valence state of 0, 4, or 5, and these types of complexes may be involved in chromium binding to nucleic acids.

Trivalent chromium forms many coordination complexes, most of which are hexacoordinate. In aqueous solutions, those complexes are characterized by relative kinetic inertness such that ligand-displacement reactions have half-times in the range of several hours. Thus chromium is unlikely to be involved as the metal catalyst at the active site of enzymes, where the rate of exchange would need to be rapid. However, such relatively inert Cr complexes could function as structural components; for example, by binding ligands in proper orientation to function in catalysis in enzymes, or in tertiary structures of proteins or nucleic acids.

In aqueous solutions, chromium is coordinated to water and exists as octahedral hexaquo ion, $[Cr(H_2O)_6]^{3+}$. Hexaquo chromium occurs in salts (e.g., $[Cr(H_2O)_6]Cl_3$) and alums (e.g., $MCr(SO_4)_2 \cdot 12H_2O$, where M is any monoatomic cation except lithium).

At neutral pH, as in biological tissues, hydrolysis of the coordinated water of hexaquo chromium occurs, which leads to formation of bridges between hydroxyl groups, or olation. This results in the formation of polynucleate chromium complexes that ultimately precipitate and are biologically inert. Olation is enhanced by alkali and heat to 120° (Schroeder, 1970). Strong ligands, such as oxalate ions, can prevent and even reverse olation, but weaker ligands can only prevent the reaction. In biological systems, chromium is able to function because it is kept soluble by weaker organic and inorganic ligands. Bridging between chromium complexes may also occur with other ligands such as thiocyanate or amino groups. For example, Rollinson et al. (1967) found that the following naturally occurring ligands inhibit olation of Cr(III) under biological conditions: pyrophosphate, methionine, serine, glycine, leucine, lysine, and proline.

8.3 Biologically Active Chromium

Conversion of inorganic chromium to a biologically active form is essential for the physiological functions of Cr. The biological activity of Cr is measured *in vitro* by the potentiation of the action of insulin in the glucose oxidation of chromium-deficient rat adipose tissue (Mertz and Roginski, 1971) or more efficiently, in isolated epididymal fat cells from Cr-deficient rats (Anderson et al., 1978). The biologically active form of chromium isolated from brewer's yeast is postulated to contain nicotinic acid and the amino acids glycine, glutamic acid, and cysteine. Synthetic chromium compounds derived from these components or from Cr, nicotinic, and glutathione have been shown to display biological activity (Toepfer et al., 1977; Anderson et al., 1978). Tuman et al., (1978) found that the biologically active Cr preparation isolated from brewer's yeast and synthetic biologically active Cr complexes but not simple Cr compounds lowered plasma glucose and triglycerides of diabetic mice. A synthetic complex of Cr, nicotinic acid, and glutathione binds tightly to insulin *in vitro* (Anderson et al., 1978). The exact structure of the biologically active form(s) of chromium is not yet known.

8.4 Absorption and Transport of Chromium

Chromium appears to be absorbed primarily in the jejunum, based on *in vitro* perfusion of rat intestinal segments with $^{51}CrCl_3$ (Chen et al., 1973). Absorption of orally administered ^{51}Cr in normal humans was 0.69% and was not significantly different for elderly or maturity-onset diabetics; however, chromium absorption of insulin-requiring diabetics is almost three-fold higher (Doisy et al., 1976). Anderson et al. (1983a) found that chromium absorption for normal subjects was 0.4% both for nonsupplemented and for subjects supplemented with 200 μg of inorganic chromium per day.

Absorption of chromium is influenced by the presence of chelating agents. Chen et al. (1973) found that in rats, oxalate significantly increased and phytate significantly decreased chromium absorption *in vitro* and *in vivo*. However, citrate and EDTA had no effect. Ligands that increase absorption may function by preventing olation and precipitation of chromium that may occur in the near neutral milieu of the intestine.

Absorption and metabolism of chromium are also affected by interactions with other metals, especially zinc and iron. Hahn and Evans (1975) reported that the intestinal-wall and whole body contents of an oral dose of radioactive chromium were greater in zinc-deficient rats than in controls. Orally administered zinc decreased ^{51}Cr absorption in zinc-deficient rats and chromium decreased ^{65}Zn absorption. Both zinc and chromium eluted in the same low-molecular-weight fraction when mucosal supernatant extracts were separated by gel filtration, suggesting that in the intestine similar ligands bind both metals. Chromium and iron may also share a common gastrointestinal transport mechanism, since iron-deficient animals appear to absorb more Cr than iron-supplemented controls (Hopkins and Schwarz, 1964). Oral administration of iron to the deficient animals inhibited Cr absorption. Chromium absorption also appears to be influenced by vanadium. Hill (1975) reported that high intakes of chromium prevent the growth depression and mortality of chicks associated with feeding high levels of vanadate. Vanadate inhibits the uptake of chromate by respiring mitochondria, and similarly, chromium inhibits the uptake of vanadate.

After absorption, trivalent chromium binds to the β-globulin fractions of serum proteins, specifically to transferrin. Transferrin appears to be the protein involved in the transport of chromium to body tissues. Under normal physiological conditions, transferrin is only 30% saturated with iron, and *in vitro* under conditions of saturation of transferrin with iron, chromium competes with iron for the same binding sites (Hopkins and Schwarz, 1964). More recent work indicates that transferrin has two binding sites that have different affinities for iron *in vitro,* depending on pH. At lower levels of iron saturation, iron and chromium preferentially occupy sites A and B, respectively. However, at higher iron concentrations, iron and chromium compete for binding mainly to site B. This may explain why patients with hemochromatosis, with greater than 50% saturation of transferrin by iron, appear to retain less whole-body or blood ^{51}Cr than iron-depleted patients or normal subjects (Sargent *et al.*, 1979). Since diabetes is a frequent complication associated with hemochromatosis, the effect of high iron saturation of transferrin on the transport of chromium may play a role. In addition to binding to transferrin, at higher than physiological concentrations, Cr binds nonspecifically to several plasma proteins (Hopkins and Schwarz, 1964).

8.5 Chromium Concentrations in Blood, Tissues, and Hair

Reported values for the chromium concentrations in blood and urine have decreased precipitously in the past two decades. Values greater than 1 ppb for the chromium concentration of normal human urine, serum, plasma, and milk samples may be in error and, unless verified by independent means, should not be considered accurate. Presently accepted normal concentrations for urine and serum have been verified by three different methods: atomic absorption, gas chromatography/mass spectrometry, and neutron activation analysis. Data for the chromium concentrations of human tissues should be accepted with caution unless the reported values have been verified in samples that have not contacted stainless steel and have been analyzed using improved laboratory techniques and instrumentation. There appears to be reasonable agreement among laboratories for the Cr concentration of foods, water, and hair. Therefore, reported values usually appear to be indicative of the true Cr content of these materials.

Standard reference materials, with certified Cr concentrations similar to those of the samples to be analyzed, should be used for analyzing Cr in biological samples. More than one certified reference material should be utilized, since chromium from different sources responds differently during sample digestion and analysis.

8.5.1 Blood

The reported values for the concentration of chromium in serum have decreased in recent years because of improved analytical techniques and increased awareness of problems associated with Cr contamination. The decrease in the reported values for Cr content of serum or plasma of healthy subjects is shown in Table 8-1. The lowest and probably most reliable, by present standards, value for fasting serum chromium concentration of normal healthy subjects is about 0.14 ng/ml, determined by atomic absorption spectrophotometry (Kayne *et al.*, 1978) and neutron activation analysis (Versieck *et al.*, 1979).

Chromium concentration in tissues is 10–100 times higher than blood. Since blood chromium is not in equilibrium with tissue stores, the concentration of chromium in blood is not a good indicator of chromium nutritional status (Mertz, 1969).

Table 8-1. Reported Plasma or Serum Cr Concentration of Healthy Individuals

Concentration (ng/ml)	Year
520	1962
170	1962
180	1959
150	1959
28	1974
24	1971
28	1960
23	1968
17	1969
14	1972
7	1976
5	1972
3.1	1974
1.7	1978
1.6	1974
0.73	1974
0.50	1975
0.16	1978
0.14	1978
0.14	1983

Source: Adapted from Anderson, 1981b.

8.5.2 Tissues

The distribution of chromium in body tissues has been determined directly in human tissues at autopsy in animal tissues after injection or oral ingestion of radiochromium. As with serum and urine, determination of the chromium concentration in body tissues is affected by newer analytical techniques and avoidance of contamination. For example, a common source of contamination for human tissue samples, usually obtained by hospital personnel, is stainless steel which contains about 18% chromium.

The chromium concentration in human tissues has been reported to vary considerably with geographic location; however, this work needs confirmation. Schroeder et al. (1962) reported wide variability in liver and kidney samples obtained in the United States and other countries. In general, tissue concentrations of chromium tend to be lower in areas of the world where the incidence of maturity-onset diabetes and atherosclerosis is high. For example, chromium in the aorta is significantly lower in samples from the United States than from other countries (Schroeder et al., 1970). However, values obtained using newer methods indicate that serum chromium concentrations for individuals from different geographic regions are similar (see Section 8.4.1). Tissue Cr content of individuals residing in different geographical regions needs to be redetermined using newer methods and increased awareness of Cr contamination beginning at the collection of the samples and continuing throughout the preparation and analysis.

Factors associated wtih age also affect tissue chromium concentrations in humans. Chromium concentrations of the lung, aorta, heart, and spleen decrease considerably within the first few months of life, while the liver and kidney retain neonatal concentrations until after 10 years of age (Schroeder et al., 1962). This decline in tissue chromium with age is less pronounced in tissue samples from countries other than the United States (Schroeder et al., 1970). The lungs are the only tissue in which subsequent increase in chromium concentration is universally observed after age 20, probably because of inhalation of chromium pollutants in air.

Insulin-dependent diabetes also affects the chromium content of human tissues. Morgan (1972) found that hepatic chromium content of diabetics was less than that of controls, and Schroeder et al. (1962) reported the pancreas of diabetic subjects had less chromium than that of nondiabetic subjects. In contrast, Eatough et al. (1978) found no difference in liver, pancreas, or spleen concentrations of chromium between diabetic and nondiabetic Pima Indians. However, since the incidence of diabetes is 50% in this population, the nondiabetic group may have included subjects predisposed to diabetes, especially since the mean age of the nondiabetic group was 40, compared to 61 years in the diabetic group.

The tissue distribution of radiochromium is affected by chemical form, age, species and the presence of diabetes. Visek et al., (1953) found that after intravenous (IV) administration, almost 100% of ^{51}Cr administered as sodium chromite was localized in the reticuloendothelial system but 55% of a dose of ^{51}Cr administered as $CrCl_3$ was present in the liver. However, less than 5% of $^{51}CrCl_3$ buffered with acetate or citrate reached the liver and most was excreted in the urine. About 25% of the dose of ^{51}Cr administered as sodium chromate reached the liver, but the radioactivity of the liver decreased rapidly. Except for oral administration, the route of administration, whether IV, intraperitoneal (IP), or intratracheal, did not affect tissue Cr distribution. Kraintz and Talmage (1952) found that 24 hours after IV injection of $^{51}CrCl_3$ in rats, 40% of the dose was excreted by the kidneys and the bone marrow had the highest concentration of remaining activity, indicating ^{51}Cr was deposited in the reticuloendothelial system in a manner similar to small colloids. In rabbits, however, the highest concentration of ^{51}Cr was in the spleen. After IP injection of Na_2CrO_4 in mice, the liver and kidney incorporated the greatest amount of ^{51}Cr, followed by the epididymal fat pads and spleen (Vittorio *et al.*, 1962). In mice, age leads to decreased ^{51}Cr uptake in certain tissues, specifically liver, testes, and epididymal fat pads. Hopkins (1965) also observed differences in tissue retention with age after injection of physiological amounts of ^{51}Cr as $^{51}CrCl_3$. Mature rats retained less ^{51}Cr in bone but more in spleen, kidney and testes than immature rats. There was no difference in tissue distribution associated with sex, dose level, or previous deficiency or adequacy of chromium in diet. Mathur and Doisy (1972) reported similar tissue distribution of ^{51}Cr between diabetic and normal rats, but the liver of diabetic rats had more ^{51}Cr in nuclear and supernatant fractions and less in mitochondrial and microsomal fractions. The pattern was similar in normal rats fed high-fat diets. However, when the rate of hepatic lipogenesis was normal or elevated, chromium appeared to move from the nuclear to the microsomal fraction of liver cells.

Various researchers have attempted to characterize the kinetics of chromium metabolism by compartment analysis. Mertz *et al.* (1965) found that elimination of $^{51}CrCl_3$ that was administered IV was independent of amount injected and previous dietary status of the animals and occurred in three distinguishable phases with half-times for removal of 0.5, 5.9, and 83.4 days. By analyzing both total tissue chromium and ^{51}Cr distribution, Jain *et al.* (1981) also proposed a model of extracellular chromium in equilibrium with rapid and slow cellular exchange pools and an inner tissue chromium pool of very slow exchange.

8.5.3 Hair

The analysis of chromium in hair has been suggested as a method of assessing chromium nutritional status. The advantages of hair chromium determinations are as follows: (1) analytical methods are less difficult because hair

Cr concentrations are relatively high compared to serum or urine; (2) hair chromium concentrations are not subject to rapid fluctuations because of diet and other variability and therefore would more likely reflect long-term nutritional status; (3) sample collection is noninvasive; (4) samples are stable at room temperature. Sample contamination may be a problem but Hambidge *et al.* (1972) concluded that with careful washing and the absence of a history of dying, bleaching, or other specific environmental exposures, the Cr in hair reflects endogenous Cr available to cells in the hair follicle.

Hair chromium concentrations decrease with age, similar to other body tissues. The fetus accumulates Cr, especially in the last months before birth, since premature infants have lower hair Cr than full-term babies (Hambidge, 1971). A full-term baby at birth has hair concentrations $2\frac{1}{2}$ times that of the mother. Hambidge (1971) reported the hair Cr concentration of 25 full-term infants to be 900 ppb, compared to 440 ppb for 20 children 2–3 years of age. Similar decreases after age 2 were observed by Gurson *et al.*, (1975).

Pregnancy and the time between pregnancies affects hair Cr concentrations. Nulliparous women have higher hair Cr than parous women. However, hair Cr concentration does not appear to decrease further with increasing parity and actually increases significantly between pregnancies separated by 4 or more years (Mahalko and Bennion, 1976). Hair chromium concentration is also affected by diabetes and arteriosclerosis. Hambidge *et al.* (1968) reported that insulin-dependent diabetic children have lower hair chromium than normal children. A similar trend was seen in adult female diabetics compared to normal females but adult male diabetics had hair Cr concentrations similar to normals (Rosson *et al.*, 1979). A greater number of subjects with arteriosclerotic heart disease had hair chromium concentrations in the lowest quartile of values observed compared to healthy subjects of the same age (Cote *et al.*, 1979). Subjects with arteriosclerotic heart disease had hair concentrations similar to those with both arteriosclerosis and diabetes and the median hair concentrations of both groups was 449 ppb, compared to 819 ppb for normal subjects.

8.5 Chromium Excretion

Absorbed chromium is excreted primarily in the urine, with small amounts lost in hair, perspiration, and bile (Hopkins, 1965; Doisy *et al.*, 1971). Since absorption of orally ingested Cr is less than 1%, the rest is lost in the feces. However, minimal amounts of absorbed Cr may also be excreted via the gastrointestinal tract. In rat studies, detection of injected radiolabeled Cr in feces is difficult because of contamination by the urine. Kraintz and Talmage (1952) detected no radioactive chromium in the feces 1 day after injection, but Visek *et al.* (1953) found up to 20% of labeled Cr in feces 4 days after injection.

Hopkins (1965) found 0.5–1.7% of an IV dose of Cr in the feces removed from the intestine of a rat sacrificed 8 hours after injection. This latter observation has been confirmed in our laboratory, thus verifying that not all the Cr found in feces after injection of labeled Cr is due to contamination from urine. However, the physiological significance of Cr excretion after IV injection is not known.

The reported values for the concentration of Cr in urine of humans have decreased with improved methods of sampling, prevention of contamination, and improved instrumentation (Veillon et al., 1982). Table 8-2 lists the reported concentrations of urinary Cr excretion. Urinary chromium concentrations of less than 1 ppb for normal subjects in the United States and Western countries have been obtained by several investigators and are generally accepted as accurate by present standards.

The most recent values for urinary Cr excretion are in better agreement with other parameters of chromium metabolism than the 10-fold greater values reported earlier. The chromium content of several typical U.S. diets was found to be less than 100 µg/day (62 ± 28 µg/day in "high"-fat diets and 89 ± 56 µg/day in "low"-fat diets) (Kumpulainen et al., 1979). This agrees with reports from earlier studies (Schroeder et al., 1962; Levine et al., 1968). Thus absorption of about 1% or less would account for less than 1.0 µg per day, which would be in approximate balance with an excretion of less than 1.0 µg/day.

Urinary chromium excretion can be expressed in relation to creatinine excretion. Daily urinary output of endogenous creatinine is correlated with muscle mass and varies little from day to day in healthy individuals, regardless of diet

Table 8-2. Reported 24-Hour Urinary Cr Excretion of Normal Subjects

µg/day	Number of subjects	Year
150	2	1966
115	2	1969
18	3	1964
11.3	8	1978
8.4	20	1971
7.2	9	1975
6.6	316	1978
4.3	20	1980
3.6; 3.7	234	1977
3.1	5	1975
2.7	1	1977
0.8	12	1978
0.16	9	1982
0.22	42	1983

Source: Adapted from Anderson et al., 1983a.

or diuresis, although it may increase with physical exercise. Gurson and Saner (1978) found that chromium excretion expressed as the ng chromium/mg creatinine ratio (Cr/Cre) eliminated the diurnal variation seen in chromium excretion as well as corrected for variation in urine volume. Although their values for urinary Cr excretion were high compared to currently accepted values, they found no significant difference between 24-hour chromium excretion calculated from 4-hour samples and that measured in a 24-hour sample. The Cr/Cre ratios based on 4-hour samples were not significantly different from Cr/Cre ratios of 24-hour samples. The 4-hour samples were collected after an overnight fast.

The exact mechanisms of metabolism of Cr by the kidney are not known. Stable Cr complexes such as Cr–EDTA are not reabsorbed and have been suggested as good indicators of glomerular filtration rates. The ultrafilterable Cr in plasma, as a measure of Cr in glomerular filtrate, was determined by Rabinowitz *et al.* (1980) to be 6–28% of total plasma chromium and to have a molecular weight of less than 10,000 daltons. Davidson *et al.* (1974) reported plasma ultrafilterable Cr in dogs to be 41%. Recently, Wu and Wada (1981) have characterized a low-molecular-weight substance (MW = 1500) found in rat and human urine that apparently binds trivalent Cr chemically and is similar to a chromium binding substance found in dog and rabbit liver and other organs. Donaldson *et al.* (1982) reported that plasma ultrafilterable ^{51}Cr was greater when the radiolabeled Cr was given orally than when injected IP or IV, suggesting different binding of Cr to serum proteins, depending on route of administration.

Estimates of renal tubular reabsorption of filtered Cr have ranged from 63% in dogs to 80–97% in man (Davidson *et al.*, 1974; Rabinowitz *et al.*, 1980). Similarly, Vanderlinde *et al.* (1979) reported that renal tubular reabsorption for normal subjects and insulin-dependent and diet-controlled diabetics was 94–95%. However, preliminary evidence of Donaldson *et al.* (1982) indicates that ultrafilterable Cr is excreted similar to creatinine, and tubular reabsorption may not be as great as previously reported.

Based on the fractional excretion of Cr (the ratio of Cr clearance to creatinine clearance), urinary excretion of Cr in humans and dogs was reported to increase with water diuresis (Davidson *et al.*, 1974; Rabinowitz *et al.*, 1980). However, evidence for this should be reexamined in light of both the recent observations of tubular reabsorption and present techniques of Cr analysis. Data from our laboratory (Anderson and Bryden, 1983) indicate that a more than two-fold increase in total urine output did not affect 24-hour urinary Cr excretion.

8.7 Functions of Chromium and Signs of Chromium Deficiency

Chromium may have a role in activating enzymes and in maintaining the structural stability of proteins and nucleic acids, but the primary physiological role of chromium in a biologically active complex is to potentiate the action of insulin. The biological effects of chromium are observed in carbohydrate, protein, and fat metabolism in insulin-dependent processes.

The role of chromium in carbohydrate metabolism was first observed when impaired glucose tolerance of rats was restored by a postulated "glucose tolerance factor," or GTF, in brewer's yeast (Schwarz and Mertz, 1957). The "factor" was subsequently isolated from brewer's yeast and demonstrated to contain trivalent chromium. In the presence of insulin, chromium increases the oxidation of glucose to carbon dioxide when incubated with epididymal fat tissue from rats fed low-chromium diets (Mertz et al., 1961). The effects of chromium on other insulin-dependent processes have been studied in the rat epididymal fat pad. Chromium enhances the action of insulin in glucose uptake and the incorporation of glucose into fat. The incorporation of acetate into fat, which is not an insulin-dependent process, is not stimulated by chromium (Mertz, 1969). The glucose uptake of isolated eye lens (rats) and the formation of glycogen from glucose are also stimulated by chromium in the presence of insulin.

Chromium affects the action of insulin in protein metabolism, as indicated by rats fed chromium-deficient diets repleted with chromium (Roginski and Mertz, 1969). Insulin-mediated amino acid transport into tissues was enhanced and incorporation of labeled glycine, serine, and methionine into heart protein was greater in chromium-supplemented animals.

Evidence of a role for chromium in lipid metabolism and chromium deficiency in the development of atherosclerosis is accumulating from animal and human studies. Early animal studies indicated that with increasing age, rats fed a low-chromium diet have increased serum cholesterol, elevated fasting glucose, impaired glucose tolerance, and increased aortic lipids and plaques (Schroeder, 1965). In rabbits fed a high-cholesterol diet to induce atherosclerotic plaques, subsequent administration of chromium (20 μg potassium chromate per day injected intraperitoneally) reduced the size of aortic plaques and decreased aortic cholesterol content significantly (Abraham et al., 1980). Chromium administered concurrently with a high-cholesterol diet resulted in significantly less aortic cholesterol and plaque area (Abraham et al., 1982).

Schroeder et al. (1970) reported that individuals who died from coronary artery disease had significantly lower chromium concentration in aortic tissue compared to subjects dying from accidents, although other tissues analyzed were similar in chromium content. A more recent study indicates that subjects with coronary artery disease have lower serum chromium than subjects without symptoms of disease (Newman et al., 1978). Decreased serum chromium correlated highly ($P < 0.01$) with the appearance of coronary artery disease, whereas elevated serum triacylglycerol correlated less significantly ($P < 0.05$), and in this study there was no correlation between other risk factors, such as serum cholesterol, blood pressure, or body weight. Chromium supplementation of 12 adult men (200 μg daily, 5 days a week for 12 weeks) resulted in significant decreases in serum triglycerides and increases in high-density lipoprotein cholesterol compared to controls (Riales and Albrink, 1981).

Elevated circulating insulin levels are characteristic of many subjects who either have developed or might develop atherosclerosis. In response to chronic exposure to high insulin concentrations, insulin-sensitive arterial tissue may develop lipid-filled lesions resembling those seen in early atherosclerosis (Stout, 1977). Since chromium potentiates the action of insulin and thus is involved in regulating insulin levels, the link of chromium deficiency with the development of atherosclerosis may be due to inadequate chromium to maintain normal levels of circulating insulin.

The role of chromium in the function of nucleic acids is indicated by the high concentration of chromium in nuclear proteins relatve to other metals, the stability of chromium attachment to nucleic acid fractions, and the protection chromium offers against heat-induced conformational changes (Mertz, 1969). Chromium chloride significantly stimulated RNA synthesis when incubated with mouse liver DNA or chromatin prior to the addition of RNA polymerase but inhibits RNA synthesis with simultaneous incubation with polymerase (Okada *et al.*, 1981). On the other hand, chromium deficiency may depress nucleic acid synthesis as indicated by the significantly lower sperm count and decreased fertility of male rats fed chromium-deficient diets compared to chromium-supplemented controls (Anderson and Polansky, 1981).

Experimental animals fed chromium-deficient diets can manifest impaired glucose tolerance, fasting hyperglycemia, impaired growth, decreased longevity, elevated serum cholesterol, and related cardiovascular abnormalities (Table 8-3). Specific signs and symptoms of chromium deficiency in humans have been reported in two patients receiving long-term total parenteral nutrition (Jeejeebhoy *et al.*, 1977; Freund *et al.*, 1979). They included peripheral neuropathy, weight

Table 8-3. Signs of Chromium Deficiency

Impaired glucose tolerance
Elevated circulating insulin
Glycosuria
Fasting hyperglycemia
Impaired growth
Decreased longevity
Elevated serum cholesterol and triglycerides
Increased incidence of aortic plaques
Peripheral neuropathy
Brain disorders
Decreased fertility and sperm count

Source: Anderson, 1981a.

loss despite adequate caloric intake, and glucose intolerance refractory to the addition of insulin. In both cases, these diabetic-like symptoms were reversed within a few days of infusion of 150–250 μg chromium per day. In the general population, these overt signs of chromium deficiency have not been observed but signs of marginal chromium deficiency characterized primarily by impaired glucose tolerance and elevated serum lipids that are improved by chromium supplementation are observed frequently (see section on chromium supplementation).

8.8 Factors Affecting Chromium Metabolism

It has been hypothesized that chromium is mobilized into the plasma from body stores in reponse to a glucose challenge and acts in relation with insulin in glucose metabolism. Chromium is then excreted in the urine (Mertz, 1969). Early studies support this mechanism of chromium mobilization. In 1966, Glinsmann *et al.* (1966) reported that following an oral glucose load, plasma chromium levels increased. This was not evident in two diabetics with impaired glucose tolerance until after dietary chromium supplementation, which also led to improved glucose tolerance.

Liu and Morris (1978) suggested that the serum chromium concentration after an oral glucose load is a function of chromium status. They measured "relative chromium response" (RCR), which is equal to the 1-hour serum chromium level divided by fasting serum chromium times 100. In normal and hyperglycemic women, the serum chromium 1 hour after an oral glucose load was lower than fasting (RCR = 95%). After chromium supplementation, the fasting serum chromium was lower, and the 1-hour serum chromium levels were higher (RCR = 144%). Urinary chromium excretion was not measured.

In contrast to these studies, Davidson and Burt (1973) reported decreases in plasma chromium in normal women administered an oral glucose load and in both men and women administered glucose intravenously. However, in pregnant women fasting plasma chromium levels were lower than those of nonpregnant women. The differences in chromium metabolism were postulated to be related to the changes in carbohydrate metabolism observed in late pregnancy.

Davidson *et al.* (1974) found that urinary chromium excretion decreased significantly after an oral glucose load in eight healthy subjects. They hypothesized that this was a result of chromium being mobilized from the plasma to peripheral tissues, the site of insulin action. The hypothesized decrease in plasma chromium would result in decreased glomerular filtration and renal excretion (not determined in that study).

In contrast, urinary chromium was reported to increase in 10 normal adults after an oral glucose load (Gurson and Saner, 1978). However, in 13 adults or adolescents from families with overt diabetes and in 9 juvenile diabetics, there was no mean increase in urinary chromium excretion. Reported values of urinary excretion of that study were high, compared to currently accepted values. Using current analytical techniques, Anderson *et al.* (1982a) found that 90 minutes after an oral glucose load, urinary Cr excretion exceeded fasting levels for 61 of 76 subjects. For all 76 subjects, the mean fasting chromium–creatinine ratio was 0.12 ng Cr/mg Cre, which increased to 0.19 ng Cr/mg Cre at 90 minutes after the glucose load. They concluded that although the elevation in urinary chromium excretion after a glucose load is observed in most healthy individuals, this is not an accurate indicator of chromium status.

Thus researchers differ as to the effect of glucose on chromium metabolism. These differences may be due to differences in analytical methods, some of which are outdated, or they may be real, especially when differences in subjects are considered (e.g., pregnant women or diabetics). In addition, the time of sampling after the glucose challenge may be a factor, as an initial decrease in serum chromium may indicate mobilization of serum chromium to peripheral tissues for insulin action but subsequent increases may indicate chromium released to the blood from body stores.

Chromium is metabolized differently by diabetics than by normal individuals. There appear to be differences in chromium metabolism, depending upon the type of diabetes (i.e., whether the subject is a juvenile-onset, insulin-dependent diabetic or an adult-onset non-insulin-dependent diabetic). Chromium absorption in juvenile-onset diabetics is two to four times greater than that in normal subjects (Doisy *et al.*, 1976; Nath *et al.*, 1979). Insulin-dependent diabetics have higher serum chromium than normal subjects (Vanderlinde *et al.*, 1979) and other diabetics (Rabinowitz, 1980). The urinary chromium excretion of diabetics is also greater than that of normal subjects (Doisy *et al.*, 1976; Vanderlinde *et al.*, 1979). The exact reasons for these observations are not known, but Doisy *et al.* (1976) hypothesized that diabetics cannot utilize chromium normally and may be unable to convert inorganic chromium to a biologically active form.

8.9 Chromium and Stress

Under normal physiological conditions, blood glucose concentrations are regulated primarily by the hormones insulin, glucagon, and growth hormone. During different forms of stress, such as trauma, infection, surgery, intense heat or cold, elevated secretion of several other hormones alters glucose metabolism. For example, after trauma, plasma cortisol and catecholamines are elevated. Glucagon secretion is stimulated by elevated catecholamines, plasma growth hormone concentrations are elevated but insulin secretion is not increased. The changed pattern of hormone concentrations has the net effect of stimulating gluconeogenesis by the liver so that hyperglycemia prevails.

Stress also appears to affect chromium metabolism. In experimental animals, stress induced by a low protein diet, controlled exercise, acute blood loss or infection aggravated the symptoms of depressed growth and survival caused by low chromium diets (Mertz and Roginski, 1969). Pekarek *et al.* (1975) reported that subjects infected with sandfly fever had lower fasting plasma chromium than healthy subjects. Serum chromium concentrations decreased after an intravenous glucose load in controls and infected subjects. They postulated that the availability of circulating chromium may be decreased during infection, and this may act in conjunction with hormonal patterns of stress to cause the relative glucose intolerance despite elevated insulin levels.

A similar pattern of increased serum levels of glucagon, growth hormone, catecholamines and cortisol, and decreased serum insulin is also observed during the stress of strenuous exercise. Anderson *et al.* (1982b) found that the increased glucose utilization of exercise (running) affects chromium excretion. In nine adult males, a 6-mile run resulted in elevated blood glucose and glucagon but no increase in serum insulin. The mean urinary chromium was increased fivefold 2 hours after running and the total chromium excretion on the day of running was more than two times greater than excretion on the following rest day. Severely traumatized patients also excreted several times more chromium than normal subjects (Borel *et al.*, 1983). However, some of the additional chromium excreted in the urine was due to the high levels of chromium present in the blood components administered in the treatment of the injuries.

8.10 Dietary Requirements of Chromium

In 1980 the recommended safe and adequate intake for chromium was established at 50–200 µg/day for adults (National Research Council, 1980). This recommended range is for total chromium and not for specific organic forms. However, chromium must be converted by the body to a biologically active form to function physiologically.

There have not been any comprehensive studies determining the Cr content of foods eaten in the United States. However, Finnish workers completed a large study to determine the nutrient content, including Cr, of numerous foods from each of the basic food groups (Koivistoinin, 1980). Foods that are high in chromium include brewer's yeast, meats, cheeses, whole grains, mushrooms, black pepper, nuts, and asparagus. In a survey of a large number of foods, Toepfer *et al.* (1973) found that total chromium and biologically active chromium often do not correlate. It has been postulated that the dietary form of chromium is important, since some forms of chromium were reported to be absorbed much better than inorganic chromium (Mertz, 1969; Mertz and Roginski, 1971). However, since absorbed chromium is excreted primarily in the urine and newer analytical techniques indicate urinary losses are more than 10-fold less than previous values, the importance of the dietary role of biologically active forms of chromium needs to be reevaluated (Anderson, 1981a).

Studies that have determined the chromium content of diets consumed in the U.S. indicate that, at best, intake is in the lower end of the recommended range and may be much less. Levine et al. (1968) reported a daily intake of 5–115 µg/day for elderly institutionalized subjects. The mean chromium content of meals served in 50 U.S. colleges ranged from 33–125 µg/day, with a mean of 77 µg/day. Kumpulainen et al. (1979) found that low-fat (25%) diets contained more chromium than high-fat (43%) diets (89 ± 50 µg/day compared to 62 ± 28 µg/day). These meals were planned by dietitians to provide the recommended dietary allowance of vitamins and minerals, but approximately one third of the diets contained less than the minimum safe and adequate intake for Cr (50 µg). It is likely that many people consume meals that are less nutritionally balanced and thus are consuming considerably less than the minimum recommended dietary intake of Cr.

The amount of chromium in foods decreases with processing. For example, molasses contains 0.26 µg Cr/g; unrefined sugar, 0.16 µg/g; and refined sugar 0.02 µg/g (Wolf et al., 1974). Similarly, whole wheat contains 1.75 µg Cr/g, white flour has 0.60 µg Cr/g, and white bread 0.14 µg Cr/g (Schroeder, 1968). However, the amount of chromium in some processed foods may be due to nonspecific contamination during processing, which may even contribute to dietary intake, as was observed in the analysis of chromium in many varieties of beer (Anderson and Bryden, 1983). Chromium is released from stainless steel cooking vessels into water in a range of 16–59 ppb at high temperatures (90–100°C) and low pH (2.5–3.0), whereas no chromium is transferred to unacidified water. Similar release of chromium from stainless steel vessels by fruit juices was also observed (Offenbacher and Pi-Sunyer, 1983). In general, however, the chromium content of highly processed foods is low compared to less processed food sources. The widespread tendency toward increased consumption of highly processed foods in Western countries, particularly refined sugar, which stimulates urinary losses of chromium, may result in a marginal intake of chromium and depletion of chromium stores. The long-term effects of this suboptimal intake of chromium may be related to the decrease in tissue chromium with age and the increased incidence of diabetes and atherosclerosis observed in developed countries.

8.11 Effects of Chromium Supplementation

In light of the marginal dietary intake of Cr and the role of Cr in glucose and lipid metabolism and consequent link to diabetes and atherosclerosis, many investigators have examined the effect of chromium supplementation on glucose metabolism of various populations susceptible to abnormal glucose tolerance. These include children with protein-calorie malnutrition (PCM), the elderly, diabetics, patients receiving total parenteral nutrition (TPN) as well as normal subjects.

Brewer's yeast, which is high in total and biologically active Cr, has been used as a source of Cr for supplementation studies. In contrast, Torula yeast is low in total Cr but similar in other nutrients and has therefore been used as a control for brewer's yeast supplementation studies. Chromium as $CrCl_3$, administered in solution or as a pill, is usually the inorganic form used for supplementation, including studies involving intravenous administration.

The effect of Cr supplementation on children with PCM has been studied in Turkey, Jordan, Nigeria, and Egypt. In Turkey, an increased glucose removal rate after an intravenous glucose tolerance test was seen in marasmic children who were given 250 µg Cr as $CrCl_3 \cdot 6H_2O$ orally (Gurson and Saner, 1971). Hopkins et al. (1968) observed similar results in children given the same dose in Jordan and Nigeria. However, not all children with PCM had low Cr status. Children from an area with a three-fold greater chromium concentration in the drinking water had normal glucose removal rates compared to those from an area of low Cr in the drinking water. In Egypt, studies with 34 children with kwashiorkor showed no effect of Cr supplementation on fasting blood glucose or glucose tolerance (Carter et al., 1968). However, the Cr content of many foods consumed by these children prior to admission was high, and plasma Cr was high relative to levels considered normal at that time.

The effect of Cr supplementation of elderly subjects is of interest because of the reported decline in tissue Cr levels, the abnormal glucose tolerance, and the increased incidence of diabetes and atherosclerosis observed with increasing age (Schroeder et al., 1962). In elderly subjects supplemented with inorganic Cr (150 µg Cr as $CrCl_3 \cdot 6H_2O$ per day), Levine et al. (1968) reported that the subjects who responded to Cr supplementation with improved oral glucose tolerance tests originally had mild abnormalities in their ability to utilize glucose. The "nonresponders" had more serious diabetic-like abnormalities in glucose tolerance that were not improved by Cr supplementation in the time period studied. Doisy et al. (1976) reported similar results in 12 elderly subjects who had abnormal glucose tolerance. After supplementation with brewer's yeast for 1–2 months, the glucose tolerance of 50% of the subjects was normal. In a more recent study (Offenbacher and Pi-Sunyer, 1980), elderly subjects supplemented with 9 g brewer's yeast daily (10.8 µg Cr/day) were compared to elderly subjects supplemented with Torula yeast (< 0.45 µg Cr/day). After 8 weeks, the brewer's yeast group had significantly improved glucose tolerance and decreased insulin output after an oral glucose load. Cholesterol and total lipids also decreased significantly. The Torula yeast group showed no significant changes in these parameters. Supplementation of six elderly subjects with 200 µg of chromium as chromium chloride led to a significant improvement in oral glucose tolerance and using the "glucose clamp" procedure on these same subjects a significant improvement in glucose utilization and β-cell sensitivity to glucose (Potter et al., 1983).

Chromium supplementation appears to improve glucose utilization and decrease exogenous insulin requirement in some diabetics. These effects depend on the type of diabetes, form and amount of Cr supplemented, and duration of supplementation. Glinsmann and Mertz (1966) found that three out of six diabetics had improved glucose tolerance after long-term supplementation with inorganic Cr but not after short-term supplementation (1–7 days). Sixty days of supplementation with 500 μg inorganic Cr per day resulted in significantly lowered glucose and insulin after a glucose challenge in 12 maturity-onset diabetics (Nath et al., 1979). Doisy et al. (1976) reported decreased need for exogenous insulin in insulin-dependent diabetics supplemented with brewer's yeast. Glucose tolerance was improved to normal in offspring of an insulin-dependent diabetic with similar supplementation.

Anderson et al. (1983b) recently reported that 200 μg inorganic Cr per day given to normal adults living at home improved the glucose tolerance of subjects with higher levels of serum glucose 90 minutes after an oral glucose load. In those subjects that had 90-minute serum glucose values greater than 100 mg/dl prior to supplementation, the 90-minute serum glucose decreased significantly after 2 and 3 months of Cr supplementation. Subjects whose 90-minute serum glucose was less than fasting responded to Cr supplementation with an increase in serum glucose at 90 minutes (from 71 to 81 mg/dl). Cr supplementation had no effect on subjects whose 90-minute serum glucose was greater than fasting but less than 100 mg/dl. These data suggest that Cr may normalize blood glucose levels in subjects with tendencies toward high and low blood sugar but has no effect on subjects with near optimal ability to utilize glucose.

The most dramatic effects of Cr supplementation were observed in two patients receiving long-term total parenteral nutrition (TPN). Jeejeebhoy et al. (1977) observed weight loss, peripheral neuropathy, and glucose intolerance in a 37-year-old woman who had been receiving TPN for 3 years following a complete enterectomy. These symptoms were not improved with increased caloric intake or 45 units of exogenous insulin per day. Blood and hair Cr concentrations were low according to the analytical methods used. Supplementation of the TPN solutions with 250 μg/day Cr for 2 weeks led to lowered plasma glucose, weight gain, and reversal of the diabetic-like symptoms; the 45 units of exogenous insulin were no longer required. Similar symptoms plus metabolic encephalopathy were reported by Freund et al. (1979) in a patient who received TPN for 5 months after a complete bowel resection. Supplementation of 150 μg/day Cr as $CrCl_3$ was given intravenously and within 3–4 days, blood glucose levels were normal, 20–30 units of exogenous insulin were no longer required, the encephalopathy cleared, and the patient began to gain weight.

Some intravenous solutions may contain significant amounts of chromium as an unintentional contaminant. This varies, depending on manufacturer and lot (Hauer and Kaminski, 1978; Borel *et al.*, 1983). Chromium is considerably higher in protein or protein hydrolysates and albumin or plasma protein solutions compared to synthetic amino acid mixtures, fat, carbohydrate, or electrolyte solutions (Seeling *et al.*, 1979; Fell *et al.*, 1979; Hauer and Kaminski, 1978; Borel *et al.*, 1983). Thus the amount of Cr a patient may receive varies, depending on the specific intravenous solutions administered.

The American Medical Association (1979) recommends supplementation of TPN solutions with 10–15 µg/day Cr for stable adult patients and 20 µg/day for stable adults with intestinal fluid losses. No amount is recommended for an adult in a more acute catabolic state. These values are based primarily on the case report of Jeejeebhoy *et al.* (1977), where 20 µg/day was calculated to balance observed losses in urine and gastrointestinal drainage. However, urinary and other losses of Cr as well as Cr in intravenous solutions of TPN patients need to be determined according to more current analytical techniques, to establish optimum levels of Cr supplementation for TPN patients.

It is clear that dietary supplementation of chromium leads to improved glucose tolerance in some cases in normal subjects, children with PCM, elderly and diabetics, and subjects with marginally elevated blood glucose. It has been suggested that an improvement in glucose tolerance after chromium supplementation is a valid indicator of Cr deficiency. The most definitive cases of this were observed when inorganic Cr administered intravenously to patients receiving TPN alleviated the diabetic-like symptoms of apparent Cr deficiency. Supplementation studies with brewer's yeast introduce many other nutrients that could influence glucose metabolism. Glucose tolerance may be influenced by infections and other nutrient deficiencies observed in PCM. Thus chromium supplementation appears to be effective in improving glucose metabolism in cases where previous dietary intake was not adequate, but it cannot be used pharmacologically to treat all abnormalities of glucose metabolism.

8.12 Toxicity of Chromium

Information on the toxicity of hexavalent chromium compounds has been reviewed extensively (Towill *et al.*, 1978; Industrial Health Foundation, 1981; National Research Council, 1974). Hexavalent chromium is more toxic than trivalent. Chromium poisoning in humans is limited to accidental ingestion of chromic acid or chromates. Toxicity to kidney, liver, nervous system, and blood are the major causes of death (Langard, 1980). Chromium compounds have a wide variety of industrial uses, including production of stainless steel and other alloys, high-melting refractory materials, pigments and mordants for paints and dyes, tanned leather goods, and chrome plating. Thus the potential occupational exposure exists for many workers in these industries. Chronic occupational exposure to chromates causes skin allergies and ulceration, perforation of nasal septum, and bronchial asthma (Langard, 1980).

Trivalent Cr, the form of Cr found in foods, is poorly absorbed (see Section 8.3); therefore exceedingly high oral intakes are necessary to attain toxic levels. In our laboratory, we have not been able to determine any toxic levels of orally administered trivalent chromium compounds.

8.13 Summary

Chromium is an essential trace element required for normal carbohydrate and lipid metabolism. The biological function of chromium is closely associated with that of insulin and most chromium-potentiated reactions are also insulin dependent. Sufficient dietary chromium leads to a decreased requirement for insulin and an improved blood lipid profile. Dietary intake of chromium is marginal and marginal Cr status is exacerbated by increased urinary losses due to pregnancy, strenuous exercise, infection, physical trauma, and other forms of stress. Chromium functions *in vivo* as an organic chromium complex that is postulated to contain, in addition to Cr, nicotinic acid and glutathione or its constituent amino acids, glycine, cysteine, and glutamic acid. This organic form of chromium, which potentiates insulin activity, can be measured *in vitro* by determining the increase in apparent insulin activity due to organic forms of Cr in the breakdown of glucose by adipose tissue or cells. *In vitro* insulin potentiation by Cr is relatively specific since inorganic Cr compounds and most organic complexes do not potentiate insulin activity. Research involving the nutritional significance of dietary Cr has been hampered by the lack of suitable reproducible methods for total Cr analysis of foods, tissues, and biological fluids. Chromium concentrations in the parts per billion range, ubiquitous presence of contaminating Cr, extreme matrix effects, possible volatility, and the inherent property of Cr to bind nonspecifically to reaction vessels, pipettes, graphite tubes, and so on, have made the accurate determination of Cr in biological materials extremely difficult. However, current instrumentation and methodology, coupled with the use of standard reference materials, permit the determination of total Cr in biological materials to be completed accurately and reproducibly. However, numerous reports are still appearing in which researchers have not taken the extreme precautions involved in the accurate determination of Cr and consequently their values are often erroneous. The only meaningful measure of Cr status is retrospective, involving Cr supplementation of subjects and determining improvements in glucose tolerance and blood lipids.

References

Abraham, A. S., Sonnenblick, M., Eini, M., Shemash, O., and Batt, A. P., 1980. The effect of chromium on established atherosclerotic plaques in rabbits, *Am. J. Clin. Nutr.* 33:2294–2298.

Abraham, A. S., Sonnenblick, M., and Eini, M., 1982. The action of chromium on serum lipids and on atherosclerosis in cholesterol-fed rabbits, *Atherosclerosis* 42:185–195.

American Medical Association Department of Foods and Nutrition, 1979. Guidelines for essential trace element preparations for parenteral use. A statement by an expert panel, *J. Am. Med. Assoc.* 241:2051–2054.

Anderson, R. A., 1981a. Nutritional role of chromium, *Sci. Total Environ.* 17:13–29.

Anderson, R. A., 1981b. Chromium as a naturally occurring chemical in humans. Proceedings of Chromate Symposium-80, Industrial Health Foundation, Inc., Pittsburgh, pp. 332–345.

Anderson, R. A., and Bryden, N. A., 1983. Concentration, insulin potentiation and absorption of chromium in beer, *J. Agric. Food Chem.* 31:308–311.

Anderson, R. A., and Polansky, M. M., 1981. Dietary chromium deficiency: Effect on sperm count and fertility in rats, *Biol. Trace Element Res.* 3:1–5.

Anderson, R. A., Brantner, J. H., and Polansky, M. M., 1978. An improved assay for biologically active chromium, *J. Agric. Food Chem.* 26:1219–1221.

Anderson, R. A., Polansky, M. M., Bryden, N. A., Roginski, E. E., Patterson, K. Y., Veillon, C., and Glinsmann, W., 1982a. Urinary chromium excretion of human subjects: Effect of chromium supplementation and glucose loading, *Am. J. Clin. Nutr.* 36:1184–1193.

Anderson, R. A., Polansky, M. M., Bryden, N. A., Roginski, E. E., Patterson, K. Y., and Reamer, D. C., 1982b. Effects of exercise (running) on serum glucose, insulin, glucagon, and chromium excretion, *Diabetes* 31:212–216.

Anderson, R. A., Polansky, M. M., Bryden, N. A., Patterson, K. Y., Veillon, C., and Glinsmann, W. A., 1983a. Effect of chromium supplementation on urinary Cr excretion of human subjects and correlation of Cr excretion with selected clinical parameters, *J. Nutr.* 113:276–281.

Anderson, R. A., Polansky, M. M., Bryden, N. A., Roginski, E. E., Mertz, W., and Glinsmann, W., 1983b. Chromium supplementation of human subjects: Effects on glucose, insulin and lipid parameters, *Metabolism.* 32:894–899.

Borel, J. S., Majerus, T. C., Polansky, M. M., Moser, P. B., Anderson, R. A., 1983, Chromium intake and urinary chromium excretion of trauma patients, *Fed. Proc.* 42:925.

Carter, J. P., Kattab, A., Abd-El-Hadi, K., Davis, J., El Gholmy, A., and Patwardhan, V. N., 1968. Chromium(III) in hypoglycemia and in impaired glucose utilization in kwashiorkor, *Am. J. Clin. Nutr.* 21:195–202.

Chen, N. S. C., Tsai, A., and Dyer, I. A., 1973. Effect of chelating agents on chromium absorption in rats, *J. Nutr.* 103:1182–1186.

Collins, R. J., Fromin, P. O., and Collings, W. D., 1961. Chromium excretion in the dog, *Am. J. Physiol.* 201:795–798.

Cote, M., Munan, L., Gagne-Billon, M., Kelly, A., Di Pietro, O., and Shapcott, D., 1979, Hair chromium concentration and arteriosclerotic heart disease, in *Chromium in Nutrition and Metabolism*, D. Shapcott and J. Hubert (eds.), Elsevier/North-Holland, Amsterdam, pp. 223–228.

Davidson, I. W. F., and Burt, R. L., 1973, Physiologic changes in plasma chromium of normal and pregnant women: Effect of a glucose load, *Am. J. Obstet. Gynecol.* 116:601–608.

Davidson, I. W. F., Burt, R. L., and Parker, J. C., 1974. Renal excretion of trace elements; Chromium and copper, *Proc. Soc. Exp. Biol. Med.* 147:720–725.

Doisy, R. J., Streeten, D. H. P., Souma, M. L., Kalafer, M. E., Rekant, S. L., and Dalakos, T. G., 1971. Metabolism of chromium 51 in human subjects, in *Newer Trace Element in Nutrition*, W. Mertz and W. E. Cornatzer (eds.), Dekker, New York, pp. 155–168.

Doisy, R. J., Streeten, D. H. P., Freiberg, J. M., and Schneider, A. J., 1976. Chromium metabolism in man and biochemical effects, in *Trace Elements in Human Health and Disease, Vol. II, Essential and Toxic Elements*, A. S. Prasad and D. Oberleas (eds.), Academic Press, New York, pp. 79–104.

Donaldson, D. L., Anderson, R. A., Veillon, C., and Mertz, W. E., 1982. Renal excretion of orally and parenterally administered chromium-51, *Fed. Proc.* 41:391. (Abstract.)

Eatough, D. J., Hansen, L. O., Starr, S. E., Astin, M. S., Larsen, S. B., Izatt, R. M., and Christensen, J. J., 1978. Chromium in autopsy tissues of diabetic and non-diabetic American (Pima) Indians, in *Trace Element Metabolism in Man and Animals*, Vol. III, M. Kirchgessner (ed.), Institut fur Ernahrungspysislogie, Freising-Weihenstephan, pp. 259–263.

Fell, G. S., Halls, D., and Shenkin, A., 1979. Chromium requirements during intravenous nutrition, in *Chromium in Nutrition and Metabolism*, D. Shapcott and J. Hubert (eds.) Elsevier/North Holland, Amsterdam, pp. 105–112.

Freund, H., Atamian, S., and Fischer, J. E., 1979, Chromium deficiency during total parenteral nutrition, *J. Am. Med. Assoc.* 241:496–498.

Glinsmann, W. H. and Mertz, W., 1966. Effect of trivalent chromium on glucose tolerance, *Metabolism* 15:510–520.

Glinsmann, W. H., Feldman, F. J., and Mertz, W., 1966. Plasma chromium after glucose administration, *Science* 152:1243–1245.

Gurson, C. T. and Saner, G., 1971. Effect of chromium on glucose utilization in marasmic protein-calorie malnutrition, *Am. J. Clin. Nutr.* 24:1313–1319.

Gurson, C. T., and Saner, G., 1978. The effect of glucose loading on urinary excretion of chromium in normal adults, in individuals from diabetic families and in diabetics, *Am. J. Clin. Nutr.* 31:1158–1161.

Gurson, C. T., and Saner, G., 1971. Effect of chromium on glucose utilization in marasmic protein calorie malnutrition, *Am. J. Clin. Nutr.* 24:1313–1316.

Gurson, C. T., Saner, G., Mertz, W., Wolf, W. R., and Sokucu, S., 1975. Nutritional significance of chromium in different chronological age groups and in populations differing in nutritional backgrounds, *Nutr. Rep. Int.* 12:9–17.

Hahn, C. J., and Evans, G. W., 1975. Absorption of trace metals in the zinc-deficient rat, *Am. J. Physiol.* 228:1020–1023.

Hambidge, K. M., 1971. Chromium nutrition in the mother and the growing child, in *Newer Trace Elements in Nutrition*, W. Mertz and W. E. Cornatzer (eds.), Dekker, New York, pp. 169–194.

Hambidge, K. M., Franklin, M. L., and Jacobs, M. A., 1972. Hair chromium concentration: Effect of sample washing and external environment, *Am. J. Clin. Nutr.* 25:384–389.

Hauer, E. C., and Kaminski, M. V., 1978. Trace metal profile of parenteral nutrition solutions, *Am. J. Clin. Nutr.* 31:264–268.

Hill, C. H., 1975. Mineral interrelationships, in *Trace Elements and Human Disease*, Vol. 2, A. S. Prasad (ed.), Academic Press, New York, pp. 281–300.

Hopkins, L. L., Jr., 1965. Distribution in the rat of physiological amounts of injected $Cr^{51}(III)$ with time, *Am. J. Physiol.* 209:731–735.

Hopkins, L. L., Jr., and Schwarz, K., 1964. Chromium(III) binding to serum proteins, specifically siderophilin, *Biochem. Biophys. Acta* 90:484–491.

Hopkins, L. L., Jr., Ransome-Kuti, O., and Majaj, A. S., 1968. Improvement of impaired carbohydrate metabolism by chromium(III) in malnourished infants, *Am. J. Clin. Nutr.* 21:203–211.

Jain, R., Verch, R. L., Wallach, S., and Peabody, R. A., 1981. Tissue chromium exchange in the rat, *Am. J. Clin. Nutr.* 34:2199–2204.

Jeejeebhoy, K. N., Chu, R. C., Marliss, E. B., Greenberg, G. R., and Bruce-Robertson, A., 1977, Chromium deficiency, glucose intolerance, and neuropathy reversed by chromium supplementation in a patient receiving long-term total parenteral nutrition, *Am. J. Clin. Nutr.* 30:531–538.

Kayne, F. J., Komar, G., Laboda, H., and Vanderlinde, R. E., 1978. Atomic absorption spectrophotometry of chromium in serum and urine with a modified Perkin-Elmer 603 atomic absorption spectrophotometer, *Clin. Chem.* 24:2151–2154.

Koivistoinen, P. (ed.), 1980. Mineral element composition of Finnish foods: N, K, Ca, Mg, P, S, Fe, Cu, Mn, Zn, Mo, Co, Ni, Cr, F, Se, Si, Rb, Al, B, Br, Hg, As, Cd, Pb and Ash, *Acta Agric. Scand., Suppl.* 22, Stockholm.

Kraintz, L., and Talmage, R. V., 1952. Distribution of radioactivity following intravenous administration of trivalent chromium[51] in the rat and rabbit, *Proc. Soc. Exp. Biol. Med.* 81:490–492.

Kumpulainen, J. T., Wolf, W. R., Veillon, C., and Mertz, W., 1979. Determination of chromium in selected United States diets, *J. Agric. Food Chem.* 27:490–494.

Langard, S., 1980. Chromium, in *Metals in the Environment*, H. H. Waldron (ed.), Academic Press, London, pp. 111–132.

Levine, R. A., Streeten, D. H. P., and Doisy, R. J., 1968. Effects of oral chromium supplementation on the glucose tolerance of elderly subjects, *Metabolism* 17:114–125.

Liu, V. J. K., and Morris, J., 1978. Relative chromium response as an indicator of chromium status, *Am. J. Clin. Nutr.* 31:972–976.

Mahalko, J. R., and Bennion, M., 1976. The effect of parity and time between pregnancies on maternal hair chromium concentration, *Am. J. Clin. Nutr.* 29:1069–1072.

Mathur, R. K., and Doisy, R. J., 1972. Effect of diabetes and diet on the distribution of tracer doses of chromium in rats, *Proc. Soc. Exp. Biol. Med.* 139:836–838.

Mertz, W., 1969. Chromium occurrence and function in biological systems, *Physiol. Rev.* 49:163–239.

Mertz, W., and Roginski, E. E., 1969. Effects of chromium(III) supplementation on growth and survival under stress in rats fed low protein diets, *J. Nutr.* 97:513–536.

Mertz, W., and Roginski, E. E., 1971. Chromium metabolism: The glucose tolerance factor, in *Newer Trace Elements in Nutrition,* W. Mertz and W. E. Cornatzer (eds.), Dekker, New York, pp. 123–153.

Mertz, W., Roginski, E. E., and Schwarz, K., 1961. Effect of trivalent chromium complexes on glucose uptake by epididymal fat tissue of rats, *J. Biol. Chem.* 236:318–322.

Mertz, W., Roginski, E. E., and Reba, R. C., 1965. Biological activity and fate of trace quantities of intravenous chromium(III) in the rat, *Am. J. Physiol.* 209:489–494.

Morgan, J. M., 1972. Hepatic chromium content of diabetic subjects, *Metab. Clin. Exp.* 21:313–316.

Nath, R., Minocha, J., Lyall, V., Sunder, S., Kumar, V. Kapoor, S., and Dhar, K. L., 1979. Assessment of chromium metabolism in maturity onset and juvenile diabetes using chromium-51 and therapeutic response of chromium administration on plasma lipids, glucose tolerance and insulin levels, in *Chromium in Nutrition and Metabolism,* D. Shapcott and J. Hubert (eds.), Elsevier/North Holland, Amsterdam, pp. 213–222.

National Research Council, 1974. *Medical and Biological Effects of Environmental Pollutants: Chromium,* National Academy of Sciences, Washington, D.C.

National Research Council, 1980. *Recommended Dietary Allowances,* National Academy of Sciences, Washington, D.C.

Newman, H. A. I., Leighton, R. F., Lanese, R. R., and Freedland, N. A., 1978. Serum chromium and angiographically determined coronary artery disease, *Clin. Chem.* 24:541–544.

Offenbacher, E. G., and Pi-Sunyer, F. X., 1980. Beneficial effect of chromium-rich yeast on glucose tolerance and blood lipids in elderly subjects, *Diabetes* 29:919–925.

Offenbacher, E. G., and Pi-Sunyer, F. X., 1983. Temperature and pH effects on the release of chromium from stainless steel into water and fruit juices, *J. Agric. Food Chem.* 31:89–92.

Okada, S., Hiroshi, O., and Taniyama, M., 1981. Alterations in ribonucleic acid synthesis by chromium(III), *J. Inorg. Biochem* 15:223–331.

Onkelinz, C., 1977. Compartment analysis of metabolism of chromium(III) in rats of various ages, *Am. J. Physiol.: Endocrinol. Metab. Gastrointest. Physiol.* 1:E478–E484.

Pekarek, R. S., Hauer, E. C., Rayfield, E. J., Wannemacher, R. W., and Beisel, W. R., 1975. Relationship between serum chromium concentrations and glucose utilization in normal and infected subjects, *Diabetes* 24:340–353.

Potter, J. F., Levin, P., Anderson, R. A., Freiberg, J. M., Andres, R., and Elahi, D., 1983. Glucose metabolism in glucose intolerant older people during chromium supplementation, *Metabolism*. Submitted.

Rabinowitz, M. B., Levin, S. R., and Gonick, H. C., 1980. Comparisons of chromium status in diabetic and normal men, *Metabolism* 29:355–364.

Riales, R., and Albrink, M. J., 1981. Effect of chromium chloride supplementation on glucose tolerance and serum lipids including high-density lipoprotein of adult men. *Am. J. Clin. Nutr.* 34:2670–2678.

Roginski, E. E., and Mertz, W., 1969. Effects of chromium supplementation on glucose and amino acid metabolism in rats fed a low protein diet, *J. Nutr.* 97:525–530.

Rollinson, C. L., Rosenbloom, E., and Lindsay, L., 1967, Reactions of chromium(III) with biological substances, in *Proc. 7th International Congress of Nutrition,* Vol. 5, Pegamon, New York, pp. 692–698.

Rosson, J. W., Foster, K. J., Walton, R. J., Monro, P. P., Taylor, T. G., and Alberti, K. G. M. M., 1979. Hair chromium concentrations in adult insulin-treated diabetics, *Clin. Chim. Acta* 93:299–304.

Sargent, T., Lim, T. H., and Jensen, R. L., 1979. Reduced chromium retention in patients with hemochromatosis, a possible basis of hemochromatotic diabetes, *Metabolism* 28:70–79.

Schroeder, H. A., 1965. Serum cholesterol levels in rats fed thirteen trace elements, *J. Nutr.* 94:475–480.

Schroeder, H. A., 1968. The role of chromium in mammalian nutrition, *Am. J. Clin. Nutr.* 21:230–244.

Schroeder, H. A., 1970. *Chromium Air Quality Monograph,* #70–15, American Petroleum Institute, Washington, D.C., 28 pp.

Schroeder, H. A., Balassa, J. J., and Tipton, I. H., 1962. Abnormal trace metals in man—chromium, *J. Chronic Dis.* 15:941–964.

Schroeder, H. A., Nason, A. P., and Tipton, I. H., 1970. Chromium deficiency as a factor in atherosclerosis, *J. Chronic Dis.* 23:123–142.

Schwarz, K. and Mertz, W., 1957. A glucose tolerance factor and its differentiation from factor 3, *Arch. Biochem. Biophys.* 72:515–518.

Seeling, W., Ahnefeld, F. W., Grunert, A., Kienle, K. H., and Swobodnik, M., 1979. Chromium in parenteral nutrition, in *Chromium in Nutrition and Metabolism,* D. Shapcott and J. Hubert (eds.), Elsevier/North Holland, Amsterdam, pp. 95–104.

Stout, R. W., 1977. The relationship of abnormal circulating insulin levels to atherosclerosis, *Atherosclerosis* 27:1–13.

Toepfer, E. W., Mertz, W., Roginski, E. E., and Polansky, M. M., 1973. Chromium in foods in relation to biological activity, *J. Agric. Food. Chem.* 21:69–73.

Toepfer, E. W., Mertz, W., Polansky, M., Roginski, E. E., and Wolf, W. R., 1977. Preparation of chromium-containing material of glucose tolerance factor activity from brewer's yeast extracts and by synthesis, *J. Agric. Food Chem.* 25:162–166.

Towill, L. E., Shriner, C. R., Drury, J. S., Hammons, A. S., and Holleman, J. W., 1978. *Reviews of the environmental effects of pollutants: III. Chromium.* Oak Ridge National Laboratory and U.S. Environmental Protection Agency, Cincinnati, Ohio.

Tuman, R. W., Bilbo, J. T., and Doisy, R. J., 1978. Comparison and effects of natural and synthetic glucose tolerance factor in normal and genetically diabetic mice, *Diabetes* 27:49–56

Vanderlinde, R. E., Kayne, F. J., Komar, G., Simmons, M. J., Tsou, J. Y., and Lavine, R. L., 1979. Serum and urine levels of chromium, in *Chromium in Nutrition and Metabolism*, D. Shapcott and J. Hubert (eds.), Elsevier/North-Holland, Amsterdam, pp. 49–57.

Veillon, C., Patterson, K. Y., and Bryden, N. A., 1982. Direct determination of chromium in human urine by electrothermal atomic absorption spectrometry, *Anal. Chim. Acta.* 136:233–241.

Versieck, J., de Rudder, J., Hoste, J., Barbier, F., Lemey, G., and Vanballenberghe, L., 1979. Determination of the serum chromium concentration in healthy individuals by neutron activation analysis, in *Chromium in Nutrition and Metabolism*, D. Shapcott and J. Hubert (eds.), Elsevier/North-Holland, Amsterdam, pp. 49–57.

Visek, W. J., Whitney, I. B., Kuhn, U. S. G., III, and Comar, C. L., 1953. Metabolism of Cr^{51} by animals as influenced by chemical state, *Proc. Soc. Exp. Biol. Med.* 84:610–615.

Vittorio, P. V., Wight, E. W., and Sinnott, B. E.. 1962, The distribution of chromium-51 in mice after intraperitoneal injection, *Can. J. Biochem. Physiol.* 40:1677–1683.

Wolf, W., Mertz, W., and Masironi, R., 1974. Determination of chromium in refined and unrefined sugars by oxygen plasma ashing flameless atomic absorption, *J. Agr. Food Chem.* 22:1037–1042.

Wu, G. Y., and Wada, O., 1981. Studies on a specific chromium binding substance (a low molecular weight chromium binding substance) in urine, *Jpn. J. Ind. Health,* 23:505–512.

Selenium 9

Raymond J. Shamberger

9.1 Introduction and History

Selenium was first discovered in 1818 by Berzelius in the sediment found in the lead chamber of a sulfuric acid plant in Gripsholm, Sweden. Because of its chemical similarity to tellurium, which in 1782 was named for the earth (Latin, meaning *tellus*, "earth"), selenium was named after the Greek word for moon.

Selenium is found in VIa of the periodic table between sulfur and tellurium. Selenium exists in several allotropic forms: amorphous, crystalline, and metallic. Amorphous selenium has three forms—vitreous, red amorphous, and colloidal selenium. The amorphous forms range from a dark-red to black powder that softens at 50–60°C and becomes elastic at 70°C. Monoclinic (red) selenium has two crystalline forms. The dark-red crystalline form has the only sharp melting point of these allotropic forms, m.p. 144°C. These red crystals are metastable and change when heated to the gray form, which is more stable. In the gray metallic form, selenium appears as gray to black lustrous hexagonal crystals, which melt at 217°C.

Because the electrical conductivity of metallic selenium increases when light strikes it and because it can convert light directly into electricity, the element is used in photoelectric cells, solar cells, and photographic exposure meters. Metallic selenium is also used extensively in rectifiers because of its ability to convert alternating current to direct current. Some of the properties of elemental selenium are included in Table 9-1.

The abundance of selenium in the earth's crust is 0.000009%. There are no important ores that are direct sources of selenium. Selenium is more commonly

---

Raymond J. Shamberger • The Cleveland Clinic, 9500 Euclid Avenue, Cleveland, Ohio 44106.

Table 9-1. Properties of the Selenium Atom

Atomic weight	78.96
Atomic number	34.0
Covalent radius, Å	1.16
Atomic radius, Å	1.40
Ionic radius, Å	1.98 (−2)
	0.42 (+6)
Electronegativity	2.4
Electronic structure	$(Ar)3d^{10}4s^24p^4$
Masses of stable isotopes	74,76,77,78,80,82
Melting point	
Amorphous	50°C
Gray	217°C
Boiling point	685°C
Density	
Amorphous	4.28 g/cm^3
Gray	4.79 g/cm^3
Oxidation states	−2, 0, +4, +6

associated with the sulfides of metal ores, and it is usually recovered as a byproduct in the flue dusts from the refining of the sulfide ores of other metals, such as copper, iron, lead, and silver. Selenium is also extracted from the slime deposited at the anode in the electrolytic refining of copper. Selenium occasionally occurs uncombined, usually in conjunction with free sulfur.

The naturally occurring isotopes of selenium include ^{80}Se (49.8%), ^{78}Se (23.5%), ^{76}Se (9.0%), ^{82}Se (9.2%), ^{77}Se (7.6%) and ^{74}Se (0.87%). Because none of the naturally occurring isotopes of selenium are radioactive, the radioisotopes are manufactured using a nuclear reactor and neutron activation technology. Selenium-75 has proved to be one of the more useful radionuclides. Selenium-75 is widely used in biological tracer experiments and diagnostic procedures because of its convenient gamma ray for counting and its relatively long half-life of 120 days.

Selenium, like sulfur, exists in several oxidation states, which are -2, 0, +4, and +6. Because of their similarity in oxidation states, these elements have some analogous properties even though there are also some prominent differences.

Selenium and sulfur have similar configurations of electrons in their outermost valence shells, even though the third shell of selenium is completely filled. The bond energies, the ionization potentials, the sizes of the atoms whether they are in the covalent or ionic state, the electronegativities, and the polarizabilities are essentially identical.

Even though these two elements are very similar chemically and physically, they cannot always substitute for one another *in vivo*. One possible reason for these differences may be found in the following equation:

$$H_2SeO_3 + 2H_2SO_3 \rightarrow Se + 2H_2SO_4 + H_2O \qquad (9\text{-}1)$$

This reaction states that the quadrivalent selenium in selenite has a tendency to undergo reduction, whereas the quadrivalent sulfur in sulfite tends to undergo oxidation. The chemical difference in the ease of reduction of selenite versus the ease of oxidation of sulfite is also reflected in mammalian metabolism of these compounds. In mammals selenium compounds also generally tend to be reduced, whereas sulfur compounds have a general tendency to be oxidized.

Even though the analogous oxyacids of selenium and sulfur are of similar strength, H_2Se is a much stronger acid than H_2S. In addition, the hydrides of selenium and sulfur also have a different acid strength. This difference may be important in the dissociation behavior of the selenohydryl group of selenocysteine (pK 5.24) compared to that of the sulfhydryl group of cysteine (pK 8.25). This difference is likely to be biologically important because at physiological pH, the sulfhydryl group in cysteine (or other thiols) exists mainly in the protonated form; the selenohydryl group in selenocysteine (or other selenols) exists largely in the dissociated form. Even though several superficial similarities in the metabolism of selenium and sulfur do exist, the major physiologic role of selenium may well be found in some of the small chemical and physical differences of these two elements.

Some of the known selenium compounds are H_2Se, metallic selenides, SeO_2, H_2SeO_3, SeF_6, Se_2Cl_2, and H_2SeO_4 (selenic acid). Selenic acid possesses the unusual ability to dissolve gold. One of the more important selenium compounds is the oxychloride ($SeOCl_2$), a yellowish corrosive liquid that attacks most metals and is a good industrial solvent for rubber, Bakelite, gums, resins, Celluloid, glue, asphalt, and other materials. In the rubber industry, selenium is also used along with sulfur and tellurium to increase the tensile strength, abrasion resistance, and the life of rubber. Selenium is also able to form a large number of organic compounds that are analogous to those of sulfur.

The greatest amounts of selenium are used for the manufacture of photoelectric cells. Some selenium is used for xerography; in addition, the electronics industry uses the element, as do the ceramics and rubber industry. A small amount of selenium corrects the green color of glass, and a large amount gives the ruby color to glass that is widely used in taillights and signals. Red enamelware contains selenium, and the element is used as a catalyst and in flameproofing electric cable.

Table 9-2. Naturally Occurring Organic Selenium Compounds

Compound	Formula
Selenocysteine	$HSe-CH_2CH(NH_2)COOH$
Selenocystine	$HOOCCHNH_2CH_2-Se-Se-CH_2CH(NH_2)COOH$
Selenohomocystine	$HOOCCH(NH_2)CH_2CH_2-Se-Se-CH_2CH_2-CH(NH_2)COOH$
Se-methylselenocysteine	$CH_3-Se-CH_2CH(NH_2)COOH$
Selenocystathionine	$HOOCCH(NH_2)CH_2-Se-CH_2CH_2CH(NH_2)COOH$
Selenomethionine	$CH_3-Se-CH_2CH_2CH(NH_2)COOH$
S-methyl selenomethionine	$(CH_3)_2-Se-CH_2CH_2-CH(NH_2)COOH$
Dimethylselenide	$CH_3-Se-CH_3$
Dimethyldiselenide	$CH_2-Se-Se-CH_3$
Trimethyl selenonium	$(CH_3)_3Se^+$
Selenotaurine	$H_2NCH_2CH_2-SeO_3H$

9.2 Selenium and Its Compounds in Cells and Tissues

The forms of selenium that occur in living systems (Table 9-2) and the forms that are biologically available to the organism depend on the form of selenium supplied to the organism, the amount of selenium supplied, and the species of plant or animal. In this presentation the forms of selenium have been divided into low- and high-molecular-weight compounds.

9.2.1 Low-Molecular-Weight Compounds

Selenocysteine

Protein hydrolysates of corn and wheat seeds have been subjected to one-dimensional paper chromatography. The selenium locations were compared to positions of synthetic seleno-amino acids. The R_f migrations were found to be similar to that of selenocysteine (Table 9-2).

Wool hydrolysates from sheep injected with $H_2^{75}SeO_3$ have also been analyzed by both column and paper chromatography. Peaks of standard selenocysteine were similar to those found in the wool hydrolysates.

Selenocystine

Peterson and Butler (1962) have fractionated the 80% ethanol extracts of red clover, white clover, and rye grass, which were given radioactive selenite. Radioactive tracings of paper chromatograms showed substantial amounts of selenocystine (Table 9-2), selenomethionine, and their oxidation products in the soluble fraction. Two-dimensional paper chromatograms of alcoholic onion extracts (*Allium copa*) showed that two of the five radioactive spots observed were similar to selenocystine and selenomethionine. One of the five major radioactive spots was probably propenyl β-amino-β-carboxyethyl selenoxide, the selenium analog of the lachrymotory precursor.

Selenohomocystine

Leaves from *Astragalus crotalariae* metabolize ^{75}Se-selenomethionine to selenohomocystine (Table 9-2) (Virupaksha et al., 1966). Dowex 50-column chromatography followed by further fractionation by paper chromatography and electrophoresis gave a compound that yielded selenomethionine after reduction and methylation.

Se–methylselenocysteine

Aqueous extracts (Trelease et al., 1960) were analyzed from soil-grown plants of *Astragalus bisulcatus*. Eighty percent of the total selenium was present as Se–methylselenocysteine (Table 9-2), which was isolated as a crystalline solid with small amounts of S-methylcysteine.

Selenocystathionine

Hot water extracts of soil-grown *Astragalus pectinatus* were analyzed and selenocystathionine (Table 9-2) was isolated as a crystalline solid (Horn and Jones, 1941). Selenocystathione has been identified as the cytotoxic compound in seeds of coco de mono (*Lecythis ollaria*) or monkey nut tree, which is distributed in Central and South America.

Selenomethionine

Selenomethionine (Table 9-2) has been found in protein hydrolysates from livers of dogs injected with Se^{75} and from wool hydrolysates from sheep that were injected with $H_2Se^{75}O_3$. Selenomethionine has also been detected in hydrolysates from *E. coli* grown with $Na_2Se^{75}O_3$ on a sulfur-deficient medium. In another experiment methionine was replaced by selenomethionine in β-galactosidase isolated from a strain of *E. coli* grown on a medium high in selenate and low in sulfate. About 80 of its 150 methionine residues have been replaced by selenomethionine, but there was no apparent replacement of cystine by selenocystine. Rumen microorganisms can also incorporate inorganic selenium into seleno-amino acids. In ruminant fluid selenate also appears to undergo rapid reduction to selenite. In addition, selenite was rapidly and extensively bound to protein by a nonenzymic process.

Seleno-amino acids were also found in protein hydrolysates from corn and wheat seeds of soil-grown plants. Extensive incorporation of selenium into selenomethionine has also been observed in the protein portion of ryegrass, wheat, red clover, and white clover.

Se-methylselenomethionine

Se-methylselenomethionine (Table 9-2) was first detected as a radioactive compound of high electrophoretic mobility in aqueous ethanol extract of clover and ryegrass roots that had been grown on radioactive selenite. This compound was also found to be the predominant soluble organoselenium compound synthesized from selenite by several species of *Astragalus* that are unable to accumulate selenium. The occurrence of this compound in nonaccumulators may distinguish these species from the accumulator species, where Se-methylselenocysteine is the predominant compound.

Dimethyl Selenide

Rats injected with radioactive selenate exhaled a radioactive compound that was isolated with carrier dimethyl selenide and derivatized with mercuric chloride and recrystallized to a constant specific activity. Dimethyl selenide (Table 9-2) has also been isolated from *Scopulariopis brevicaulis* and *Aspergillus niger* grown with Na_2SeO_4 or Na_2SeO_3. The methyl groups of dimethyl selenide originated from methionine. Using gas chromatography the presence of dimethyl selenide was detected as a respiratory product in rats given selenite. Dimethylselenide was also found in *Astragalus* plants and seeds.

Dimethyl Diselenide

Four volatile compounds from *Astragalus racemosus* were trapped on active carbon followed by extraction from the carbon in various solvents. One of the four compounds was identified by gas chromatography as dimethyl diselenide.

Trimethyl Selenonium

One of the urinary selenium compounds excreted by rats was identified as the trimethyl selenonium ion $(Ch_3)_3Se^+$. Since the compound was not formed when normal rat urine was mixed with selenite and could also be detected in rat tissues, the trimethyl selenonium ion appears to be an actual metabolite and not an artifact of the isolation procedure.

Elemental Selenium

Microorganisms have been reported to reduce selenium salts to elemental selenium. Amorphous selenium has been identified as the end product of selenite reduction by *Salmonella heidelberg*. Deposits of elemental selenium have been reported in the roots of plants given toxic selenium levels.

Selenotaurine

After sheep were injected with $H_2{}^{75}SeO_3$, the selenotaurine derivative of bile cholic acid was separated by celite column chromatography.

Other Compounds

After rats were injected with $Na_2{}^{75}SeO_3$, seleno-coenzyme A was identified chromatographically. Evidence for a seleniferous wax containing selenoesters of the type $R-\overset{\overset{\displaystyle O}{\|}}{C}-Se-R$ was obtained after *Stanleya bipinnati* was grown in the presence of ^{75}Se-selenite. Two isomers of glutamylselenocystathionine have been isolated from *Astragalus rasei* (a nonaccumulator) when this plant was given selenite or selenate. Dimethyl selenone $(CH_3)_2SeO_2$ was produced by soil and sewage sludge samples after they were incubated with sodium selenite or elemental selenium. Dimethyl selenide and dimethyl diselenide were also produced by the soil and sewage sludge samples. These results suggest that microbial methylation of selenium may commonly occur. Thus biomethylation might be a pathway for the mobilization of selenium to the atmosphere.

9.2.2 Macromolecular Weight Compounds

Formate Dehydrogenase

In a purified culture medium, traces of selenite were needed for the production of formate dehydrogenase in *E. coli*. Molybdenum and iron were also required for the activity of this enzyme. Protein synthesis was essential in order to show the response to selenium. Selenocystine and selenite were shown to be about equally effective in stimulating the synthesis of formate dehydrogenase, but selenomethionine was much less active. Similar effects were observed with selenite and molybdate on the activity of the formate dehydrogenase of *Clostridium thermoaceticum*.

Studies on the formate dehydrogenase activity of *M. vannielii* revealed that this organism produces both a selenium-dependent and a selenium-independent formate dehydrogenase. *M. vannielii* is strictly an anaerobic microorganism that grows on formate as the sole organic substrate and converts it into carbon dioxide and methane.

$$4HCOOH \rightarrow 3CO_2 + CH_4 \qquad (9\text{-}2)$$

The immediate electron acceptor for formate oxidation by both formate dehydrogenases of *M. vannielii* is an 8-hydroxy-5-deazaflavin, which is an abundant cofactor in methane bacteria. Reduction of FAD, FMN, and tetrazolium dyes but not pyridine nucleotides was observed with the isolated enzymes. A second protein, an 8-hydroxy-5-deazaflavin-dependent $NADP^+$ reductase is required to link formate oxidation to $NADP^+$ reduction.

The chemical form of selenium in the selenium-containing formate dehydrogenase of *M. vannielii* is selenocystine. This is also likely true for the *E. coli* formate dehydrogenase. There should be four selenocysteine residues in the *E. coli* enzyme or one per 100,000 subunit.

Glycine Reductase

When selenite was added to the medium used for *C. sticklandii*, the activity of glycine reductase was markedly increased. After this initial observation, the low-molecular-weight "protein A" of the Clostridial glycine reductase was found to be a selenoprotein. Protein A is a heat stable, acidic protein with a low molecular weight of about 12,000. The glycine reductase selenoprotein A contains one gram atom of selenium per mole, which is present within the polypeptide chain as a single seleno-cysteine residue. The clostridial glycine reductase system participates in a specific electron transfer process that is coupled to the esterification of orthophosphate and synthesis of ATP.

$$NH_2CH_2COOH + R(SH)_2 + Pi + ADP \rightarrow CH_3COOH$$
$$+ NH_3 + RS_2 + ATP \qquad (9\text{-}3)$$

Arsenate can replace phosphate in the enzyme system *in vitro* without affecting the reaction rate. With arsenate an adenylate acceptor is not required. The soluble glycine reductase system consists of at least three or four different proteins. Two of these have been isolated in purified form and the third protein has been extensively purified. In addition to selenoprotein A—the 12,000-dalton, acidic, heat-stable protein, which has been extensively purified—there is also protein

B, approximately 200,000 daltons, which contains one or more essential carboxyl groups and a fraction C protein of approximately 250,000 daltons. All three components are essential for the reduction of glycine by a dithiol to acetate and ammonia.

Nicotinic Acid Hydrolase

When supplemental selenium is added to the medium in which *Clostridium barkeri* is cultured, there is about a 16-fold elevation in nicotinic acid hydroxylase. Nicotinic acid hydroxylase catalyzes an anaerobic reaction in which water is added to nicotinic acid followed by the removal of two hydrogen atoms and generation of NADPH (Figure 9-1). The purified 300,000-dalton enzyme contains 6 acid-labile sulfides, 1.5 FAD equivalents per mole, and 11 iron equivalents per mole. When nicotinic acid hydrolase was purified from *C. barkeri* cells labeled with ^{75}Se, both the enzyme activity and ^{75}Se were enriched in parallel. These results suggest that selenium may indeed be an essential cofactor for the enzyme with the possibility of serving as another redox center in addition to the bound flavin and the iron sulfur centers.

Xanthine Dehydrogenase

Selenite supplementation of growth media also increased the activity of xanthine dehydrogenase, another clostridial enzyme that also contains multiple redox centers. FAD and molybdenum and acid-labile sulfide are also present on the enzyme. The reduction of uric acid to xanthine is considered to be the major role of xanthine dehydrogenase in the purine-fermenting anaerobic bacteria:

$$\text{uric acid} + \text{reduced acceptor} \rightarrow \text{xanthine} + H_2O + \text{acceptor} \quad (9\text{-}4)$$

The purified enzyme can oxidize a variety of purines, substituted purines and aldehydes using dyes, ferricyanide, or oxygen as electron acceptor. The form of selenium in the enzyme has not yet been elucidated.

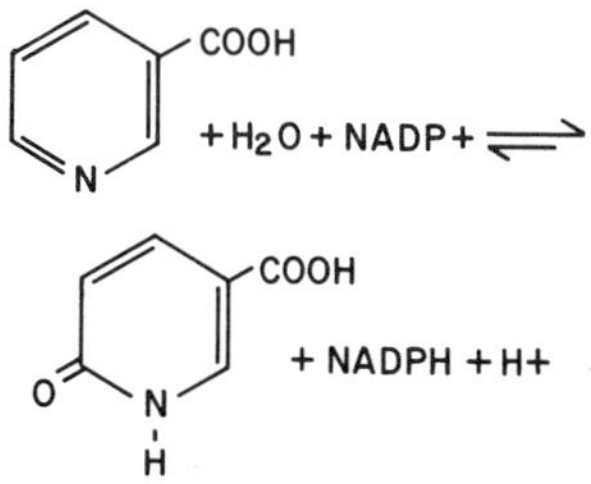

Figure 9-1. Reaction catalyzed by nicotinic acid hydrolase.

Thiolase

Even though the first selenium-dependent enzymes were observed to function as redox catalysts, thiolase seems to be an exception to this trend. Thiolase catalyzes the coenzyme A-dependent cleavage of acetoacetyl-CoA to form 2 acetyl-CoA.

$$\text{acetoacetyl-COA} + \text{COASH} \rightarrow 2 \text{ acetyl-CoA} \qquad (9\text{-}5)$$

Two thiolases have been elucidated that catalyze the CoA-dependent cleavage of aceto acetyl-CoA, but only one contains selenium. The condensation reaction should be considerably facilitated if an enzyme-bound selenolacyl ester rather than a thiolacyl ester, took part as an intermediate, since a selenol group is more reactive than a thiol group. The native selenoenzyme of about 155,000 daltons is composed of four subunits of 38,000–40,000 daltons. The thiolase, which lacks selenium, is of similar molecular weight and is less acidic. The factors that control the relative activity of the two enzymes in the cell have not been identified.

Hydrogenase

A ^{75}Se-labeled hydrogenase was purified to near homogeneity from extracts of *Methanococcus vannielii* cells grown in the presence of (^{75}Se) selenite. The molecular weight of the enzyme was estimated as 340,000 by gel filtration. A value of 3.8 g-atoms of selenium/mole of enzyme was determined by atomic absorption analysis. The chemical form of selenium in the enzyme was shown to be selenocysteine.

Glutathione Peroxidase

Glutathione peroxidase is the first selenium-dependent enzyme that has been studied extensively in mammals. The enzyme was first discovered in 1957, the same year selenium was found to be essential, but it was not until 1971 that the relationship of selenium to the enzyme became known. Rotruck *et al.* (1971) reasoned that because several of the effects of selenium deficiency in animals could be prevented by vitamin E or antioxidants as well as by selenium, selenium might play a role in preventing oxidative damage. Rotruck *et al.* (1971) chose erythrocytes as a model system for investigation. The membrane lipids of the red blood cell can be damaged by oxidants, resulting in rupture of the cell and release of its hemoglobin through hemolysis. This hemolysis is prevented by vitamin E and has long been used as an *in vitro* test for vitamin E deficiency.

Hemolysis of vitamin E-deficient animals was prevented by selenium, but only if glucose was present in the incubation medium. The glucose-dependent effect of selenium indicated that some step was involved that linked glucose oxidation to the destruction of peroxides by reduced glutathione. Because the concentration of reduced glutathione (GSH) was effectively maintained during an *in vitro* incubation, the defect was not in the generation of GSH, but in its utilization to protect the cell, possibly by means of glutathione peroxidase. Subsequent experiments confirmed this observation. Erythrocytes as well as other tissues from animals deficient in selenium had much lower levels of glutathione peroxidase.

Glutathione peroxidase was purified from sheep erythrocytes. During the purification process the selenium content rose progressively and reached a value of 0.34% in the pure enzyme, equivalent to nearly 4 g-atoms per mole. The same value was found in cows, humans, and rat liver. The molecular-weight values reported for the enzymes were between 76,000 and 92,000. The enzyme is composed of four almost identical 19,000–23,000-dalton subunits. Each subunit contains selenium in the form of a single selenocysteine residue.

Glutathione peroxidase is capable of reducing a wide variety of organic peroxides in addition to H_2O_2, but reduced glutathionine (GSH) is the only reductant of physiologic significance.

$$2GSH + H_2O_2 \rightarrow GSSG + H_2O \tag{9-6}$$

$$2GSH + ROOH \rightarrow GSSG + ROH + H_2O \tag{9-7}$$

Other thiol compounds have also been accepted as donor substrates, if they show structural similarities to GSH. Specificity studies have revealed that both carboxylic groups of the GSH molecule are essential for substrate binding.

Conventional tissue fractionation of rat liver showed that glutathione peroxidase can be characterized as a soluble enzyme present in the cytosol and the mitochondrial matrix. Cell fractionation demonstrated that 25.9% of the enzyme activity was in the high- and low-density mitochondria and 73.3% of the enzyme activity was located in the soluble fraction. Almost no activity was found in the nuclei, liposomes, peroxisomes, and the high- and low-density microsomes. No glutathione peroxidase was found in the inner or outer membranes of mitochondria. The matrix space contains 92% of the activity and the intermembrane space contains 4% of the activity.

Assay of glutathione peroxidase levels in blood is often used as a way of detecting selenium deficiency because of the frequently observed correlation between the level of this enzyme and the selenium nutritional status of humans and animals. In some cases, however, assays using organic peroxides (e.g., cumene hydroperoxide) as substrate may give misleading results, especially if tissues such as liver were analyzed. The lack of correlation in certain deficiency conditions was attributed to the presence of nonselenium-dependent glutathione peroxidase, which was subsequently identified as glutathione transferase. Because detoxification enzymes of this class account for as much as 5% of the soluble protein of liver, their reactivity with organic peroxides, even though having much less total activity when compared to glutathione peroxidase, may also be an important factor in protecting liver from damage resulting from peroxides in selenium-deficient animals.

Miscellaneous Selenoproteins

There are several selenium-containing proteins whose biochemical roles are still unknown, but they have been found in various biological samples. One such example is a 10,000 dalton molecular-weight selenoprotein that has been isolated from normal semitendinous muscles and hearts of lambs but is lacking in selenium-deficient animals. White muscle disease of lambs and calves, gizzard myopathy of domestic birds, and pancreatic fibrosis of chicks are known selenium deficiency syndromes that appear to be a type of nutritional muscular dystrophy in which collagen replaces damaged tissue. It is not known whether the muscle small-molecular-weight selenoprotein that has been reported to possess a cytochrome c chromophore is the component that is the critical factor in all these conditions.

Fertility in males has been shown by farmers and animal nutritionists to be directly related to the availability of selenium in the diet. A selenium-containing protein of about 15,000 daltons appears in the testis of the rat concomitant with the onset of sexual maturity and may be of special importance in regard to fertility. A 17,000-dalton selenoprotein isolated from rat sperm tail may be identical to the testis selenoprotein.

Another selenoprotein, called ^{75}Se-P, has been isolated from rat liver. Selenium deficiency has decreased ^{75}Se incorporation by glutathione peroxidase at 3 and 72 hours after ^{75}Se injection but increased ^{75}Se incorporation by ^{75}Se-P. The apparent molecular weights of ^{75}Se-P from liver and plasma were determined by gel filtration and were found to be 83,000 and 79,000. ^{75}Se-P may account for some of the physiological effects of selenium.

A novel enzyme that exclusively decomposes L-selenocysteine into L-alanine and H_2Se in various mammalian tissues has been named selenocysteine lyase. The enzyme from pig liver has been purified and has a molecular weight of 85,000 and contains pyridoxal 5'-phosphate as a coenzyme. Even though the enzyme is not a selenoprotein, the enzyme is the first proven enzyme that specifically acts on selenium compounds.

Seleno-tRNAs

An example of the natural occurrence of a different type of macromolecule that contains selenium is provided by the observation that a few bacterial aminoacyl transfer nucleic acids (tRNAs) are specifically modified with selenium in the polynucleotide portions of the molecules. *C. sticklandii* has been shown to synthesize three readily separable ^{75}Se-labeled tRNAs when incubated with ^{75}Se-selenite or ^{75}Se-selenocysteine. In this system seleno-tRNA formation is highly specific for selenium and is not decreased in the presence of 1000-fold molar excesses of sulfur analogs. One of the four seleno-tRNAs isolated from *C. sticklandii* copurified with L-prolyl-tRNA species. Another tRNA has been isolated in a highly purified form and has been shown to be isoaccepting L-glutamyl-tRNA. The four different tRNAs account for 5–8% of the total tRNA in *C. sticklandii*. *M. vannielii* and several other bacteria synthesize selenium-containing tRNAs, but in each of these only a few of the tRNAs contain selenium. In many cases a specifically modified base in a particular tRNA appears to play a role in some type of regulatory process. In *E. coli* selenium can be incorporated into tRNA, and a ^{75}Se-labeled nucleoside that chromatographed with 4-selenouridine has been isolated from the enzymic digests of ^{75}Se-labeled tRNA.

Transfer RNA has also been isolated from the selenium indicator plant *A. bisulcatus*. This material had a high guanosine to cytidine ratio and showed a major and modified nucleoside composition characteristic of plant transfer RNAs and exhibits chromatographic and electrophoretic properties similar to transfer RNAs from other well-studied plant and bacterial systems.

9.3 Selenium Deficiency and Function

9.3.1 Dietary Liver Necrosis and Factor 3

Dietary liver necrosis is a rapidly developing condition that can cause an apparently healthy rat to become ill and die within a day or two. Upon autopsy, massive hepatic necrosis is found. Relatively early it became apparent the necrotic degeneration of the liver was produced by dietary means.

Sulfur amino acids exerted some protective effect, and vitamin E was found to be an agent that prevented dietary liver necrosis. It also became clear that another unidentified dietary factor occurred in nutrients that independently prevented this disease. This dietary factor was designated Factor 3 and was present in crude casein, kidney and liver powder, and in brewer's yeast. However, it was absent from torula yeasts. Because of the absence of Factor 3 from torula yeast the latter has been used for years as a protein source in the standard diets for producing dietary liver necrosis and related diseases. Some of the selenium vitamin E deficiency diseases are summarized in Table 9-3.

At first Factor 3 appeared to be a new vitamin. The discovery of organically bound selenium in kidney powder as the essential constituent of Factor 3 was made after seven years of attempts to isolate this substance (Schwarz and Foltz, 1957). Factor 3 was also isolated from American brewer's yeast and liver.

Table 9-3. Vitamin E and Selenium Deficiency Disease of Animals

Disease	Animal	Tissue affected	Vit. E	Se	Antioxidants
Reproductive failure					
Fetal death, resorption	Rat	Embryonic vascular system	X		X
	Cow, ewe		X	X	
Testicular degeneration	Rooster, rat, rabbit, hamster, dog, pig, monkey	Germinal epithelium	X	X	
Nutritional myopathies					
Nutritional muscular dystrophy (NMD)	Chick,[a] rat, guinea pig, rabbit, dog, monkey	Striated muscle	X	[b]	
"Mulberry heart" disease	Pig	Cardiac muscle	X		
NMD	Mink	Striated cardiac muscle	X	X	
Gizzard myopathy	Turkey,	Gizzard muscle	X		
	duck		X		
"Stiff lamb" disease	Newborn lamb	Striated muscle		X	
NMD	Sheep, goat, calf	Striated muscle	[c]	X	
Creatinuria	Rat, rabbit, guinea pig, monkey	Plasma	X		
Erythrocyte hemolysis	Chick, rat, rabbit	Erythrocyte	X	X	X
Incisor Depigmentation	Rat	Incisor enamel	[c]		
Systemic Disorders					
Liver necrosis	Rat, mouse, pig	Liver	X	X	X
Membrane lipid peroxidation	Chick, rat	Hepatic mitochondria and microsomes	X	X	
Accumulation of ceroids[d]	Rat, mink, calf, lamb, dog, chick, turkey	Adipose	X	[e]	[e]
Plasma protein loss	Chick, turkey	Serum albumin	X	X	
Kidney degenerations	Mouse, rat, pig	Renal tubule contorti	X		X
Anemia	Monkey, pig	Bone marrow	X		
Encephalomalacias[d]	Chick	Cerebellum	X		X
Exudative diathesis	Chick	Capillary walls	X	X	[f]
Pancreatic fibrosis	Chick	Pancreas		X	
Lack of growth	Rat, monkey	Body size, mass	X	X	
Lack of hair growth	Rat, monkey	Hair	X		
Lack of feather growth	Turkey	Feathers	X		

Prevented by dietary: Vit. E, Se, Antioxidants

[a] Responsive to sulfur-containing amino acids.
[b] May partially reduce severity.
[c] Syndrome not easily produced in absence of dietary polyunsaturated fatty acids.
[d] Accelerated by polyunsaturated fatty acids.
[e] Involvement proposed but not confirmed.
[f] Active only in presence of selenium.

Fractionation of Factor 3 led to a semicrystalline preparation that developed a characteristic "garlic like" odor upon alkali addition. This led to the discovery that selenium is a part of Factor 3. Previously, it had been reported that the breath of cattle consuming high-selenium plants from seleniferous soils had a garlic like odor. Some of the very best Factor 3 preparations were about 10,000-fold as active as the starting material. However, because of the extreme instability of the purified agent, efforts to isolate the active compound from kidney powder were halted.

Inorganic compounds (selenite, selenate, and selenocyanate) were found to be about one-third as potent as the naturally occurring form of Factor 3 from kidney powder. Schwarz et al. (1972) have tested several hundred organoselenium compounds in an effort to match the potency of Factor 3. Several compounds were about as potent as selenite; others were less potent, and some were not at all effective. Attempts to identify Factor 3 in this way also failed.

Schwarz (1965) has postulated that there are three important phases in the development of dietary liver necrosis. These phases may also be important in other deficiency diseases. In the first phase it is likely that there is an induction period during which the animal becomes deficient. In this phase the stores of the two primary protective agents—vitamin E and selenium—are gradually depleted but are nonetheless present in adequate amounts to maintain normal function. In the next phase marked disturbances of intermediary metabolism as well as the subcellular, electron microscopic structure are detectable. No gross pathological lesions are detectable. However, in the last phase there are severe anatomical and macroscopic changes, which are often irreversible in this acute phase. The gross macroscopic changes of the liver develop extremely fast, usually over a period of only a few hours. These changes are often fatal within a few hours or days in the majority of animals.

The exact biochemical lesion of dietary necrosis is uncertain, but it is likely that this disease is due to lipid peroxidation. When rats are dying of apparent liver necrosis, large amounts of ethane are exhaled. Ethane has been established as an end product of peroxidation. One could postulate that in the absence of vitamin E and of selenium-dependent glutathione peroxidase, the defensive mechanisms against lipid peroxidation are inadequate to hold the process in check.

Death from liver necrosis in rats is usually preceded by a latent phase of about seven days. There is a characteristic impairment of energy metabolism that can be observed in liver slices and homogenates during this last phase. This disorder has been designated "respiratory decline" and is characterized by the sudden failure of oxygen consumption. Electron microscopy has shown that the respiratory decline is related to swelling of the mitochondria as well as a cystic degeneration of the elongated profiles of the ergastoplasmic reticulum. It is likely that selenium in the active site of glutathione peroxidase helps protect against the swelling of mitochondria.

9.3.2 Hepatosis Dietetica

When the liver necrosis occurs in pigs it is called hepatosis dietetica. There are two clinical patterns of hepatosis dietetica in pigs. In the more severe type, apparently healthy, rapidly growing 3-month-old piglets become ill and die within hours after the onset of the disease. Upon autopsy massive liver necrosis is observed. This first pattern markedly resembles dietary liver necrosis in rats. In other animals acute liver failure may not occur. Instead, the disease runs a subacute course with the development of jaundice and ascites. When the liver is examined, it shows areas of massive necrosis over the capsular and cut surfaces and fibrosis is also present. The affected lobules may be swollen with hemorrhages or may show nonhemorrhagic coagulative necrosis. Other conditions frequently found are effusions of fluids into the body cavities, skeletal muscle degeneration, and esophageal ulcers.

In addition to rats and pigs, hepatic necrosis can also occur in monkeys. Seven adult squirrel monkeys were fed a low-selenium semipurified feed with *Candida utilis* as the protein source and adequate vitamin E. A variety of problems, such as alopecia, loss of body weight, and listlessness, developed after 9 months. Some of the monkeys became moribund and died. Upon necropsy several lesions were observed, including hepatic necrosis, skeletal muscle degeneration, myocardial degeneration, and nephrosis.

9.3.3 Nutritonal Muscular Dystrophy

Several important sheep-raising areas have a low soil and plant selenium content. High incidences of selenium and vitamin E deficiency have been observed in New Zealand and in western Oregon. These deficiencies of vitamin E and selenium bring about a nutritional muscular dystrophy (NMD). This disease has also been called white muscle disease and stiff lamb disease. Lambs and calves about 3 months old are usually affected. After the lambs lose strength and become weak, they develop a stiff appearance, which results in an abnormal gait. Their weakness may interfere with foraging or suckling and may lead to death from starvation. Secondary infections such as pneumonia are frequently found. If cardiac involvement occurs sudden death could result. Even though NMD occurs mostly in the age range of 1–3 months, NMD also occurs in newborns and in adults. NMD seems to be increased by muscular activity, and the symptoms of the disease are sometimes confused with heavy metal poisoning or other conditions that involve the central nervous system.

When the muscles are examined on necropsy, they reveal a pattern of symmetrical involvement, with the pale areas of several skeletal muscles indicating the areas of greatest degeneration. The deep muscles overlying the cervical vertebrae are especially affected with typical chalky-white lesions, which are particularly characteristic of chronic dystrophy.

Experimental NMD has been shown to respond to either selenium or vitamin E administration. Blood selenium and glutathione peroxidase have been shown to predict NMD's occurrence. Abnormally elevated levels of serum glutamic oxalacetic transaminase, lactic acid dehydrogenase, and creatinine have been observed with NMD. Prevention of NMD has been successful using several different methods. These include administration of selenium by drench, injection, or heavy pellets in the rumen.

NMD also occurs in pigs and chickens. In pigs, NMD often occurs where hepatosis dietetica and mulberry heart disease are also present. In most farm animals, NMD is considered to be a disease of the young, but in swine the disease occurs in piglets, sows, and fattening pigs. In cases that have been reported in the United States, the pigs that have NMD are frequently weak and unsteady. If the hindlegs are affected, this produces a condition that is termed "splay-leg" or "spraddle-legged" pigs. Lesions are not usually grossly apparent, but in some cases pale areas are grossly visible in skeletal muscle. In general, NMD does not assume the clinical importance in pigs that it does in sheep because the pigs are more susceptible to severe liver or heart disease. The animals frequently die from these problems before NMD can develop to a life-threatening extent.

In chickens, NMD is primarily due to vitamin E and sulfur-containing amino acid deficiencies; however, selenium spares vitamin E in preventing NMD. NMD in chicks is characterized by white striations in the breast and in some cases the leg muscles. Selenium deficiency in the chick causes other defects, which are more serious than NMD. These include myopathy of the smooth muscle (gizzard) and myocardial myopathy. Levels of serum glutamic oxalacetic transaminase have also been related to NMD in chicks and poults that have been maintained on selenium and vitamin E-deficient diets.

9.3.4 Exudative Diathesis

Exudative diathesis in chickens and turkeys is characterized by an edema of body tissues, with large amounts of fluid accumulation under the skin of the abdomen and breast. Between 3 and 6 weeks of age, the chicks may become dejected, lose condition, and show leg weakness. In severe cases they may become prostrate and die. Swelling of the subcutaneous tissue is frequently observed. Hemorrhages in cutaneous tissues near the areas of fluid accumulation cause a reddish color of the skin. Once animals develop this condition, they usually die within a few days. Exudative diathesis can be prevented in chicks by adding as little as 0.1 ppm of selenium as selenite to a vitamin E-deficient diet. Several controlled treatment trials have been carried out in areas of New Zealand that experienced outbreaks of exudative diathesis. If 15 mg of selenium was added to 1 gal of drinking water for only one day, immediate control of the disease occurred and most of the affected birds made a rapid recovery.

9.3.5 Pancreatic Degeneration

Chicks fed a severely selenium-deficient diet but supplemented with vitamin E and bile salts develop a severe atrophy of the pancreas. If the chicks are continued on the deficient diet, they subsequently die of pancreatic insufficiency. This disease has not been observed in other animals. Death usually results after a marked decreased absorption of lipids, including vitamin E.

As little as 0.01 mg of selenium as sodium selenite per kilogram of diet will completely prevent pancreatic degeneration even when vitamin E levels were very high (100 IU/kg or more). However, when the vitamin E is at more normal levels (10–15 IU/kg), a level of 0.02 mg Se/kg of diet is required.

Chicks produced from hens fed a low-selenium, low vitamin E diet also had low levels of the activities of glutathione peroxidase in plasma and in pancreas when they hatched. Even though mitochondrial swelling is believed to be involved in the early stages of nutritional pancreatic atrophy, no significant differences were observed in the oxygen uptake, respiratory control index, or adenosine diphosphate to oxygen ratio between pancreatic mitochondria isolated from either selenium-deficient or selenium-adequate chicks.

9.3.6 Mulberry Heart Disease

Mulberry heart disease sometimes occurs together with hepatosis dietetica in pigs but sometimes itself causes sudden death. The clinical patterns are those of an apparently healthy animal suddenly developing heart failure and dying within hours. The disease has been called "mulberry heart disease" because the animals frequently have such extensive cardiac hemorrhage that the organ has a reddish-purple gross appearance comparable to that of a mulberry. Mulberry heart disease occurs in animals deprived of selenium and vitamin E. Administration of selenium increases tissue levels and prevents mulberry heart disease.

9.3.7 Reproductive Problems

In the South Island of New Zealand there is an embryonic mortality in ewes that often runs as high as 75%. This infertility tends to occur in those areas where congenital white muscle disease is also found. Even though many of the ewes conceive, many fail to produce a lamb. Embryonic death at about three to four weeks after conception is the cause of the infertility. When selenium is administered to the ewe before mating, this infertility is almost totally eliminated. Lambing percentages were increased from 25–90%. Decreased embryonic mortality was likely responsible for the increase of lambing percentages, but improvement in the ova fertilization rate is also possible. Improvement in ova fertility in Se + vitamin E-treated ewes may be due to an increase in the number of uterine contractions migrating toward the oviduct at mating. The increase in the number of contractions should enhance sperm transport toward the oviduct. In the same grazing area calving percentages appear satisfactory for cattle grazing side by side with sheep who are experiencing selenium-responsive infertility.

When selenium has been added to low-selenium, low vitamin E diets of chicks, improved egg production, hatchability, and growth have been observed. Similar results were observed in Japanese quail. Eggs from chickens fed commercial feedstuffs had about an equal selenium content of dried egg white and yolk.

In rats fed Se-deficient (<0.02 ppm) diets, sperm morphology and motility appear to be normal at four months, but half of the animals in the 11–12-month interval produced sperm with impaired motility and a characteristic breakage of the midpiece. Poor motility of spermatazoa from the cauda epididymus was observed in males born to females on a selenium-deficient diet. Some of the sperm showed breakage of the fibrils in the axial filaments. In many first-generation rats fed selenium-deficient diets for a year or more, aspermatogenesis has been observed.

The accumulation of ^{75}Se in the testes and epididymus has been observed in rats fed a low-selenium torula yeast diet. Autoradiographs showed that the ^{75}Se concentrated in the midpiece of the sperm. This suggests that there might be a specific need for selenium in the mitochondria that are exclusively located in the sperm midpiece. Subcellular fractionation showed that the mitochondria of the testes contained the greatest amount of selenium. Selenium in rat sperm is associated with a cysteine-rich 17,000-dalton structural protein of the mitochondrial capsule.

In the bull ^{75}Se retention in the epididymis was highly correlated with spermatozoal concentration. When ejaculated bovine spermatozoa were subjected to repeated freezing and thawing in distilled water, the selenium remained with the spermatozoa. This suggests that the selenium was bound tightly to the cell structural components. The flagellum of rat spermatozoa contains a specific selenopolypeptide that has been named "selenoflagellin," which may be an important protein in the formation and function of the flagellum.

Selenium and vitamin E have a beneficial effect on the incidence of retained placentas in dairy cows on low-selenium diets.

9.3.8 Myopathy of the Gizzard

Originally, gizzard myopathy was one of the most characteristic vitamin E deficiency symptoms in turkeys. The addition of sulfur-containing amino acids to the diet had no beneficial effect on these myopathies. Scott *et al.* (1967) have used a low-selenium turkey diet and a semipurified basal diet to produce a selenium deficiency in young turkey poults. Myopathies of the heart and of the gizzard as well as poor growth and mortality were observed in young poults. When vitamin E and methionine were added to the diet, improved growth was observed, but gizzard myopathy was not prevented. These results suggest that the lesion of gizzard myopathy and that of poor growth may result from different biochemical processes.

The order of prominence of the affected organs of the "selenium-responsive" diseases of the young poult appear to be (1) myopathy of the smooth muscle (gizzard); (2) myopathy of the myocardium; and (3) myopathy of the skeletal muscle. The primary nutritional factor required appears to be selenium with vitamin E of lesser significance, and sulfur amino acids are completely ineffective in preventing these myopathies in poults.

9.3.9 Growth

Several investigators have maintained rats on selenium-deficient diets that had been supplemented with adequate vitamin E for long periods. Growth impairment was the only consistent selenium-responsive abnormality reported.

In the offspring of selenium-deficient mothers, additional signs of selenium deficiency have been observed. The offspring tended to grow poorly and had sparse or absent hair. In the second-generation selenium-deficient rats, cataracts have been reported.

With *Candida utilis* as the protein source, several adult squirrel monkeys were fed a low-selenium semipurified diet with adequate vitamin E. Body weight loss, alopecia, and listlessness developed in the monkeys after they were on the diet for nine months.

9.3.10 Selenium-Responsive Unthriftiness of Sheep and Cattle

The most widespread and economically significant of all the selenium-responsive diseases of New Zealand livestock is selenium-responsive unthriftiness. This disease is characterized by an inability to maintain optimum rates of growth. Lambs may appear normal for several months and then may show markedly reduced weight gains; others stop eating, stop growing, lose weight, become dejected, and die. Diarrhea is not a constant symptom of this disease. The only post-mortem findings are advanced emaciation and osteoporosis.

In dairy and beef cattle the symptoms vary from a subclinical growth depression to a more severe syndrome characterized by a sudden and rapid progressive loss of health, profuse diarrhea, and sometimes high mortality. In adult cattle severe outbreaks occur on occasion. The autumn and winter months are the most critical times for this disease. Selenium can prevent the unthrifty condition in both sheep and cattle. Lambs that are treated produce up to 30% more wool than do those affected with unthriftiness and untreated. The effects of vitamin E and selenium-deficient diet have been tested against experimental swine dysentery. The incubation times were much shorter and the clinical symptoms were much less pronounced in the deficient diet than in the group supplemented with vitamin E and selenium. The treatment with vitamin E and selenium apparently greatly increases resistance to swine dysentery.

9.3.11 Periodontal Diseases of Ewes

Periodontal disease has been described in both the North and South Islands of New Zealand, where two long-term controlled selenium trials have been conducted. Both studies have shown that selenium administration will greatly reduce the incidence of periodontal disease. Since the treatment did not prevent periodontal disease completely, it appears that other factors may also be involved.

9.3.12 Encephalomalacia

Encephalomalacia is an ataxia resulting from hemorrhages and edema within and/or granular layers of the cerebellum with pyknosis. Antioxidants such as vitamin E and ethoxyquin have been shown to be effective against encephalomalacia. It is believed that the protection by antioxidants against peroxidation occurs through the neutralization by the antioxidant of lipoperoxides that are produced in an early stage of the auto-oxidation of fat. During the prevention of encephalomalacia in chicks, vitamin E functions as a biological antioxidant closely related to linoleic acid metabolism. Vitamin E may protect chicks against encephalomalacia by preventing the breakdown of linoleic acid to 12-oxo-*cis*-9-octadecenoic (keto) acid. This possibility is based on observations that α-tocopherol gave protection against encephalomalacia in the presence of 1.5% methyl linoleate, but not in the presence of 0.25% of the ketoacid.

Diphenyl-*p*-phenylenediamine, ethoxyquin, and methylene blue have also been shown to be effective in preventing encephalomalacia in chicks with very low intakes of vitamin E over prolonged periods. These antioxidants might prevent encephalomalacia by either replacing vitamin E in the metabolism of the animal or preventing destruction of traces of vitamin E in the diet or animal body. In general, selenium has been found to be ineffective in preventing encephalomalacia. In one experiment, the addition of 0.13 ppm of selenium as sodium selenite to a semisynthetic diet containing 4% corn oil stripped of tocopherol, reduced the incidence of encephalomalacia. But when 8% corn oil was fed, much greater amounts of selenium had no effect on the incidence of encephalomalacia. These results indicate that the protective action of selenium was small but significant when the dietary lipid stresses were borderline. The failure of other experiments to show a protective effect by selenium against encephalomalacia may have been due to their use of larger overwhelming levels of the stressor lipid, thereby masking any possible protection that added selenium might give.

9.4 Metabolism and Toxicity of Selenium

9.4.1 Absorption

Availability

Different amounts of selenium are absorbed from the gastrointestinal tract. The amounts absorbed vary with the species and with the chemical form and the amount of the element ingested. Cantor *et al.* (1975) have studied the biological availability of selenium in feedstuffs and selenium compounds for the prevention of exudative diathesis in chicks. In these experiments graded levels of selenium supplied as sodium selenite, which was the standard control, or by test ingredients were fed for 12–21 days. In most of the feedstuffs of plant origin selenium was highly available, ranging from 60 to 90%. These included wheat (70.7%), brewer's yeast (88.6%), brewer's grain (79.8%), corn (86.3%), soybean meal (59.8%), cottonseed meal (86.4%), dehydrated alfalfa meal (210.0%), and distiller's dried grains plus solubles (65.4%).

Selenium was less than 25% available from feedstuffs of animal origin, which included tuna meal (22.4%), poultry byproduct meal (18.4%), menhaden meal (15.6%), fish solubles (8.5%), herring meal (24.9%), and meat and bone meal (15.1%). High availability values were also found for sodium selenate and selenocystine, while low values were obtained for selenomethionine, sodium selenide, selenomethionine, and selenopurine. Elemental selenium in the gray form was almost completely unavailable. The activity of plasma glutathione peroxidase in chicks fed sodium selenite or selenomethionine were highly correlated with protection against exudative diathesis. The results of Cantor *et al.* (1975) suggest that biological availability is determined by the ability of the chick to utilize the various forms of selenium for enzyme activity.

The duodenum is the main site of absorption of physiological amounts of selenium as ^{75}Se and there is no absorption from the rumen or abomasum of sheep or the stomach of pigs. At dietary levels of 0.35–0.50 ppm Se, about 35% of the ingested ^{75}Se isotope was absorbed in sheep and 85% in pigs. In three young women, the intestinal absorption of (^{75}Se) selenite was 70, 64, and 44% of the dose (Thomson and Stewart, 1974). Other investigations have also observed that monogastric animals have a higher intestinal absorption of selenium than ruminants. It is likely that selenite is reduced in the rumen to insoluble compounds that are not as available for intestinal absorption.

If animals are selenium-deficient, radioselenium is more efficiently retained than in animals on a selenium-supplemented diet. This trend has been observed in chicks, rats, and sheep. This increased retention probably reflects greater tissue demands for selenium. The retention of selenium in several tissues and the urine and fecal losses have been measured following the administration of ^{75}Se to lambs fed graded levels of dietary selenium. The body loss of isotopic selenium was inversely proportional to the dietary level of selenium. The concentration of ^{75}Se in various tissues was also inversely related to the dietary selenium level.

Blood Distribution

After selenium is absorbed, it is first carried mainly in the plasma where selenium is apparently associated with the plasma proteins. They deliver selenium to all of the tissues, including the bones, the hair, and the red blood cells and the leukocytes.

In dogs, time-distribution studies of serum proteins have demonstrated that albumin is the immediate receptor of ^{75}Se. After some time, the selenium is released and is bound by other serum proteins, namely, alpha-2 and beta-1-globulins. The distribution of ^{75}Se among various electrophoretic serum protein fractions have been studied at different time intervals after the administration of the isotope to chicks. Within the first 2 hours after intubation with ^{75}Se, about 40% and 22% of the serum protein radioactivity was located in the alpha-2 and alpha-3 globulins, respectively. During the first 24 hours the percentage of serum proteins with radioactivity bound to the gamma globulin increased rapidly. During the next 24–173-hour interval, 52–70% of the ^{75}Se was carried by the alpha-2 and the gamma globulin fractions.

When ^{75}Se, as selenite, is added to human blood, it is rapidly taken up by the cells (50–70% within 1–2 minutes) and then is released into the plasma so that most of the radioactivity is in the plasma in 15–20 minutes. Analysis of the subfractions of plasma and cells from normal individuals have indicated that the highest selenium concentrations are in the plasma alpha- and beta-globulins.

The lipid fraction of serum lipoproteins also becomes labeled with ^{75}Se in both rat and man. Selenium-75 has also been found to be incorporated into the alpha- and beta-lipoproteins of rat and dog serum. The binding of ^{75}Se-selenite human plasma proteins has indicated that the most heavily labeled lipoproteins were the beta-lipoprotein and an unidentified fraction located electrophoretically between the alpha-1 and alpha-2 globulin fractions.

Enhanced *in vitro* uptake of ^{75}Se by the red blood cells has been observed in selenium-deficient sheep and in the red blood cells of children suffering from kwashiorkor. After subtoxic amounts of ^{75}Se were administered to dogs, the isotope could be detected in various blood proteins for as long as 310 days. The greatest disappearance of radioactivity from the red cells occurred at 100–120 days after the intitial injection. Apparently, once selenium enters these cells, it remains there throughout their life span.

Tissue Distribution

The intracellular distribution of radioselenium varies with the tissue and with the level of selenium. In the liver ^{75}Se was found to be distributed rather evenly among the particulate and soluble fraction in liver. In contrast, nearly 75% of the activity in kidney cortex was found in the nuclear fraction. The microsomes appear to be the initial site for the incorporation of selenium into protein.

Selenium occurs in all the cells and tissues of the body at levels that vary with the tissue as well as the level of dietary selenium. The kidney, and especially the kidney cortex, has by far the greatest selenium concentration, followed by the glandular tissues, especially the pancreas, the pituitary, and the liver. Tissues that are low in selenium include muscle, bone, blood, and adipose tissue. Cardiac muscle is consistently higher in selenium than skeletal muscle. The kidney and liver are the most sensitive indicators of the animal's selenium status. Normal concentrations are greater than 1.0 ppm of selenium in the kidney cortex and 0.1 ppm in the liver. About one-half of this quantity indicates a marginal selenium deficiency.

When intakes of selenium are toxic (i.e., 10–100 times or more greater than those normally ingested), tissue selenium concentrations rise steadily until levels as high as 5–7 ppm in liver and kidneys and 1–2 ppm in the muscles are reached. Beyond these tissue levels, excretion begins to keep pace with absorption. Selenium was not continuously accumulative beyond a certain tissue level.

9.4.2 Excretion

Selenium is removed from the body through excretion into the feces, the urine, and the expired air. The amounts and proportions that are excreted depend on the level and form of the intake, the nature of the diet, and the species. At higher dietary intakes of the element, exhalation of selenium is an important route of excretion, but exhalation of selenium is much less at low intakes. When the protein and methionine contents of the diet were increased, pulmonary excretion of selenium was increased. Fecal excretion of ingested selenium is generally greater than urinary excretion in ruminants, but not in monogastric species. When selenium was injected into sheep or other nonruminant species, the urine was the major pathway of excretion.

Increasing dietary levels of selenium did not substantially affect volatile urine or fecal excretion of ^{75}Se when it was administered orally. In contrast, increasing dietary selenium levels markedly increased the amounts of ^{75}Se in the urine or feces. The selenium in the feces mostly consisted of selenium that had not been absorbed from the diet, together with small amounts excreted into the bowel with the biliary, pancreatic, and intestinal secretions.

The effect of selenium metabolism by combinations of diets containing high- or low-protein or phosphorus diets has been studied by Greger and Marcus (1981), who conducted two 51-day human metabolic studies in eight adult males. Subjects lost significantly less selenium in their feces, but significantly more selenium in their urine, when they were fed high-protein, less-phosphorus and high-protein, high-phosphorus diets rather than low-protein, low-phosphorus and low-protein, high-phosphorus diets. Supplementation of the diets with methionine and cystine led to even more selenium loss in the urine.

In addition, selenium metabolism and excretion can be altered appreciably by the presence of sulfate. When rats were given sodium selenate parenterally followed by sodium sulfate parenterally or in the diet, the urinary excretion of selenium was increased almost threefold. Sulfate, however, only has a slight effect on the urinary excretion of selenium administered in the form of selenite.

9.4.3 Placental Transfer

In the mouse, rat, dog, and sheep, selenium has been shown to be transmissible through the placenta to the fetus, whether supplied in inorganic or organic form. The placenta apparently presents something of a barrier to the transfer of selenium in the inorganic form. However, when ewes are injected with ^{75}Se-selenomethionine, ^{75}Se-selenocystine, the ^{75}Se concentration in the lambs is higher than when selenite is injected, and it is nearly as high as in the mother.

9.4.5 Mechanism of the Antioxidant Action of Selenium

After it was observed that several vitamin E deficiency diseases were prevented by synthetic antioxidants and that there was increased lipid peroxidation in tissues of deficient animals, the biological antioxidant hypothesis was developed. Vitamin E and selenium have several specific metabolic functions, the most important of which is the protection of the biological membranes from lipid peroxidation. Figure 9-2 outlines the function of selenium and vitamin E in the inhibition of lipid peroxidation damage.

The main opposition to the biological antioxidant hypothesis has resulted from difficulties in detecting lipid peroxides in vitamin E-deficient animals. However, the existence of peroxides *in vivo* adipose has been reported (Chrapil *et al.,* 1974). In addition, numerous *in vitro* studies have demonstrated the protection by vitamin E and selenium against peroxidation of unsaturated membrane lipids under oxidizing conditions. Many investigators believe that controlled lipid peroxidation may be a continuous metabolic process in all tissues.

Plasma membranes from erythrocytes are quite labile to lipid peroxidation because of their high content of polyunsaturated fatty acids and because of their exposure to molecular oxygen. Hemolysis of the cells results from peroxidation of the erythrocyte membrane lipids. These membrane lipids can be more sensitive to peroxidation by feeding the animals increased levels of unsaturated fats, which increase the unsaturated fatty acid content of the erythrocyte, thus making it more sensitive to oxidative hemolysis. Increasing erythrocyte fragility has been associated with the anemias of vitamin E-deficient chicks, rats, monkeys, fish, and humans with protein–calorie malnutrition and premature infants.

In addition, dietary selenium has also been observed to prevent oxidative hemolysis in the vitamin E-deficient rat. Selenium is an essential constituent of the enzyme glutathione peroxidase, which is important in the metabolic destruction of peroxides, utilizing reducing equivalents from glutathione to reduce hy-

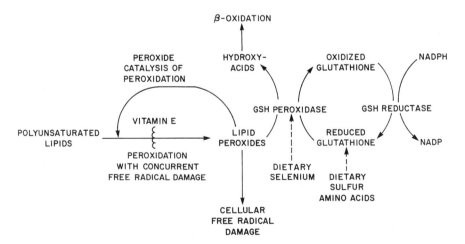

Figure 9-2. Function of selenium in the GSH peroxidation system in the inhibition of lipid peroxidation damage.

drogen peroxide as well as fatty acid hydroperoxides. In the absence of adequate glutathione peroxidase, hydrogen peroxide may react further with excess superoxide to produce the powerful hydroxyl radical, which is capable of initiating severe oxidative degradation of cellular components. Highly purified glutathione peroxidase has been shown to inhibit lipid peroxidations in hepatic mitochondria and microsomes.

The reactive superoxide anion (O_2^-) has been produced by many cellular oxidations that reduce oxygen univalently. If levels of superoxide are unchecked, lipid peroxidation is increased. Superoxide dismutase reduces superoxide levels to hydrogen peroxide, which in turn is reduced to water by catalase and peroxidases, including glutathione peroxidase.

The membranes of subcellular organelles, especially those of the mitochondria and microsomes of the liver, are labile to damage from lipid peroxidation. Lipid peroxidation in these membranes can result in structural damage that interferes with cell function. Severe disruption of normal cell function can result from peroxidative damage to lysosomal membranes.

Vitamin E and selenium have a cooperative function in protecting subcellular membranes in preparations of chick liver from lipid peroxidation. The protection of biological membranes appears to be due to (1) the action of vitamin E as a membrane-protecting antioxidant and (2) the prevention by glutathione peroxidase of hydroxyl radical formation and attack on unsaturated membrane lipids, and also possibly the destruction of any lipid hydroperoxides that may form.

Effect of Paraquat

Burk *et al.* (1980) have studied the effect of herbicide (paraquat and diquat) administration on liver necrosis and lipid peroxidation in the rat. Paraquat and diquat enhance formation of superoxide anion formation in biological systems, and lipid peroxidation has been postulated to be their mechanism of toxicity. Selenium deficiency enhances both paraquat and diquat toxicity. Both caused rapid and massive liver and kidney necrosis and very high ethane production rates in selenium-deficient rats. Injection of selenium protected against diquat poisoning and the concomitant lipid peroxidation and mortality. However, there was no increase in glutathione peroxidase activity of liver, kidney, lung, or plasma after 10 hours. Apparently, there is a selenium-dependent factor that exists in addition to glutathione peroxidase, and that factor either protects against peroxidation or some other unknown process takes place.

Effect of Cytochrome P-450

Selenium deficiency in rats has produced an increase in cytochrome P-450 content of only 70%, as compared to 150% in similarly treated controls. The induction of ethylmorphine demethylase activity by phenobarbital was also impaired in selenium-deficient rats. In another experiment, intraperitoneal injections of graded levels of selenium into adult male rats significantly increased heme oxygenase at all levels of the trace element.

Selenium and Hepatic Heme Metabolism

Maines and Kappas (1976) have found that selenium induced the mitochondrial enzyme delta-amino levulinate synthase, as well as the microsomal enzyme heme oxygenase in liver. Phenobarbital administration has been shown to stimulate heme synthesis in the liver in both the selenium-deficient and the selenium-adequate rats. The activity of microsomal heme oxygenase was increased six- to eight-fold by phenobarbital in selenium-deficient rats but not in controls. In addition to phenobarbital other agents, such as tryptophan and methemalbumin, which also raise heme content, have caused a greater stimulation of microsomal heme oxygenase activity in selenium-deficient livers than in controls.

9.4.5 Interactions of Selenium with Other Substances

Cadmium

Single subcutaneous injections of cadmium chloride in amounts much below toxic levels selectively damage the testis of rats and other laboratory animals. The site and mode of cadmium action have not been satisfactorily established, but it is believed that cadmium specifically damages the testicular artery–pampiniform plexus complex and its countercurrent exchange mechanism.

Administration of zinc salts has a marked protective effect against many of the acute lesions of cadmium toxicity, including the development of testicular damage and necrosis. Apparently, zinc and cadmium are able to compete with one another for binding sites for essential enzymes.

Administration of selenium compounds has been shown to protect effectively against toxic as well as lethal doses of cadmium. Although selenium protects effectively against cadmium-induced testicular injury, the cadmium content of the testis is actually increased, indicating that selenium causes a redistribution of cadmium in this organ. Marked diversions of cadmium from 10,000- and 30,000-MW proteins to large molecular weight proteins of 110,000 or 330,000 daltons were also observed. Since the target of the cadmium-induced testicular injury appears to be the 30,000 molecular weight protein, the diversion of the cadmium by selenium may be an important mechanism involved in the protection against cadmium-induced testicular injury. The diversion in the binding of cadmium in the soluble fraction has also been observed in the kidney and liver.

Arsenic

Arsenic given in the diet or the drinking water was first found to protect against the chronic toxicity of seleniferous grains or of selenite in rats and dogs. After injecting rats with single subtoxic doses of selenite, arsenic has been reported to increase the concentration of selenium in blood and to decrease it in the liver. In addition, the amount of selenium exhaled as volatile compounds following the injection of selenite was greatly decreased when arsenite was present. The formation of volatile compounds has long been regarded as a possible detoxification process. The compound exhaled is thought to be mainly dimethyl selenide, which suggests that this mode of selenium elimination involves a very active reduction of selenium. The reaction pathway seems to consist of reduction of selenite to the selenide oxidation state followed by methylation by methyl transferase enzymes. The microsomal fraction of the liver contains the methyl transferase, which is very sensitive to arsenite, which may account for the ability of arsenic to inhibit the production of volatile compounds.

The precise chemical mechanism by which arsenic detoxifies selenium is not known. It is possible that selenium and arsenic react in the liver to form a detoxification conjugate, possibly a selenoarsenite that is then excreted into the bile. Both arsenic and selenium increase each other's biliary excretion.

Copper

If high dietary levels of copper are added to a chick diet marginal in selenium, selenium deficiency signs result. Exudative diathesis and muscular dystrophy have been observed in chicks fed a high copper level. The exact mechanism by which copper contributes to selenium deficiency is not known. High dietary levels of copper probably reduce the availability of selenium in the tissue for synthesis of glutathione peroxidase by interfering with absorption and/or formation of insoluble intracellular selenium compounds. It is also known that copper inhibits yeast glutathione reductase. Glutathione reductase is thought to be necessary for release of reduced selenium from erythrocytes, an important step in the metabolism of intravenously injected selenite.

Silver

Administration of silver nitrate or silver lactate in the drinking water of rats fed a low vitamin E diet has resulted in muscular dystrophy, necrotic degeneration of the liver, and increased mortality. Selenium added to the diet counteracted the toxicity of silver. These results suggest that silver induced a conditional deficiency of selenium and made selenium unavailable for glutathione peroxidase activity.

Vitamin E was as effective as selenium in overcoming the growth depression produced by silver but did not prevent the depression of glutathione peroxidase. It appears that growth depression is not related to the silver-induced decrease in glutathione peroxidase.

Cobalt

Macroscopic lesions were not observed in pigs fed cobalt (500 mg/kg, as the chloride). However, evidence of selenium–vitamin E deficiency was indicated by microscopic necrosis of cardiac and skeletal muscle in 50% of the pigs fed cobalt. In addition, ducklings fed cobalt (200 or 500 mg/kg, as chloride) developed lesions characteristic of selenium–vitamin E deficiency, such as necrosis of skeletal and cardiac muscle, and of smooth muscle of the gizzard and intestine.

Lead

Chronic lead poisoning can affect the gastrointestinal, hematopoietic, or central nervous system, producing clinical signs accordingly. Autopsies of animals subjected to either chronic or acute lead toxicities showed central nervous system (CNS) and/or gastrointestinal lesions. Edema of the kidneys as well as abnormal erythrocyte metabolism and morphology have also been demonstrated. Erythrocytes from these animals have an increased mechanical fragility, decreased peroxidative fragility, osmotic fragility, and decreased filterability. Both selenium and vitamin E have been shown to be involved in decreasing the toxic effects of lead in rats. Vitamin E status, however, seems to be more important than selenium. The nature of the interaction of lead and selenium is probably not as complex or as extensive as those of mercury and selenium, cadmium and selenium, or arsenic and selenium.

Mercury

Selenite and to a lesser extent selenomethionine have markedly decreased the acute nephrotoxicity of mercuric mercury in rats, as long as the selenium compound was given after the mercury compound. If selenium was given before mercuric mercury, males showed an increased mortality.

In red blood cells, selenite accelerated the uptake of inorganic mercury. The major fractions of mercury and selenium were eluted in a high-molecular-weight fraction on gel filtration of rat plasma, as well as rabbit stroma-free hemolysate following simultaneous administration of mercuric chloride and selenite.

In a similar way, selenium has been observed to reduce the toxicity of methyl mercury when administered at the same time. Bis(methylmercuric) selenite (BMS) $(CH_3HG_2)Se$ has been reported to be formed from methylmercury and selenite in rabbit blood. BMS was also formed from the reaction of methylmercury and selenite in the presence of reduced glutathione (GSH). Since GSH exists extensively in animal tissues at high concentrations, there is a possibility that BMS is also formed in animal tissues from methylmercury and selenium. BMS formation could play an important role in the modifying effect of selenium on methylmercury toxicity.

Other Metals

Oral or parenteral administration of selenate has been reported to prevent death in rats due to thallium poisoning. The content of thallium was markedly increased in the liver, kidney, and bones by the selenate treatment.

Tellurium toxicosis resulted in the development of cardiac lesions in ducklings fed 500 ppm tellurium tetrachloride. The cardiac lesions resembled those in ducks fed diets inadequate in selenium and vitamin E. Anorexia, slowed growth, and a reluctance to stand affected many of the birds, and eventually many died. Lesions included myopathy of the gizzard, intestine, skeletal muscles, and heart. The birds fed tellurium had marked vascular injury, which included myocardial congestion, edema, and hemorrhage. The damaged hearts in these ducks fed tellurium appeared similar to the red "mulberry hearts" described in pigs with selenium–vitamin E deficiency.

9.4.6 Toxicity of Selenium

Introduction

Selenium is present at least in trace amounts in all soils and in all natural feeds, and its occurrence is not restricted to any specific area of the earth. Some soils contain an excess of selenium in forms which are available to plants. In this case plants absorb the selenium in amounts that make them toxic to animals. In acid soils the selenium is usually in a form that is not available to plants. The soils and plants with large available amounts of selenium have been referred to as "seleniferous" or "indicator plants," which implies excessive levels of the element. The indicator plants have been divided into two categories. The first category was called primary indicators and included over 20 species of Astragalus (1000 ppm Se) *Machaeranthera,* Oonoposis (800 ppm Se), *Haplopappus,* Zylorhiza (120 ppm Se), and Stanleya (700 ppm Se), which appear to require selenium for their growth. High levels of selenium normally accumulate in these species, sometimes as much as several thousand parts per million.

The second category is referred to as secondary indicators because they do not appear to require selenium for their growth. They will accumulate the element when they grow on soils of high available selenium content. They belong to a number of genera which include Aster (72 ppm Se), Atriplex (50 ppm Se), Gutierrezia (60 ppm Se), Castilleja, Comandra, and Grindelia (38 ppm Se).

Finally, there is a third group of plants that includes the grasses and grains. This group does not normally accumulate selenium in excess of about 50 ppm field conditions. When livestock, usually cattle, eat toxic plants, three types of selenium poisoning have been described: (1) acute; (2) chronic, blind staggers; (3) chronic, alkali disease.

Acute Toxicity

Acute poisoning comes about when high-selenium content (10,000 ppm Se) plants are consumed in large quantities. Usually animals avoid the indicator plants because they do not have a good taste. However, under poor grazing conditions brought about by bad weather or overgrazing of an area, cattle or sheep may sometimes be forced to eat these toxic indicator plants. The symptoms, which are severe, include abnormal movement and posture, anorexia, watery diarrhea, labored breathing, bloat, elevated temperature and pulse rate, prostration, and often death from respiratory failure. The pathological changes include hepatic necrosis, hemorrhage, edema, nephritis, and congestion and ulceration of numerous tissues. Because the highly seleniferous plants are not palatable, deaths from the acute type are quite rare. In individual herds a few cases of large doses have been observed.

Blind Staggers

Blind staggers is a more advanced type of chronic selenium poisoning that results from the consumption of primary indicator plants (100–10,000 ppm Se) in smaller amounts over a longer period of time. In Wyoming, animals are found from time to time to be acutely poisoned by selenium. In the early stages of blind staggers, cattle wander, stumble, have impaired vision, and lose their appetite for food and water. Later the front legs become weak, and if extreme enough, there is finally respiratory failure and paralysis preceding death. Other symptoms include stumbling over obstacles, weak forelegs, and paralysis of the tongue and throat. Lesions include hepatic necrosis, nephritis, hyperemia, and ulceration of the upper intestinal tract.

Alkali Disease

The first report of alkali disease in livestock was made in 1860 by Dr. Madison, an army surgeon stationed at Fort Randall, a territory of Nebraska. He described a fatal disease of horses that had grazed in a certain area near the fort. The horses lost the long hair from the mane and tail, and their feet became sore. In 1891 after farmers settled around Fort Randall, the farmers experienced difficulty with the same disease in their livestock. Experiments by Franke (1934) led to the discovery of selenium as the cause of alkali disease.

Chronic poisoning of the alkali disease type results from the continuous ingestion of feeds that contain more than 5 ppm but usually less than 40 ppm of selenium. Alkali disease is found primarily in South Dakota. In horses, cattle, and swine the most common symptoms of the poisoning have been described as follows: rough hair coat, lack of vitality, loss of appetite, malformation and sloughing off of hoofs, lameness due to joint erosion of the long bones, inflammation with swelling along the coronary band, and emaciation. In advanced cases, liver cirrhosis, atrophy of the heart, and anemia occur. A "bobtail" appearance can occur in horses if they lose the long hair from the mane and tail. In horses other lesions include alopecia and elongated and weak cracked hoofs. On occasion hogs lose their body hair and cattle lose the long hair from their tails. Sheep seem to be more tolerant and get a milder form of the disease. They do not have symptoms similar to those of cattle and horses. However, sheep may suffer a loss of appetite and show a reduced rate of weight gain. Inorganic selenium has been found to affect rats, dogs, swine, poultry, and the embryos of chicken eggs in a similar way as do seleniferous grains, causing the typical "alkali disease" symptoms.

In growing chicks, depressed rates of weight gain, roughed feathers, and characteristics of nervousness have been observed. In chickens and turkeys, egg production is delayed, and as a result of deformed embryos a lower percentage of hatchability is observed. The most common deformities are missing or short upper beaks, deformed feet and legs, edema of the head and neck, missing eyes, and wiry down. In sheep and pigs reduced reproducibility is also observed. From the economic standpoint, reduced reproductive performance may be the most significant effect of excessive selenium intake.

Effect of Diet on Toxicity

The effect of the protein content of diets containing toxic amounts of selenium has been studied. Selenium poisoning was found to be less severe on a diet with a high protein content. In general neither methionine nor cystine were themselves of value in reducing the toxicity of selenium.

Crude casein and linseed meal were found to be the best proteins in preventing the characteristic liver lesions, and both proteins provided an excellent growth response. In addition, 10% linseed oil has been shown to alleviate the toxic effect of 10 ppm selenium in white leghorn cockerel chickens. Palmer *et al*. (1980) have isolated two new cyanogenic glycosides, linustatin and neolinustatin, from linseed oil meal. Both substances gave significant protection against growth depression caused by 9 ppm sodium selenite. It is likely that there is a factor in linseed oil that may complex with selenium to form a less harmful compound in the tissues.

Biochemical Lesions

Traces of selenium will greatly reduce the fermentation of glucose by yeast and selenium has been shown to inhibit the succinic dehydrogenase of minced muscle and other tissues. Inhibition of succinic dehydrogenase seems to be an important factor in the toxicity of selenium. Selenium also exerts an inhibiting action on rat liver urease. It is likely that the inhibition of urease and succinic dehydrogenase by selenium is a result of a loss of sulfhydryl groups by oxidation. Both urease and succinic dehydrogenase are dependent on sulfhydryl groups for their activity. Methionine adenosyltransferase is also inactivated by selenium, but the exact mechanism remains to be established. It is possible that a reaction of selenite with sulfhydryl groups is involved.

Toxicity in Man

Selenium exposure may occur in the following list of occupations. This list is not meant to be all-inclusive, but it does represent most occupations in which selenium may pose an industrial threat: arc light electrode makers, copper smelter workers, electric rectifier makers, glass makers, organic chemical synthesizers, pesticide makers, makers of photographic chemicals, pigment makers, plastic workers, pyrite roaster workers, rubber makers, semiconductor makers, sulfuric acid makers, and textile workers.

Selenium enters the body when it is inhaled in the form of a dust or a vapor. Liquid solutions of selenium compounds can also pass through the skin. Finally, selenium compounds may be swallowed and exert their poisonous action by being absorbed from the digestive tract. Elemental selenium and some of its natural compounds are not particularly irritating to the body and are not well absorbed by the body. Certain chemical compounds of selenium, such as selenium oxychloride ($SeOCl_2$) and selenium chloride Se_2Cl_2, however, are strong vesicants. $SeOCl_2$ can blister and severely burn the skin and thus destroy it. Certain compounds, such as selenium dioxide and selenium oxychloride, are strong irritants, which are able to attack the upper parts of the respiratory tract and the eyes; these vesicants can also irritate the mucous lining of the stomach.

Certain selenium compounds may also cause dermatitis or inflammation of the skin if it is exposed to these compounds. An allergy to selenium dioxide may develop in some individuals; the general appearance is in the form of a rash of the urticarial type. The allergy to selenium dioxide may manifest itself by causing a pink discoloration of the eyelids and palpebral conjunctivitis. Another mode of entry for selenium dioxide is penetration under the fingernails, which can produce excruciatingly painful nail irritation and a painful condition called paronychia.

Hydrogen selenide poisoning is also known to occur, and the general effects are much like the effects seen after exposure to various other irritating gases used or released by industrial processes. The most common sign of hydrogen selenide poisoning is irritation of the mucous membranes of the nose, eyes, and upper respiratory tract. These responses are followed by a sensation of tightness in the chest. When the victim is removed from the area of exposure, these signs and symptoms disappear. If there is severe enough exposure, or if the individual has a special susceptibility to hydrogen selenide, pulmonary edema may suddenly develop.

Hydrogen selenide is considered 15 times more dangerous than hydrogen sulfide. Even though hydrogen sulfide claims its annual toll of deaths every year, hydrogen selenide has never caused a death or an illness lasting more than 10 days in a human being. There are two reasons for this lack of severe toxicity: first is the fact that in industry hydrogen sulfide is used by the tanker load, whereas hydrogen selenide is never used in quantity; the second reason is that hydrogen selenide is very easily oxidized back to the red selenium on the surface of the mucous membranes of the nose and probably in the alveoli of the lungs. It is likely that this quick breakdown of the hydrogen selenide to the harmless red elemental selenium has prevented death.

There have been no toxic incidents reported in the use of organic selenium compounds except where hydrogen selenide has suddenly been evolved during their synthesis. In addition, organic selenium compounds seem to be less toxic than the inorganic forms.

From the diagnostic point of view, the first and probably the most characteristic indication of selenium exposure and absorption by the human body is the garlic odor imparted through the breath of the victim. Excretion in the breath of small amounts of dimethyl selenide seems to be a cause of this odor. The odor disappears completely in somewhere between 7 to 10 days after the victim is removed from selenium exposure. Garlic breath is not an absolutely certain guide to the absorption of selenium. An early more subtle sign which many victims experience is a metallic taste of the mouth. There are also other fairly generalized effects of selenium exposure. They include pallor, lassitude, irritability, vague complaints of indigestion and giddiness. Vital organs do not oridinarily seem to be harmed by the absorption of selenium in man. However, results of animal experimentation suggest that the possibility of liver and kidney damage in human beings exposed to toxic levels of selenium compounds should not be completely discounted. Damage to the liver and other toxic effects are well recognized in livestock that grazed on plants containing high levels of selenium.

Permissible Limits for Selenium Exposure

The federal standards are: selenium compounds (as Se), 0.2 mg/m^2; selenium hexafluoride, 0.05 ppm, 0.4 mg/m^3; and hydrogen selenide, 0.05 ppm, 0.2 mg/m^3 (Wilber, 1980). In studies on 200–300 selenium workers over a period of 17 years, Glover (1967) found an average urinary level of 84 mcg/L. The highest figure was 490 mcg of selenium per liter for a selenium worker who had inhaled selenium dust. From these studies Glover suggested that in public, rural and industrial health situations, the selenium urinary level should be below 100 mcg/L.

9.5 Summary

This chapter on the biochemistry of selenium includes the chemical properties and occurrence of selenium; selenium and its compounds in cells and tissues; selenoproteins; and seleno-tRNAs. The vitamin E and selenium-related deficiency diseases are outlined: dietary liver necrosis, hepatosis dietetica, nutritional muscular dystrophy, exudative diathesis, pancreatic degeneration, mulberry heart disease, reproductive problems, gizzard myopathy, growth, unthriftiness of sheep and cattle, periodontal disease of ewes, and encephalomalacia.

The metabolism of selenium is also described including availability, blood distribution, tissue distribution, excretion, placental transfer, antioxidant action mechanism, and interaction of selenium with other substances. The toxicity of selenium includes acute toxicity, blind staggers, alkali disease, dietary effects, biochemical lesions, and its toxicity in man.

References

Burk, R. F., Lawrence, R. A., and Lane, J. M., 1980. Liver necrosis and lipid peroxidation in the rat as the result of paraquat and diquat administration. Effect of selenium deficiency, *J. Clin. Invest.* 65:1024–1031.

Cantor, A. H., Scott, M. L., and Noguchi, T., 1975. Biological availability of selenium in feedstuffs and selenium compounds for prevention of exudative diathesis in chicks, *J. Nutr.* 105:96–105.

Chvapil, M., Peng, Y. M., Aronson, A. L., and Zakowski, C., 1974. Effect of zinc on lipid peroxidation and metal content in some tissues of rats, *J. Nutr.* 104:434–443.

Franke, K. W., 1934. A toxicant occurring naturally in certain samples of plant foodstuffs. I. Results obtained in preliminary feeding trials, *J. Nutr.* 8:596–608.

Glover, J. R., 1967. Selenium in human urine: A tentative maximum allowable concentration for industrial and rural populations, *Ann. Occup. Hyg.* 10:3–14.

Greger, J. L., and Marcus, R. E., 1981. Effect of dietary protein, phosphorus, and sulfur amino acids on selenium metabolism of adult males, *Ann. Nutr. Metab.* 25:97–108.

Horn, M. J., and Jones, D. B., 1941. Isolation from *Astragalus pectinatus* of a crystalline amino acid complex containing selenium and sulfur. *J. Biol. Chem.* 139:649–660.

Maines, M. D., and Kappas, A., 1976. Selenium regulation of hepatic heme metabolism: Induction of delta-aminolevulinate synthase and heme oxygenase, *Proc. Nat. Acad. Sci.* 73:4428–4431.

Palmer, I. S., Olson, O. E., Halverson, A. W., Miller, R., and Smith, C., 1980. Isolation of factors in linseed oil meal protective against chronic selenosis in rats. *J. Nutr.* 110:145–150.

Peterson, P. I., and Butler, G. W., 1962. The update and assimilation of selenite by higher plants, *Aust. J. Biol. Sci.* 15:126–146.

Rotruck, J. T., Hoekstra, W. G., and Pope, A. L., 1971. Glucose dependent protection by dietary selenium against haemolysis of rat erythrocytes *in vitro Nature (Lond.) New Biol.* 231:223–224.

Schwarz, K., and Foltz, C. M., 1957. Selenium as an integral part of factor 3 against dietary necrotic liver degeneration, *J. Am. Chem. Soc.* 79:3292–3293.

Schwarz, K., 1965. The role of vitamin E, selenium and related factors in experimental liver disease, *Fed. Proc. Fed. Am. Soc. Exp. Biol.* 24:58–67.

Schwarz, K., Porter, L. A., and Fredga, A., 1972. Some regularities in the structure–function relationship of organoselenium compounds effective against dietary liver necrosis, *Ann. N.Y. Acad. Sci.* 192:200–214.

Scott, M. L., Olson, G., Krook, L., and Brown, W. R., 1967. Selenium-responsive myopathies of myocardium and of smooth muscle in the young poult., *J. Nutr.* 91:573–583.

Thomson, C. D., and Stewart, R. D. H., 1974. The metabolism of (^{75}Se) in young women, *Br. J. Nutr.* 32:47–57.

Trelease, S. F., DiSomma, A. A., and Jacobs, A. L., 1960. Seleno-amino acid found in *Astragalus bisulcatus, Science* 132:618.

Virupaksha, T. K., Shrift, A., and Tarrer, H., 1966. Biochemical differences between selenium accumulator and non-accumulator species, *Biochim. Biophys. Acta* 130:45–55.

Wilber, C. G., 1980. Toxicology of selenium: A review, *Clin. Toxicol.* 17:171–230.

Vanadium   10

Barbara J. Stoecker and Leon L. Hopkins

10.1 Introduction and History

10.1.1 Discovery and History

Vanadium was unequivocally discovered in 1830 by Sefström in Sweden. Sefström named the element after the Norse goddess of beauty, Vanadis, because of the rich colors of its derivatives. H. E. Roscoe made major contributions to the early study of vanadium and published numerous papers from 1868 to 1871 that provided the basis for subsequent work (Clark, 1973).

10.1.2 Occurrence and Distribution

The average concentration of vanadium in the earth's crust is approximately 150 mg/kg. Vanadium is found in more than 50 mineral species, and crude oils usually have high vanadium concentrations. In the soil, vanadium occurs as a relatively insoluble salt and is often in the trivalent state (Byerrum *et al.*, 1974).

Vanadium is found in both fresh and sea water. In fresh water, vanadium is usually in the pentavalent state. Usual levels have been reported to vary from 0.3 to over 20 µg/liter; however, as much as 220 µg/liter of vanadium has been reported in the water from one area in the western United States. In the sea, vanadium salts precipitate and result in marine muds very high in vanadium.

Barbara J. Stoecker and Leon L. Hopkins • Department of Food and Nutrition, Texas Tech University, Lubbock, Texas 79409.

10.1.3 Nuclear and Chemical Characteristics

Vanadium is element number 23 and has only two naturally occurring isotopes. ^{51}Vanadium occurs with 99.76% frequency and ^{50}vanadium with 0.24% frequency. Various other nuclides are available and ^{48}vanadium with a half-life of 16.1 days is most frequently used in biological research (Clark, 1973).

Vanadium is positioned in the Vb group of the first transition series. Vanadium occurs in compounds in the 0, +2, +3, +4, and +5 oxidation states but is most stable at the +4 or +5 valence. In alkaline solutions vanadate is usually bound to oxygen and forms anionic species. In neutral and acidic solutions, the cation VO_2^+ predominates (Meisch and Bielig, 1980). If the concentration of vanadium is less than $10^{-6}M$ or if ligand binding sites are plentiful, polymer formation is minimized. Both of these criteria are met in physiological systems (Rubinson, 1981).

A large proportion of the vanadium present in physiological systems is bound to coordinating groups such as $-SH$, $-SS-$, $-OH$, $-N-$, $-COO-$ and phosphate. However, extracellular vanadium must compete with very high levels of calcium and magnesium for ligand sites and intracellular vanadium competes with magnesium for binding sites. Extracellularly, vanadium appears to be in the oxidized or vanadate(V) form with intracellular vanadium in the reduced vanadyl(IV) form (Rubinson, 1981).

10.1.4 Essentiality

Vanadium has been shown to be nutritionally essential for both the chick and the rat (Hopkins and Mohr, 1971; Strasia, 1971; Hopkins and Mohr, 1974; Nielsen and Ollerich, 1973). Chicks fed diets low in vanadium (10 ppb) demonstrated reduced growth of wing and tail feathers when maintained on purified diets (Hopkins and Mohr, 1971). Rats fed purified diets low in vanadium responded to dietary vanadium supplementation with increased weight gain (Strasia, 1971) and with improved reproductive performance (Hopkins and Mohr, 1974). Essentiality of vanadium for man has been hypothesized but not demonstrated (Hopkins and Mohr, 1974).

10.2 Vanadium in Tissues

10.2.1 Vanadium in Plants and Plant Products

Vanadium has been analyzed using neutron activation analysis in several types of fruits and vegetables grown in Sweden. Most sample means ranged from less than 0.027–2.1 ng/g wet weight, although certain vegetables appeared to have particularly high concentrations of vanadium (Söremark, 1967). Vanadium levels of fruits and vegetables grown in the United States were found to be <1–5 ng/g by flameless atomic absorption (Myron *et al.*, 1977).

The vanadium content of unprocessed cereal grains has been reported to be in a similar range using both catalytic and flameless atomic absorption techniques. Vanadium levels of <6.5–20 or <1–14 ng/g were found (Welch and Cary, 1975; Myron *et al.*, 1977). Legumes tended to contain more vanadium than other plants (Byerrum *et al.*, 1974). Wheat, barley, oats, and peas grown in nutrient solutions contained 28–55 ng/g vanadium and their vanadium content was increased when vanadium was added to the solutions (Welch and Cary, 1975).

Both analytical difficulty and geographical variations may contribute to differences observed in vanadium content. Vanadium content of fruits and vegetables grown in two locations in the northeastern United States tended to be lower than for the same species grown in Sweden (Söremark, 1967). Welch and Cary (1975) also suggested that the concentrations of vanadium in seeds such as wheat may be influenced by geochemical factors and varietal differences.

10.2.2 Vanadium in Tunicates, Crustaceans, Shellfish, and Fish

Several reports have indicated that some Ascidia have extremely high blood concentrations of vanadium (3–1900 µg/g). High levels of vanadium also occur in some holothurians and mollusks (Underwood, 1977). In tunicates vanadium is held within specialized blood cells against a concentration gradient of as much as 10^{+6}- to 10^{+7}-fold. The mechanisms for vanadium accumulation apparently involve facilitated diffusion across the blood cell plasma membrane followed by the reduction of vanadate to a nontransportable cation. Electron paramagnetic resonance signals indicate that the vanadate is not bound to a protein or macromolecule within the cell (Dingley *et al.*, 1981).

Vanadium concentrations of various shellfish and one tunicate from different locations in Japan ranged from 25 to 328 ng/g wet weight when determined by flameless atomic spectrophotometry (Ikebe and Tanaka, 1979). Additionally, a level of vanadium of 7.4 µg/g was found in the viscera of one sample of top shell. The vanadium concentrations of several species of fish were lower than for most shellfish and ranged from nondetectable (<20 ng/g with their method) to 112 ng/g.

Further research is needed to clarify the extent to which geographic location and species contribute to the highly variable vanadium levels. Ikebe and Tanaka (1979) also suggested the need to clarify whether the relatively high levels of vanadium in many marine products occur naturally or whether they result from contamination due to crude oil pollution.

10.2.3 Human Intakes of Vanadium

Vanadium concentrations of 1–22 ng/g for various meat and poultry products have been estimated using flameless atomic absorption spectroscopy (Myron *et al.*, 1977). Calf liver from two locations averaged 2.4 and 10 ng/g wet weight by neutron activation analysis (Söremark, 1967). The vanadium concentration of milk has been analyzed and found to be 0.07–0.11 ng/g (Söremark, 1967) or 3 ng/g (Myron *et al.*, 1977), again with differences in location and analytical methods.

Considerable variation exists in vanadium levels reported for vegetable oils and similar products. Myron and colleagues (1977) using flameless atomic absorption found oils, butter, and margarine to contain 1–4 ng vanadium/g while Welch and Cary (1975) found levels of 14–139 ng/g using a catalytic method. Analytical techniques, geographic location, and manufacturing processes may contribute to the observed differences. Earlier reports of extremely high levels of vanadium in some oils (Schroeder *et al.*, 1963) now appear to have been due to analytical difficulty.

Certain beverages may contribute significantly to the human intake of vanadium. Drinking water has been reported to vary from 0.04 to 0.85 ng/g (Söremark, 1967) and to 1.8 ng/g (Myron *et al.*, 1977) in different locations and with different analytical techniques. The vanadium content of several prepared foods ranged from 11 to 93 ng/g suggesting the introduction of vanadium into products during processing (Myron *et al.*, 1977).

Nine institutional diets were analyzed for vanadium using an atomic absorption spectrometer with graphite furnace and deuterium-arc background correction (Myron *et al.*, 1978). Vanadium concentrations of these diets ranged from 0.019 to 0.050 µg/g dry weight. On a caloric basis the mean vanadium content of the diets was 8.9 µg/1000 kcal. The mean daily intake of vanadium from these diets was calculated to be 20.4 µg with a range of 12.4–30.1 µg/day. Myron and co-workers (1978) suggested that the vanadium content of these diets may not be adequate if the vanadium requirement of the human being is similar to that of the laboratory animal.

10.2.4 Vanadium Levels in Human Beings

Vanadium concentration in the tissues of human beings is very low. Reported values for various tissues, including liver, spleen, brain, muscle, testis, pancreas, prostate gland, lung, lymph nodes, and dental enamel, are in the range of 0.01–0.6 µg vanadium/g wet weight. The vanadium content of the lung tends to be higher than that of many other organs, probably because of air pollution (Underwood, 1977).

Estimates of vanadium levels in blood and serum have declined with better analytical capability; however, there are still severalfold differences between laboratories. Environmental factors may influence the vanadium concentrations but many of the analyses have been performed very near detection limits of the instruments or methods employed. Serum vanadium levels reported recently from several laboratories have been summarized (Cornelis and Versieck, 1982). By neutron activation analysis mean vanadium levels from 0.031 ± 0.010 µg/liter to 6.6 ± 3.0 µg/liter were estimated. By flameless atomic absorption spectroscopy, a value of 3.35 ± 0.63 µg/liter was reported. Serum vanadium levels were elevated in patients with chronic renal disease and in vanadium-exposed workers.

10.3 Vanadium Deficiency and Function

10.3.1 Growth

Chicks fed diets containing less than 10 ppb vanadium showed reduced wing and tail feather growth (Hopkins and Mohr, 1971). Reduced weight gain and abnormal bone development were subsequently observed in chicks fed diets containing 30–35 ppb vanadium (Nielsen and Ollerich, 1973). Significant growth stimulation in rats was reported when 0.5 ppm vanadium was added to a diet containing less than 100 ppb vanadium (Strasia, 1971). Similarly, in rats raised in a trace element-controlled environment, Schwarz and Milne (1971) found growth stimulation from adding 250 or 500 ppb vanadium to the highly purified basal diet, which was presumably low in vanadium.

10.3.2 Reproduction

Fertility was reduced in rats fed diets containing less than 10 ppb vanadium for three or four generations. Number of pregnancies per mating period was reduced and mortality of the pups was markedly increased in the vanadium-deficient compared to the vanadium-supplemented animals (Hopkins and Mohr, 1974).

10.3.3 Nutritional Edema

Nutritional edema has been hypothesized to be the result of sodium retention due to the effects of vanadate deficiency on the renal sodium pump (Golden and Golden, 1981). Vanadate infusions cause natriuresis and diuresis (Balfour *et al.*,

1978; Grantham, 1980). Perhaps any disease that causes vanadium depletion could cause sodium and water retention. Golden and Golden (1981) reviewed work that reported lower vanadium concentrations in the serum of children with kwashiorkor compared to control children. The reported values were higher than those obtained with improved methodology, but the relative levels of vanadium in the two groups are of interest.

10.3.4 Manic-Depressive Illness

Vanadium has been hypothesized by Naylor and co-workers to be an etiological factor in manic-depressive illness. In manic-depressive psychosis, the control of erythrocyte sodium concentration is abnormal. These changes may be secondary to decreases in (Na^+, K^+)-ATPase activity (Naylor and Smith, 1981). They assumed that the reduced sodium pump activity could be a response to elevated intracellular vanadate; therefore, treatments to reduce cell vanadate were investigated. Manic-depressive patients were treated with a 3-g dose of either vitamin C or parenteral ethylene diaminetetraacetic acid (EDTA). Decreased seriousness of illness was reported with treatment. Naylor and Smith (1981) suggested that large doses of vitamin C might alter the redox state so that more vanadyl was formed from vanadate and that EDTA could reduce absorption of vanadium.

Lithium is an established treatment for manic-depressive psychosis, and lithium has been reported to diminish the inhibition of (Na^+, K^+)-ATPase by vanadate (Naylor and Smith, 1981). However, MacDonald and co-workers (1982) suggested that the role of (Na^+, K^+)-ATPase has been overemphasized as a factor in the etiology of affective disorders. They found no protective effect of lithium against vanadate-induced inhibition of (Na^+, K^+)-ATPase and cautioned against any attempts to lower tissue vanadium levels until the "vanadate hypothesis of affective disorders" is critically evaluated.

10.3.5 Dental Caries

Data relating vanadium and dental caries remain equivocal and have been reviewed by Underwood (1977). Several epidemiological studies have reported lower caries incidence in areas with higher levels of vanadium in the drinking water. Some studies with rats, guinea pigs, and hamsters have indicated protection against caries with the administration of vanadium. However, other researchers have found no benefit or even a detrimental effect of vanadium supplementation on dental caries. Radiovanadium in the tooth was concentrated in the dentine.

10.3.6 Inotropic Effects of Vanadium

Relatively high levels of vanadate affect the force of contraction of cardiac muscle. Increasing ammonium vanadate (20–500 μM) produced a concentration-dependent increase in the force of contraction of electrically driven papillary muscle isolated from the right ventricles of cats. This effect was not due to the NH_4^+ ion nor to the stimulation of B-adrenoceptors (Hackbarth et al., 1978). However, inotropic effects of vanadium vary with the species, tissue, concentration, and form of vanadium, presence of other ions, and activity of adenylate cyclase. Negative inotropic effects have also been reported (Ramasarma and Crane, 1981). Erdmann and co-workers (1980) have reported a NADH–vanadate oxidoreductase that converts vanadate(V) to vanadyl(IV). The inhibitory effects of vanadyl ions on (Na^+, K^+)-ATPase, for example, are much less than the effects of vanadate on this ATPase. Thus, NADH–vanadate oxidoreductase could regulate vanadate-sensitive enzymes such as (Na^+, K^+)-ATPase.

10.3.7 Vanadium and Renal Function

Vanadate(V) is concentrated in the renal cells and inhibits (Na^+, K^+)-activated renal ATPase (Grantham, 1980). Therefore, possible natriuretic and diuretic effects of vanadate were investigated in rats under ether anesthesia. Intravenous injection of 1.5 μmol sodium or orthovanadate resulted in 9–25 fold increases in urine flow compared to flow rates prior to injection. An increase in urine flow of this magnitude indictated that one site of vanadium action was the proximal tubule. Increases in sodium excretion were of a similar magnitude to the increases in urine flow (Balfour et al., 1978). In rats natriuresis and diuresis were especially apparent when serum K^+ was elevated. However, vanadium infused in dogs and cats causes primarily vasoconstriction and does not produce the marked diuresis seen in rats (Grantham, 1980).

Vanadate has been suggested as a candidate for the cellular control mechanism of the sodium pump. Several factors affecting the inhibition of the pump by vanadate have been reviewed. For example, higher external K^+ intensifies inhibition by vanadium. Reduction of external sodium to below physiological levels likewise increased inhibition of the pump by vanadium. Inhibition of the pump required internal magnesium. Vanadium as vanadyl(IV) was much less effective than vanadate in inhibiting the pump and the inhibition was from the cytoplasmic side of intact cells. The balance between vanadate(V) and vanadyl(IV) may be the prime mechanism for control of the sodium pump (Grantham, 1980).

10.3.8 Vanadium and Glucose Metabolism

Vanadium compounds have been demonstrated to have effects on glucose metabolism *in vitro*. In systems of rat adipocytes the optimal concentration of vanadate ions in the medium stimulated glucose oxidation to a level equivalent to that achieved with optimal concentrations of insulin (Shechter and Karlish, 1980; Dubyak and Kleinzeller, 1980). If adipocytes were pretreated with the maximally activating concentration of insulin, no additional stimulation of glucose oxidation was seen with the addition of vanadate (Dubyak and Kleinzeller, 1980). When the glucose concentration of the medium was raised, the degree of stimulation by either insulin or vanadate decreased, indicating that they activated a glucose transport system that had become saturated (Shechter and Karlish, 1980). Inhibition of (Na^+, K^+)-ATPase was not the mechanism for the insulin-mimetic effects of vanadate. Vanadate was still effective in a K^+-free medium (Shechter and Karlish, 1980), no loss of cellular potassium was observed, and uptake of $^{86}Rb^+$ showed no inhibition of (Na^+, K^+)-ATPase function (Dubyak and Kleinzeller, 1980). The vanadyl ion appeared to be the principal form of vanadium involved in stimulation of glucose oxidation. Dubyak and Kleinzeller (1980) favored the hypothesis that vanadate entered the cell and was altered or sequestered within since vanadate transport remained linear for long periods of time. They further suggested that intracellular vanadate was reduced to vanadyl(IV). Shechter and Karlish (1980) found that addition of glutathione (GSH) to the medium lowered the concentration of vanadate required for half-maximal stimulation by almost an order of magnitude. They reported that GSH entered the adipocytes and supplemented the endogenous reduction of vanadate.

10.3.9 Vanadium and Lipid Metabolism

Several reports have indicated a role for vanadium in lipid metabolism but the results have frequently been contradictory (Hopkins and Mohr, 1974; Underwood, 1977; Ramasarma and Crane, 1981). Lowered serum cholesterol levels were initially reported in vanadium-deficient chicks at 4 weeks but at 7 weeks the serum levels of the deficient birds were elevated compared to vanadium-supplemented controls. Other scientists observed elevated serum cholesterol levels in chicks by 4 weeks on a vanadium-deficient diet. On the other hand, supplementation of diets with the near toxic level of 100 ppm vanadium increased liver and plasma cholesterol. Recently the activity of microsomal 3-hydroxy-3-methylglutaryl (HMG) CoA reductase and its inactivating enzyme were reported to be inhibited by vanadate (Menon *et al.*, 1980). The inactivating enzyme was more sensitive to vanadate and was largely inhibited by 1 mM vanadium while 10 mM vanadate was required for inhibition of the HMG CoA reductase. Since HMG CoA reductase is the rate-limiting enzyme in cholesterol biosynthesis in the liver, cholesterol synthesis could increase with low concentrations of vanadium and decrease with higher vanadium concentrations (Menon *et al.*, 1980). Relative activity of these enzymes might explain some of the contradictory results seen with lower levels of supplementation. However, interpretation of the elevated cholesterol levels seen with pharmacological levels of vanadium remains difficult.

Lipid peroxidation was accelerated by both sulfite and vanadium *in vitro*. Increased lipid peroxidation was inhibited by free-radical scavengers such as hydroquinone, α-tocopherol, and superoxide dismutase (Ramasarma and Crane, 1981).

10.3.10 Vanadium and ATPases

In 1977 a (Na^+, K^+)-ATPase inhibitor that had been noted in several laboratories was identified as vanadate (Cantley *et al.*, 1977). Interest in inhibition by vanadium has been keen since vanadate levels in mammalian tissues appear to be in the appropriate range to inhibit a significant proportion of (Na^+, K^+)-ATPase *in vivo* (Cantley *et al.*, 1977).

Using red cell ghosts, Cantley and colleagues (1978) concluded that vanadate equilibration was catalyzed by the anion-exchange system common to phosphate and that vanadate inhibited (Na^+, K^+)-ATPase from the cytoplasmic side. Subsequently it was reported that most cytoplasmic vanadate was in the $+4$ oxidation state as vanadyl ions associated with large molecules presumably hemoglobin. Vanadium in the $+4$ oxidation state was less effective than in the $+5$ state for inhibiting the (Na^+, K^+)-ATPase. Thus a reduction of vanadate could reduce the inhibition of (Na^+, K^+)-ATPase (Cantley and Aisen, 1979).

(Ca^{2+}, Mg^{2+})-ATPase from brain tissue was inhibited by vanadate and the inhibition was increased by both KCl and dimethylsulfoxide. Potentiation of vanadate inhibition by K^+ suggested similarities between the brain (Ca^{2+}, Mg^{2+})-ATPase and both plasma membrane (Na^+, K^+)-ATPase and the sarcoplasmic reticulum (Ca^{2+}, Mg^{2+})-ATPase (Robinson, 1981). Vanadium inhibits a number of other ATPases and many of these reactions have recently been reviewed by

Ramasarma and Crane (1981). Inhibition of the ATPases occurred at differing speeds in various tissues and was influenced by a range of vanadium and other ion concentrations.

10.3.11 Additional Effects of Vanadium

Vanadyl(IV), but not vanadate(V) was found to enhance oxygen binding to hemoglobin and myoglobin *in vitro*. Enhancement of oxygen has been postulated to be due to the direct redox reaction of vanadyl ion and ferric heme. The resultant ferrous form was readily bound to oxygen (Sakurai *et al.*, 1982). Sakurai and co-workers (1982) suggested that enhancement of oxygen binding to heme protein may be an important physiological role of vanadyl ion.

In erythrocytes depleted of ATP, vanadate induced a 10–15-fold increase in K^+ permeability. The effect on K^+ permeability occurred after a lag phase that would have allowed conversion of vanadate(V) to vanadyl(IV). Siemon and co-workers (1982) suggested that VO^{2+} might be the form that opens the "potassium channel" in the membrane.

Additionally, vanadium has a wide range of other biological activities. Effects of vanadium on ribonuclease, alkaline phosphatase, adenylcyclase, NADH oxidase, and phosphofructokinase are among the functions recently reviewed (Ramasarma and Crane, 1981).

10.4 Vanadium Metabolism

10.4.1 Absorption of Vanadium

Tissue absorption of vanadium from oral doses has been reported to be low. However, there is little information on the absorption and distribution of vanadium from different salts. Parker and Sharma (1978) gave 5 or 50 ppm vanadium as vanadyl sulfate or sodium orthovanadate *ad libitum* for 3 months in the drinking water of rats being fed lab chow. Tissue concentrations of rats fed 5 ppm of either of the vanadium salts were similar to control; however, the tissues of animals fed 50 ppm had markedly elevated vanadium concentrations. The highest concentration was in the kidney, followed by bone, liver, and muscle. Tissue concentrations of vanadium in animals given 50 ppm sodium orthovanadate were higher than in animals receiving similar doses of vanadyl sulfate. These differences presumably were due to differences in solubility, in absorption, or in metabolism, especially elimination, of the compounds. The vanadium content of tissues tended to plateau after 3 weeks on the feeding regimen except for the kidney where the concentration increased until the ninth week of the study.

Since Al(OH)$_3$ restricts absorption of phosphorus and since the oxy anions of vanadium and phosphorus are very similar, the effects of Al(OH)$_3$ on intestinal absorption of vanadium were examined. Accumulation of ^{48}vanadium in the tissues was reduced in rats gavaged with 1 ml of Al(OH)$_3$ containing 5 μmol Na$_3$VO$_4$ and 1 μCi ^{48}vanadium compared to animals receiving the dose in 1 ml of saline. Tissue levels of ^{48}vanadium were consistently higher in controls than in Al(OH)$_3$-treated animals. Equivalent doses of Al(OH)$_3$ had no effect on tissue levels of ^{48}vanadium injected intraperitoneally; thus it appeared that Al(OH)$_3$ reduced intestinal VO$_3$ absorption (Wiegmann et al., 1982).

10.4.2 Tissue Distribution of Vanadium

Trace amounts of vanadium injected intravenously as either the (III), (IV), or (V) oxidation state accumulated in the liver, kidney, spleen, and testis up to 96 hours. Vanadium concentration in other tissues measured was declining at this time. Subcellular analyses of the liver indicated that 43% of the vanadium was found with the nuclei and debris, 37% with the mitochondria, 10% with the microsomes, and 11% with the supernatant (Hopkins and Tilton, 1966).

Accumulation of vanadium in the tissues of rats seemed to be directly related to the level of intraperitoneal dose of vanadyl trichloride in the range of 0.1 to 2 mg/kg body weight. At the high level of 8 mg/kg body weight, vanadium accumulated more rapidly than predicted in deposition organs such as bone, kidney, and liver. Overall, ^{48}vanadium was distributed in the following order: bone > kidney > liver > spleen > intestine > stomach > muscle > testis > lung > brain. Subcellular distribution analysis indicated that vanadium was associated with the high-molecular-weight proteins in the soluble fraction of liver and was also associated with nuclei, mitochondria, and microsomes (Sharma et al., 1980).

10.4.3 Effects of Hormones on Vanadium Metabolism

Vanadium metabolism was disturbed in hypophysectomized (HYPOX) or thyroid-parathyroidectomized (TPTX) rats. In HYPOX rats, tissue–serum ^{48}vanadium ratios of pancreas, kidney, and testis were increased while the ratios in liver, heart, and bone decreased compared to controls. TPTX caused an increase in tissue–serum ^{48}vanadium ratios in pancreas and kidney of similar magnitude to the change induced by HYPOX. Adrenal ablation did not appear to influence ^{48}vanadium distribution. The changes in the kidney and pancreas seemed to be due to thyroid deficiency while changes in other tissues were due to hypopituitarism (Peabody et al., 1976).

In a subsequent study, the effects of supplementation with bovine growth hormone (BGH), BGH plus thyroxine (T_4), or T_4 alone were compared in HYPOX rats and controls. BGH alone had a partial restorative effect on the tissue–serum ratio of ^{48}vanadium in kidney and bone of HYPOX rats. T_4 alone partially restored the tissue–serum ratio in kidney and pancreas and returned the ratio in bone to control values. The combination of BGH and T_4 reduced ratios in both kidney and pancreas to levels statistically indistinguishable from controls. Neither hormone reduced the elevated ^{48}vanadium levels in the testis of HYPOX rats to control values.

10.4.4 Excretion of Vanadium

The main route of excretion of intravenously injected ^{48}vanadium was the urine. At 96 hours, 46% of the dose had been excreted in the urine and 8.6% in the feces (Hopkins and Tilton, 1966). Bile seemed to contribute to the intestinal elimination of vanadium when plasma vanadium levels were high; however, in the period 0–6 hours after injection of pentavalent ^{48}vanadium, the renal excretion was 5–10 times higher than biliary excretion (Sabbioni et al., 1981).

10.5 Vanadium Toxicity

10.5.1 Factors Affecting Toxicity of Vanadium

Vanadium toxicity has been reported in experimental animals and man. The degree of toxicity of vanadium depends on route of administration, valence, chemical form and interactions, and to some extent species. Toxicity of ingested vanadium is low compared to the toxicity elicited by parenteral administration. The most commonly reported cases of toxicity are due to industrial exposure via the respiratory system (Byerrum et al., 1974; Underwood, 1977; Dailey et al., 1981).

10.5.2 Toxicity in Chicks, Rats, and Sheep

Signs of vanadium toxicity including depressed growth, increased susceptibility to *Salmonella gallinarum,* and uncoupling of oxidative phosphorylation have been reported in chicks fed diets supplemented with 25 ppm or more vanadium (Byerrum et al., 1974; Underwood, 1977). Several studies, however, have indicated that different combinations of dietary components increase or decrease the tolerance for high levels of vanadium (Underwood, 1977). Recently,

Hill (1979) reported that vanadium toxicity was decreased when dietary protein levels were 20 or 30% rather than 10%. Hafer and Kratzer (1976) found that high levels of ammonium metavanadate (50, 100, or 200 ppm) inhibited growth to a much greater degree when added to a semipurified diet than when added to their practical diet containing natural ingredients. Weight gain of chicks consuming 50 ppm vanadium incorporated into the practical diet was slightly but insignificantly less than the gain of the group to which no supplementary vanadium was fed. They also reported that the addition of disodium ethylenediaminetetraacetate (EDTA) to the diets at six times the molar concentration of vanadium significantly reduced the growth depression seen with 50 and 200 ppm supplementary vanadium. Addition of lactose to diets containing 200 ppm vanadium increased growth depression significantly compared to the control diet. Addition of 1000 or 2000 ppm chromium as chromium chloride to diets containing 100 and 200 ppm vanadium reduced the growth depression otherwise seen with the high vanadium levels. Nielsen and co-workers (1980), however, have reported difficulty in confirming their studies of combined dietary chromium and vanadium effects. They state that without further studies "discussion of interaction between vanadium and chromium in biological systems seems inappropriate."

10.5.3 Toxicity in Human Beings

The toxicity of vanadium in man via the oral route tends to be low (Byerrum et al., 1974); likewise, serious toxicity caused by the presence of vanadium in implant structural members has not been demonstrated (Crews and Hopkins, 1981). However, human beings exposed to vanadium via the respiratory system have exhibited numerous signs of toxicity (Byerrum et al., 1974).

The effects of ingestion of vanadium have been reviewed (Schroeder et al., 1963; Byerrum et al., 1974; Underwood, 1977; Crews and Hopkins, 1981). Chronic doses greater than 22.5 mg vanadium per day (55 ppm based on consumption of 400 g dry matter daily) have not been reported. With a daily intake of 22.5 mg vanadium, some cramps and diarrhea were noted and a green tongue was observed in 5 to 12 patients.

Vanadium in varying amounts is found in the alloys used to make medical implants, but the amount of vanadium released with chronic exposure to the physiological fluids is not thought to present a toxicity problem (Crews and Hopkins, 1981).

Exposure of personnel to vanadium-containing dust or fumes has produced toxicity. The high levels of vanadium in residual fuel oils may be a matter of concern. Many of the symptoms of vanadium intoxication appear to be caused by local irritation of the mucous membranes of the respiratory tract and the conjunctiva. Symptoms include cough, rhinitis, conjunctivitis, sputum production, chest pain, wheezing, bronchospasm, lassitude, nausea, and green coating of the tongue (Byerrum et al., 1974; Crews and Hopkins, 1981). In 1972, a threshold limit of value of 0.5 mg/m^3 for vanadium dust and a maxiumum concentration of 0.05 mg/m^3 for vanadium fumes was stipulated by the American Conference of Governmental Industrial Hygienists (Byerrum et al., 1974).

Lees (1980) recently observed changes in lung function in workers exposed, due to faulty respirators, to high levels of vanadium compounds in fuel oil ash. Reductions in forced vital capacity, forced expiratory volume, and forced mid-expiratory flow had occurred within 24 hours of exposure to the dust. These parameters had not returned to preexposure levels by 8 days; but 4 weeks after exposure no residual effects were observed. Kiviluoto (1980), however, found no difference in the ventilation tests of a reference group and a group of 63 workers in a vanadium factory. The vanadium workers were chronically exposed to vanadium concentrations of 0.01–0.04 mg/m^3. Their previous exposure, however, had been in the range of 0.1–3.9 mg/m^3 for an average of 11 years.

10.6 Summary

Vanadium, a transition element, occurs in several oxidation states but is most stable at the +4 or +5 valence. Much of the vanadium present in physiological systems is bound to coordinating groups such as $-SH$, $-SS-$, $-OH$, $-N-$, $-COO-$, and phosphate.

Vanadium has been demonstrated to be nutritionally essential for the chick and the rat but to date essentiality has not been shown in man. Vanadium concentrations of most foods were low (<100 ppb); however, certain marine species contained higher concentrations of vanadium. Since toxicity of ingested vanadium is low, vanadium from foods does not create toxicity problems. Toxicity has been reported due to industrial exposure via the respiratory system.

Several physiological functions for vanadium have been hypothesized. In addition to effects on growth and reproduction, vanadium has been suggested to be an etiological factor in nutritional edema and manic-depressive illness. Various mechanisms for an effect of vanadium at the cellular level have been hypothesized such as involvement in the control mechanism of the sodium pump. Additionally, vanadium compounds have been reported to have an insulin-mimetic effect *in vitro*. Many studies, however, have been equivocal and additional research will be necessary to clarify the physiological role of vanadium.

References

Balfour, W. E., Grantham, J. J., and Glynn, I. M., 1978. Vanadate-stimulated natriuresis, *Nature* 275:768.

Byerrum, R. U., Eckardt, R. E., Hopkins, L. L., Libsch, J. F., Rostoker, W., Zenz, C., 1974. *Vanadium*, National Academy of Sciences, Washington, D.C.

Cantley, L. C., Jr., and Aisen, P., 1979. The fate of cytoplasmic vanadium. Implications on (Na, K)-ATPase inhibition, *J. Biol. Chem.* 254:1781–1784.

Cantley, L. C., Jr., Josephson, L., Warner, R., Yanagisawa, M., Lechene, C., and Guidotti, G., 1977. Vanadate is a potent (Na, K)-ATPase inhibitor found in ATP derived from muscle, *J. Biol. Chem.* 252:7421–7423.

Cantley, L. C., Jr., Resh, M. D., and Guidotti, G., 1978. Vanadate inhibits the red cell (Na^+, K^+) ATPase from the cytoplasmic side, *Nature* 272:552–554.

Clark, R. J. H., 1973. Vanadium, in *Comprehensive Inorganic Chemistry*, J. C. Bailar, H. J. Emeléus, S. R. Nyholm, and A. F. Trotman-Dickenson (eds.), Vol. 3, Pergamon Press, Oxford, pp. 491–551.

Cornelis, R., and Versieck, J., 1982. Determination of vanadium in tissues and serum, *Clin. Chem.* 28:1708.

Crews, M. G., and Hopkins, L. L., 1981. Metabolism and toxicity in vanadium, in *Systemic Aspects of Biocompatibility*, Vol. I, D. F. Williams (ed.), CRC Press, Inc., Boca Raton, Fl., pp. 179–186.

Dailey, N. S., Mashburn, S. A., and Winslow, S. G., 1981. Toxicity of selected vanadium compounds: A bibliography with abstracts, 1965–1980, Federation of American Societies for Experimental Biology, Bethesda, Md.

Dingley, A. L., Kustin, K., Macara, I. G., and McLeod, G. C., 1981. Accumulation of vanadium by tunicate blood cells occurs via a specific anion transport system, *Biochim. Biophys. Acta* 649:493–502.

Dubyak, G. R., and Kleinzeller, A., 1980. The insulin-mimetic effects of vanadate in isolated rat adipocytes. Dissociation from effects of vanadate as a (Na^+-K^+)ATPase inhibitor, *J. Biol. Chem.* 255:5306–5312.

Erdmann, E., Werdan, K., Krawietz, W., Lebuhn, M., and Christl, S., 1980. Significance of NADH-vanadate-oxidoreductase of cardiac and erythrocyte cell membranes, *Basic Res. Cardiol.* 75:460–465.

Golden, M. H., and Golden, B. E., 1981. Trace elements. Potential importance in human nutrition with particular reference to zinc and vanadium, *Br. Med. Bull.* 37:31–36.

Grantham, J. J., 1980. The renal sodium pump and vanadate, *Am. J. Physiol.* 239:F97–106.

Hackbarth, I., Schmitz, W., Scholz, H., Erdmann, E., Krawietz, W., and Philipp, G., 1978. Positive inotropism of vanadate in cat papillary muscle, *Nature* 275:67.

Hafez, Y. S., and Kratzer, F. H., 1976. The effect of diet on the toxicity of vanadium, *Poultry Sci.* 55:918–922.

Hansard, S. L., Ammerman, C. B., Henry, P. R., and Simpson, C. F., 1982. Vanadium metabolism in sheep. 1. Comparative and acute toxicity of vanadium compounds in sheep, *J. Anim. Sci.* 55:344–349.

Hill, C. H., 1979. The effect of dietary protein levels on mineral toxicity in chicks, *J. Nutr.* 109:501–507.

Hopkins, L. L., Jr., and Mohr, H. E., 1971. The biological essentiality of vanadium, in *Newer Trace Elements in Nutrition*, W. Mertz and W. E. Cornatzer (eds.), Dekker, New York, pp. 195–213.

Hopkins, L. L., Jr., and Mohr, H. E., 1974. Vanadium as an essential nutrient, *Fed. Proc.* 33:1773–1775.

Hopkins, L. L., Jr., and Tilton, B. E., 1966. Metabolism of trace amounts of vanadium 48 in rat organs and liver subcellular particles, *Am. J. Physiol.* 211:169–172.

Ikebe, K., and Tanaka, R., 1979. Determination of vanadium and nickel in marine samples by flameless and flame atomic absorption spectrophotometry, *Bull. Environ. Contam. Toxicol.* 21:526–532.

Kiviluoto, M., 1980. Observations on the lungs of vanadium workers, *Br. J. Ind. Med.* 37:363–366.

Lees, R. E. M., 1980. Changes in lung function after exposure to vanadium compounds in fuel oil ash, *Br. J. Med.* 37(3):253–256.

MacDonald, E., LeRoy, A., and Linnoila, M., 1982. Failure of lithium to counteract vanadate-induced inhibition of red blood cell membrane Na^+, K^+-ATPase, *Lancet* ii:774.

Meisch, H. U., and Bielig, H. J., 1980. Chemistry and biochemistry of vanadium, *Basic Res. Cardiol.* 75:413–417.

Menon, A. S., Rau, M., Ramasarma, T., and Crane, F. L., 1980. Vanadate inhibits mevalonate synthesis and activates NADH oxidation in microsomes, *FEBS Lett.* 114:139–141.

Myron, D. R., Givand, S. H., and Nielsen, F. H., 1977. Vanadium content of selected foods as determined by flameless atomic absorption spectroscopy, *J. Agric. Food. Chem.* 25:297–300.

Myron, D. R., Zimmerman, T. J., Shuler, T. R., Klevay, L. M., Lee, D. E., and Nielsen, F. H., 1978. Intake of nickel and vanadium by humans. A survey of selected diets, *Am. J. Clin. Nutr.* 31:527–531.

Naylor, G. J., and Smith, A. H., 1981. Vanadium: A possible aetiological factor in manic depressive illness, *Psychol. Med.* 11:249–256.

Nielsen, F. H., and Ollerich, D. A., 1973. Studies on a vanadium deficiency in chicks, *Fed. Proc.* 32:929.

Nielsen, F. H., Hunt, C. D., and Uthus, E. O., 1980. Interactions between essential trace and ultratrace elements, *Ann. N.Y. Acad. Sci.* 355:152–164.

Parker, R. D., and Sharma, R. P., 1978. Accumulation and depletion of vanadium in selected tissues of rats treated with vanadyl sulfate and sodium orthovanadate, *J. Environ. Pathol. Toxicol.* 2:235–245.

Peabody, R. A., Wallach, S., Verch, R. L., and Kraszeski, J., 1976. Metabolism of vanadium-48 in normal and endocrine-deficient rats, in *Trace Substances in Environmental Health-X*, D. D. Hemphill (ed.), University of Missouri, Columbia, pp. 441–450.

Peabody, R. A., Wallach, S., Verch, R. L., and Lifschitz, M. L., 1977. Effect of thyroxin and growth hormone replacement on vanadium metabolism in hypophysectomized rats, in *Trace Substances in Environmental Health-XI*, D.D. Hemphill (ed.), University of Missouri, Columbia, pp. 297–304.

Ramasarma, T., and Crane, F. L., 1981. Does vanadium play a role in cellular regulation? *Current Topics in Cellular Regulation* 20:247–301.

Robinson, J. D., 1981. Effect of cations on (Ca^{2+} + Mg^{2+})-activated ATPase from rat brain, *J. Neurochem.* 37:140–146.

Rubinson, K. A., 1981. Concerning the form of biochemically active vanadium, *Proc. Roy. Soc. Lond. (Biol.)* 212:65–84.

Sabbioni, E., Marafante, E., Rade, J., Gregotti, C., Di Nucci, A., and Manzo, L., 1981. Biliary excretion of vanadium in rats, *Toxicol. Eur. Res.* 3:93–98.

Sakurai, H., Goda, T., and Shimomura, S., 1982. Vanadyl (IV) ion dependent enhancement of oxygen binding to hemoglobin and myoglobin, *Biochem. Biophys. Res. Comm.* 107:1349–1354.

Schroeder, H. A., Balassa, J. J., and Tipton, I. H., 1963. Abnormal trace metals in man—vanadium, *J. Chron. Dis.* 16:1047–1071.

Schwarz, K., and Milne, D. B., 1971. Growth effects of vanadium in the rat, *Science* 174:426–428.

Sharma, R. P., Oberg, S. G., and Parker, R. D., 1980. Vanadium retention in rat tissues following acute exposures to different dose levels, *J. Toxicol. Environ. Health* 6:45–54.

Shechter, Y., and Karlish, S. J., 1980. Insulin-like stimulation of glucose oxidation in rat adipocytes by vanadyl (IV) ions, *Nature* 284:556–558.

Siemon, H., Schneider, H., and Fuhrmann, G. F., 1982. Vanadium increases selective K^+-permeability in human erythrocytes, *Toxicology* 22:271–278.

Söremark, R., 1967. Vanadium in some biological specimens, *J. Nutr.* 92:183–190.

Strasia, C. A., 1971. Vanadium: Essentiality and toxicity in the laboratory rat, Ph.D. Dissertation, University Microfilms, Ann Arbor, Mich.

Underwood, E. J., 1977. Vanadium, *Trace Elements in Human and Animal Nutrition*, 4th ed., Academic Press, New York, pp. 388–397.

Welch, R. M., and Cary, E. E., 1975. Concentration of chromium, nickel, and vanadium in plant materials, *J. Agric. Food Chem.* 23:479–782.

Weigmann, T. B., Day, H. D., and Patak, R. V., 1982. Intestinal absorption and secretion of radioactive vanadium ($^{48}VO_3^-$) in rats and effect of Al(OH)$_3$, *J. Toxicol. Environ. Health* 10:233–245.

Silicon

11

Edith Muriel Carlisle

11.1 Introduction

Although interest in the silicon content of animal tissues and the effect of siliceous substances on animals was expressed over half a century ago (King and Belt, 1938), emphasis has been placed until recently on the toxicity of silicon, mainly its involvement in silicosis. The occurrence of silicon in living systems, its physiological and biochemical roles, and its toxicology have attracted increasing research interest over the last 10 years (for an extensive review, see Bendz and Lindquist, 1978) because it is only within the last decade that silicon has been recognized as participating in the normal metabolism of higher animals and as an essential trace element. Silicon has been shown to be required in bone, cartilage, and connective tissue formation as well as participating in several other important metabolic processes.

In terms of the relationship of silicon to human disease, several relatively new entities have attracted interest. Moreover, the pathogenesis of one of the "old" diseases—silicosis—has become better understood. A relatively newly recognized disease, asbestosis, is of great concern, because asbestos is carcinogenic.

11.1.1 Discovery and History

Berzelius is generally credited with the discovery in 1823 of silicon as an element, preparing elementary silicon from potassium fluosilicate. He correctly claimed it to be an element, studied its properties and converted it to SiO_2 by

---

Edith Muriel Carlisle • Division of Nutrition, School of Public Health, University of California, Los Angeles, California, 90024.

combustion, thus proving that silica was a compound. He named the new element silicium (from the Latin *silex,* "quartz").

Silicon is one of man's most useful elements—as a semiconductor in electronics, as a major component in ceramics, building materials, glasses, and the organo-metallic silicones among others.

11.1.2 Occurrence and Distribution

Silicon is the second most abundant element in the earth's crust. In the hydrosphere, on the other hand, silicon is a rather scarce element and extremely scarce in sea water. Nevertheless, diatoms and other microorganisms are able to extract dissolved silica or silicic acid from sea water and to deposit it in their skeletons in an insoluble form. Silicon is not found free in nature, but occurs chiefly as the oxide and as silicates. SiO_2 occurs mainly in crystalline form as quartz, rarely as tridymite and cristobalite, and as the amorphous mineral opal. Asbestos, tremolite, the feldspars, clays, and micas are but a few of the silicate minerals. Silicon is prepared commercially by heating silica and carbon in an electric furnace, using carbon electrodes.

11.1.3 Chemistry

Silicon is a relatively inert element, but it is reactive toward halogens and dilute alkali. Most acids, except hydrofluoric, do not affect it. Silicon has three stable isotopes—^{28}Si, ^{29}Si, and ^{30}Si—and two radioisotopes that can be used as tracers, ^{31}Si and ^{32}Si. ^{31}Si has the disadvantage of a half-life of only 2.6 hours making this a useful tracer only for short-term experiments. ^{32}Si has a longer half-life but is difficult to produce and is not available commercially. It also has the disadvantage of decaying to ^{32}P as the daughter nuclide.

Silicon is similar to carbon in several respects, the two elements being adjacent in Group IV of the periodic table. However, their differences are important. For example, because of greater electronegativity, the hydrogen on silicon has hydride reactivity unlike the hydrogen on carbon. The C–C bond is more stable than the Si–Si bond. However, the Si–O bond is more stable than a C–O bond because silicon bonding is both ionic and covalent in character. The high energy of this bond renders the formation of silica and the great variety of oxygen-containing silicon compounds thermodynamically favorable. For the same reason, silicic acid is only sparingly soluble and has a tendency to polymerize.

Silicon compounds are both inorganic and organic. Organo-silicon compounds are characterized by the presence of one Si–C bond. Their chemistry is exemplified by the organo-siloxanes, better known as the commercially important silicones. These compounds are not found in nature, but are readily synthesized in the laboratory and behave chemically and biologically as organic compounds. So far, no bonds, other than Si–O, have been demonstrated in biological systems, and silicon metabolism, therefore, is not known to involve organo-silicon chemistry.

11.1.4 Essentiality

Although silicon had long been suspected to be physiologically important and a constituent of connective tissue in mammals (for a brief historical review see Schwarz, 1978), it had been classed along with 20–30 other elements as one of the environmental contaminants that occur in highly variable concentrations in living tissues but do not meet criteria for essentiality (Underwood, 1977). As late as 1967, in an extensive review of the biological properties of silicon compounds, it was stated that "although traces of silica are found in all animal tissues, there is no evidence that there is any biological need for silicon in the higher animals" (Fessenden and Fessenden, 1967).

A series of experiments has contributed to the establishment of silicon as an essential element. The first of these were *in vitro* studies (Carlisle, 1969, 1970a) that showed that silicon is localized in active growth areas in bones of young mice and rats, suggesting a physiological role for silicon in bone calcification processes. These were followed by *in vivo* studies (Carlisle, 1970b; Carlisle, 1974) showing that silicon affects the rate of bone mineralization. Of critical importance, it was subsequently demonstrated that silicon deficiency is incompatible with normal growth and skeletal development in the chick and that these abnormalities could be corrected by a silicon supplement (Carlisle, 1972). During the same year Schwarz and Milne (1972) showed that silicon deficiency in the rat results in depressed growth and skull deformations. Later studies, both *in vitro* (Carlisle and Alpenfels, 1978) and *in vivo* (Carlisle, 1980a, b, 1981), emphasize silicon's importance in bone formation and connective tissue metabolism and confirm the postulate that silicon is involved in an early stage of bone formation. These studies are discussed further in the present chapter.

11.2. Silicon in Tissues

11.2.1 Primitive Organisms

Silicon compounds play a particularly important role in certain primitive organisms (see Simpson and Volcani, 1981); silicate bacteria, the simplest algae, spore plants, and two groups of animals, notably the radiolaria (belonging to the Protozoa), and some sponges (Porifera). There have been some interesting reports concerning silicate bacteria (*Bacillus siliceous*), which are capable of breaking down the insoluble potassium aluminosilicates in soil and may even be cultured in a granite bowl or in a medium of powdered glass (Fessenden and Fessenden, 1967). The mode of silicon uptake by some related soil bacteria has been studied by Heinen (1965). Diatoms, unicellular microscopic plants, have been shown to have an absolute requirement for silicon, in the form of monomeric silicic acid, for normal cell growth. It is incorporated into the silica shell of the cell (Lewin, 1955) and is essential for the synthesis of DNA, through facilitating the synthesis of the two nuclear DNA polymerases. It is also involved in the formation of thymidylate kinase (Simpson and Volcani, 1981). Evidence has been presented for a relationship between silicon fixation and intermediary metabolism in diatoms. Werner (1977) has reported that diatoms require $Si(OH)_4$ for a number of metabolic processes prior to wall formation, including protein synthesis, the regulation of respiration and chrysolaminarin utilization, and chlorophyll synthesis. The structural relationship of silica to the organic constituents of the cell wall has also been established (Reimann *et al.*, 1965).

11.2.2 Higher Plants

Whether silicon is essential, in a strict sense, to the growth of higher plants has remained questionable. The fact that all soil-grown plants contain silicon and that large quantities may be present in such species as rice, other grasses, and horsetails, has led to the suggestion that silicon has beneficial effects on plant growth or crop yield. It has been suggested that in some circumstances, the element may promote resistance to insect and fungal attack. A good review has been prepared by Lewin and Reimann (1969).

Less is known on the function of silicon in plants than on the forms and concentrations in which it occurs. Variation between species is considerable. For example, cereal grains high in fiber are much richer in silicon than low-fiber grains. The element is present mainly as silica (silica gel), as soluble silicates, and partly in organic combination, bound to the cellulosic structure of the cell.

Plant uptake of silicon is governed by its concentration in the soil solution among other factors, and by the nature of the plant. Jones and Handreck (1965) grew oats in soil and soil-oxide mixtures that had a wide range of concentrations of silicon in solution. Their data showed that total silica in the plant tops was directly proportional to the concentration of monosilicic acid in the soil solution. Abundant data show that there are marked differences in the tendencies of various plants to accumulate silicon, although the reasons for these differences are not completely clear (Jones and Handreck, 1967; Lewin and Reimann, 1969). Some plants, such as legumes, have levels as low as those of animal tissues (see the following discussion). Other plants, such as grasses, oats, and rice, accumulate silicon in very large amounts. The flowering part of the plant commonly has the highest concentration, while the endosperm is quite low in silicon. Silicon in grass varies widely, depending on the region in which the grasses are grown and on the species. Rice straw commonly contains more than 5% silicon, and the silicon content of *Equisetum* is similar (Jones and Handreck, 1967; Lewin and Reimann, 1969). Plants can be sorted into those that take up silicon passively, those that take up silicon actively (e.g., rice), and those that restrict the uptake of silicon, although, as with many elements, this may vary according to the concentration of silicon in the substrate.

Silicon is deposited in plant tissue as the solid, hydrated oxide $SiO_2 \cdot nH_2O$, known as opaline silica or as silica gel, following polymerization of the water-soluble monosilicic acid, $Si(OH)_4$ entering the plant. The deposits are most commonly in the form of particles (phytoliths), the shapes of which are characteristic of a given plant and vary enormously between plant species. Very little is known about the occurrence of silicon compounds other than free silica in plants. Reports have described the occurrence of organic complexes (Heinen, 1965). Van Soest and Lovelace (1969), for example, suggested that the organically bound silica is complexed by carbohydrate. Lovering and Engel (1967), on the other hand, suggested that it occurs in soluble complexes with polyphenols in a kind of silicone linkage. The complexes, if they occur in plants, are very unstable, particularly in the presence of traces of acid (Van Soest and Lovelace, 1969) and have defied isolation and characterization thus far. An exception may be the recent isolation and identification of a silicon chelate of thujaplicine, an isopropyl tropolone in the conifer *Thuja plicata* (Weiss and Herzog, 1978). Even in this case, however, there is still some question whether the silicon complex might not have been formed in the course of the extraction.

Moreover, it is not completely clear in what soluble form silicon is transported or otherwise enters into cells or occurs in tissue fluids or the cytosol. In diatoms, the transport of silicic acid into the cell has been found to be an energy-dependent, membrane phenomenon, but in higher plants, once silicon is within the conduction tissues it may move passively up the shoot of the plant. Biochemically, silicic acid polymerization per se apparently can occur in two very different localities: extracellularly, in association with a structure that is otherwise not known to be specialized for silicon metabolism (namely, the cell wall in higher plants) or intracellularly, within a discrete specialized cytoplasmic membrane, the silicalemma, as in diatoms, chrysophytes, and so on.

The question of whether there is an active metabolism of silicon in higher plants is still not answered. Most of the accumulations in plants might be explained by passive mechanisms. The saturation concentration of orthosilicic acid might merely be exceeded by evaporation of the solvent, thus leading to polycondensation and to formation of insoluble aggregates of silica.

Deposits of silicon in *Gramineae, Bamboo,* or *Equisetum* needs further confirmation, mainly with respect to mechanical occlusion. Deposits other than silica are unknown in higher plants up to now. Also transport forms other than orthosilicic acid are uncertain, though there are discussions about interaction of silicic acids with sugars. However, all the sugar–silicon derivates studied up to now are extremely unstable in the presence of water.

11.2.3 Animals and Man

Earlier data on the silicon content of living tissues have varied greatly, and in general, reported values were considerably higher before the advent of plastic laboratory ware and the development of suitable methods. Even with more recent methods, most of which are modifications of the molybdate reaction, considerable variance exists in reported tissue concentrations of silicon, probably due mainly to limitations of the colorimetric determination by this technique (Schwarz, 1978).

The concentration of silicon in blood is much lower than previously assumed. Normal whole blood levels in man (McGavack et al., 1962), monkeys (LeVier, 1975), and rats (McGavack et al., 1962; Carlisle, 1982) average around 100 μg/dl except for ovines, which is higher (500 μg/dl). Normal human serum has a narrow range of silicon concentration averaging 50 μ/dl (Carlisle, 1982). The range is similar to that found for most of the other well-recognized trace elements in human nutrition. The silicon is present almost entirely as free soluble monosilicic acid (Baumann, 1960), and the concentration of silicon in other examined body fluids has been found to be similar to that of normal serum, indicating that silicon is freely diffusible throughout tissue fluid. No correlations of age, sex, occupation, or pulmonary condition with blood silicon concentrations were found as a result of measurements on hundreds of people, although the level increased when silicon compounds were specifically administered (Worth, 1952). Moderate increases have been obtained in rat's blood after feeding silicon metasilicate. Much higher levels have been reached, however, after feeding organic silicates (Carlisle, 1982).

Connective tissues such as the aorta, trachea, tendon, bone, skin, and its appendages are unusually rich in silicon, as shown by studies in several animal species (Carlisle, 1974). In the rat, for example (Figure 11-1), the aorta, trachea, and tendon are four to five times richer in silicon than liver, heart, and muscle. Among the human tissues, epidermis and hair have been reported to contain unusually large amounts of localized silicon. The element accumulates in the cornified epidermis on the surface of skin and in the epicuticle of hair as well as the wool and feathers of other animals in an alkali-insoluble component constituting only 0.4–1.7% of the total tissue weight. It has been suggested (Fregert, 1959) that this small alkali-insoluble component with its high silicon

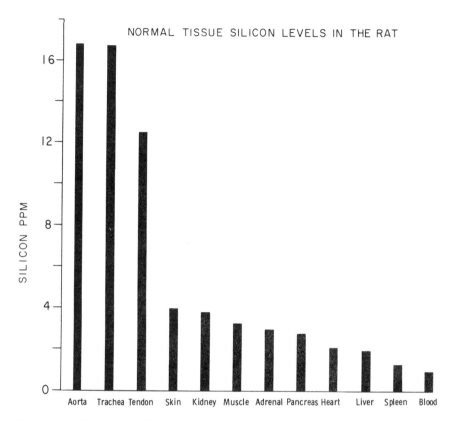

Figure 11-1. Normal tissue silicon levels in the adult male rat. Values represent mean silicon levels of 20 animals (4 months of age) expressed as parts per million wet weight of tissue (Carlisle, 1974).

content may contribute to the solidity and great chemical resistance of keratinous tissues and may also play a role as a barrier of absorption. High silicon levels have also been reported in human dental enamel (Losee *et al.*, 1973) and in the head of the monkey femur containing the epiphyses (Le Vier, 1975).

The high silicon content of connective tissues appears to arise mainly from its presence as an integral component of the glycosaminoglycans and their protein complexes that contribute to the structural framework of this tissue. Fractionation procedures reveal that connective tissues, such as bone, cartilage, and skin, yield complexes of high silicon content. Silicon is also found as a component of glycosaminoglycans isolated from these complexes.

The consistently low concentrations of silica in most organs do not appear to vary appreciably during life. Parenchymal tissue, such as heart and muscle, for example, ranges from 2 to 10 μ of silicon/g dry weight (Carlisle, 1982).

The lungs are an exception. Similar levels of silicon in rat tissues have been reported by McGavack et al. (1962) and in both rat and rhesus monkey by Le Vier (1975), where soft tissue levels in both species varied from 1 to 33 μg of silicon/g dry weight excepting the primate lung and lymph modes, which averaged 942 ppm and 101 ppm, respectively.

In order to gain information about silicon's function in soft tissues, the subcellular distribution of silicon in whole rat liver was determined in terms of percent of homogenate total. The element was found to be equally distributed in the supernatant, mitochondria, and nuclei/debris. Little silicon was associated with the microsomal fraction (Le Vier, 1975). Silicon has also been detected in the centriole of guinea pig renal tubule cells using the electron microprobe; the significance of this finding remains unresolved (Schafer et al., 1970).

11.3 Silicon Deficiency and Functions

11.3.1 Growth and Development

As mentioned earlier silicon was shown to have a marked effect on growth. Feeding a low-silicon diet based on an optimal mixture of L-amino acids for the chick and using special trace element techniques, it was possible to show that silicon is required for normal growth and development (Carlisle, 1972). Increases of nearly 50% in growth rates in chicks were observed upon the feeding of silicon supplied as sodium metasilicate. The chicks on the deficient diet appeared stunted. On subsequent examination all organs appeared relatively atrophied. Macropathologic examination showed that the skin and mucous membranes were somewhat anemic. The deficient chick had no wattles and the comb was severely attenuated. Significantly retarded skeletal development was also evident in the long bones. The skulls were also smaller and abnormally shaped. This effect of silicon on skeletal development supports the earlier findings that silicon is involved in an early stage of bone formation (Carlisle, 1969, 1970a,b).

Silicon deficiency in rats also resulted in depressed growth and skull deformities (Schwarz and Milne, 1972). As with chicks, chemically defined diets based on amino acids in place of protein were used. The skulls were shorter and the bone structure surrounding the eye appeared distorted. Pigmentation of the incisors also was affected. The addition of 50 mg of silicon per 100 g of diet produced a 25–34% increase in growth rates. Silicon was only partly effective in preventing the impairment of pigment deposition. Significant effects on incisor pigmentation also were produced by physiological levels of tin, vanadium, or fluorine.

11.3.2 Calcification

In Vitro Studies

Concern for the problem of osteoporotic bone loss and the fact that small amounts of fluoride have been known for some time to change markedly the properties of bone apatite led to a study of the possible effects of other less-known elements in bone metabolism. Techniques were developed for application of the electron microprobe to bone and related tissues during various stages of development and, as indicated earlier, silicon, was shown to be uniquely localized in active growth areas in young bone (Carlisle, 1969, 1970a). The amount of silicon present in specific small sites within the active growth areas appeared to be uniquely related to "maturity" of the bone mineral. Furthermore, in the earliest stages of calcification in these sites, when the calcium content of osteoid tissue is very low, a direct relationship exists between silicon and calcium. In more advanced stages the amount of silicon in the sites falls markedly and, as calcium approaches the proportions present in bone apatite, silicon is present only at the detection limit; the more "mature" the bone mineral, the smaller the amount of measurable silicon. Likewise, maximal amounts of silicon are present at Ca/P molar ratios of approximately 0.7, but at Ca/P ratios approaching that of hydroxyapatite (1.67), silicon falls below the detection limit.

This relationship is demonstrated in Figure 11-2 in a typical traverse across the periosteal region of young tibia as obtained by electron microprobe techniques. For example, it was found that in the fibrous layer of the periosteum, both the calcium and silicon values are invariably low, whereas in the adjacent osteoid layer, silicon-rich sites appear, containing up to 25 times as much silicon and 9 times as much calcium as in the fibrous layer. The silicon content in the sites falls again to the original extremely low value as calcium approaches the proportions in bone apatite.

Neither the initiating nor the limiting factor in the mineralization of bone in the living animal is known. Several investigators have regarded calcium binding as a most important and first event in calcification. The data presented previously suggest that silicon is associated with calcium in such a process.

In Vivo Studies

Subsequent experiments showed that silicon has a demonstrable effect on *in vivo* calcification (Carlisle, 1970b, 1974); that is, a relationship between the level of dietary silicon and bone mineralization, as measured by bone ash content, was established. Conventional trace element procedures to keep silicon contamination at a minimum were used (Carlisle, 1972). Weanling rats, from mothers

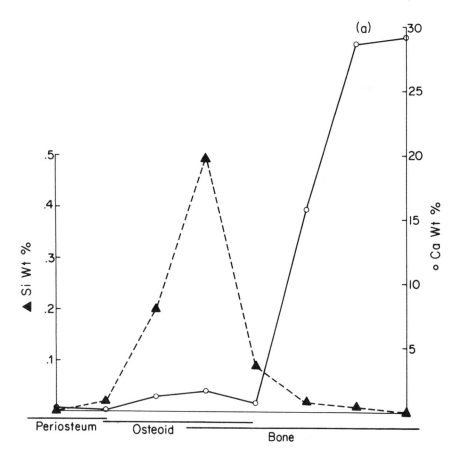

Figure 11-2. A spatial relation between silicon (▲) and calcium (○) composition (% by weight) in a typical traverse across the periosteal region of young rat tibia (cross section), using microprobe techniques (Carlisle, 1970a, with permission).

Table 11-1. Effect of Silicon Intake on Absolute Ash Content of Tibia of Rats on Low-Calcium (0.08%) Diet

Diet silicon (ppm)	Ash content[a]	
	2 weeks (mg)	5 weeks (mg)
250	88[b]	89
25	69[b]	90
10	66[b]	85

[a] Mean ash values of 12 animals per treatment group.
[b] Indicates a difference at the 5% confidence level.

placed on a low-silicon diet prior to mating, were maintained on diets containing three levels of calcium (0.08, 0.40, 1.20%) at three levels of silicon (10, 25, 250 ppm). Increasing silicon in the low-calcium diet resulted in a highly significant (35%) increase in the percentage of ash contained in the tibia during the first three weeks of the experiment. Calcium content of the bone also increased with increased dietary silicon, substantiating the theory of a relationship between mineralization and silicon intake.

The effect of the three levels of dietary silicon on mineralization rate was then tested in five bone groups (tibia, humerus, radius, ulna, femur). On the low-calcium diet, the percentage ash content at all silicon levels fell markedly in all bone groups by the end of the third or fourth week, but absolute ash content (Table 11-1), as well as rate of mineralization, rose with increased silicon intake. Thus, in the tibia, for example, at the end of week 2 on the low-silicon (10 ppm) diet, the mean total ash content was 60 mg rising to 85 mg at the end of week 5. At 250 ppm, however, total ash was already 88 mg by week 2, and increased to 89 by the end of the fifth week. Thus, although total ash content was about the same at all silicon levels in all bones tested at the end of the experimental period, it was clear that the bones had reached maximal mineralization much more rapidly on the high-silicon diet. The tendency of silicon to accelerate mineralization, especially on the low-calcium diet, was also demonstrated by its effect on bone maturity, as indicated by the [Ca]/[P] ratio. At the end of week 3 on the 10 and 25 ppm silicon diet, the ratio was 1.32 and 1.31, respectively, whereas at 250 ppm, it was 1.41. The concept of an agent that affects the speed of chemical maturity of bone is not new. Muller *et al.* (1966) found that the chemical maturity of vitamin D-deficient bone (measured by the amount of heat-produced pyrophosphate), although inferior to control bone during the period of maximum growth, approached the control level at the end of the experiment.

11.3.3 Bone Formation

The earliest studies suggesting a role for silicon in bone formation were those mentioned earlier. Most significant, however, was the establishment of a silicon deficiency state incompatible with normal skeletal development. In the chick this is evidenced by reduced circumference, thinner cortex, and less flexible leg bones, as well as by smaller and abnormally shaped skulls, with the cranial bones appearing flatter (Carlisle, 1972). Independently, silicon deficiency in rats was also shown to result in skull deformations (Schwarz and Milne, 1972). Subsequent *in vivo* studies with chicks demonstrated a relationship between silicon, magnesium, and fluorine in growing bone. A relationship between silicon, endocrine balance, and mineral metabolism in rats has been reported by Charnot and Pères (1978).

Recent studies further emphasize the importance of silicon in bone formation. Skull abnormalities associated with reduced collagen content have been produced in silicon-deficient chicks under conditions promoting optimal growth using a semisynthetic diet containing a natural protein in place of the crystalline amino acid diet used in earlier studies (Carlisle, 1980a). Additionally, a striking difference between the silicon-deficient and silicon-supplemented chicks was observed in the appearance of the skull matrix, the matrix of the deficient chicks totally lacking the normal striated trabecular pattern of the control chicks. The deficient chicks showed a nodular pattern of bone arrangement, indicative of a primitive type of bone. Trabecular bone formed later as the nodules coalesced.

Still more recently, using the same conditions as the preceding and by introducing three different levels of vitamin D, it has been shown that the effect exerted by silicon on bone formation is substantially independent of the action of vitamin D (Carlisle, 1981). All chicks on silicon-deficient diets, regardless of the level of dietary vitamin D had gross abnormalities of skull architecture, and furthermore, the silicon-deficient skulls showed considerably less collagen at each vitamin D level (Figure 11-3). As in the previous study in the groups receiving adequate vitamin D, the bone matrix of the deficient chicks totally lacked the normal striated trabecular pattern of the control chicks. In the rachitic groups of chicks, the appearance of the bone matrix was quite different from

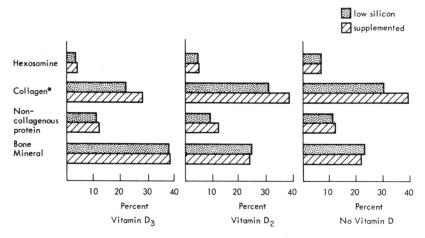

Figure 11-3. Effect of silicon intake on skull frontal bone composition of chicks fed different levels of vitamin D. Total duration of experiment, 4 weeks; 6 animals per group mean values. *Significantly different from the supplemented animals at $P < 0.05$. Vitamin D_2 = 6000 IU: vitamin D_3 = IU (Carlisle, 1980c).

the groups receiving adequate vitamin D, being considerably less calcified and more transparent, enabling the cells and underlying structure to be seen more easily. The deficient chicks appeared to have a marked reduction in number of osteoblasts compared to the controls. In these two studies, the major effect of silicon appears to be on the collagen content of the connective tissue matrix, and this is independent of vitamin D.

11.3.4 Cartilage and Connective Tissue Formation

In addition to bone growth, silicon deficiency is manifested by abnormalities involving glycosaminoglycans and the formation of cartilage matrix and connective tissue (Carlisle, 1976a). Chicks in the silicon-deficient group had thinner legs and smaller combs in proportion to their size. The tibial-metatarsal and tibial-femoral joints appeared markedly smaller in the silicon-deficient chicks. The epiphyseal (proximal) ends of the tibia and metatarsus had less articular cartilage and were narrower and less shaped. Joints at the distal extremities of the bones also were smaller and less well formed. Analysis of articular cartilage removed from the tibia of the experimental animals revealed that both the amount of cartilage and the glycosaminoglycan content measured by hexosamines was considerably less in the silicon-deficient chicks (Table-11-2).

The relationship established between silicon and glycosaminoglycans in cartilage formation was also confirmed in another type of connective tissue, the cock's comb. The comb of the cockerels was examined because of the difference noted in comb size and because cockerel comb is a so-called target connective tissue. Here too the amount of connective tissue, the total amount of hexosamines, and the proportion of hexosamine in the comb were found to be larger in the supplemented group. Additional analysis has revealed a significantly higher silicon content as well.

In studies using the semisynthetic diet containing a natural protein in place of crystalline amino acids, long bone abnormalities similar to those reported

Table 11-2. Effect of Silicon Intake on Articular Cartilage Composition[a]

Diet	Tissue (wet wt, mg)	Total hexosamine (wet wt, mg)	Percent hexosamine (wet, %)
Low silicon	63.32 ± 8.04[b]	0.187 ± 0.023[b]	0.296 ± 0.009[b]
Supplemented	86.41 ± 4.82	0.310 ± 0.031	0.359 ± 0.0911

[a] There were 12 chicks per group. All values reported as mean ± SD.
[b] Significantly different from the supplemented animals at $P < 0.001$.

earlier have been produced in silicon-deficient chicks, and a requirement for silicon in articular cartilage formation was again demonstrated (Carlisle, 1980b). Tibia from silicon-deficient chicks had significantly less glycosaminoglycans and collagen, the difference being greater for glycosaminoglycans than for collagen. Previously undocumented, tibia from silicon-deficient chicks showed rather marked pathology, profound changes being demonstrated in epiphyseal cartilage. The disturbed epiphyseal cartilage sequences resulted in defective endochondral bone growth indicating that silicon is involved in a metabolic chain of events required for normal growth of bone.

11.3.5 Connective Tissue Matrix

Although silicon may participate in the mineralization process itself, its primary effect in bone and cartilage appears to be on formation of the matrix. The preceding *in vivo* studies have shown silicon to be involved in both collagen and glycosaminoglycan formation. However, formation of the organic matrix, whether of bone or cartilage, appears to be more severely affected by silicon deficiency than is the mineralization process.

Organ Culture

The *in vivo* findings have been corroborated by studies of bone (Carlisle and Alpenfels, 1978) and cartilage (Carlisle and Alpenfels, 1980) in organ culture. Studies using embryonic skull bones further demonstrate the dependence of bone growth on the presence of silicon. Most of the increase in growth appears to be due to a rise in collagen content; silicon-supplemented bones showed a 100% increase in collagen content over silicon-low bones after 12 days. Silicon is also shown to be required for formation of glycosaminoglycans; at day 8, the increase in hexosamine content of supplemented bones was nearly 200% more than in silicon-low bones, but by day 12 it was the same in both groups.

A parallel effect has been demonstrated in the growth of cartilage and is especially marked in cartilage from 14-day embryos as compared with 10- and 12-day embryos. Silicon's effect on collagen formation was also especially striking in cartilage from 14-day embryos, appearing to parallel the rate of growth. Similarly, matrix hexosamines (glycosaminoglycans) were formed more rapidly by silicon-supplemented cartilage, the most striking difference in this case being in cartilage from 12-day embryos. The requirement for silicon in collagen and glycosaminoglycan formation thus proves not to be limited to bone matrix but applies also to cartilage.

Further studies on cartilage formation in culture have shown an interaction between silicon and ascorbate (Carlisle and Suchil, 1983). Silicon's effect on cartilage formation was investigated both in the presence and absence of ascorbate. No significant effect on hexosamine content occurred in the absence of ascorbate. However, silicon supplementation resulted in significant increases in wet weight, hexosamine and proline content in the presence of ascorbate. The greatest effect was on hexosamine content; at day 8 in the presence of ascorbate the hexosamine content in silicon-supplemented media had increased 100% more than that in low-silicon media. Furthermore, silicon and ascorbate interact to give maximal production of hexosamines. Silicon also appears to increase hydroxyproline, total protein, and noncollagenous protein beyond the effects of ascorbate.

Cell Culture

An effect of silicon on formation of extracellular cartilage matrix components by chondrocytes in culture has been demonstrated (Carlisle and Garvey, 1982). Chondrocytes isolated from chick epiphyses were cultured under low-silicon and silicon-supplemented conditions for 18 days. Cultures were assayed for DNA, hexosamines, hydroxyproline, proline, and noncollagenous protein. The major effect of silicon appeared to be on collagen. Silicon-supplemented cultures demonstrated a 243% ($p < 0.01$) increase in cartilage matrix procollagen hydroxyproline over low-silicon cultures. Silicon also had a pronounced stimulatory effect on matrix polysaccharides; matrix polysaccharide content of silicon-supplemented cultures increased 152% ($p < 0.01$) more than that of low-silicon cultures. Silicon's effect on collagen and glycosaminoglycan formation is not due to cellular proliferation but to some system in the cell participating in their formation. These findings extend the concept of a requirement for silicon in collagen and glycosaminoglycan formation to the cellular level.

11.3.6 Enzyme Activity

A dependence on silicon for maximal prolyl hydroxylase activity has been demonstrated (Carlisle *et al.*, 1981). Prolyl hydroxylase was obtained from frontal bones of 14-day-old chick embryos incubated for 4 to 8 days under low-silicon conditions with 0, 0.2, 0.5, or 2.0 mM silicon added to the media. Prolyl hydroxylase activity was measured as specific tritium release as tritium water from a [3,4-^{3}H] proline-labeled unhydroxylated collagen substrate prepared from chick calvaria. The results show lower enzyme activity in 0.2, 0.5, and 2.0 mM cultures. The greatest increase occurred in 8-day bones, where enzyme activity was stimulated sixfold by the addition of 0.2 mM silicon, seven to eightfold by 0.5 mM silicon and tenfold by 2.0 mM silicon. These results support the earlier *in vivo* and *in vitro* findings of a requirement for silicon collagen biosynthesis, the activity of prolyl hydroxylase being a measure of the rate of collagen biosynthesis.

11.3.7 Connective Tissue Cellular Component

Additional support for silicon's metabolic role in connective tissue is provided by evidence of its presence in connective tissue cells as demonstrated in the active bone-forming cell, the osteoblast, and its subcellular localization in mitochondria (Carlisle, 1975). X-ray microanalysis of active growth areas in young bone and isolated osteoblasts, and further studies including ultracentrifugation, demonstrate that silicon is concentrated in the cytoplasm of the osteoblast in the mitochondria. In subsequent studies (Carlisle, 1976) where the total silicon, calcium, phosphorus, and magnesium stores of single osteogenic cells were successfully measured, silicon was shown to be a major ion that had been heretofore neglected. The amounts of silicon were in the same range as that of calcium, phosphorus, and magnesium. Moreover, silicon appeared to be especially high in the metabolically active state of the osteoblast.

Clear evidence that silicon occurs in the osteoblast and is localized in the mitochondria adds strong support to the proposition that silicon is required for connective tissue matrix formation. It also supports the original proposal that silicon plays a fundamental role in the early stages of the bone calcification process. It is generally agreed that during bone formation the osteoblast synthesizes the organic matrix—collagen, the glycosaminoglycans, and probably the lipids—and thus presumably modifies the calcification process.

A role for mitochondria in silicon metabolism has also been suggested by the finding of a relative abundance of silicon in mitochondria from diatoms (Mehard *et al.*, 1974) and in the liver, kidney, and spleen of rats, where silicon-containing granules were observed in the matrices of isolated mitochondria (Mehard and Volcani, 1976). Silicon has also been detected in a subcutaneous connective tissue cell (Takaya, 1975).

11.3.8 Structural Component

Although the preceding discussion indicates that silicon plays an important metabolic role in connective tissue, a structural role has also been proposed, mainly supported by the finding that in connective tissue silicon is a component of animal glycosaminoglycans and their protein complexes. In higher animals, the glycosaminoglycans, hyaluronic acids, chondroitin sulfates, and keratan sulfate are found to be linked covalently to proteins as components of the extracellular amorphous ground substance that surrounds the collagen, elastic fibers, and cells. By extraction and purification of several connective tissues, silicon has been shown to be chemically combined in the glycosaminoglycan fraction. The silicon content of the glycosaminoglycan protein complex extracted in this laboratory from bovine nasal septum, for example, is 87 ppm, compared to 13 ppm in the original dried cartilaginous tissue (Carlisle, 1976b). From this complex, smaller molecules considerably richer in silicon were isolated. Silicon was found to be associated with the larger, purer polysaccharide and smaller protein moieties. Purified chondroitin sulfate A, the major component of glycosaminoglycans in bovine nasal septum, was found to contain significantly more silicon than chondroitin sulfate C, a minor component. Certain other purified glycosaminoglycans also contain amounts of silicon in the same range as those of chondroitin sulfate A, namely, hyaluronic acid and dermatan sulfate extracted from skin (pig and rat) and hyaluronic acid extracted from rooster comb and umbilical cord. Chondroitin sulfate C and keratan sulfate from nasal septum cartilage contain lesser amounts. Hyaluronic acid from vitreous humor appears to be low in silicon (Carlisle, 1974).

Similar results on isolated glycosaminoglycans, which included some reference research standards, have been reported (Schwarz, 1973). Relatively large amounts of bound silicon were obtained in notably purified hyaluronic acid from umbilical cord, chondroitin 4-sulfate, dermatan sulfate, and heparan sulfate. Lesser amounts occurred in chondroitin 6-sulfate, heparin, and keratan sulfate-2 from cartilage, while hyaluronic acids from vitreous humor and keratan sulfate-1 from cornea were silicon-free. More recently, however, it has been reported (Schwarz, 1978) that many of the earlier observations on the occurrence of bound silicon in glycosaminoglycans were in error because they were based partially on results obtained with materials contaminated by silica or polysilicic acid, and that the hypothesis that silicon acts generally as a cross-linking agent may have to be modified, to silicon acting as a cross-linking agent in some special situations. Work in this laboratory shows that silicon is indeed a component of the glycosaminoglycan–protein complex; however, the amount of silicon in these complexes is less than the values reported by Schwarz (1973) for isolated glycosaminoglycans.

Glycosaminoglycans of connective tissues probably represent one of the most primitive and most simple of the high polymers in which regularly repeating anionic groups give rise to an astonishing variety of polymeric carbohydrates—compounds with distinct chemical, physical, and biological properties. Significantly, polymeric carbohydrates are perhaps best known in plant cells in which silicon is also shown to be a component (Carlisle, 1974). Cell walls of angiosperms provide one example, made up of cellulose and calcium pectate, a salt of polygalacturonic acid. In cell walls of brown algae the corresponding substance is calcium alginate, a salt of polymannuronic acid. Evidence has been presented (Schwarz, 1976) that silicon is chemically bound in pectin and alginic acid.

Silicon, then, in addition to being a component of glycosaminoglycans in animals, is apparently a component of plant polysaccharides, of which polygalacturonic acid (pectic acid), alginic acid, fucoidin, and carrageenan are known examples. In addition, of course, many plants contain large amounts of unbound silicon or silica. Furthermore, in invertebrates, silicon is a component of chitin, probably the most abundantly distributed polysaccharide in nature, particularly well known as forming the exoskeleton of crustaceans and insects (Carlisle, 1974).

Although hyaluronic acid and polygalacturonic acid, two of the ester-linked polymers that have been investigated, contain significant quantities of silicon, glucuronic acid, and galacturonic acid monomers derived from them do not (Carlisle, 1974). Silicon, therefore, appears to be involved in the ester linkage. A series of investigations concerned with silicic acid uptake by the bacterium *Proteus mirabilis* (Heinen, 1965) may be of significance in this respect. In spite of the fact that this organism does not require silicon for growth, it is reported to be able to integrate silicon into its metabolism when incubated in a phosphorus-deficient medium. Analysis by infrared spectra revealed that this bacterium forms carbohydrate-silicon esters, C–O–Si bonds have been similarly demonstrated in bacterial extracts. Earlier laboratory synthesis studies (Henglein *et al.*, 1956) have shown that polygalacturonic acids (pectic acid) could be linked by silicic acid either via metal ions or directly via silicic acid esters of galacturonic acid. Moreover, the hydroxyl groups of silicic acid can condense with those of sugars (and other molecules) just as the hydroxyl groups of phosphoric, sulfuric, and boric acids do, forming silicic acid esters C–O–Si (Needham, 1965). In further experimental work (Carlisle, 1974) disaccharides enzymatically derived from chondroitin sulfate A were found to contain considerably more silicon than disaccharides obtained from chondroitin sulfate C. The finding that silicon is a constituent of disaccharide units suggests that it may be added at the stage of formation of the polysaccharide chain from smaller units.

The preceding data indicate that silicon is not merely involved in glycosaminoglycan formation but that, in animal glycosaminoglycans at least and quite probably in plant polysaccharides and in chitin, silicon is a structural component.

The unique properties of silicon atoms, which lend themselves to macromolecular structure (Needham, 1965), appear to be manifested in connective tissues where silicon may contribute to the structural framework by forming links or bridges within and between individual polysaccharide chains and perhaps linking the polysaccharide chains to proteins. In this way silicon may aid in the development of the architecture of the fibrous elements of connective tissue and contribute to its structural integrity by providing strength and resilience. Significant here is the fact that the Si–O bond is very stable in terms of energy levels and chemical nonreactivity. The value reported for the bond energy of the Si–O bond is 108 kcal/mole, compared with the value of 85.5 kcal/mole for the C–O and 82.6 kcal for the C–C bond (Fessenden and Fessenden, 1967).

11.3.9 Aging

Connective tissue changes are prominent in aging so that it is not surprising to find a relationship between silicon and aging in certain tissues. The silicon content of the aorta, other arterial vessels, and skin was found to decline with age, in contrast with other analyzed tissues, which showed little or no change (Carlisle, 1974). The decline in silicon content was significant and was particularly dramatic in the aorta, commencing at an early age. This relationship occurred in several animal species, including the rabbit, rat, chicken, and pig. For example, in the rabbit between 12 weeks and 18–24 months of age there was a decrease of silicon in the aorta of 84% and in the skin of 83% (Figure 11-4). In the pig, the silicon content of mature pig skin decreased 90%, compared with fetal pig skin. During fetal development silicon was shown to increase with age.

Leslie et al. (1962) also found a decrease in silicon content in rat skin with age, in contrast to other tissues analyzed that showed an increase, such as brain, liver, spleen, lung, and femur. The increase seen in the lung is probably due to dust inhalation. In muscle and tendon, no significant changes in silicon were found. In skin, there was a 60% decrease between 5 weeks and 30 months. The dermis samples of the 30-month-old rat contained less silicon than the corresponding whole skin.

Similarly, in human skin, the silicon content of the dermis has been stated to diminish with age (MacCardle et al., 1943). It has also been reported by French investigators (Loeper et al., 1966; 1978) that the silicon content of the normal human aorta decreases significantly with age, in contrast to earlier findings (Kvorning, 1950), and, furthermore, that the level of silicon in the arterial wall decreases with the development of atherosclerosis. The possible involvement of silicon in atherosclerosis has also been suggested by others (Schwarz, 1978; Dawson et al., 1978). Of possible significance here, a relationship has been reported (Charnot and Pères, 1971) between silicon, age, and endocrine balance as a result of finding changes in absorption and resulting levels of silicon in the blood and intestinal tissues of rats in relation to age, sex, and various endocrine glands. It is suggested that the decline in hormonal activity with age may be responsible for the changes in silicon levels in senescence.

The precise relationship of silicon with the aging process remains to be determined. In contrast to the decrease in silicon content with age found in certain connective tissues mentioned earlier, the accumulation of silicon with

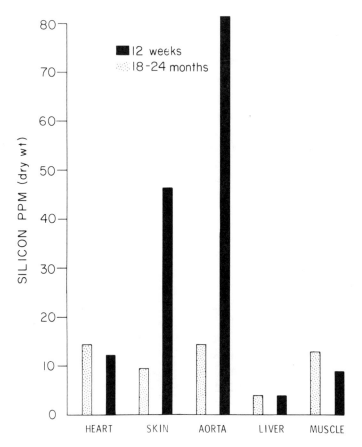

Figure 11-4. Effect of age on rabbit tissue silicon levels. Values represent mean silicon levels of 18 rabbits expressed as parts per million dry weight of tissue (Carlisle, 1974).

age in certain other tissues, mainly due to environmental influences, raises the possibility that a failure to dispose of silicon may also affect the aging process. In humans, it was shown in an earlier study (King and Belt, 1938) that silicon levels gradually increase with age in the human peribronchial lymph nodes, even in subjects who have no history of unusual dust exposure. Probably less well known is the fact, demonstrated by postmortems, that most urban dwellers harbor silicon in the form of asbestos fibers in their lungs. More recently, in Alzheimer's disease (Nikaido et al., 1972), a presenile condition characterized pathologically by the presence of glial plaques in the brain, an unexpectedly high increase of silicon has been reported in the cores and rims of the senile plaques. The effect of silicon on the aging process deserves further investigation.

11.4 Metabolism

4.1 Absorption

It has been estimated that man assimilates 9 to 14 mg. of silicon daily. This figure correlates well with the report of Goldwater (1936) that man excretes 9 mg of silicon in urine daily. In a recent balance study (Kelsay *et al.*, 1979) the silicon intake of men on a high-fiber diet was about double that on a low-fiber diet, and although urinary excretions ranged between 12 and 16 mg/day for the low- and high-fiber diets, the differences are not statistically significant. These values appear to be in the same range as those estimated earlier.

Balance trials in animals indicate that almost all ingested silicon is unabsorbed, passing through the digestive tract to be lost in the feces. Moreover, most of the small proportion that is absorbed is excreted in the urine. The proportion of absorbed silicon actually retained in the body is not known. Little is also known of the extent or mechanism of silicon absorption from the products entering the alimentary tract derived from food sources such as silica, monosilicic acid, and silicon found in organic combination, such as pectin and mucopolysaccharides.

In man silicic acid in foods and beverages has been reported to be readily absorbed across the intestinal wall and rapidly excreted in the urine (Baumann, 1960). In guinea pigs absorption is found to occur mainly as monosilicic acid (Sauer *et al.*, 1959), some of which comes from the silica of the plant materials, which is partly dissolved by the fluids of the gastrointestinal tract. In sheep the extent of absorption of silica as monosilicic acid varies with the silica content of the diet. Jones and Handreck (1969) found that the amounts excreted in the urine increased with increasing silica content of the diet from 0.10 to 2.84%, but reached a maximum of 205 mg SiO_2 perday, this amount representing less than 4% of the total intake.

The form of dietary silicon has been shown to be an important factor affecting its absorption appearing to correlate with the rate of production of soluble or absorbable silicon in the gastrointestinal tract (Benke and Osborn, 1978). Other factors have been reported to influence silicon absorption, the dietary fiber content of the diet has been implicated in studies with humans (Kelsay *et al.*, 1979) and changes in silicon absorption in rats have been found to be related to age, sex, and the activity of various endocrine glands (Charnot and Pères, 1971).

As is the case for many other elements, silicon availability is also probably affected by excess amounts of certain other mineral elements in the diet that may result in a diminution of silicon absorption through a reduction in the production of soluble silicon. Several elements suggest themselves as likely candidates based on their ability to depress greatly the amounts of silicic acid yielded in solution from various forms of silica suspended in body fluids (ascitic fluid and serum), especially aluminum and including Fe_2O_3, $Ca(OH)_2$, MgO, and SrO (King and McGeorge, 1938). The solubility depressor effect is believed to be a simple precipitation of the insoluble silicate by the metal. On the other hand, manganese availability may be influenced by silicon. In poultry rations, differences in manganese availability among various inorganic manganese sources have been demonstrated, but only two, the carbonate and silicate, are relatively unavailable (Underwood, 1977). Furthermore, silicon has been used as a therapeutic agent to alleviate manganese toxicity in plants (e.g., wheat, oats, and rice) (Okuda and Takahashi, 1964). A discussion of the interrelationship between silicon and molybdenum follows.

Phytoliths are minute bodies of amorphous silica (opal) that are formed within a great variety of plants. Many of the smaller phytoliths are able to pass the intestinal barrier of animals, including man, and enter the lymphatic and blood vascular systems, thus to be distributed throughout the body. In addition to possibly producing pathological effects, the presence of phytoliths throughout many tissues, especially lungs, liver, lymph nodes, and spleen, makes it difficult to determine the amount of silicon in animal tissues that is actually contributing to physiologic mechanisms. The gut wall of man is also permeable to particles the size of diatoms. The capacity of these particles to travel in the blood and to penetrate membranes, including the placenta, is illustrated further by their presence in the organs of stillborn and premature infants.

11.4.2 Transport

Silicon is found to be freely diffusible throughout tissue fluid. In earlier studies Baumann(1960) showed that monomeric $Si(OH)_4$ penetrates all body liquids and tissues at concentrations less than its solubility (0.01%) and is readily excreted. These findings are supported by later studies where the concentration of silicon in those body fluids examined was found to be similar to that of normal human serum except for urine. This indicates that silicon is freely diffusible throughout tissue fluid, with the higher levels and wider range encountered in urine suggesting that the kidney is the main excretory organ. Policard *et al.*, (1961) have suggested that polymeric molecules of silicic acid containing four to five silicon-oxygen units characterize the transport form of silicic acid in blood.

11.4.3 Excretion

In man, rats, guinea pigs, cows, and sheep increased urinary silicon output with increasing intake up to fairly well-defined limits has been demonstrated. The upper limits of urinary silicon excretion do not seem to be set by the ability of the kidney to excrete more because much greater urinary excretion can occur after peritoneal injections (Sauer *et al.*, 1959). These limits are determined by the rate and extent of silicon absorption from the gastrointestinal tract into the blood. Once silicon has entered the bloodstream it must pass rapidly into the urine and tissues, because even at widely divergent intakes, the silicon level in the blood remains relatively constant (Jones and Handreck, 1969). Jones and Handreck (1969) reported that in sheep the sum of the amounts of silica excreted in the urine and feces was within 1% of the amounts ingested. The amounts excreted in the urine represent only small proportions of the amounts ingested, while in the only known silicon balance study in humans (Kelsay *et al.*, 1979) urinary excretion represented a large portion of the silicon ingested and more silicon was excreted than was taken in. The recovery of ingested silica in steers and sheep has also been found often to exceed the calculated intake.

11.4.4 Interaction with Molybdenum

An interrelationship between silicon and molybdenum, a newly studied element, has recently been established (Carlisle, 1979). A marked interaction was demonstrated with both semisynthetic and amino acid diets. Plasma silicon levels were strongly and inversely affected by molybdenum intake; silicon-supplemented chicks on a liver-based diet (Mo 3 ppm) had a 348% lower plasma silicon level than chicks on a casein diet (Mo 1 ppm) (Figure 11-5). Molybdenum supplementation also reduced silicon levels in those tissues examined. Conversely, plasma molybdenum levels are also markedly and inversely affected by the inorganic silicon intake. Silicon supplementation of an amino acid diet reduced the plasma molybdenum levels of the Si-supplemented chicks by 280% and the red blood cell molybdenum by 425% compared to the low-silicon group. Reduction in molybdenum tissue retention by silicon also occurred, in liver by 63% and other tissues examined by 220–300%.

11.4.5 Enzyme Interaction

The effect of ionic silicate and colloidal silica on some enzymes has received the attention of several workers, who were guided by the hypothesis that silica is the pathogenic agent in silicosis. King *et al.* (1956) observed an inhibition of

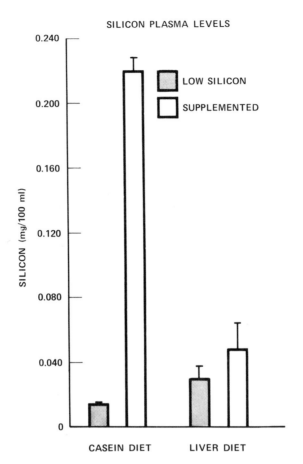

Figure 11-5. Effect of silicon supplementation on plasma silicon levels of chicks fed a low-silicon basal diet with casein as the protein source and with liver as the protein source. Total duration of experiment, 4 weeks; 12 animals per group. Plasma silicon values of silicon supplemented diets significantly different at $P < 0.001$ (Carlisle, 1979).

acetylcholinesterase, succinic dehydrogenase, liver esterase, and hyaluronidase by colloidal silica. Rowsell and Leonard (1957) reported a positive effect of colloidal silica on the respiration of rat liver homogenates.

An enzyme, silicase, has been discovered in pancreas, stomach, and kidney, which liberates silicic acid from a synthetic substrate, trimethoxy dodecanoxy silane (Schwarz, 1978). The enzyme is bound both to mitochondria and microsomes, from which it can be removed by treatment with suitable, nonionic detergents. Assuming that the enzyme has a function in silicate metabolism, it is of interest that it is membrane-bound and occurs in mitochondria and microsomes. The natural substrate of silicase is at present unknown.

11.5 Toxicity

Investigations in the area of silicon toxicity are almost invariably associated with the silicosis problem. When abnormal amounts of silica enter the body by inhalation some type of toxicity commonly results, mainly affecting the lungs and pleura. In addition, however, as a result of some recent studies, it may be necessary to consider inhaled silica dust as a rare cause of renal disease.

Silicon is generally considered to be an element of a relatively low order of toxicity. The ingestion of small amounts of siliceous materials and inhalation of siliceous dusts are harmless and of common occurrence for most animals. The majority of the ingested material passes unchanged through the alimentary canal. However, a small part of the silicon is absorbed and eliminated in the urine by the kidney, which is capable, as mentioned previously, of excreting much larger doses of silicic acid than are normally absorbed. Nevertheless, occurrence of siliceous uroliths gives clear evidence that the quantity of silicon absorbed, and excreted in the urine, under conditions of excessively high intake can be harmful.

A report concerning the health aspects of using the Generally Recognized as Safe (GRAS) silicates as food ingredients concludes that although silicates vary considerably in physical properties and solubility in aqueous solvents, most of the silicates added to foods as anticaking and antifoaming agents are insoluble in water and relatively inert (Select Committee on GRAS substances, 1979). Amorphous silicates are considered safe additions in foods, therefore, and their use as anticaking agents, for example, is permitted in amounts up to 2% by weight. The water-soluble silicates are also of low toxicity; studies of the effects of feeding various silicon compounds to laboratory animals have generally shown the substances to be innocuous under the test conditions.

11.5.1 Pneumoconioses in Man

Most published work on the pathologic effects of the element silicon in human beings concern the pneumoconioses, a group of diseases characterized by a chronic fibrous reaction in the lungs caused by dust inhalation. Inhalation of finely particulate silica (especially in the range of 0.3–3µ) and the group of fibrous silicates collectively termed asbestos cause the most marked reactions. These are known as silicosis and asbestosis respectively. Asbestos comprises fibrous forms of five amphiboles—amosite, $(Mg,Fe)_7Si_8O_{22}(OH)_2$; crocidolite, $Na_2Fe_3Fe_2Si_8O_{22}(OH)_2$; tremolite, $Ca_2Mg_5Si_8O_{22}(OH)_2$; anthophyllite, $(Mg,Fe)_7Si_8O_{22}(OH,F)_2$; and actinolite, $Ca_2(Mg,Fe)_5(Si_8O_{22})(OH,F)_2$—all of which comprise less than 5% of commercial asbestos in the United States and one serpentine—chrysotile, $Mg_3(Si_2O_5)(OH)_4$, which represents over 95% of potential asbestos exposure. Talc is a flaky, very soft hydrated magnesium silicate that may contain in nature up to 30% or more of admixed fibrous tremolite.

In addition to being characterized by a widespread fibrosis, silicosis is clinically complicated by a greatly increased susceptibility to tuberculosis. This is not seen with asbestosis. However, asbestos fibers have been shown to produce cancer in man and animals. Cancer of the lung, pleura and peritoneum (mesothelioma), gastrointestinal sites and several other sites may frequently occur. This is now identified as an important public health problem. Silica (quartz) does not reproduce these carcinogenic effects.

11.5.2 Silicosis

Foreign particles, taken into the human body by inhalation, are usually innocuous, like the carbon particles that remain in phagocytic cells of the lungs, more or less indefinitely. However, certain minerals, such as silica, stimulate a severe fibrogenic reaction. Many hypotheses have been advanced to explain the fibrogenic properties of silica, among which the solubility theory was most widely accepted until several years ago. Detailed consideration of the disease silicosis, which occurs in certain classes of workers, such as miners and sandblasters, because of the confirmed inhalation of silica particles into the lungs is outside the scope of this chapter and is covered extensively in the literature.

The *in vivo* response to the accumulation of crystalline silicon dioxide may be envisaged as having two main components. The initial step is phagocytosis of inhaled particles by alveolar macrophages and their subsequent uptake into lysosomes, where damage of lysosomal membranes occurs through H-bonding reactions, rendering them permeable and allowing its enzymes to leak into the cytoplasm and destroy the cell. Collagen synthesis by neighboring fibroblasts is stimulated by death of these macrophages. Heppleston and Styles (1967) have shown that the viable macrophage–silica interaction results in the release of a nonlipid factor of unknown nature that stimulates collagen formation by fibroblasts as judged by synthesis of hydroxyproline. This is a finding of great interest in light of the involvement of silicon in collagen synthesis disclosed by recent research presented earlier in this paper.

In the second step, which occurs at the same time, silica evidently stimulates the production of lipids, especially phospholipids, by the alveolar epithelium and the interaction with macrophages. Despite the evidence against participation of lipid as a fibrogenic agent within silicotic nodules themselves, it nevertheless appears that lipid, when absorbed from or diffusing in the lesions, may possess a systemic and even a local role designed to maintain a supply of macrophages to replace those destroyed under the local action of silica. This second aspect requires further investigation.

11.5.3 Asbestosis

Less information is available on asbestosis than on silicosis, principally because asbestos has been recognized more recently. Moreover, the causative agent is much more complex than silica. The lung scarring that follows the inhalation of asbestos fibers is clinically different from that which follows silica inhalation, with the scarring diffuse rather than nodular. There are also other important differences between the pneumoconiosis caused by silica and that caused by asbestos fibers. For example, much larger particles are involved in asbestosis; furthermore, the fibrosis develops more rapidly (with heavy exposure) and produces greater pulmonary disability, and there is a clearly discernible causal relationship between asbestosis and cancer.

Since the asbestos minerals are all silicates, it might be presumed that loss of metal cations leaving needlelike pseudomorphs of silica might be the causative agent. However, this cannot be so according to Selikoff (1978) because needlelike particles of alumina or nonsiliceous glass can cause similar effects, at least in test animals, where this type of cancer develops more rapidly. In addition, the effect of asbestos is not inhibited by PVPNO, which is effective against silica (Davis, 1972).

Physical shape and size of the carcinogenic particles appears to be more important than their chemical composition in the induction of asbestosis and cancer; for example, the particles must be an accurate size to produce the disease. Thus, mesothelioma cannot be produced by injecting fibrils less than $\frac{1}{2}$ μm long into the pleural space, whereas longer fibers will produce the disease in a high percentage of animals. On the other hand, fibrous asbestos that is definitely carcinogenic can be rendered quite innocuous by grinding it into roughly spherical shapes. The reasons for this are not known.

As with cancer, in general, there is no clear understanding of the cellular mechanisms that underlie the development of cancer associated with silicon particulate matter. There is good reason to believe, however, that such an effect is not related simply to the presence of silicon as an element. Since this disease appears to be due to mechanical damage to tissues and is not related to the chemical nature of the microacicular dust particles, it is proposed that a broader and more descriptive term than *asbestosis* should be used (Iler, 1979).

11.5.4 Renal Toxicity

Two major types of functional renal toxicity have been observed: obstructive (urolithiasis) and toxic (nephropathy).

Urolithiasis

Normally, urinary-silica is readily excreted, but under some conditions a part of it is deposited in the urinary tract to form calculi (uroliths). Small calculi may be excreted harmlessly, whereas large calculi may block the passage of urine and cause death of the animal. Silica urolithiasis, a problem in range animals in western Australia, western regions of Canada, and the northwestern parts of the United States is related to a diet of pasture plants containing a high silicon content in the form of opaline silica. This is in contrast to urolithiasis in man where oxalates, urates, or phosphates play the predominant role.

The silica of calculi from both sheep and cattle has been specifically identified as amorphous opal, most of which has been derived from the absorbed monosilicic acid along with accessory elements—notably Mg, Ca, and P—and appreciable amounts of organic material, mainly glycoprotein. A minute fraction of the opal consists of phytoliths from plants with occasional fragments of sponge spicules and diatoms embedded in the calculi. The exact nature of the organic material, or matrix, in siliceous calculi has not been determined, but chemical studies (Keeler, 1963) indicate the presence of a glycoprotein containing a neutral carbohydrate moiety.

Factors leading to the formation of urinary calculi are poorly defined. Attempts to produce them in sheep and cattle by raising the intake of silica or lowering the intake of water have not been successful. It is apparent that high dietary intakes and high silica outputs in the urine, associated with supersaturation of the urine, are insufficient to explain the polymerization of the monosilicic acid, deposition of silica, and formation of calculi.

It has been suggested that among the accessory elements magnesium might be involved in the formation of calculi (Baker *et al.*, 1961). However, it is not possible to state whether these elements are simply occluded in the silica or whether they play a role in the aggregation of urolithic material.

The glycoprotein component of the organic matrix has been assigned a critical role in the formation of urinary calculi in man, through acting as a primary matrix that becomes secondarily mineralized. A similar theory has been adopted to explain the formation of siliceous calculi in cattle (Keeler, 1963) and phosphatic calculi in sheep, cattle, and dogs. This theory may be questioned on the basis that the main mechanism involved is precipitation of the inorganic components, which, in turn, depends upon both supersaturation and nucleation.

Calculi have been produced in rats with oral ingestion of magnesium trisilicate and sodium metasilicate. Supplementation of diets with tetraethoxysilane was particularly effective in producing calculi, however, presumably because of its greater degree of absorption (Sauer *et al.*, 1959).

Siliceous urinary calculi have been produced in dogs fed an atherogenic diet high in silicate bulking constituents, containing 11% magnesium silicate and 44% silicic acid gel as part of the 27% nonnutritive bulk (Ehrhart and McCullaug, 1973).

Relatively few cases of siliceous calculi have been reported in the urinary tracts of human beings. Herring (1962) reported finding no silica stones among 10,000 urinary calculi analyzed by crystallography and X-ray diffraction. Urinary siliceous calculi were first documented by Herman and Goldberg (1960). As in most of the reported cases, the source of silicon was attributed to the ingestion of magnesium trisilicate, an antacid and the most extensively used silicon medicinal agent today. Silicon levels have been shown to be increased by its ingestion, since approximately only 7% of the contained silicon is eliminated in the urine (Keeler and Lovelace, 1963). On the other hand, magnesium trisilicate has been ingested by human subjects for many years in amounts of several grams per day without apparent serious adverse effects. Nevertheless, occasional reports of urinary calculi containing silicon have appeared in which prolonged use of magnesium trisilicate was a common feature. As of 1964, nine such cases were recorded in the world literature. No renal malignancies were noted and no renal parenchymal damage was observed in this group. Lagergren (1962) reported urinary calculi, consisting mainly of silica, in five human patients, all of whom had been ingesting magnesium trisilicate tablets over several years. A recent case is described by Joekes *et al.* (1973). The original analysis did not include silicon, but showed the presence of calcium, oxalate, phosphorus, and magnesium; subsequent analysis showed the calculus to be predominantly amorphous silica. Stone analysis in most hospital laboratories is qualitative and not quantitative so that trace constituents can be erroneously reported as main ingredients. In contrast to the well known occurrence of calculi composed chiefly of silica in some animals is a sparcity of information on urologic problems in man associated with silica.

Nephropathy

Although the pulmonary effects of silicon are well documented, there are relatively few studies in the literature that indicate an effect of silicon on the kidney in either man or experimental animals. A causal relationship of free silica inhalation and nephropathy in humans is probable but not yet established.

A moderate increased frequency of renal pathology in individuals suffering from silicosis has been noted. Kolev *et al.* (1970) presented the first evidence of distinct renal morphologic alteration in silicosis, describing glomerular and tubular lesions in 23 (51%) out of 45 patients who died of advanced silicosis. More recently, three patients were reported developing a nephropathy after inhalation of free silica. The diagnosis of silicon nephropathy was supported by determination of the silicon content in kidney tissue. In the first patient, a bricklayer presenting with proteinuria and hypertension, renal biopsy showed a focal glomerulonephritis and degenerative changes in the proximal tubules, although no change in tubular function could be demonstrated (Saldanha et al., 1975). The silicon content of renal tissue was 200 ppm dry weight, some 14 times normal. The second report (Giles et al., 1978) describes a sandblaster, with acute pulmonary silicoproteinosis, who died in acute renal failure after a period of massive proteinuria. Again, excessive amounts of silicon were found in the kidney and histological examination revealed a mild proliferative glomerulonephritis, not judged severe enough to explain the proteinuria and renal failure. A third patient also with excessive industrial exposure to silicon presented proteinuria with no obvious tubular dysfunctions (Hauglustaine et al., 1980). Renal biopsy again disclosed a mild focal proliferative glomerulonephritis and an elevated silicon content in kidney tissue.

It would be of interest to obtain data on the renal silicon content in subjects with silicosis but without glomerulopathy. The pathogenesis of silicon nephropathy is unclear. There are speculations of an immunologic injury, but so far the evidence is not conclusive. A simple toxic effect of silicon on the kidney, perhaps on renal enzymes, has also been postulated, but the fact that tubular function is unaffected suggests a major difference from the toxic nephropathies associated with many metals, such as cadmium, lead, mercury, and copper. The difference could result from silicon's lack of influence on the Na-K-ATPase transport enzyme system, in contrast to the other nephrotoxic metals.

Nephrotoxicity of silicon has been suggested in animal studies. Newberne and Wilson (1970) fed rats and dogs sodium silicate, magnesium trisilicate, silicon dioxide, and aluminum silicate for 4 weeks at an equivalent dose level of 0.8 g/kg/day for all compounds. No pathology was noted in rats but renal tubular damage, including inflammatory changes, was produced in dogs fed the first two compounds. Long-term studies of feeding silica and silicates to rats in amounts comparable to antiacid medications have revealed no nephropathy. It has been suggested there may be a species-related susceptibility to renal damage from ingestion of silicic acid. The dose of magnesium trisilicate fed to rats and dogs in the preceding study is approximately 25 times greater than that of a patient taking 5 g magnesium trisilicate daily. Other investigators have demonstrated similar renal changes in guinea pigs by administering orally a quartz suspension over a 6-month period (Markovic and Arambasic, 1971) and magnesium trisilicate over a 4-month period (Dobbie and Smith, 1982) at a level more closely approximating that consumed by a patient. Primary proliferation glomerular lesions were not described in any of these animal studies.

An association between silicates and a condition known as endemic or

Balkan nephropathy has been suggested. This disorder is characterized by a chronic interstitial nephritis that inevitably leads to renal failure, affecting individuals living in specific locations in Yugoslavia, Bulgaria, and Rumania. Histologically, there is marked interstitial fibrosis with tubular damage and in one third of the cases at autopsy carcinoma of the upper urinary tract. The search for etiological agents has focused on several possible local factors, including silicate minerals leached into the drinking water of the region, immunity, and infection. Endemic nephropathy is described in a recent editorial (Lancet, 1977).

Although it appears likely that the etiology of endemic nephropathy is due to causes other than silicon, possibly infection, the potential for nephrotoxicity of long-term consumption of silicon compounds, including those that might be added to food and drugs, should not be overlooked.

11.6 Summary

Silicon, considered heretofore as an environmental contaminant, is one of the most recent trace elements established as "essential" for higher animals and a mechanism and site of action have been identified. Silicon has been demonstrated to perform an important role in connective tissue, especially in bone and cartilage. Abnormalities of both bone and cartilage have been produced in silicon-deficient animals. These skeletal abnormalities were associated with a reduction in matrix components, resulting in the establishment of a requirement for silicon in collagen and glycosaminoglycans formation. Silicon's primary effect in bone and cartilage appears to be on the matrix, formation of the organic matrix appearing to be more severely affected by silicon deficiency than the mineralization process. Additional support for silicon's metabolic role in connective tissue is provided by the finding that silicon is a major ion of osteogenic cells, especially high in the metabolically active state of the cell, and furthermore, that silicon reaches high levels in mitochondria of these cells, indicating that silicon participates in the biochemistry of the subcellular, enzyme-containing structures. Although it is clear from the body of recent work that silicon performs a specific metabolic function, a structural role has also been proposed for silicon in connective tissue. A relationship established between silicon and aging is probably related to glycosaminoglycan changes.

The subject of silicon toxicity is almost invariably associated with the silicosis problem. As with many other essential elements, certain chemical forms of silicon may be toxic if inhaled or ingested in large amounts although the chronic oral ingestion of small amounts of many siliceous materials is generally considered safe as evidenced by the number of silicates on the FDA GRAS list. However, the potential for nephrotoxicity of long-term consumption of certain silicon compounds should not be overlooked.

References

Baker, G., and Jones, L. H. P., 1961. Opal uroliths from a ram, *Aust. J. Agric. Res.* 12:473–482.
Baumann, H., 1960. Verhatten der Kieselsaure im menschlichen Blut und Harn, *Hoppe-Seyler's Z. Physiol. Chem.* 320:11–20.
Bendz, G. and I. Lindqvist (eds.), 1981. *Biochemistry of Silicon and Related Problems*, Plenum Press, New York.
Benke, G. M., and Osborn, T. W., 1978. Urinary silicon excretion by rats following oral administration of silicon compounds, *Food Cosmet. Toxicol.* 17:123–127.
Carlisle, E. M., 1979. Silicon localization and calcification in developing bone, *Fed. Proc.* 28:374.
Carlisle, E. M., 1970a. Silicon: A possible factor in bone calcification, *Science* 167:179–280.
Carlisle, E. M., 1970b. A relationship between silicon and calcium in bone formation, *Fed. Proc.* 29:265.
Carlisle, E. M., 1971. A relationship between silicon, magnesium and fluorine in bone formation in the chick, *Fed. Proc.* 30:462.
Carlisle, E. M., 1972. Silicon an essential element for the chick, *Science* 178:619–621.
Carlisle, E. M., 1974. Silicon as an essential element, *Fed. Proc.* 33:1758–1766.
Carlisle, E. M., 1975. Silicon in the osteoblast, *Fed. Proc.* 34:927.
Carlisle, E. M., 1976a. *In vivo* requirement for silicon in articular cartilage and connective tissue formation in the chick, *J. Nutr.* 106:478–484.
Carlisle, E. M., 1976b. Bone cell silicon stores and interrelationships established with other elements, *Fed. Proc.* 35:256.
Carlisle, E. M., 1979. A silicon–molybdenum interrelationship *in vivo*, *Fed. Proc.* 38:553.
Carlisle, E. M., 1980a. A silicon requirement for normal skull formation, *J. Nutr.* 10:352–359.
Carlisle, E. M., 1980b. Biochemical and morphological changes associated with long bone abnormalities in silicon deficiency, *J. Nutr.* 10:1046–1056.
Carlisle, E. M., 1981. Silicon: A requirement in bone formation independent of vitamin D, *Calc. Tissue Intern.* 33:27–34.
Carlisle, E. M., 1982. The nutritional essentiality of silicon, *Nutr. Rev.* 40:193–198.
Carlisle, E. M., and Alpenfels, W. F., 1978. A requirement for silicon for bone growth in cultures, *Fed. Proc.* 37:404.
Carlisle, E. M., and Alpenfels, W. F., 1980. A silicon requirement for normal growth of cartilage in culture, *Fed. Proc.* 39:787.
Carlisle, E. M., and Garvey, D. L., 1982. The effect of silicon on formation of extracellular matrix components by chondrocytes in culture, *Fed. Proc.* 41:461.
Carlisle, E. M., and Suchil, C., 1983. Silicon and ascorbate interaction in cartilage formation in culture, *Fed. Proc.* 42:398.
Carlisle, E. M., Berger, J. W., and Alpenfels, W. F., 1981. A silicon requirement for prolyl hydroxylase activity, *Fed. Proc.* 40:866.
Charnot, Y., and Pères, G., 1971. Contribution à l'etude de la regulation endocrinienne du metabolisme silicique, *Anal. Endocrinol.* 32:397–402.
Charnot, Y., and Pères, G., 1978. Silicon, endocrine balance and mineral metabolism, in *Biochemistry of Silicon and Related Problems*, G. Bendz and I. Lindquist (eds.), Plenum Press, New York, pp. 269–280.
Davis, J. M. G., 1972. Effects of polyvinylpyridine-N-oxide (P204) on the cytopathogenic action of chrysotile asbestos in vivo and in vitro, *Br. J. Exp. Pathol.* 53:652–658.
Davison, E. B., Frey, M. J., Moore, T. D., and McGanity, 1978. Relationship of metal metabolism to vascular disease mortality rates in Texas, *Am. J. Clin. Nutr.* 31:1188–1197.
Dobbie, J. W., and Smith, M. J. B., 1982. Silicate nephrotoxicity in the experimental animal: The missing factor in analgesic nephropathy, *Scot. Med. J.* 27:10–16.

Ehrhart, L. A., and McCullaug, K. G., 1973. Silica urolithiasis in dogs fed an atherogenic diet, *Proc. Soc. Exp. Biol. Med.* 143:131–132.
Fessenden, R. J., and Fessenden, J. S., 1967. The biological properties of silicon compounds, *Adv. Drug. Res.* 4:95–132.
Fregert, S., 1959. Studies on silicon in tissues with special reference to skin, *Acta Derm. Venereol. Suppl.* 39:1–91.
Giles, R. D., Sturgill, B. C., Suratt, P. M., and Bolton, W. K., 1978. Massive proteinuria and acute renal failure in a patient with acute silicoproteinosis, *Am. J. Med.* 64:336–342.
Goldwater, L. J., 1936. The urinary excretion of silica in non-silicotic humans, *J. Ind. Hyg. Toxicol.* 18:163–166.
Hauglustaine, D., Van Damme, B., Daenens, P., and Michielsen, P., 1980. Silicon nephropathy: A possible occupational hazard, *Nephron* 26:219–224.
Heinen, W., 1965. Time-dependent distribution of silicon in intact cells and cell-free extracts of *Proteuns mirabilis* as a model of bacterial silicon transport, *Arch. Biochem. Biophys.* 110:137–149.
Henglein, F. A., and Scheinost, K., 1956. Substituierte Silylderivative des Pektins und der Glucose, *Makromol. Chem.* 20:59–73.
Heppleston, A. G., and Styles, J. A., 1967. Activity of a macrophage factor in collagen formation by silica, *Nature* 214:521–522.
Herman, J. R., and Goldberg, A. S., 1960. New type of urinary calculus caused by antacid therapy, 1960. *J. Am. Med. Assoc.* 174:128–129.
Herring, L. C., 1962. Observations on the analysis of ten thousand urinary calculi, *J. Urol.* 88:545–562.
Joekes, A. M., Rose, G. A., and Sutor, J., 1973. Multiple renal silica calculi, *Br. Med. J.* 1:146–147.
Iler, R., 1979. *The Chemistry of Silica,* Wiley, New York.
Jones, L. H. P., and Handreck, K. A., 1967. Silica in soils, plants, and animals, *Adv. Agron.* 19:107–149.
Jones, L. H. P., and Handreck, K. A., 1969. The relation between the silica content of the diet and the excretion of silica by sheep, *J. Agric. Sci.* 65:129–134.
Keeler, R. F., 1963. Silicon metabolism and silicon–protein matrix interrelationship in bovine urolithiasis, *Ann. N.Y. Acad. Sci.* 104:592–611.
Kelsay, J. L., Behall, K. M., and Prather, E. S., 1979. Effect of fiber from fruits and vegetables on metabolic response of human subjects. II. Calcium, magnesium, iron and silicon balances, *Am. J. Clin. Nutr.* 32:1876–1880.
King, E. J., Schmidt, E., Roman, W., and Kind, P. R. N., 1956. The effect of silica and silicic acid on enzymes, *Enzymologia* 17:341–351.
King, E. J., and Belt, T. H., 1938. The physiological and pathological aspects of silica, *Physiol. Rev.* 18:329–365.
King, E. J., and McGeorge, M., 1938. The biochemistry of silicic acid. V. The solution of silica and silicate dusts in body fluids, *Biochem. J.* 32:417–433.
Kolev, K., Doitschinov, D., and Todorov, D., 1970. Morphologic alterations in the kidney by silicosis, *Med. Lav.* 61:205–210.
Kvorning, S. A., 1950. The silica content of the aortic wall in various age groups, *J. Gerontol.* 5:23–25.
Lagergren, C., 1962. Development of silica calculi after oral administration of magnesium trisilicate, *J. Urol.* 87:994–996.
Lancet (editorial), 1977. Balkan nephropathy, *Lancet* 1:683–684.
Leslie, J. G., Kao, K. T., and McGavack, T. H., 1962. Silicon in biological material. II. Variaions in silicon contents in tissues of rats at different ages, *Proc. Soc. Exp. Biol. Med.* 110: 218–220.
Le Vier, R. R., 1975. Distribution of silicon in the adult rat and rhesus monkey, *Bioinorg. Chem.* 4:109–115.

Lewin, J. C., 1955. Silicon metabolism in diatoms. II. Sources of silicon for growth of *Navicula pelliculosa, Plant Physiol.* 30:129–134.
Lewin, J., and Reimann, B. E. F., 1969. Silicon and plant growth, *Ann. Rev. Plant Physiol.* 20:289–304.
Loeper, J., Loeper, J., and Fragny, M., 1978. The physiological role of silicon and its antiatheromatous action, in *Biochemistry of Silicon and Related Problems,* G. Bendz and I. Lindquist (eds.), Plenum Press, New York, pp. 281–296.
Losee, F., Cutress, T. W., and Brown, R., 1973. *Trace Substances in Environmental Health,* Vol. VII, D. D. Hemphill (ed.), University of Missouri Press, Columbia, p. 19.
Markovic, B. L., and Arambosic, M. D., 1971. Experimental chronic interstitial nephritis compared with endemic human nephropathy, *J. Pathol.* 103:35–40.
MacCardle, R. C., Engman, M. F., Jr., and Engman, M. F., Sr., 1943. XCIV, Mineral changes in neurodermatitis revealed by microincineration, *Arch. Dermatol. Syphilol.* 47:335–372.
McGavack, T. H., Leslie, J. G., and Kao, K. T., 1962. Silicon in biological material. I. Determinations eliminating silicon as a contaminant, *Proc. Soc. Exp. Biol. Med.* 110:215–218.
Mehard, C. W., Sullivan, C. W., Azam, F., and Volcani, B. E., 1974. Role of silicon in diatom metabolism. IV. Subcellular localization of silicon and germanium in *Nitschia alba* and *Cylindrotheca fusiformis, Physiol. Plant.* 30:265–272.
Mehard, C. W., and Volcani, B. E., 1976. Silicon-containing granules of rat-liver, kidney and spleen mitochondria. Electron probe X-ray microanalysis, *Cell Tissue Res.* 174:315–328.
Muller, S. A., Posner, H. S., Firschein, H. E., 1966. Effect of vitamin D-deficiency on the crystal chemistry of bone mineral, *Proc. Soc. Exp. Biol. Med.* 121:844–846.
Needham, A. E., 1965. *The Uniqueness of Biological Materials,* Pergamon Press, Oxford.
Newberne, P. M., and Wilson, R. B., 1970. Renal damage associated with silicon compounds in dogs, *Proc. Natl. Acad. Sci.* 65:872–875.
Nikaido, T., Austin, J., Trueb, L., and Rinehart, R., 1972. Studies in aging of brain. II. Microchemical analysis of the nervous system in Alzheimer patients, *Arch. Neurol.* 27:549–554.
Okuda, A., and Takahashi, E., 1964. *In the Mineral Nutrition of the Rice Plant,* Symp., Intern. Rice Res. Inst., Johns Hopkins Press, Baltimore, Maryland, pp. 123–146.
Policard, A., Collet, A., Moussard, D. H., and Pregermain, S., 1961. Deposition of silica in mitochondria: An electron microscopic study, *J. Biophys. Biochem. Cytol.* 9:236–238.
Reimann, B. E. F., Lewin, J. C., and Volcani, B. E., 1965. Studies on the biochemistry and fine structure of silica shell formation in diatoms. I. The structure of the cell wall of *Cylindrotheca fusiformis,* Reiman and Lewin, *J. Cell. Biol.* 24:39–55.
Rowsell, E. V., and Leonard, R. A., 1957, The effect of soluble silica on the respiration of rat liver homogenates, *Biochem. J.* 66:3 P.
Saldanha, L. F., Rosen, V., and Gonick, H. C., 1975. Silicon nephropathy, *Am. J. Med.* 59:95–103.
Sauer, F., Laughland, D. H., and Davidson, W. M., 1959. Silica metabolism in guinea pigs, *Can. J. Biochem. Physiol.* 37:183–191.
Schafer, P. W., and Chandler, J. A., 1970. Electron probe x-ray microanalysis of a normal centriole, *Science* 170:1204–1205.
Schwarz, K., 1973. A bound form of silicon in glycosaminoglycans and polyuronides, *Proc. Natl. Acad. Sci.* 70:1608–1612.
Schwarz, K., 1978. Significance and functions of silicon in warm-blooded animals, in *Biochemistry of Silicon and Related Problems,* G. Bendz and I. Lindquist (eds.), Plenum Press, New York, pp. 207–230.
Schwarz, K., and Milne, D. B., 1972. Growth-promoting effects of silicon in rats, *Nature* 239:333–334.
Select Committee on GRAS Substances, 1979. Evaluation of the health aspects of certain silicates as food ingredients (SCOGS-61). Life Sciences Research Office, Federation of American Societies for Experimental Biology, Bethesda, Md.

Selikoff, I. J., 1977. Carcinogenic potential of silica compounds, in *Biochemistry of Silicon and Related Problems*, G. Benz and I. Lindquist (eds.), Plenum Press, New York, pp. 311–335.

Simpson, T. L., and Volcani, B. E. (eds.), 1981. *Silicon and Siliceous Structures in Biological Systems*, Springer-Verlag, New York.

Takaya, K., 1975. Intranuclear silicon detection in a subcutaneous connective tissue cell by energy-dispersive x-ray microanalysis using fresh air-dried spread, *J. Histochem. Cytochem.* 23:681–685.

Underwood, E. J., 1977. *Trace Elements in Humans and Animal Nutrition*, Academic Press, New York.

Van Soest, P. J., and Lovelace, F. E., 1969. Solubility of silica in forages, *J. Anim. Sci.* 29:182.

Weiss, A., and Herzog, A., 1978. Isolation and characterization of a silicon–organic complex from plants, in *Biochemistry of Silicon and Related Problems*, G. Bendz and I. Lindquist (eds.), Plenum Press, New York, pp. 109–127.

Werner, D., 1977. *The Biology of Diatoms Botanical Monograph*, Vol. 13, Blackwell Press, Oxford.

Worth, G., 1952. Der Kieselsaürespiegel im menschlichen Blut, *Klin. Wochenschr.* 30:82–83.

Nickel

12

Forrest H. Nielsen

12.1 Introduction and History

Historically, the first study of the biological action of nickel was reported in 1826 when the oral nickel toxicity signs exhibited by rabbits and dogs were described. Between 1853 and 1912 numerous other studies on the pharmacologic and toxicologic actions of various nickel compounds were described. The findings from these studies were summarized by Nriagu (1980a). The first reports on the presence of nickel in plant and animal tissues appeared in 1925 (Berg, 1925; Bertrand and Macheboeuf, 1925). Although Bertrand and Nakamura (1936) first suggested that nickel may be an essential element, conclusive evidence for essentiality did not appear until 1970–1975. Thus, most of the studies on the biochemical, nutritional, and physiological roles of nickel were done subsequent to 1970.

12.2 Nickel and Its Compounds in Cells and Tissues

Divalent nickel is apparently the important oxidation state of nickel in biochemistry. Divalent nickel forms a large number of complexes encompassing coordination numbers 4, 5, and 6, and all main structural types, which include square planar, square pyramidal, tetrahedral, octahedral, and trigonal pyramidal. Moreover, in complicated equilibria Ni^{2+} complexes often exist between these structural types. Like other ions of the first transition series, Ni^{2+} has the ability

---

Forrest H. Nielsen • U.S. Department of Agriculture, Agricultural Research Service, Grand Forks Human Nutrition Research Center, Grand Forks, North Dakota 58202.

to complex, chelate, or bind with many substances of biological interest. Thus, it is not surprising that nickel is a ubiquitous element found in all biological material. Various authors have tabulated the nickel content in numerous plants, microorganisms, and animals in a recently published book (Nriagu, 1980b). Except for some nickel-accumulating plants and marine species, nickel levels in nearly all biological materials are in the range of nanograms per gram (ppb) to a few micrograms per gram (ppm).

The binding of nickel *in vitro* to numerous molecules isolated from cellular materials may have important counterparts *in vivo*. Sunderman (1977) suggested that ultrafilterable Ni^{2+} binding ligands play an important role in extracellular transport of nickel, intracellular binding of nickel, and excretion of nickel in urine and bile. In human serum, amino acids were found to be components of the low-molecular-weight Ni^{2+}-binding fraction and L-histidine was found to be the main Ni^{2+}-binding amino acid (Lucassen and Sarkar, 1979). At physiological pH, nickel coordinates with histidine via the imidazole nitrogen. In rabbit serum, cysteine, histidine, and aspartic acid may be involved in the binding of Ni^{2+} (Sunderman, 1977). The binding of Ni^{2+} by cysteine probably occurs at the N (amino) and S (sulfhydryl) sites. Computer approaches have predicted that the predominant interaction with naturally occurring low-molecular-weight ligands would occur with histidine and cysteine (Jones *et al.*, 1980).

The binding of nickel by some proteins has been suggested to be of physiological significance. Albumin is the principal Ni^{2+}-binding protein in human, bovine, rabbit, and rat serum. Nickel apparently is bound to albumin by a square planar ring formed by the terminal amino group, the first two peptide nitrogen atoms at the N-terminus, and the imidazole nitrogen of the histidine residue which is located at the third position from the N-terminus. Canine and porcine albumins, which contain tyrosine instead of histidine at the third position, have less affinity for Ni^{2+} than albumins from other species (Sunderman, 1977). Another serum protein, histidine-rich glycoprotein (HRG), apparently has the ability to bind Ni^{2+} (Morgan, 1981). Both albumin and HRG have been suggested to have a role in the transport and homeostasis of nickel in serum. Another histidine-rich protein, purified from newborn rat epidermis, also binds Ni^{2+}. Takeda *et al.* (1981) suggested that in addition to contributing to the formation of the matrix material of cornified cells, the histidine-rich protein regulates the utilization of trace metals, including nickel, in granular cells. They also suggested that the carboxyl groups of aspartic acid and glutamic acid are more likely to be involved in the binding of Ni^{2+} than histidine in the epidermal histidine-rich protein.

The binding of nickel to nucleotides also may be of physiological significance. Nickel, via a structural role, may be a stabilizer for DNA and RNA. This role may be more important in some species than others. Nickel was found in relatively high amounts in chromatin of dinoflagellates (Kearns and Sigee, 1980). In these organisms, metal ions, including nickel, may be of particular importance for chromatin structure because the chromatin is permanently condensed with no associated histones. Although the phosphate oxygens are a major binding site for nickel, there is considerable evidence that Ni^{2+} interacts with the purine base adenine at the N-7 site in the nucleotide ATP.

Organic acids apparently are major bioligands in the uptake and translocation of nickel in plants. Nickel translocates in plants as a stable anionic organic complex. Early suggestions were that these complexes were formed by amino acids. However, recent phytochemical studies showed that in nickel-accumulating plants, nickel was contained as anionic citrate or malate complexes (Kersten et al., 1980).

Nickel is an integral component, not just bound, to some biological substances. To date, four nickel-containing substances have been, or are tentatively, identified.

A nickel-containing macroglobulin, nickeloplasmin, has been found in human and rabbit serum (Sunderman, 1977). Characteristics of nickeloplasmin include an estimated molecular weight of 7.0×10^5, nickel content of 0.90 g-at/mol, and esterolytic activity. Nickel in nickeloplasmin is not readily exchangeable with $^{63}Ni^{2+}$ in vivo or in vitro. Nickeloplasmin has been suggested to be a ternary complex of serum α_1-macroglobulin with a nickel constituent of serum. Sunderman (1977) noted that a 9.55 α_1-glycoprotein that strongly bound Ni^{2+} has been isolated from human serum and thus suggested that nickeloplasmin might represent a complex of the 9.55 α_1-glycoprotein with serum α_1-macroglobulin. Unfortunately, there is no clear indication of the physiological significance or function of nickeloplasmin.

Urease from several plants and microorganisms was found to be a nickel metalloenzyme. Urease from the jack bean *(Canavalia ensiformis)* was the first example of this natural nickel metalloenzyme. Jack bean urease (E.C. 3.5.1.5) contains stoichiometric amounts of nickel, 2.00 ± 0.12 g-at/mol of 96,600-dalton subunits (Dixon et al., 1980a,b). Nickel in jack bean urease is part of the active site and is tightly bound, being similar to the zinc ion in yeast alcohol dehydrogenase (E.C. 1.1.1.1) and manganous ion in chicken liver pyruvate carboxylase (E.C. 6.4.1.1). Jack bean urease is stable and fully active in the presence of 0.5 mM EDTA at neutral pH. The nickel ion can be removed only upon exhaustive dialysis in the presence of chelating agents or by EDTA at low pH, and then it is not possible to restore nickel with reconstitution of enzymatic activity. The findings of Dixon et al. (1980b) were consistent with an octahedral coordination of Ni^{2+} with urease as seen in model complexes and Ni^{2+}-phosphoglucomutase and Ni^{2+}-carboxypeptidase. Jack bean urease has relatively low reactivity of the active-site sulfhydryl groups, thus suggesting coordination of the active-site nickel with the unreactive cysteine. Of interest, jack bean urease was the first enzyme protein to be crystallized. It took 50 years to show that this enzyme contains nickel in its structure.

More than 50% of the nickel taken up by methanogenic bacteria is incorporated into a low-molecular-weight compound with an absorption maximum at 430 nm (Thauer et al., 1980). This nickel-containing compound, which has been found in every methanogenic bacterium examined to date, has been named factor F_{430}. Preliminary experiments indicate that nickel is tightly bound to factor F_{430} because $^{63}Ni^{2+}$ is not exchanged with factor F_{430} nickel even when the incubation mixture is 6 N HCl. The exact structure of factor F_{430}, which has a mass per mol nickel of approximately 1500 daltons, has not been elucidated. However, findings to date suggest that factor F_{430} contains a nickel tetrapyrrole structure. A nickel-containing degradation product of factor F_{430} has an absorption spectrum in the visible region resembling that of vitamin B_{12}. Also, biosynthetic studies indicate that, per mol of nickel, 8 mol of δ-aminolevulinic acid are incorporated into factor F_{430}. Like nickeloplasmin, there is no clear indication as to the physiological significance or function of factor F_{430}.

In acetogenic bacteria, the reductive carboxylation of methyl tetrahydrofolate to acetate is catalyzed by a multienzyme complex with one portion having carbon monoxide dehydrogenase activity. The synthesis of the moiety with carbon monoxide dehydrogenase activity requires nickel (Thauer et al., 1980). Furthermore, analyses of the multienzyme complex indicate that the moiety is a protein with a nickel-containing prosthetic group with properties similar to those of vitamin B_{12}.

12.3 Nickel Deficiency

For animals, the first description of possible signs of nickel deprivation appeared in 1970. However, those findings, and others that followed shortly thereafter, were obtained under conditions that produced suboptimal growth in the experimental animals (Nielsen, 1980a). Also, some reported signs of nickel deprivation appeared inconsistent. However, retrospective analyses of the methodology used in those studies indicated that much of the inconsistency was probably caused by variance in the iron status of, and environmental conditions for, the experimental animals. Thus, most of the early findings were true nickel deprivation signs in animals under certain dietary and environmental conditions. The importance of iron in nickel nutrition will be described *(vide infra)*.

Since 1975, diets and environments that allow for apparently optimal growth and survival of experimental animals have been used in the study of nickel metabolism and nutrition. To date, signs of nickel deprivation have been described for six animal species—chick, cow, goat, minipig, rat, and sheep.

Nielsen (1980a) summarized the signs of nickel deprivation in chicks. The signs included depressed hematocrits, oxidative ability of the liver in the presence of α-glycerophosphate and yellow lipochrome pigments in the shank skin, and ultrastructural abnormalities in the liver. The abnormalities were characterized by pyknotic nuclei and dilated cisternal lumens of the rough endoplasmic reticulum. The cisternae appeared to be draped around the mitochondria, bringing the ribosomes into close proximity to the outer mitochondrial membrane. There was a general impression that the ribosomes were more irregular in their spacing on the rough endoplasmic reticulum and that large amounts of membrane were devoid of ribosomes. Mitochondria were numerous and closely packed and elongate forms were common. The matrices of the mitochondria were reduced in density and appeared hydrated.

Two independent laboratories have described signs of nickel deprivation for the rat. In the laboratory of Nielsen (1980a), successive generations of rats were exposed to a low-nickel diet throughout fetal, neonatal, and adult life. Signs of nickel deprivation found in the rat included elevated perinatal mortality, unthriftiness characterized by a rough coat and/or uneven hair development in pups, pale livers, and ultrastructural changes in the liver. The most obvious effect of nickel deprivation on liver ultrastructure was a reduced amount of rough endoplasmic reticulum which appeared disorganized in that the normal "stacking" of the cisternae was partially or totally absent. Nickel deprivation appeared to depress growth and hematocrits of rats in these experiments in which an adequate, relatively available form of dietary iron was used, but these signs were not consistently significant. Subsequent to those studies, Nielsen (1980a) found that the iron status of the rat has a major influence on the extent and severity of the signs of nickel deprivation. Thus, when relatively unavailable ferric sulfate at marginally adequate levels was the dietary source of iron, depressed growth and hematopoiesis were consistently found in nickel-deprived offspring. Furthermore, Nielsen (1980a) found that when the iron content and form of the diet was properly manipulated, nickel deprivation could be induced in rats that were not from nickel-deprived dams. Apparent signs of deficiency in weanling rats fed a diet deficient in nickel and marginally adequate in levels of relatively unavailable ferric sulfate for 9–11 weeks included depressed hematopoiesis and liver iron, and elevated plasma lipids and liver copper.

In a series of reports summarized recently, Kirchgessner and Schnegg (1980) described the following nickel deprivation signs for offspring of nickel-deprived dams. At age 30 days, nickel-deprived pups exhibited significantly depressed growth, hematocrit, hemoglobin, and erythrocyte counts; those signs were still evident, but less marked, at age 50 days. However, at age 120 days, depressed growth and hematopoiesis were not evident in the nickel-deprived offspring. Thus, like those of Nielsen (1980a), the findings on growth and hematopoiesis were inconsistent. Perhaps the findings of Kirchgessner and Schnegg (1980) were also influenced by a changing iron status of the animal. Although the initial form of iron used in their studies was ferrous sulfate, the method of adding the ferrous sulfate to the diet could have converted some Fe^{2+} to Fe^{3+}, possibly in varying amounts for each diet. The ferrous sulfate was put into distilled water with several other mineral salts, then added to a moist diet mixture, and heated at 50° to remove water. Also, the diet may have contained residual EDTA (used to prepare the casein), which can affect the availability of iron. Kirchgessner and Schnegg (1980) found that the severity of the nickel deprivation signs in young pups was reduced by increasing dietary iron from 50 µg/g to 100 µg/g.

Other apparent nickel deprivation signs in the young pups described by Kirchgessner and Schnegg (1980) included: (1) At age 30 days, the activities of the liver enzymes malate dehydrogenase, glucose-6-phosphate dehydrogenase, isocitrate dehydrogenase, lactate dehydrogenase, glutamate-oxalacetate transaminase, and glutamate-pyruvate transaminase were depressed whereas the activities of alkaline phosphatase and creatine kinase were elevated. At age 50 days, nickel deprivation did not significantly affect liver malate dehydrogenase, or glucose-6-phosphate dehydrogenase activity. (2) At age 30 days, the activities of the kidney enzymes glutamate dehydrogenase, glutamate-oxalacetate transaminase, and glutamate-pyruvate transaminase were depressed. (3) The levels of urea, ATP, and glucose in serum were reduced. (4) The levels of triglycerides, glucose, and glycogen in liver were reduced. (5) Iron absorption was impaired. (6) Levels of iron, copper, and zinc were depressed in the liver, kidney, and spleen. (7) Levels of calcium and phosphorus were depressed and the level of magnesium was elevated in the femur. (8) Activities of proteinase and leucine arylamidase increased, and the activity of α-amylase decreased in the pancreas. Kirchgessner and Schnegg (1980) suggested that the depression in digestion of starch by α-amylase might be indirectly responsible for their observed large depressions in the activities of hepatic enzymes and in the concentration of hepatic metabolites in nickel-deprived rats.

Anke *et al.* (1980) described the signs of nickel deprivation for goats and minipigs. The signs in these animals included depressed growth, delayed estrus, elevated perinatal mortality, unthriftiness characterized by a rough coat and scaly and crusty skin, depressed levels of calcium in the skeleton and of zinc in liver, hair, rib, and brain. Nickel-deprived goats also showed depressed hematocrit, hemoglobin, iron content in the liver, and triglycerides, β-lipoproteins, glutamate-oxaloacetate transaminase activity, and glutamate dehydrogenase activity in serum. The level of α-lipoproteins was elevated in serum.

Spears and Hatfield (1980) described nickel deprivation signs for sheep and cattle. Nickel-deprived lambs exhibited depressed growth, total serum protein, erythrocyte counts, ruminal urease activity, and total lipids, cholesterol, and copper in liver. Iron contents were elevated in liver, spleen, lung, and brain.

Signs of nickel deprivation in cattle fed a low-protein diet included depressed ruminal urease, serum urea, nitrogen, and growth.

Upon casual inspection, there appears to be some divergency in the described signs of nickel deprivation. Species differences might explain some disagreements, but, most likely, the iron status and age of the animals and the length of experiments probably were major determinants in the direction, extent, and severity of the signs of nickel deprivation obtained by various investigators. Taking the latter parameters into consideration, the signs of nickel deprivation agree relatively well. The major finding apparently is that nickel deprivation leads to abnormal iron metabolism.

In organisms other than animals, depressed growth is the predominant sign of nickel deprivation. Nickel is required for chemolithotropic growth of a number of hydrogen-oxidizing (Knallgas) bacteria (Tabillion et al., 1980), including five strains of *Alcaligenes eutrophus,* two strains of *Xanthobacter autotrophicus, Pseudomonas flava, Arthrobacter* spec. 11X, and *Arthrobacter* strain 12x. Apparently, nickel is necessary for the synthesis of active soluble and membrane-bound hydrogenase in Knallgas bacteria. Growth of methanogenic bacteria, such as *Methanobacterium thermoautotrophism,* which utilize the formation of methane from H_2 and CO_2 as their energy source, is dependent on nickel (Thauer et al., 1980). Depression in the formation of nickel-containing factor F_{430} *(vide supra)* may be responsible for the poor growth during nickel deprivation in these bacteria. Growth of microorganisms dependent on the nickel metalloenzyme urease is also depressed in nickel deprivation. These microorganisms include rumen bacteria (Spears and Hatfield, 1980) and urease-deficient mutant of *Aspergillus nidulans* (Mackay and Pateman, 1980). Nickel may also be required for growth of the Legionnaires' disease bacteria *Legionella pneumophilia* (Reeves et al., 1981), a blue-gree algae (cyanobacteria) *Oscillatoria* sp. (Van Baalen and O'Donnell, 1978), and acetogenic bacteria such as *Clostridium thermoaceticium, C. formicoaceticium, C. aceticium,* and *Acetobacterium Woodii* (Thauer et al., 1980). Carbon monoxide dehydrogenase activity apparently depends on nickel in the acetogenic bacteria that reduce CO_2 to acetate as their energy source.

The biological significance of nickel for plants was recently reviewed by Welch (1981). Nickel apparently is required by some plants for optimum growth, germination rate, and nitrogen utilization. Plants reported to require nickel for growth include *Chlorella vulgaris* (green algae), *Pinus radiata* (Monterey pine), and some nickel-accumulating species of *Alyssum.* Plant species that show improved growth upon nickel supplementation include wheat, grape vines, cotton, paprika, tomato, chinese hemp, potatoes, and soybean. Germination rates of seeds of peas, beans, wheat, castor beans, white lupine, soybeans, timothy, and rice apparently were improved by pretreating with, or germinating in the presence of a low concentration of, nickel. Plant growth dependent on urea-N, thus dependent on the nickel-containing enzyme urease, is also depressed by nickel deprivation. With urea as the sole source of nitrogen, growth was depressed by nickel deprivation in plants such as duckweed (*Lemna paucicostata, Spirodela polyrrhiza,* and *Wolffia globosa*) and rice, tobacco, and soybean callus grown in suspension cultures. Most likely, any plant that requires urease in growth and metabolism would be adversely affected by nickel deprivation. Those plants and the point in their life cycle in which urease may be of importance are described by Welch (1981).

12.4 Nickel Function

To date, the most firmly established biological function of nickel is its *vide supra* described role in the enzyme urease. Thus, nickel is needed for the reaction:

$$(NH_2)_2CO + H_2O \xrightarrow[\text{urease}]{\text{Ni}} CO_2 + 2NH_3$$

Apparently, binding of the substrate urea to a nickel ion in urease is an integral part of the mechanism in the hydrolysis reaction. A carbamato-enzyme intermediate involving active-site nickel has been proposed. Nucleophilic attack or general base catalysis by a suitable active-site group would then lead to an active-site nickel–ammonia complex.

Although nickel is required for synthesis of active hydrogenases for the Knallgas reaction

$$2H_2 + O_2 \rightarrow 2H_2O$$

in Knallgas bacteria, the specific function in the synthesis is unknown. Nickel might be a constituent of both hydrogenase proteins or a regulatory protein essential at the transcriptional or translational level for the synthesis of the hydrogenases, or a factor needed to convert the hydrogenases into a catalytically active configuration (Friedrich *et al.,* 1981).

In acetogenic bacteria, the energy source reaction

$$8[H] + 2CO_2 \rightarrow CH_3COOH + 2H_2O$$

proceeds via formate, formyl tetrahydrofolate, methenyl tetrahydrofolate, methylene tetrahydrofolate, and methyl tetrahydrofolate. The reductive carboxylation of the methyl tetrahydrofolate to acetate is catalyzed by a multienzyme complex, part of which has carbon monoxide dehydrogenase activity. The synthesis of this part of the multienzyme complex is dependent on nickel. Most likely, nickel is a constituent of the carbon monoxide dehydrogenase (Thauer *et al.,* 1980).

In higher animals, the evidence showing that nickel is essential does not clearly define its metabolic function. The finding of nickel metalloenzymes in plants and microorganisms suggests that a similar function for nickel in animals may be found. Nickel can activate many enzymes *in vitro*, but its role as a specific cofactor for any animal enzyme has not been shown. Other possible metabolic functions have been described (Nielsen, 1980b). Perhaps the most promising possibility is that nickel functions as a bioligand cofactor facilitating the intestinal absorption of the Fe^{3+} ion. This hypothesis is supported by findings from factorially arranged experiments which showed that nickel enhanced the absorption of iron present in the diet in less than adequate levels and in a relatively unavailable Fe^{3+} form. Nickel apparently had little or no effect on the absorption of the Fe^{2+} ion when it was present in the diet in less than adequate, but not severely inadequate levels. Possible mechanisms whereby nickel could enhance Fe^{3+} absorption were described by Nielsen (1980b).

12.5 Biological Interactions between Nickel and Other Trace Elements

In animals, plants, and microorganisms, nickel interacts with at least 13 essential minerals: Ca, Cr, Co, Cu, I, Fe, Mg, Mn, Mo, P, K, Na, and Zn. These interactions were recently reviewed by Nielsen (1980c). The interactions that appeared to be of most biological significance were those with iron, copper, and zinc.

The interaction between nickel and iron can be either synergistic or antagonistic. The synergistic interaction apparently occurs between nickel and ferric iron (Nielsen *et al.*, 1982). In factorially arranged experiments, hematopoiesis was affected by an interaction between iron and nickel when the iron supplement was ferric sulfate only. When the dietary ferric sulfate level was low, hemoglobin and hematocrit were lower in nickel-deprived than in nickel-supplemented rats. There was no evidence for an interaction when dietary iron was supplied as a mixture of ferric–ferrous sulfates. When the interaction between nickel and iron affected other parameters, such as the copper content of liver, it generally did so in a manner similar to that found with hemoglobin and hematocrit. That is, signs of nickel deprivation were more severe when dietary iron, as ferric sulfate, was low; or the signs of moderate iron deficiency were more severe when dietary nickel was deficient.

The antagonistic interaction apparently occurs between nickel and ferrous iron (Nielsen *et al.*, 1982). Severe iron deficiency was more detrimental to nickel-supplemented than to nickel-deficient rats, as growth was more severely depressed and perinatal mortality was higher in nickel-supplemented rats. This suggests nickel impaired the utilization of the small amount of dietary iron apparently in the ferrous form.

The form of dietary iron also apparently affects the nature of the signs of nickel deficiency. When only ferric sulfate was supplemented to the diet, plasma and liver total lipids were elevated, and liver iron content was depressed in nickel-deprived rats. On the other hand, when a ferric–ferrous mixture was supplemented to the diet, nickel deprivation depressed plasma total lipids did not affect liver total lipids and elevated the liver content of iron.

Most of the interaction between nickel and iron probably occurs during absorption. There is evidence that both active and passive transport mechanisms play a role in iron absorption. Active transport of iron to the serosal surface is relatively specific for the divalent cation so the ferric ion is absorbed by passive transport. Becker *et al.* (1980) reported that the transport of nickel across the mucosal epithelium apparently is an energy-driven process rather than simple diffusion and suggested that nickel ions use the iron transport system. Because different mechanisms are involved in the absorption of the two forms of iron, and nickel is involved in both mechanisms, it seems likely that iron nutrition should affect nickel absorption and requirement, and vice versa. Evidence for the competitive interaction between nickel and iron during absorption includes the findings that the transfer of nickel from the mucosal to the serosal side of iron-deficient rat intestinal segments was elevated (Becker *et al.*, 1980), and that moderately anemic iron-deficient rats absorbed approximately 2.5 times as much $^{63}Ni^{2+}$ administered via gavage, as did those fed a control diet (Ragan, 1978). Thus, in nickel deficiency, because of the lack of competition with the Ni^{2+} ion, more of the active transport system could be utilized for iron absorption. This would explain the finding of an increased iron content in liver of nickel-deprived rats fed dietary iron in the ferrous form. On the other hand, nickel apparently promotes the passive transport of Fe^{3+} because in nickel deficiency the absorption of Fe^{3+} is depressed. This would explain the finding of a decreased iron content in liver of nickel-deprived rats fed dietary iron as ferric sulfate only. These changes in iron absorption, leading to changes in levels of iron and other trace elements in tissue, probably explain the apparent divergent effects of nickel deprivation on plasma and liver total lipids. In other words, dietary nickel affects these parameters indirectly, with the direction and magnitude of the changes determined by the extent and the mechanism through which nickel and iron interact during their absorption.

The competitive interaction between nickel and iron is possible because they both can form the same type of complex. Nickel ions can possess outer orbital bonding in which the coordination number is 6 and form octahedral complexes. Both ions of iron (Fe^{2+}, Fe^{3+}) can have a coordination number of 6 and form octahedral complexes regardless of whether there is inner or outer orbital bonding. Thus, the same orbitals could be involved in bonding both nickel and iron when they form outer orbital complexes.

Cuprous and cupric ions have a preferred coordination number of 4 and form tetrahedral and square coplanar complexes, respectively. Nickel can possess inner orbital bonding, in which the preferred coordination number is 4, and form either tetrahedral or square coplanar complexes. Thus, if nickel is present in biological systems with a coordination number of 4, nickel and copper could have similar chemical parameters and a nickel–copper interaction probably would be competitive. Supporting this suggestion are findings from factorially arranged experiments that show an antagonistic interaction between nickel and copper (Nielsen et al., 1982). In rats deficient in copper, with significant but not too severe anemia, the copper deficiency signs of elevated heart weight and plasma cholesterol, and depressed hemoglobin, were exacerbated by nickel supplementation. The effect was greater when dietary nickel was 50 µg/g rather than 5 µg/g.

Nickel supplementation did not depress the level of copper in liver, or plasma, of copper-deficient rats. This finding indicates that nickel did not exacerbate copper deficiency signs by interfering with the absorption of copper. The antagonism between copper and nickel was probably due to the isomorphous replacement of copper by nickel at various functional sites. If nickel did not perform, or less efficiently performed, the functions of copper, the end result would be less copper function at various physiological sites and a more severe copper deficiency.

Zn^{2+} has a preferred coordination number of 4 and forms tetrahedral complexes. As mentioned *vide supra,* nickel ions can have inner orbital bonding, in which the preferred coordination number is 4, and form either tetrahedral or square coplanar complexes. Thus if nickel forms tetrahedral complexes in biological systems, a competitive interaction between zinc and nickel should occur. However, to date, most of the findings indicate that the interaction between nickel and zinc is noncompetitive. Instead of interacting directly at sites of zinc function, deficient or toxic levels of nickel apparently act indirectly to shift slightly the distribution of zinc in the body. For example, Anke et al. (1980) found that nickel deprivation depressed the zinc content of liver, hair, rib, and brain in minipigs and goats. Because some signs of nickel deprivation were similar to those of zinc deficiency, Anke et al. (1980, 1981) suggested nickel deficiency disturbs zinc metabolism. Kirchgessner and Pallauf (1973) found that depressed growth and other signs of zinc deficiency in the rat were not changed by supplementation with 50 µg of nickel/g of diet. Nickel supplementation did reduce serum zinc and increased the zinc concentration in the liver. Spears et al. (1978) found that in the rat 50 µg of nickel/g of diet partially alleviated some signs of zinc deficiency, but did not affect others. They noted that nickel depressed tibia zinc but elevated liver zinc in the zinc-deficient animals. This suggests that nickel, through an indirect means, redistributed zinc in the animal body so that some signs of deficiency were alleviated.

Because nickel competitively interacts with copper and iron, but apparently not with zinc, nickel probably forms square coplanar or octahedral complexes in biological systems. The formation of tetrahedral complexes by nickel *in vivo* probably occurs infrequently.

12.6 Nickel Metabolism and Toxicity

In higher animals there are four entry routes into the body—oral intake, inhalation, percutaneous absorption, and parenteral administration. In metabolism, oral intake is of primary importance, and thus is emphasized here. Parenteral administration of nickel, at present, is used only to study nickel distribution, metabolism, and toxicity, and will be discussed only when it indicates possible mechanisms in handling oral nickel. Percutaneous absorption and inhalation entry routes are of major importance in nickel toxicity.

Most ingested nickel remains unabsorbed by the gastrointestinal tract and is excreted in the feces (Sunderman, 1977). Some fecal nickel may come from the bile as nickel has been found in the bile of rats and rabbits injected with ^{63}Ni(II). Limited studies indicate that less than 10% of ingested nickel is normally absorbed. However, a higher percentage may be absorbed during gravidity (Kirchgessner *et al.*, 1981). As mentioned *vide supra*, the transport of nickel across the mucosal epithelium appears to be an energy-driven process rather than simple diffusion. Nickel ions apparently use the iron transport system located in the proximal part of the small intestine.

Although fecal nickel excretion is 10–100 times as great as urinary excretion, the small fraction of nickel absorbed from the intestine and transported to the plasma is excreted primarily via the urine as low-molecular complexes believed to include histidine and aspartic acid. Sweat may also be important in nickel metabolism because in healthy adults it contains several times the amount found in serum. This suggests an active secretion of nickel by sweat glands. However, excretion of nickel by sweat apparently is unresponsive to acute elevated doses of oral nickel (Christensen *et al.*, 1979). Thus, measurements of serum and urinary, but not sweat, nickel may be used to detect variations in the oral intake of nickel.

Transport of nickel in blood is accomplished by serum albumin and by ultrafilterable serum amino acid ligands. No tissue except possibly fetal tissue significantly accumulates nickel. The kinetics of ^{63}Ni^{2+} metabolism in rodents apparently fits a two-component model. A summary of the tissue retention and clearance of ^{63}Ni^{2+} administered by all routes of entry has been given by Kaspizak and Sunderman (1979). This summary shows that kidney retains significant levels of nickel shortly after ^{63}Ni^{2+} is given. The retention probably reflects the role of the kidney in nickel excretion. The level of ^{63}Ni^{2+} in kidney falls quickly over time. Also, studies with $^{63}Ni^{2+}$ show that nickel readily passes through the placenta. Embryonic tissue retains greater amounts of parenteral administered nickel than does that of the dam (Jacobsen *et al.*, 1978). Also, amniotic fluid retains relatively high amounts of orally administered nickel (Kirchgessner *et al.*, 1981). The level of nickel in the fetus does not fall quickly after parenteral administration to the dam, thus suggesting retention or inhibited clearance by the fetus.

The preceding discussion shows that there are mechanisms for the homeostatic regulation of nickel. Thus, it is not surprising that life-threatening toxicity of nickel through oral intake is low, ranking with such elements as zinc, chromium, and manganese. Nickel salts exert their toxic action mainly by gastrointestinal irritation and not by inherent toxicity. Large oral doses of nickel salts are necessary to overcome the homeostatic control of nickel. Generally, 250 μg or more of nickel/g of diet is required to produce signs of nickel toxicity in rats, mice, chicks, rabbits, and monkeys (Nielsen, 1977). The ratio of the minimum toxic dose and the minimum dietary requirement for chicks and rats is apparently near 5000. If animal data can be extrapolated to humans, this translates into a daily dose of 250 mg of soluble nickel to produce toxic symptoms in humans.

Recent findings, however, suggest that oral nickel in not particularly high doses can adversely affect health under certain conditions. The effects of relatively low levels of dietary nickel on copper deficiency and severe iron deficiency were described *vide supra*. Another important condition occurs in humans with an allergy to nickel. Finally, the tendency of the fetus to retain nickel suggests that elevated levels of nickel in the blood should be avoided during pregnancy.

Nickel dermatitis is a relatively common form of nickel toxicity in humans. Several surveys have shown that incidence of sensitivity to nickel is between 4 and 13% (Nielsen, 1977). Until recently, nickel dermatitis has been thought to be caused mainly by the percutaneous absorption of nickel. However, Christensen and Möller (1975) presented evidence that suggested the ingestion of small amounts of nickel may be of greater importance than external contacts in maintaining hand eczema. Cronin *et al.* (1980) observed that an oral dose of 0.6 mg of nickel as nickel sulfate ($NiSO_4$) produced a positive reaction in some nickel-sensitive individuals. That dose is only 12 times as high as the human daily requirement postulated from animal studies.

Lu *et al.* (1981) found that the injection of $NiCl_2$ intraperitoneally in mice caused teratology. They found also that the kinetics of nickel chloride in fetal tissues was different from that in maternal tissues and suggested anomalies caused by elevated nickel levels could occur in embryo without recognizable adverse effects in maternal mice.

The metabolism and toxicity of the carcinogen nickel carbonyl differs markedly from that of Ni^{2+}. Nickel carbonyl is highly volatile and is absorbed readily by the lungs. The inhalation route is the most important in respect to nickel carbonyl toxicity. Also, the metabolism and toxicity of relatively insoluble nickel compounds such as Ni_3S_2 (usually administered intramuscularly or intrarenally) differ from Ni^{2+} in their metabolism. A review on the toxicity of nickel carbonyl and relatively insoluble nickel compounds was given by Sunderman (1977), and thus, will not be done here.

12.7 Summary

Interest in the biochemistry of nickel has been stimulated by recent discoveries of its essentiality to various microorganisms, plants, and animals and of the existence of several nickel metalloenzymes in plants and microorganisms. Signs of nickel deprivation have been described for six animal species—chick, cow, goat, minipig, rat, and sheep. Included among the more consistent signs of deficiency in mammals are depressed growth, unthriftiness characterized by rough hair coat, and an altered iron metabolism leading to depressed hematopoiesis. The predominant sign of nickel deficiency for microorganisms is depressed growth, and for plants is depressed nitrogen utilization. In plants and microorganisms, nickel is known to function in several metalloenzymes including urease, several hydrogenases, and carbon monoxide dehydrogenase. In higher animals, the evidence showing that nickel is essential has not defined its metabolic function. The finding of nickel metalloenzymes in lower forms of life suggests that a similar function for nickel in animals may be found.

Divalent nickel is the apparent important oxidation state in the metabolism of nickel. Ultrafilterable Ni^{2+} binding ligands, perhaps histidine and cystine, apparently play important roles in the extracellular transport of nickel, intracellular binding of nickel, and excretion of nickel in urine and bile. Two Ni^{2+} binding proteins suggested to have a role in the transport and homeostasis of nickel in serum are albumin and histidine-rich glycoprotein. The transport of nickel across the mucosal epithelium apparently occurs as Ni^{2+} and is an energy-driven process, rather than one driven by simple diffusion, and is probably connected with the iron-transport system. Only recently another oxidation state of nickel has been indicated to be important in biochemistry. Ni^{3+} apparently is important for the activity of bacterial hydrogenases.

In conclusion, emerging evidence indicates that nickel is a dynamic trace element in living organisms. However, knowledge of the biochemistry of nickel is very limited. Thus, further research on nickel biochemistry is needed to help evaluate the nature and importance of the physiologic, pharmacologic, and toxicologic actions of nickel.

References

Anke, M., Kronemann, H., Groppel, B., Hennig, A., Meissner, D., and Schneider, H.-J., 1980. The influence of nickel-deficiency on growth, reproduction, longevity and different biochemical parameters of goats, in *3. Spurenelement-Symposium Nickel*, M. Anke, H.-J. Schneider, and Chr. Brüchner (eds.), Friedrich-Schiller-Universität, Jena, GDR, pp. 3–10.

Anke, M., Grün, M., Hoffmann, G., Groppel, B., Gruhn, K., and Faust, H., 1981. Zinc metabolism in ruminants suffering from nickel-deficiency, in *Mengen- und Spurenelement*, M. Anke and H.-J. Schneider (eds.), Karl-Marx-Universität, Leipzig, GDR, pp. 189–196.

Becker, G., Dörstelmann, U., Frommberger, U., and Forth, W., 1980. On the absorption of cobalt(II)- and nickel(II)-ions by isolated intestinal segments in vitro of rats, in *3. Spurenelement-Symposium Nickel*, M. Anke, H.-J. Schneider, and Chr. Brückner (eds.), Friedrich-Schiller-Universität, Jena, GDR, pp. 79–85.

Berg, R., 1925. Das Vorkommen seltener Elemente in den Nahrungsmitteln und menschlichen Ausscheidungen, *Biochem. Z.* 165:461–462.

Bertrand, G., and Macheboeuf, M., 1925. Sur la presence der nickel et du cobalt chez les animaux, *C.R. Acad. Sci. (Paris)* 180:1380–1383.

Bertrand, G., and Nakamura, H., 1936. Recherches sur l'importance physiologique du nickel et due cobalt, *Bull. Soc. Sci. Hyg. Aliment.* 24:338–343.

Christensen, O. B., and Möller, H., 1975. External and internal exposure to the antigen in the hand eczema of nickel allergy, *Contact Derm.* 1:136–141.

Christensen, O. B., Möller, H., Andrasko, L., and Lagesson, V., 1979. Nickel concentration of blood, urine and sweat after oral administration, *Contact Derm.* 5:312–316.

Cronin, E., Di Michiel, A. D., and Brown, S. S., 1980. Oral challenge in nickel-sensitive women with hand eczema, in *Nickel Toxicology*, S. S. Brown and F. W. Sunderman, Jr. (eds.), Academic Press, New York, pp. 149–152.

Dixon, N. E., Gazzola, C., Asher, C. J., Lee, D. S. W., Blakeley, R. L., and Zerner, B., 1980a. Jack bean urease (EC 3.5.1.5). II. The relationship between nickel, enzymatic activity, and the "abnormal" ultraviolet spectrum. The nickel content of jack beans, *Can. J. Biochem.* 58:474–480.

Dixon, N. E., Blakeley, R. L., and Zerner, B., 1980b. Jack bean urease (EC 3.5.1.5). III. The involvement of active-site nickel ion in inhibition by β-mercaptoethanol, phosphoramidate, and fluoride, *Can. J. Biochem.* 58:481–488.

Friedrich, B., Heine, E., Finck, A., and Friedrich, C. G., 1981. Nickel requirement for active hydrogenase formation in *Alcaligenes eutrophus*, *J. Bacteriol.* 145:1144–1149.

Jacobsen, N., Alfheim, I., and Jonsen, J., 1978. Nickel and strontium distribution in some mouse tissues passage through placenta and mammary glands, *Res. Comm. Chem. Pathol. Pharmacol.* 20:571–584.

Jones, D. C., May, P. M., and Williams, D. R., 1980. Computer stimulation models of low-molecular-weight nickel(II) complexes and therapeuticals in vivo, in *Nickel Toxicology*, S. Brown and F. W. Sunderman, Jr. (eds.), Academic Press, New York, pp. 73–76.

Kaspizak, K. S., and Sunderman, F. W., Jr., 1979. Radioactive ^{63}Ni in biological research, *Pure Appl. Chem.* 51:1375–1389.

Kearns, L. P., and Sigee, D. C., 1980. The occurrence of period IV elements in dinoflagellate chromatin: An x-ray microanalytical study, *J. Cell Sci.* 46:113–127.

Kersten, W. J., Brooks, R. R., Reeves, R. D., and Jaffre, T., 1980. Nature of nickel complexes in *Psychotria douarrei* and other nickel-accumulating plants, *Phytochemistry* 19:1963–1965.

Kirchgessner, M., and Pallauf, J., 1973. Zum Einfluss von Fe-, Co-, bzw. Ni-Zulagen be Zinkmangel, *Zeit. Tierphysiol. Tierernaehr. Futtermittelkd.* 31:268–274.

Kirchgessner, M., and Schnegg, A., 1980. Biochemical and physiological effects of nickel deficiency, in *Nickel in the Environment*, J. O. Nriagu (ed.), Wiley, New York, pp. 635–652.

Kirchgessner, M., Roth-Maier, D. A., and Schnegg, A., 1981. Progress of nickel metabolism and nutrition research, in *Trace Element Metabolism in Man and Animals (TEMA-4)*, J. McC. Howell, J. M. Gawthorne, and C. L. White (eds.), Australian Academy of Sciences, Canberra, pp. 621–624.

Lu, C.-C., Matsumoto, N., and Iijima, S., 1981. Placental transfer and body distribution of nickel chloride in pregnant mice, *Toxicol. App. Pharmacol.* 59:409–413.

Lucassen, M., and Sarkar, B., 1979. Nickel(II)-binding constituents of human blood serum, *J. Toxicol. Environ. Health* 5:897–905.

Mackay, E. M., and Pateman, J. A., 1980. Nickel requirement of a urease-deficient mutant in *Aspergillus nidulans*, *J. Gen. Microbiol.* 116:249–251.

Morgan, W. T., 1981. Interactions of the histidine-rich glycoprotein of serum with metals, *Biochemistry* 20:1054–1061.

Nielsen, F. H., 1977. Nickel toxicity, in *Advances in Modern Toxicology, Vol. 2, Toxicology of Trace Elements*, R. A. Goyer and M. A. Mehlman (eds.), Wiley, New York, pp. 129–146.

Nielsen, F. H., 1980a. Evidence of the essentiality of arsenic, nickel, and vanadium and their possible nutritional significance, in *Advances in Nutritional Research*, Vol. 3, H. H. Draper (ed.), Plenum, New York, pp. 157–172.

Nielsen, F. H., 1980b. Possible functions and medical significance of the abstruse trace metals, in *Inorganic Chemistry in Biology and Medicine, ACS Symposium Series No. 140*, A. E. Martell (ed.), American Chemical Society, Washington, D.C., pp. 23–42.

Nielsen, F. H., 1980c. Interactions of nickel with essential minerals, in *Nickel in the Environment*, J. O. Nriagu (ed.), Wiley, New York, pp. 611–634.

Nielsen, F. H., Uthus, E. O., and Hunt, C. D., 1982. Interactions between the "newer" trace elements and other essential nutrients, in *New Zealand Workshop on Trace Elements in New Zealand Proceedings*, J. V. Dunckley (ed.), University of Otago, Dunedin, N.Z., pp. 165–173.

Nriagu, J. O. (ed.) 1980a. *Nickel in the Environment*, Wiley, New York.

Nriagu, J. O., 1980b. Global cycle and properties of nickel, in *Nickel in the Environment*, J. O. Nriagu (ed.), Wiley, New York, pp. 1–26.

Ragan, H. A., 1978. Effects of iron deficiency on absorption of nickel, *Northwest Lab. Annual Rep. DOE Assist. Secr. Environ., PNL-2500-PT. 1*, pp. 1.9–1.10.

Reeves, M. W., Pine, L., Hunter, S. H., George, J. R., and Harrell, W. K., 1981. Metal requirements of *Legionella pneumophilia*, *J. Clin. Microbiol.* 13:688–695.

Spears, J. W., and Hatfield, E. E., 1980. Role of nickel in ruminant nutrition, in *3. Spurenelement-Symposium Nickel*, M. Anke, H.-J. Schneider, and Chr. Brückner (eds.), Friedrich-Schiller-Universität, Jena, GDR, pp. 47–53.

Spears, J. W., Hatfield, E. E., and Forbes, R. M., 1978. Interrelationship between nickel and zinc in the rat, *J. Nutr.* 108:307–312.

Sunderman, F. W., Jr., 1977. A review of the metabolism and toxicology of nickel, *Ann. Clin. Lab. Sci.* 7:377–398.

Tabillion, R., Weber, F., and Kaltwasser, H., 1980. Nickel requirement for chemolithotrophic growth in hydrogen-oxidizing bacteria, *Arch. Microbiol.* 124:131–136.

Takeda, A., Fukuyama, K., Ohtani, O., and Epstein, W. L., 1981. Regulation of RNase activity by interaction of trace elements with histidine-rich protein from newborn rat epidermis, *Biol. Trace Element Res.* 3:317–326.

Thauer, R. K., Diekert, G., and Schönheit, P., 1980. Biological role of nickel, *Trends Biochem. Sci. (Pers. Ed.)* 5:304–306.

Van Baalen, C., and O'Donnell, R., 1978. Isolation of a nickel-dependent blue-green alga, *J. Gen. Microbiol.* 105:351–353.

Welch, R. M., 1981. The biological significance of nickel, *J. Plant Nutr.* 3:345–356.

Tin

13

David B. Milne

13.1 Introduction

Tin is widely distributed in nature and has been economically important for man since the Bronze Age. However, from a biological point of view, tin and its compounds have excited relatively little interest. Most research has been related to the toxicity of tin and its compounds, mainly because of man's exposure to tin from canned foods and various organotin compounds used as plasticizers and as fungicides. A review on tin and man and foods by Schroeder et al. (1964) treated tin as an abnormal trace element, concluding that "measurable tin is not necessary for life or health." This conclusion was based mainly on the fact that with the then current, but inadequate, methods of analysis, zero levels of tin were found in the newborn and in organs of natives of some foreign countries. This thinking was changed when Schwarz et al. (1970) reported that various tin compounds stimulated growth in rats fed highly purified diets if trace element contamination from the environment was rigidly excluded. They went on to suggest that tin may be essential. This work, however, remains to be confirmed by other investigators.

13.2 Tin in Cells and Tissues

13.2.1 Chemical Properties

The chemical nature of tin compounds in tissues is unknown at this time. However, there are a number of chemical properties of this element and its

---

David B. Milne • Department of Agriculture, Agricultural Research Service, Grand Forks Human Nutrition Research Center, Grand Forks, North Dakota 58202.

compounds that could give some insight as to its possible function. Tin shares group 4A of the Periodic Table with carbon, silicon, germanium, and lead. Thus, tin has a tendency to form covalent bonds rather than being purely ionic (Cotton and Wilkinson, 1966). Tetravalent tin has a strong tendency to form coordination complexes with four, five, six, and possibly eight ligands. It forms links with anthocyanin and other pigments and combines with halogens, dithiols, and proteins. A large number of synthetic organotin compounds are known that are used as bactericides and fungicides. Some organotin compounds also act as catalysts in polymerization, transesterification, and olefin condensation reactions. Divalent tin is readily oxidized in solution. The oxidation-reduction potential of $Sn^{2+} \rightarrow Sn^{4+}$ is -0.154 V as compared to the standard hydrogen electrode (Remey, 1956). This is within the physiological range, and is close to the oxidation-reduction potential of many of the flavin enzymes.

Table 13-1. Distribution of Tin in Human Tissues

	Kehoe (1940) (ppm wet)	Tipton and Cook (1963) (ppm dry)	Teraoka (1981) (ppm dry)
Liver	0.60	0.85	1.1 ± 0.41[a]
Kidney	0.20	0.74	0.72 ± 0.25
Lung	0.45	1.78	1.5 ± 0.80
Spleen	0.22	0.55	0.55 ± 0.37
Heart	0.22	0.20	0.59 ± 0.25
Pancreas	—	0.18	0.43 ± 0.17
Adrenal	—	0.25	0.82 ± 0.48
Thyroid	—	0.48	0.37 ± 0.25
Hylar lymph node	—	—	9.8 ± 1.3
Stomach	0.50	0.51	0.88 ± 0.55[b]
Intestine	0.16	1.09	—
Tongue	—	—	0.40 ± 0.02
Trachea	—	0.65	0.72 ± 0.54
Aorta	—	0.53	0.74 ± 0.39
Brain	n.d.	<0.30	—
Muscle	0.11	<0.21	—
Diaphragm	—	0.17	—
Esophagus	—	0.68	—
Ovary	—	0.66	—
Prostate	—	0.63	—
Skin	—	0.36	—
Testes	—	0.63	—
Whole blood	0.14	—	—

[a] Mean ± SD.
[b] Includes both stomach and intestine.

13.2.2 Distribution in Mammalian Tissues

The occurrence of tin in human tissues was surveyed by Kehoe *et al.* (1940), Tipton and Cook (1963), and Teraoka (1981) by spectographic methods (Table 13-1). Tin was found in most samples of all tissues surveyed, except brain. Highest levels occurred in the hylar lymph node. Whole blood contained 14 µg of tin per 100 g (Kehoe, 1940) with most of it concentrated in the erythrocytes. Earlier analyses demonstrated tin in almost all tissues of the ox, horse, and sheep at concentrations of 0.5–4.0 ppm dry weight (Boyd and De, 1933). The exact nature of tin in tissues is unknown at this time.

The amount of tin found in tissues is apparently age dependent (Schroeder *et al.*, 1964). Little or no detectable tin (less than 5 ppm in the ash) was found in fetal or newborn tissues. It is deposited in tissues rapidly during the first year or so of life but thereafter does not normally accumulate with age.

It should be noted that much of the preceding data was obtained with analytical methods that may prove to be inadequate for determining tin in tissues (Schwarz *et al.*, 1970; Beeson *et al.*, 1977). Most of the data were obtained from samples that had been dried at 110°C and then dry-ashed at 450°C. Under these conditions, much of the tin may be lost. Most organic tin derivatives evaporate below 200°C. Also, some inorganic tin salts such as stannic chloride and stannic acetate have low boiling points. In experiments with dry ashing, only 20% of tin as $SnCl_4$ and none of the organic tin as triphenyltin chloride were recovered (Milne, unpublished observation; see Beeson *et al.*, 1977).

13.3 Deficiency and Function

There has been little interest in tin as a possible essential element. Prior to 1970, tin had been considered as an "environmental contaminant" (Underwood, 1962). A review on tin in man and foods (Schroeder *et al.*, 1964) treated tin as an abnormal trace element and concluded that tin was not essential for life or health. This conclusion was based mainly on the inability to detect tin by the prevailing, but inadequate, methods of analysis in tissues of newborns or in organs of natives of some foreign countries. However, Schwarz *et al.* (1970) reported that several tin compounds stimulated growth in rats if trace element contamination from the environment was rigidly excluded.

Supplements of various tin compounds produced significant growth responses in rats maintained on purified amino acid diets inside trace element controlled isolators. Trimethyltin hydroxide, dibutyltin maleate, stannic sulfate, and potassium stannate enhanced growth at dose levels supplying 1 ppm of tin to the diet. As stannic sulfate, 0.5, 1.0, and 2.0 ppm of tin increased growth by 24, 53, and 59% respectively (Schwarz et al., 1970). It was shown in follow-up studies that tin as stannic sulfate or in the form of trialkyltin derivatives (i.e., R_3SnX where X is Cl or OH) fed at levels of 1–2 ppm Sn also improved growth rates and significantly increased pigmentation of rat incisors (Milne et al., 1972; Schwarz, 1974). It was estimated from the preceding data that the rats requirement for tin would be between 1 and 2 ppm in the diet. However, it is desirable that these findings be confirmed in other laboratories and extended to other species before tin can be considered an essential element.

The effort to repeat these studies has been difficult because of an apparent interaction between dietary tin and riboflavin (Nielsen et al., 1982). Apparently, much of the suboptimal growth rates, ragged greasy fur, and alopecia seen in the rats in isolators and fed purified diets (Schwarz et al., 1970; Schwarz, 1971) was caused by diets that were marginal or deficient in riboflavin (Moran and Schwarz, 1978). Diets used for the preceding studies initially contained 0.25 mg of riboflavin per 100 g of diet. However, much of the riboflavin apparently was rapidly destroyed since diets fed to rats in the isolators were exposed to high levels of direct light. This did not occur with the diets of rats maintained in conventional caging. Large improvements in growth and coat condition were noted when additional riboflavin was fed (Milne and Schwarz, Nielsen and Milne, unpublished observations). These observations may be critical when considering tin biochemistry since the oxidation-reduction potential for $Sn^{2+} \rightarrow Sn^{4+}$ of -0.15 V is the same as that of many flavoproteins.

Addition of 5 ppm of tin to drinking water had little effect on growth and longevity of mice (Schroeder and Balassa, 1967) or rats (Schroeder et al., 1968). No significant differences in weights at any age or in survival rates were seen between mice receiving or not receiving tin in drinking water (Schroeder and Balassa, 1967). Rats receiving the tin supplement exhibited increased growth in both males and females during the first 30 days of the study (Schroeder et al., 1968), but the weight differences between the tin-fed and control rats disappeared after 60 days. No significant differences due to tin were noted in mean survival rates or life spans in these rats.

13.4 Metabolism and Toxicity

13.4.1 Inorganic Tin

Dietary Intake, Absorption, and Excretion of Tin

Dietary intakes of tin are extemely variable. Tipton et al. (1969) reported a daily intake in the United States ranging from 0.10 to 100 mg of tin, with an average of 5.8 mg per day. Earlier, Schroeder et al. (1964) found the average

Table 13-2. Excretion and Retention of Dietary Tin by Humans (mg per day)

Intake	Feces	Urine	Apparent retention	Reference
17.14 ± 1.9	22.88 ± 1.9	0.14		Kehoe, 1940
1.49	2.13	0.11	−0.75	Tipton et al., 1966
2.48	1.55	0.08	0.85	Tipton et al., 1966
5.8 ± 0.7	3.6 ± 0.7	0.085	2.1 ± 1.8	Tipton et al., 1969
8.8 ± 1.1	3.6 ± 0.5	0.058	5.1 ± 2.4	Tipton et al., 1969
0.11	0.66 ± 0.02	0.029	0.03	Johnson and Greger, 1982
49.7	48.2 ± 1.5	0.122 ± 0.05	1.3	Johnson and Greger, 1982

content of an institutional diet to be 1.41 µg of tin per gram on a wet basis. They calculated that a normal human intake might range from 1 to 40 mg of tin per day, with an average of about 4 mg per day. Tin-coated cans and utensils contribute to the wide variation of daily intake of tin.

Inorganic tin is apparently poorly absorbed, with the bulk being excreted in the feces (Table 13-2). Most investigators (Kehoe, 1940; Tipton et al., 1966; Calloway and McMullen, 1966; Tipton, 1969; and Johnson and Greger, 1982) found that fecal excretion of tin approximated dietary intakes of tin when human subjects were fed foods that contained 8–190 mg of tin per day. Hiles (1974) found that at least 97% of a single oral dose of 20 mg per kilo body weight of tin fed to rats was excreted in the feces. Smaller amounts of tin were excreted in feces by human subjects fed low (<5 mg tin per day) levels of tin. Fecal losses of two human subjects who consumed from 0.33 to 5.37 mg of tin per day during a 30-day period ranged from 0.41 to 5.04 mg of tin per day (Tipton et al., 1966). Fecal excretion of tin ranged from 0.03 to 0.12 mg per day in four subjects fed 0.11 mg of tin per day for a 20-day period (Johnson and Greger, 1982).

Urinary excretion of tin was generally found to be small (4–246 µg per day) regardless of dietary intake (Kehoe, 1940; Tipton et al., 1966; Tipton, 1969; Johnson and Greger, 1982). It was hypothesized that human subjects excrete only small amounts of tin in urine because tin is poorly absorbed and/or because the body may excrete excess tin through the bile, pancreatic juice, or sweat (Johnson and Greger, 1982).

Tissue Distribution of Absorbed or Injected Tin

Distribution or retention of tin in tissues is dependent on the nature of exposure. Rats fed a single dose of radiolabeled tin (Hiles, 1974) or fed high levels of tin for several weeks (Greger and Johnson, 1981; Yamaguchi et al.,

1980) accumulated tin in the kidneys, liver, and bones. The accumulation of tin in bone appeared to be directly related to the level of tin fed (Yamaguchi et al., 1980; Hiles, 1974). Tin tended also to accumulate in liver and kidney when doses greater than 2 mg per day were fed (Greger and Johnson, 1981; Yamaguchi, 1980; Hiles, 1974). Chmielnicka et al., (1981) found 95% of the ^{113}Sn located in the skin and hair of rats after subcutaneous injections. About 4.4% was found in muscles and bone. The rest of the dose was distributed between the intestine, liver, and kidneys. The data of Durbin (1960) on the distribution of tin in the rat indicate that after intramuscular injection, tin accumulates mainly in the bones (29.4%), liver (1%), and kidney (3%). These observed differences are probably due to the type of exposure and route of administration.

Toxicity of Inorganic Tin: Interactions with Other Nutrients

Metallic tin and its salts are considered to be of low oral toxicity (DeGroot et al., 1973), mainly because they are poorly absorbed from the alimentary tract (Schroeder et al., 1964). The oral LD_{50}'s of stannous chloride in rats and mice were reported to be 700 and 1200 mg/kg, respectively (Calvary, 1942). Soluble salts of tin are gastric irritants. Canned foods and drinks containing from 250 to 700 ppm of tin have been held responsible for outbreaks of nausea, vomiting, and diarrhea in large numbers of people in several countries (Benoy et al., 1971). In a series of subacute studies, DeGroot et al. (1973) noted growth retardation, reduced hematocrits, and hemoglobin levels in rats fed 0.3% or more of several tin salts such as stannous chloride, sulfate, oxalate, or tartrate. Other tin salts such as stannic oxide, stannous sulfide, or oleate had no effect on growth or iron status at levels of up to 1% added to the diet.

Injected tin interferes with porphyrin biosynthesis and enhances heme breakdown. Stannous chloride injected into rabbits increased the concentration of coproporphyrin in blood and urine (Chiba et al., 1980). Tin severely inhibited the activity of 5-aminolevulinate dehydratase in blood, but not in liver. No effect of tin on 5-aminolevulinic acid concentrations was observed, however (Chiba et al., 1980). In contrast to the action of lead, the inhibition of 5-aminolevulinate dehydratase was reversed after cessation of tin treatment. Other workers (Kappas and Maines, 1976) have shown that injected stannous chloride rapidly accelerates heme breakdown by inducing heme oxygenase activity in both liver and kidney. Consequently, there is an impairment of heme dependent functions such as cytochrome P_{450} mediated drug metabolism (Kappas and Maines, 1976) and the occurrence of anemia (DeGroot et al., 1973) when animals receive toxic amounts of tin.

Dietary tin can influence the metabolism of several other minerals. Yamaguchi et al., (1980) found that tin altered calcium metabolism in rats. When 3.0 mg of tin as stannous chloride was administered per kilogram of body weight twice a day for 90 days, significant decreases were observed in relative femur weights, calcium concentration, lactate dehydrogenase, and alkaline phosphatase activities in the liver and calcium content and acid phosphatase activity in the femoral diaphysis and epiphyses. No effect was seen with a 0.3 mg/kg dose. On the other hand, Johnson and Greger (1982) noted that a diet containing about 50 mg of tin per day fed men for 20 days had no effect on serum calcium and on fecal or urinary losses of calcium.

Addition of about 200 µg of tin to diets also had some adverse effects on the zinc and copper nutriture of rats (Greger and Johnson, 1981). Rats fed the high-tin-containing diet lost significantly more zinc in their feces and retained lower levels of zinc in their tibias and kidneys than rats fed a diet containing 1 µg of tin per gram. Rats fed the tin-supplemented diet also had lower levels of copper in their kidneys than their controls. Human subjects fed about 50 mg of tin daily lost significantly more zinc in their feces and significantly less in the urine (Johnson *et al.*, 1982). Overall zinc retention was lower in the tin-fed subjects. Tin supplementation had no effect on fecal or urinary losses of Cu, F, Mn, or Mg.

13.4.2 Organotin Compounds

Organotin compounds are used industrially as stabilizers to prevent the thermal degradation of many chlorinated compounds such as certain types of transformer oils, PVC, chlorinated rubbers, paraffins, and modified plastics. They are also effective catalytic agents in polyurethane production and esterification reactions (Piver, 1973). Because of their toxic properties, many trialkyltin compounds are used extensively as bactericides and fungicides.

When compared with inorganic tin, organotin compounds, particularly the trialkyl derivatives, are extremely toxic. For example, the oral lethal dose of triethyltin sulfate is 10 mg/kg body weight for rabbits, rats, and guinea pigs. LD_{50}'s are about 5 mg/kg (Stoner *et al.*, 1955) for this compound.

The toxicology of organotin compounds has been thoroughly reviewed by Barnes and Stoner (1959) and Piver (1973). Trialkyl-substituted tin derivatives are the most toxic of the organic tin compounds. They apparently attack the central nervous system and cause impaired movements of the hind limbs, general ataxia and unsteadiness, then finally complete paralysis and death. Mono and dialkyl derivatives are relatively nontoxic.

Tetraalkyltin compounds are converted enzymatically *in vivo* to trialkyltin and exert delayed but similar actions to the trialkyltin (Barnes and Stoner, 1959).

Triethyltin is a powerful inhibitor of mitochondrial oxidative phosphorylation (Aldridge and Street, 1970; Rose and Lock, 1970) and inhibits dinitrophenol-stimulated ATPase activity as well as P_i–ATP exchange activity. Binding studies by Aldridge and Street (1970) showed that the inhibitory effects of triethyltin in rat liver mitochondria can be correlated with the presence of a high-affinity binding site. Cain *et al.* (1977) have also shown that dibutyltin dichloride is also a potent inhibitor of oligomycin-sensitive ATPase. The exact mechanism of this action is unknown at this time.

13.5 Summary

From a biological point of view, tin and its compounds have excited relatively little interest except for their toxicology. Little is known about the forms of tin that occur in tissues, even though its presence has been detected in most tissues surveyed except brain.

The presence of tin in the body has been treated as an "environmental contaminant." However, in 1970 it was reported that several tin compounds stimulated growth in rats if trace element contamination from the environment was rigidly excluded. Efforts to repeat this study have been difficult because of an apparent interaction between tin and riboflavin. In retrospect the diets used in the preceding study were marginal or deficient in riboflavin. Deficiency symptoms attributed to tin were improved with addition of riboflavin to the diet. This may be important when considering tin biochemistry since the reduction-oxidation potential for $Sn^{2+} \leftrightarrow Sn^{4+}$ of -0.15 mV is similar to that of many flavoproteins.

Inorganic tin is poorly absorbed, which accounts for its relatively low toxicity. However, it has been shown to interact with other essential elements and interferes with their metabolism. Organotin compounds, on the other hand, are extremely toxic when compared with inorganic tin salts. Trialkyltin derivatives, in particular, attack the central nervous system, ultimately causing paralysis and death.

In conclusion, except for toxic actions of various tin compounds, relatively little is known about the biochemistry of tin and, except for one study that has not been reproduced, its potential role as an essential trace element.

References

Aldridge, W. N., and Street, B. W., 1978, Oxidative phosphorylation. The specific binding of trimethyltin and triethyltin to rat liver mitochondria, *Biochem. J.* 118:171–179.

Barnes, J. M., and Stoner, H. B., 1959, The toxicology of tin compounds, *Pharmacol. Rev.* 11:211–231.

Beeson, K. C., Griffitts, W. R., and Milne, D. B., 1977, Tin, in *Geochemistry and the Environment*, Vol 2. The relation of other selected trace elements to health and disease. National Academy of Sciences, Washington, D.C., pp. 88–92.

Benoy, C. J., Hooper, P. A., and Schneider, R., 1971, The toxicity of tin in canned fruit juices and solid foods, *Food Cosmet. Toxicol.* 9:645–656.

Boyd, T. C., and De, N. C., 1933, some applications of the spectroscope in medical research, *Indian J. Med. Res.* 20:789–795.

Cain, K., Partis, M. D., and Griffiths, D. E., 1977, Triphosphate synthetase complex, *Biochem. J.* 166:593–602.

Calloway, D. H., and McMullen, J. J., 1966, Fecal excretion of iron and tin by men fed stored canned foods, *Am. J. Clin. Nutr.* 18:1–6.

Calvary, H. O., 1942, Trace elements in foods, *Food Res.* 7:313–320.

Chiba, M., Ogihara, K., and Kikuchi, M., 1980, Effect of tin or porphyrin biosynthesis, *Arch. Toxicol.* 45:189–195.

Chmielnicka, J., Symanska, J. P., and Sniee, J., 1981, Distribution of tin in the rat and disturbances in the metabolism of zinc and copper due to repeated exposure to Sn Cl_2, *Arch. Toxicol.* 47:263–268.

Cotton, I. A., and Wilkinson, G., 1966, *Advanced Inorganic Chemistry,* 2nd ed., Interscience, New York, pp. 309–338.

DeGroot, A. P., Feron, V. J., and Til, H. P., 1973, Short term toxicity studies on some salts and oxides of tin in rats, *Food Cosmet. Toxicol.* 11:19–30.

Durbin, P. W., 1960, Metabolic characteristics within a chemical family, *Health Phys.* 2:225–258.

Fritch, P., Blanquat, G. D. S., Derache, R., 1977, Effect of various dietary components on absorption and tissue distribution of orally administered inorganic tin in rats, *Food Cosmet. Toxicol.* 15:147–149.

Greger, J. L., and Johnson, M. A., 1981, Effect of dietary tin on zinc, copper and iron utilization by rats, *Food Cosmet. Toxicol.* 19:163–166.

Hiles, R. A., 1974, Absorption, distribution and excretion of inorganic tin in rats, *Toxicol. Appl. Pharmac.* 27:366–379.

Johnson, M. A., and Greger, J. L., 1982, Effects of dietary tin on tin and calcium metabolism of adult males, *Am. J. Clin. Nutr.* 35:655–660.

Johnson, M. A., Baier, M. J., and Greger, J. L., 1982, Effects of dietary tin on zinc, copper, iron, manganese, and magnesium metabolism of adult males, *Am. J. Clin. Nutr.* 35:1332–1338.

Kappas, A., and Maines, M. D., 1976, Tin: A potent inducer of heme oxygenase in kidney, *Science,* 192:60–62.

Kehoe, R. A., Cholak, J., and Story, R. V., 1940, A spectrochemical study of the normal ranges of concentrations of certain trace metals in biological materials, *J. Nutr.* 19:579–592.

Milne, D. B., Schwarz, K., and Sognnaes, R., 1972, Effect of newer essential trace elements in rat incisor pigmentation, *Fed. Proc.* 31:700.

Moran, J. K., and Schwarz, K., 1978, Light sensitivity of riboflavin in amino acid diets, *Fed. Proc.* 37:671.

Nielsen, F. H., Milne, D. B., and Zimmerman, T. J., 1982, Dietary tin affects riboflavin nutriture of the rat, *Proc. N.D. Acad. Sci.* 36:62.

Piver, W. T., 1973, Organotin compounds: Industrial applications and biological investigations, *Environ. Health Per.* 4:61–79.

Remy, H., 1956, *Treatise on Inorganic Chemistry,* Vol. 1, Elsevier, New York, pp. 524–537.

Rose, M. S., and Lock, E. A., 1970, The interaction of triethyltin with a component of guinea pig liver supernatant. Evidence for histidine in the binding sites, *Biochem. J.* 120:151–157.

Schroeder, H. A., and Balassa, J. J., 1967, Arsenic, germanium, tin and vanadium in mice: Effects on growth, survival and tissue levels, *J. Nutr.* 92:245–252.

Schroeder, H. A., Balassa, J. J., and Tipton, I. H., 1964, Abnormal trace elements in man: Tin, *J. Chron. Dis.* 17:483–502.

Schroeder, H. A., Kanisawa, M., Frost, D. V., and Mitchener, M., 1968, Germanium, tin and arsenic in rats: Effects on growth, survival, pathological lesions and life span, *J. Nutr.* 96:37–45.

Schwarz, K., 1971, Tin as an essential growth factor for rats, in W. Mertz and W. E. Cornatzer (eds.), *Newer Trace Elements in Nutrition,* Dekker, New York, pp. 313–326.

Schwarz, K., 1974, New essential trace elements (Sn, V, F, Si): Progress report and outlook, in W. G. Hoekstra, J. W. Suttie, H. E. Ganther, and W. Mertz (eds.), *Trace Element Metabolism in Animals—2,* University Park Press, Baltimore, pp. 355–380.

Schwarz, K., Milne, D. B., and Vinyard, E., 1970, Growth effects of tin compounds in rats maintained in a trace element controlled environment, *Biochem. Biophys. Res. Comm.* 40:22–29.

Stoner, H. B., Barnes, J. M., and Duff, J. I., 1955, Studies on the toxicity of alkyl tin compounds, *Brit. J. Pharmacol.* 10:16–25.

Teraoka, M., 1981, Distribution of 24 elements in the internal organs of normal males and the metallic workers of Japan, *Arch. Environ. Health* 36:155–165.

Tipton, I. H., and Cook, M. J., 1963, Trace elements in human tissue. Part II. Adult subjects from the United States, *Health Phys.* 9:103–145.

Tipton, I. H., Stewart, P. L., and Martin, P. G., 1966, Trace elements in diets and excreta, *Health Phys.* 12:1683–1689.

Tipton, I. H., Stewart, P. L., and Dickson, J., 1969, Patterns of elemental excretion in long term balance studies, *Health Phys.* 16:455–462.

Underwood, E. J., 1962, *Trace Elements in Human and Animal Nutrition,* 2nd ed., Academic Press, New York, 429 pp.

Yamaguchi, M., Saito, R., and Okada, S., 1980, Dose-effect of inorganic tin on biochemical indices in rats, *Toxicology,* 16:267–273.

Arsenic

14

Forrest H. Nielsen and Eric O. Uthus

14.1 Introduction and History

Since ancient times, arsenicals have been characterized by actions both benevolent and malevolent. The sulfides of arsenic—realgar (tetraarsenic tetrasulfide) and orpiment (arsenic trisulfide)—were mentioned by the Greeks and Romans prior to the Christian era. However, pure arsenic apparently was unknown to these ancient civilizations. Because arsenic has such a long history, only a few highlights will be presented here. These highlights were obtained from reviews by Vallee *et al.* (1960), Schroeder and Balassa (1966), Frost (1970), Klevay (1976), and Bruckner and Dietze (1980).

Originally, the Greek name *arsenikon* (bold, valiant, masculine) was given to the sulfides of arsenic. This name was transferred to the element itself when it was isolated.* There is no definite knowledge of who first isolated elemental arsenic and when. At one time or another, the isolation of elemental arsenic has been accredited to Albert the Great, in 1250; Paracelsus, in 1520; and Schroeder, in 1649.

Very early in the history of arsenic it was found that some arsenic compounds were convenient scentless and tasteless instruments for homicidal purposes. Thus for about 1100 years, up to the last century, arsenicals reigned as the king of poisons. Even today arsenic is often thought of as being synonymous with poison. Nonetheless, the bad reputation of arsenic did not prevent it from becoming an important pharmaceutical agent. Medicinal virtues of arsenicals were acclaimed by Hippocrates (460–357 B.C.), Aristotle (384–322 B.C.), Theophrastus (370–288

---

* In German, *arsenik* refers to arsenic trioxide, whereas *arsen* is the name for the element.

Forrest H. Nielsen and Eric O. Uthus • U.S. Department of Agriculture, Agricultural Research Service, Grand Forks Human Nutrition Research Center, and Department of Biochemistry, University of North Dakota, Grand Forks, North Dakota 58202.

B.C.), and Pliny the Elder (23–70 A.D.). In 1905, Ehrlich first synthesized organic arsenicals and demonstrated their chemotherapeutic action against trypanosomes. Subsequently, the pharmacology of arsenicals was extended to more than 8000 compounds by 1937. Arsenicals constituted two-thirds of all metal–carbon compounds known at that time. Arsenicals still are used as spirochetocides, amebicides, and trypanocides. Arsenicals were considered at various times to be specific remedies in the treatment of anorexia, other nutritional disturbances, neuralgia, rheumatism, asthma, chorea, malaria, tuberculosis, diabetes, skin diseases, and in numerous hematologic abnormalities. The use of arsenic for these disorders has fallen into disrepute or has been replaced by more effective alternatives. In addition to medicinal use, arsenicals were, and in some cases still are, used as insecticides, herbicides, and fungicides.

Early attempts to prove the essentiality of arsenic were unsuccessful. Hove et al. (1938) found diets furnishing as little as 2 µg of arsenic per day per rat had no significant effect on growth, hemoglobin concentration, and red blood cell number or fragility. They observed, however, that dietary arsenic caused a delay in the rate of fall of hemoglobin in rats fed whole milk without mineral supplementation subsequent to being fed a mineral-supplemented milk diet compared to controls not given arsenic. Schroeder and Balassa (1966) found rats and mice grew and developed normally on diets containing as little as 0.053 µg of arsenic/g. They observed, however, that arsenic supplementation resulted in a healthier appearance of the skin and hair of rats and mice. The first conclusive evidence, which will be described (*vide infra*), for arsenic essentiality appeared in 1975–1976. Nielsen et al. (1975) and Anke et al. (1976) described arsenic deprivation signs for rats, minipigs, and goats. Thus it is only recently that arsenic has been studied from the biochemical, nutritional, and physiological points of view, not the toxicological.

14.2 Arsenic and Its Compounds in Cells and Tissues

Because arsenic has long been considered a toxic element, its content in many biological materials has been determined. Tissue analyses indicate that retained arsenic in most animals is widely distributed in low concentrations (<1.0 µg/g fresh weight) under normal conditions. Arsenic, like other Group V elements, apparently has a predilection for fat. For example, Larsen et al. (1979) found that arsenic was associated with the lipid phase of brain tissue. Marine organisms generally contain more arsenic than do humans and higher animals. Although some marine organisms may contain over 100 µg of arsenic/g, most contain a few micrograms of arsenic/g. A tabulation of the arsenic content of numerous plant, microorganism, and animal tissue has been compiled recently (National Academy of Sciences, 1977).

Both the trivalent and pentavalent states of arsenic apparently exist in biological material. Which oxidation state is most important in the biochemistry, nutrition, and physiology of living organisms is not known. The organic arsenic compounds that are probably of most biochemical importance are those that contain methyl groups. Those methylated arsenicals that may be important in biologic cycles are given in Figure 14-1.

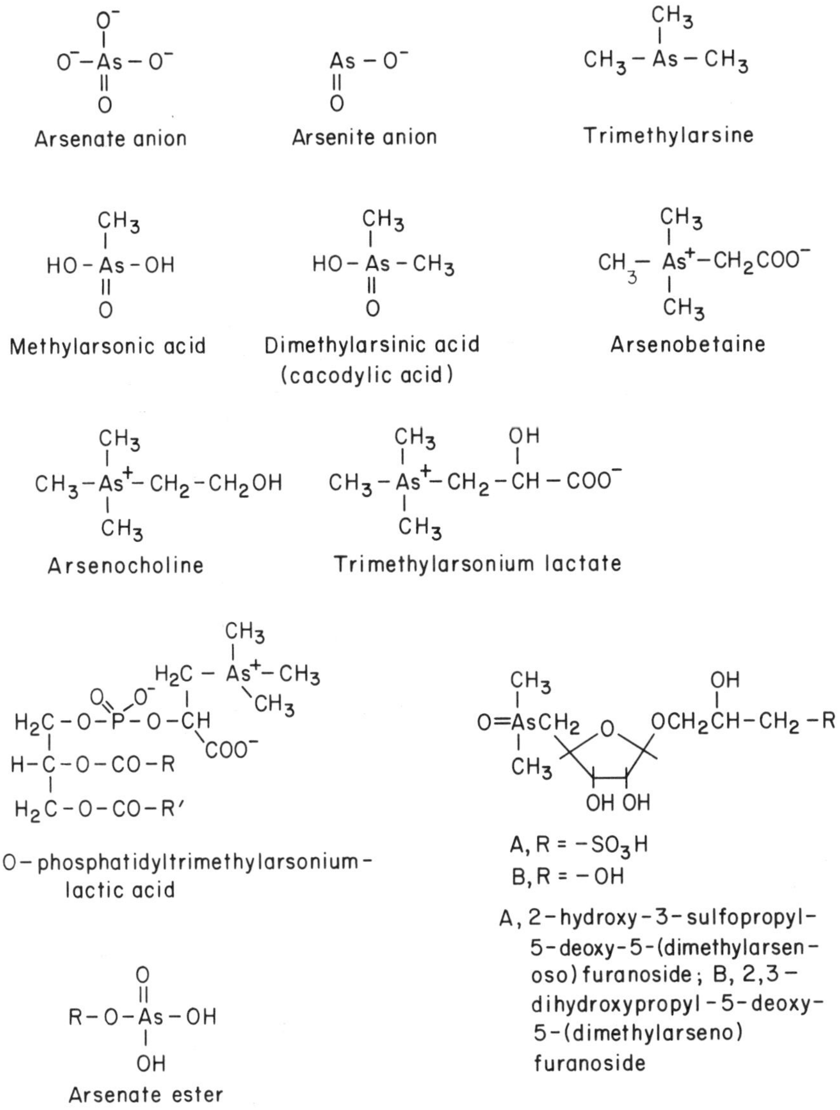

$$O^- - \underset{\underset{O}{\|}}{\overset{\overset{O^-}{|}}{As}} - O^-$$
Arsenate anion

$$\underset{\underset{O}{\|}}{As} - O^-$$
Arsenite anion

$$CH_3 - \underset{\underset{CH_3}{|}}{As} - CH_3$$
Trimethylarsine

$$HO - \underset{\underset{O}{\|}}{\overset{\overset{CH_3}{|}}{As}} - OH$$
Methylarsonic acid

$$HO - \underset{\underset{O}{\|}}{\overset{\overset{CH_3}{|}}{As}} - CH_3$$
Dimethylarsinic acid (cacodylic acid)

$$CH_3 - \underset{\underset{CH_3}{|}}{\overset{\overset{CH_3}{|}}{As^+}} - CH_2COO^-$$
Arsenobetaine

$$CH_3 - \underset{\underset{CH_3}{|}}{\overset{\overset{CH_3}{|}}{As^+}} - CH_2 - CH_2OH$$
Arsenocholine

$$CH_3 - \underset{\underset{CH_3}{|}}{\overset{\overset{CH_3}{|}}{As^+}} - CH_2 - \overset{\overset{OH}{|}}{CH} - COO^-$$
Trimethylarsonium lactate

O-phosphatidyltrimethylarsonium-lactic acid

A, R = −SO$_3$H
B, R = −OH

A, 2-hydroxy-3-sulfopropyl-5-deoxy-5-(dimethylarsenoso)furanoside; B, 2,3-dihydroxypropyl-5-deoxy-5-(dimethylarseno)furanoside

$$R - O - \underset{\underset{OH}{|}}{\overset{\overset{O}{\|}}{As}} - OH$$
Arsenate ester

Figure 14-1. Some Biologically Important Forms of Arsenic

Challenger (1978) reviewed the early history of studies on the biomethylation of arsenic. The methylated compound first identified was the poisonous gas, trimethylarsine, which was synthesized by the mold *Penicillium brevicaulis* (now designated *Scopulariopsis brevicaulis*) from arsenic oxide. Subsequently, it was found that when several mono- or dialkylarsonic acids $RAsO(OH)_2$ and $RR'AsO \cdot OH$ or their sodium salts were added to bread cultures of *S. brevicaulis*, methylation occurred, giving ethyldimethylarsine, *n*-propyldimethylarsine, allyldimethylarsine, and methylethyl-*n*-propylarsine. Other molds were also found to methylate arsenicals. *A. niger, P. notatum,* and *P. chrysogenum* in bread cultures produced trimethylarsine from methyl- and dimethylarsonic acids as sodium salts, but not from arsenite. *A. versicolor* and *A. glaucus* produced trimethylarsine from both arsenite and sodium methylarsonate.

McBride and Wolfe (1971) found that arsenate was reduced and methylated under anaerobic conditions by *Methanobacterium* when methylcobalamin (methyl-B_{12}) was present as a methyl donor. *Methanobacterium* apparently enzymatically synthesized both dimethylarsine and methylarsonic acid. Dimethylarsinic acid and methylarsonic acid are very important in the metabolic handling of arsenic by a variety of organisms, including humans. These arsenic compounds are discussed in more detail (*vide infra*).

Only recently, methylated arsenicals have been identified in a number of other organisms. Lunde (1977) reviewed some of the early studies on the form of arsenic in marine organisms. In the review, Lunde concluded that marine organisms are able to convert inorganic arsenic into organic arsenic compounds. Subsequently, in aquatic organisms, arsenic was found to be present as both lipid-soluble and water-soluble compounds. Since 1977, some of the lipid-soluble organic arsenicals were found to be closely related to some phospholipids. Irgolic *et al.* (1977) found that an arsenic-containing substance formed by *Daphia magna* and *Tetraselmis chuii* (marine algae) migrated with phosphatidylethanolamine upon thin layer chromatographic analyses. Hydrolysis of the phosphatidylethanolamine fraction by phospholipases A, C, and D resulted in the finding of an arsenocholine moiety. Benson and co-workers (1980) concluded that the final product synthesized by most marine algae is *O*-phosphatidyltrimethylarsoniumlactate. This arsenolipid differs from an arsenolecithin in that it possesses a carboxyl group which contributes to the physical and chemical properties of the hydrophilic moiety of the amphipathic lipid. A similar compound, *O*-glycerophosphoryltrimethylarsoniumlactate was found in the giant clam (*Tridacna maxima*) kidney (Benson *et al.*, 1980). Wrench and Addison (1981) reported that the unicellular marine phytoplankton, *Dunaliella tertiolecta* synthesizes arsenolipids distinctly different from those identified by Benson and co-workers. According to Wrench and Addison (1981), this organism synthesizes three acid-labile arsenolipids, one of which behaved as an anionic phosphatide similar to a compound synthesized from phosphatidylinositol and arsenite. Thus they proposed that one of the major arsenolipids in *D. tertiolecta* was a complex between arsenite and phosphatidylinositol. Wrench and Addison (1981) suggested that, because arsenic has rapid access to the lipid pool by a single biochemical process of reduction and because a wide range of organisms, from bacteria to mammals, carry out this reductive step, arsenite–lipid complexes may be widely distributed in nature.

Some of the water-soluble organic arsenicals in marine organisms may be precursors, or degradation products, of the lipid-soluble organic arsenicals. Degradation of O-phosphatidyltrimethylarsoniumlactate gives trimethylarsoniumlactate a neutral zwitterionic molecule with a structure closely related to arsenocholine and arsenobetaine (Benson et al., 1980). Trimethylarsoniumlactate has been found in giant kelp (*Macrocystis pyrifera*) (Herrera-Lasso and Benson, 1982), giant clam (*Tridacna maxima*) kidney (Benson et al., 1980), and lobster (*Homarus americanus*) (Cooney and Benson, 1980). Arsenobetaine has been isolated from several marine species including western rock lobster (*Panulirus cygnus* George), flesh of the dusky shark (*Carcharhinus obscurus* Le Sueur), American lobster (*Homarus americanus*), muscle and liver of *Prionace glaucus* (shark), and shrimp (Cannon et al., 1981; Edmonds and Francesconi, 1981a; Kurosawa et al., 1980; Norin and Christakopoulos, 1982). Shrimp may also contain arsenocholine.

Edmonds and Francesconi (1981b) suggested that arsenobetaine may evolve in some higher organisms, such as western rock lobster, dusky shark, and school whiting (*Sillago bassensis*), through the cycling in the marine ecosystem of arseno-sugars synthesized by the brown kelp (*Ecklonia radiata*). These arseno-sugars are 2-hydroxy-3-sulfopropyl-5-deoxy-5-(dimethylarsenoso) furanoside and a 2,3-dihydroxy-propyl-5-deoxy-5-(dimethylarsenoso) furanoside. Bacterial action on the arsenic-containing sugars produce dimethyloxarsylethanol a possible precursor of arsenobetaine (Edmonds et al., 1982).

Arsenic compounds other than those of the methylated form that are of interest in biochemistry are those formed when arsenate replaces phosphate in biological molecules. This occurs because arsenates and phosphates are both tetrahedral in structure and often isomorphous in crystals. Examples of these compounds are glucose-6-arsenate (Lagunas, 1980) and ADP-arsenate (Gresser, 1981). Generally, the phosphoryl compounds are more stable than the arsenyl compounds. However, under certain conditions arsenate esters are quite stable. Thus arsenylated compounds may be the form of arsenic that performs an essential function; so they should not be ignored.

14.3 Arsenic Deficiency and Interaction with Other Nutrients

In 1975–1976 the first findings showing that arsenic is essential came from two laboratories. The studies on arsenic essentiality apparently were done in each laboratory without knowledge that similar studies were being done in the other laboratory. As a result of those investigations, signs of arsenic deprivation were described for the chick, goat, minipig, and rat. The signs of deficiency for minipigs and goats were reviewed by Anke *et al.* (1978, 1980) and those for chicks and rats, by Uthus *et al.* (1983).

Anke *et al.* (1976, 1980) fed a semisynthetic diet containing less than 50 ng of arsenic/g to growing, pregnant, and lactating goats and minipigs, and to their first and second generation offspring. Controls were fed a supplemental 350 ng of arsenic/g of diet. With the initial animals, only about 60% of the deficient goats and minipigs conceived offspring, compared to nearly 100% for the respective controls. The offspring exhibited depressed growth and an elevated mortality rate. The slower growth was most marked in the second-generation deficient goats. Some lactating deficient goats died, apparently from myocardial abnormalities. Arsenic-deficient goats exhibited depressed skeletal ash and manganese and elevated levels of copper in several organs. Tissue levels of zinc and iron apparently were unaffected by arsenic deprivation. Also unaffected were several blood parameters, including hemoglobin, hematocrit, and mean corpuscular hemoglobin concentration. In serum, triglycerides were 39% lower in arsenic-deficient than in arsenic-supplemented goats. Also in serum, the α-lipoprotein fraction, rich in cholesterol, apparently was elevated, and the β-lipoprotein fraction, rich in triglyceride, apparently was depressed in arsenic-deficient goats.

While investigating the essentiality of nickel, Nielsen *et al.* (1978) observed that in third-generation pregnant rats fed a diet based on skim milk powder, whether nickel-deficient or -supplemented, perinatal death of pups was 90%. A group of ten nickel-supplemented third-generation dams were rebred and five were fed a dietary supplement of 4.0 μg of arsenic as sodium arsenate/g plus 0.5 μg of arsenic as sodium arsenite/g. Loss of pups was 80% in the group not receiving the arsenic supplement and 40% in the group receiving the arsenic supplement. In the arsenic-supplemented group most of the deaths occurred in one litter, apparently as a result of maternal neglect. At weaning (24 days) mean body weight was 26 g for the six surviving arsenic-deprived pups and 42 g for the 26 surviving arsenic-supplemented pups.

In two subsequent experiments (Nielsen *et al.*, 1975), the offspring of dams fed an arsenic-deficient (30 ng/g) diet from day 3 of gestation were examined for signs of deficiency. Controls were fed 4.5 μg of arsenic/g of diet. During the suckling period, pups did not differ in mortality or appearance. Possibly, perinatal mortality was not elevated, and growth was not depressed, because the dams had not been on the arsenic-deficient regimen for a sufficient time. After weaning, the growth of the arsenic-deprived offspring was slower than that of the supplemented offspring. At 40–44 days, control males weighed significantly more than arsenic-deprived males. At this time, the deprived rats appeared less thrifty than the supplemented rats; their coats were rougher and yellowish. Other reported signs of arsenic deprivation were elevated erythrocyte osmotic fragility, elevated spleen iron, and splenomegaly.

As described by Uthus *et al.* (1983), there are findings that show that arsenic is essential for growing chicks. Initial studies with chicks used a diet based on dried-skim milk, acid-washed ground corn and high-protein casein. The basal diet contained approximately 15–25 ng arsenic/g. Controls were fed the basal diet supplemented with 1 μg arsenic/g. Some of the arsenic-deprived and -supplemented chicks were also fed a dietary supplement of 20 g arginine/kg. During the first 3 weeks in these studies, there was no obvious difference in growth or appearance between the arsenic-deprived and -supplemented chicks. In week 4 the arsenic-deprived chicks drank and excreted more water and exhibited slower growth. After 4 weeks the arginine-supplemented, arsenic-deprived chicks weighed significantly less than comparable controls. The difference in weight at 4 weeks did not always reach significance in the groups not fed the additional arginine. Other findings in the arginine-supplemented groups included an elevated level of zinc and depressed levels of arsenic, iron, and manganese in the livers of arsenic-deprived chicks. In the groups not supplemented with arginine, similar changes, but of less magnitude, occurred.

After the initial experiments, the unavailability of low-arsenic skim milk powder and concern about the environmental conditions for the chicks prompted the development of a new diet and an improved environment. Experiments using the new diet, based on high-protein casein and containing normal levels of arginine, and an improved environment, did not give findings that agreed completely with initial findings. The signs of arsenic deficiency in these latter studies were mainly depressed growth and leg abnormalities.

Because all the signs of deficiency found in the initial experiments were not obtained in the subsequent experiments, studies were done in which arsenic-deprived chicks were stressed with high dietary levels of arginine and with zinc deficiency (Uthus *et al.*, 1983). Findings from factorially arranged experiments showed that the signs of arsenic deprivation are greatly influenced by other nutrients, such as arginine and zinc.

The interaction between arsenic and zinc affected growth, hematocrit, plasma alkaline phosphatase activity, and kidney arginase activity. The effect of dietary arsenic on growth and hematocrits was dependent upon the zinc status of the chick. When dietary zinc was 25–40 μg/g (marginally adequate), arsenic-deprived chicks exhibited depressed growth and slightly elevated hematocrits. During severe zinc deficiency, however, growth was apparently more markedly depressed and hematocrits more markedly elevated in arsenic-supplemented than in arsenic-deprived chicks. In a similar fashion, the effects of dietary zinc on plasma alkaline phosphatase activity and kidney arginase activity apparently depended upon the arsenic status of the chick. Plasma alkaline phosphatase activity was elevated in zinc-deficient chicks that were fed supplemental arsenic.

On the other hand, zinc deficiency slightly depressed plasma alkaline phosphatase activity in arsenic-deprived chicks. Zinc deficiency elevated kidney arginase of arsenic-supplemented chicks. When the chicks were arsenic-deprived, zinc deficiency did not markedly affect kidney arginase activity.

Another finding in the arsenic–zinc interaction studies was that when dietary zinc was adequate, the arsenic-deprived chicks exhibited leg abnormalities that differed from those of the zinc-deficient chicks. The legs were twisted and appeared weak but were not extremely short and thick with swollen hocks (the latter are characteristics of zinc deficiency leg abnormalities). If arsenic supplementation had any effect on leg abnormalities of zinc-deficient chicks, it was that it made them slightly worse.

The finding that arsenic deprivation apparently elevated, or depressed, kidney arginase activity with the direction depending on the zinc status of the chicks was an impetus to do three-way, two-by-two-by-two arranged experiments with supplemental arsenic, arginine, and zinc as the variables. The most interesting findings from these experiments were the following: When dietary arginine was increased to 34 mg/g by a supplement of 20 mg/g, kidney arginase and plasma urea were substantially elevated. However, the elevation of these two parameters was markedly influenced by dietary arsenic and zinc. Zinc deficiency alleviated the elevation in plasma urea and kidney arginase activity in arsenic-deprived chicks. On the other hand, zinc deficiency exacerbated the elevation in these two parameters in arsenic-supplemented chicks.

In addition to arginine and zinc, arsenic metabolism may be influenced by other nutrients, or vice versa. Two such nutrients are selenium and iodine. However, most findings concerned with the interaction between these two nutrients and arsenic were obtained when toxic levels were used.

The interaction between arsenic and selenium has been summarized by Levander (1977). Arsenic is an antagonist of selenium, and perhaps vice versa. In 1938 Moxon (1938) discovered that arsenic protected against selenium toxicity. Since that time, the protective effect of arsenic against selenium poisoning has been demonstrated in many different animal species under a wide variety of conditions. Similarly, it was found that selenium decreased the teratogenicity of arsenic in hamsters when salts of these two elements were injected simultaneously. The finding that arsenic and selenium each increase the biliary excretion of the other led Levander (1977) to suggest that these two elements react in the liver to form a detoxication conjugate that is then excreted into the bile. Because selenium excretion is altered by arsenic, findings that arsenic affects tissue retention of selenium are not surprising. Mangal and Singh (1980) found that arsenate diminished the retention of ^{75}Se-selenate in all rat organs except the kidney and gastrointestinal tract. These findings did not agree completely with those of Diplock and Mehlert (1980), who used arsenite and selenite instead of arsenate and selenate in their studies. They found that in vitamin E-deficient rats, arsenite depressed the uptake of ^{75}Se in the heart, testis, liver, and kidney of rats, but this effect was not seen when rats were supplemented with 100 mg of vitamin E/kg of diet.

Because selenate and arsenate are two oxyanions with similar chemical parameters, the antagonistic interaction between arsenic and selenium may be other than just formation of a conjugate that is readily excreted via the bile. For example, an explanation for the finding that selenate partially inhibited the uncoupling of oxidative phosphorylation by arsenate is that these two anions, because of their similarity, reciprocally inhibit the uptake of each other by mitochondria. Another finding that may be explained by the chemical parameter concept is that, while arsenite has little effect on methylmercury toxicity in Japanese quail, arsenite alters the ability of selenite to modify methylmercury toxicity (El-Begearmi *et al.*, 1982).

As reviewed by Levander (1977), there are experimental conditions in which arsenic increases selenium toxicity. The toxicity of trimethylselenonium chloride is markedly elevated by simultaneous injection with arsenite. Likewise, when injected together, arsenite potentiates the toxicity of dimethylselenide, a rather innocuous compound considered to be a detoxification product of selenium metabolism.

The interaction between iodine and arsenic is exemplified by the findings of Lindgren and Dencker (1980). They found that dietary arsenic had a goitrogenic effect in mice. Arsenic depressed the iodine level in, and elevated the weight of, the thyroid. They suggested that arsenic antagonized mechanisms of iodine uptake and utilization in the thyroid gland, which in turn caused compensatory goiter.

14.4 Arsenic Function

The evidence showing that arsenic is essential does not clearly define its metabolic function. Thus the mode of action of arsenic is open to conjecture. The findings of Uthus *et al.* (1983) suggest that arsenic does strongly influence arginine metabolism. As described (*vide supra*), arsenic deprivation apparently affects the reaction:

$$\text{Arginine} \xrightarrow{\text{arginase}} \text{urea} + \text{ornithine}$$

However, the extent and direction of the effect depends on the arginine and zinc status of the animal, indicating that arsenic indirectly affects arginase activity. Perhaps the indirect effect is a reflection of a role of arsenic in the utilization of one or more amino acids in the synthesis of protein, or in control of the degradation of some protein. Findings that suggest the possibility that arsenic has a role in amino acid or protein metabolism include:

1. Arsenic deprivation affected chick plasma uric acid levels (Uthus et al., 1983), which indicates amino acid metabolism is altered because excess amino acid nitrogen is usually eliminated from the chick as uric acid.
2. Preliminary findings show that arsenic deprivation depressed the level of arginine, and elevated the levels of cysteine, glutamine, glycine, and histidine, in the plasma of chicks (Uthus, unpublished observations).
3. Arsenic deprivation depressed the raw protein level in minipigs and goats (Anke et al., 1978).

Arsenic may function in some enzyme system. Although arsenic can either activate or inhibit enzymes, it has not been shown to be a specific activator or inhibitor of any enzyme. In its role as activator, arsenic, as arsenate, apparently replaces phosphate in phosphorylation reactions. For example, the mechanism by which inorganic arsenate facilitates the enzymatic reduction of dihydroxyacetone by α-glycerophosphate dehydrogenase (E.C. 1.1.1.8) apparently involves the formation of an esterlike compound between dihydroxyacetone and arsenate. The resulting analog is acted upon by the enzyme in a manner similar to that for the phosphorylated substrate (Jaffe and Apitz-Castro, 1977). However, an alternate mode of action for arsenic on one enzyme has been suggested. Huang and Mitchell (1971) suggested that arsenate, rather than acting as an alternate substrate in place of phosphate, stimulated the ATP-driven, energy-linked reduction of NAD^+ by succinate by acting as an enzyme modifier to change the kinetic parameters of the reaction. As an inhibitor, arsenic, as arsenite, apparently exerts its toxic effect on enzymes by reacting with vicinal-SH groups.

Finally, arsenic might have a role related or similar to lipid phosphorus in biological systems. The finding of arsenocholine, arsenobetaine, and the novel membrane phospholipid O-phosphatidyltrimethylarsoniumlactic acid in marine organisms support this suggestion (*vide supra*). Further support is that in higher animals, arsenocholine can replace choline in some of its functions (Almquist and Grau, 1944). Arsenocholine is antiperotic and growth promoting in the choline-deficient fowl.

14.5 Arsenic Metabolism

In microorganisms the methylation of inorganic arsenic apparently is one of the more important steps in arsenic metabolism. The formation of small nonpolar complexes, such as trimethylarsine which is dissipated by diffusion-controlled processes, probably prevents toxicologic manifestations of local cellular concentrations of inorganic arsenate or arsenite. Challenger (1978) proposed the following metabolic pathway for the production of trimethylarsine from arsenite:

$$AsO_3^{3-} \xrightarrow{[CH_3^+]} CH_3AsO_3^{2-} \xrightarrow[b\,[CH_3^+]]{a\,2e^-} (CH_3)_2AsO^{2-}$$
$$\text{arsenite} \qquad \text{methylarsonate} \qquad \text{dimethylarsinate}$$

$$\xrightarrow[b\,[CH_3^+]]{a\,2e^-} (CH_3)_3AsO \xrightarrow{2e^-} (CH_3)_3As$$
$$\text{trimethylarsine}$$

Findings of Cullen *et al.* (1977) indicated that *S*-adenosylmethionine or some related sulphonium compound is involved in the methylation of arsenic. This supports Challenger's proposed use of the carbonium ion (CH_3^+) rather than the carbanion (CH_3^-) or the radical (CH_3) in the methyl transfer. However, other findings of Cullen *et al.* (1979) indicated that the proposed pathway is an oversimplification of the many processes involved. Using *Candida humicola* as the test organism, they found that methylarsonate apparently does not occur as a free intermediate in the arsenate-to-trimethylarsine pathway.

Methylation of arsenic probably is the primary step of arsenic metabolism in the sea also. According to Benson *et al.* (1980), phosphate concentrations in the sea are often near those of arsenate. Because algae or phytoplankton are unable to discriminate between phosphate and arsenate ions, these organisms apparently convert the excess arsenate, absorbed during the process of absorbing the more needed phosphate, to a nontoxic methylated form. This biomethylation probably follows a pathway similar to that proposed by Challenger (1978), except the end product apparently is not the toxic gas trimethylarsine, but nontoxic compounds like arsenolipids. Benson *et al.* (1980) found that *Dunaliella tertiolecta* maintained 90%, and some other algae maintained 5–20%, of their arsenic in lipid compounds. Benson *et al.* (1980) suggested that, in *D. tertiolecta*, the ubiquitous metabolite, trimethylarsine, condenses with phosphoenolpyruvate to produce trimethylarsoniumlactate. Then this trimethylarsoniumlactate is used to make *O*-phosphatidyltrimethylarsoniumlactate (Figure 14-1), an arsenophospholipid of algae, by a mechanism analogous to that used for the production of phosphatidylserine or other plant lipids.

Benson and co-workers also suggested that the arsenophosphatide made by marine algae is degraded via lysoarsenophatide by phospholipases and other lipid-reactive enzymes. The phosphodiester, *O*-glycerophosphoryltrimethylarsoniumlactate, is cleared by bacterial phosphodiesterase to give trimethylarsoniumlactate, which is further degraded to dimethylarsinate (cacodylate). This compound apparently accumulates in considerable concentration in bacterial-contaminated algae and corals.

The findings and suggestions of Benson and co-workers, however, have not been completely accepted. Wrench and Addison (1981) stated that the products of methylation, monomethylarsonate and dimethylarsinic acid, do not affect diatom productivity and are efficiently excreted into the extracellular medium; thus there is no toxicological requirement to metabolize further these compounds to arsenolipids. They suggested that the synthesis of such compounds may be an adaptation to limited nitrate availability and that arsonium phosphatides may function as substitutes for choline, serine, or ethanolamine analogs in structural lipids. In contrast to Benson *et al.* (1980), Wrench and Addison (1981) could not find evidence of arsonium compounds in *D. tertiolecta*. The only phospholipid they found had properties consistent with a complex between arsenite and phosphatidylinositol.

Determination of the mechanisms through which arsenic is metabolized, and which are the most important, may be complex. Andreae and Klumpp (1979) found that significant differences exist between marine algae species in terms of uptake rates and the type and number of compounds produced. Moreover, they found 12 organoarsenicals in cell extracts of various marine planktonic algae of which none apparently were either arsenobetaine or arsenocholine, compounds found in higher trophic levels of the marine chain.

Freshwater and terrestrial plants also methylate arsenic. Nissen and Benson (1982) found that lower and higher freshwater plants (e.g., *Nitella tenuissima* and *Lemna min.*) not only methylated arsenic but also synthesized *O*-phosphatidyltrimethylarsoniumlactate. Higher terrestrial plants, such as tomato (*Lycopersicon esculentum*), which were phosphate deficient, produced methylarsonic acid, methylarsinic acid, and dimethylarsinic acid, but apparently no arsonium-phospholipid. Phosphate deficiency may not be a prerequisite for higher terrestrial plants to methylate arsenic, because Pyles and Woolson (1982) found methylarsonic acid in broccoli, lettuce, potato, and swiss chard grown in soil treated with 100 μg of arsenic as arsenic acid/g.

The metabolism of arsenic at trophic levels higher than phytoplankton in the marine food chain also is an unsettled topic. Some findings indicate that higher organisms themselves metabolize inorganic arsenic to organic forms, whereas other findings suggest that most organic arsenic that appears in these organisms comes from preformed sources. Both mechanisms probably occur, with the predominant one being dependent upon the organism and its environment.

Arsenic accumulates markedly in mollusks and ascidians bearing symbiotic algae, particularly the large *Tridacna* and *Hippopus* clams, which harbor zooxanthellae, and *Didemnum ternatanum*, which harbors the unique prokaryotic green alga *Prochloron*. Findings of Benson and Summons (1981) led them to conclude that the accumulation was caused by the symbiotic organisms assimilating arsenate from sea water and converting it into organic arsenic that is deposited in host tissues. The accumulation of the algae arsenicals in the clam kidney apparently occurs because their excretion by way of the urine is slow. Gill membranes exposed to bacterial oxidative processes apparently are the major site for arsenic excretion in *Tridacna*. Cooney and Benson (1980) found no organic forms of arsenic in lobsters (*Homarus americanus*) fed arsenate. Thus they concluded that organic arsenic found in crustaceans is the result of accumulation through the food chain rather than *de novo* synthesis. Unlike the mollusks, *Homarus americanus* apparently has the necessary lipase and diesterase activities to degrade the algal phospholipid to trimethylarsoniumlactate, but not to arsenobetaine. Cooney and Benson (1980) found mainly trimethylarsoniumlactate in hepatopancreas and muscle tissue of American lobster fed algae. On the other hand, others (Edmonds and Francesconi, 1981a; Cannon *et al.*, 1981) found arsenobetaine the major organic arsenical in the American lobster and the western rock lobster. Perhaps the food source was a major determinant of the form of organic arsenic in lobster. Edmonds and Francesconi (1981b) suggested that arseno-sugars such as those present in the kelp *Ecklonia radiata,* rather than algal phospholipid, were the origin of the arsenobetaine in some crustaceans.

The conclusion that arsenic compounds in higher trophic levels are derived from primary producers was not supported by the findings of Klumpp and Peterson (1981). They studied the marine food chain of *Fucus spiralis* to the snails *Littorina littoralis* and *Nucella lapillus*. The algae assimilated arsenic mainly (60%) as one lipid soluble compound, and 12 minor water-soluble organoarsenic compounds. On the other hand, the snails, whether labeled from water or food, produced predominantly one major water-soluble organoarsenical, indicating an ability on the part of the snails to metabolize arsenic. The snails, labeled from water or food, showed a temporal decrease in the proportion of arsenate and the algal-type components concurrent with a rise in the predominant organoarsenical. The organoarsenical apparently was not arsenobetaine, arsenocholine, methylarsonate, or dimethylarsonate.

The findings from the marine food chain studies show that, although organisms low in the chain bioaccumulate and convert absorbed arsenic into organic forms, their predators are usually able to degrade and/or excrete the organoarsenicals, so that biomagnification of arsenic in the food chain does not occur. Furthermore, most of the methylated forms of arsenic formed in the marine food chain are very nontoxic to the primary producers and predators.

In higher animals there are four entry routes into the body—oral intake, inhalation, percutaneous absorption, and parenteral administration. The dynamics of arsenic distribution in the body as well as the kinetics of its elimination in urine and feces varies substantially, depending on the mode of administration (Dutkiewicz, 1977). In normal metabolism, oral intake of arsenic is of primary importance, and thus is emphasized here.

Inorganic arsenate and arsenite are well absorbed by higher animals, including man. The form of organic arsenic determines whether it is well absorbed or not. For example, 8 days after ingestion, humans excreted 76% in urine, and only 0.33% in feces, of the arsenic in witch flounder (Tam et al., 1982). Freeman et al. (1979) also found that, 8–9 days after ingestion, humans excreted in urine about 77% of the arsenic in mixed flounder fillets. Siewicki and Sydlowski (1981) found that rats absorbed 98–99% of the arsenic in witch flounder, and most of it was rapidly excreted in the urine. In contrast, human subjects fed radioactive arsenic present in tissues of chickens fed arsanilic-^{74}As acid rapidly excreted the isotope (60–80%) via the feces (Calesnick et al., 1966). Perhaps some of the fecal arsenic represented that absorbed and excreted via the bile (Klassen, 1974; Cikrt et al., 1980) but much of the fecal arsenic probably was unabsorbed. Also, when given orally to rats or humans, more than 90% of a dose of sodium-p-N-glycolylarsanilate was recovered in the feces within 3 days, and urinary excretion accounted for only 4–5% (McChesney et al., 1962).

Only a limited number of studies have examined the mechanisms involved in the gastrointestinal absorption of arsenic. Arsenic in the form of arsenate apparently is absorbed in a manner similar to phosphorus as phosphate (Klevay, 1976). Hwang and Schanker (1973) found that some forms of organic arsenic (carbarsone, tryparsamide, and dimethylarsinic acid) were absorbed at rates directly proportional to their intestinal concentration over a 100-fold range in rat intestine. This finding indicated that some organic arsenicals are absorbed mainly by simple diffusion through lipoid regions of the intestinal boundary.

As indicated (*vide supra*), the excretion of ingested arsenic is quite rapid, with both urine and feces serving as major routes of elimination. Organic arsenic that is absorbed apparently undergoes little or no chemical change, because after the ingestion of arsenic-rich food, such as lobster, crab, and witch flounder, most of the arsenic in urine apparently was in its original organically bound form (Crecelius, 1977; Tam et al., 1982). After the parenteral administration of sodium p-arsanilate to rats and guinea pigs, Cristau et al. (1975) were unable to detect any other arsenic-containing material in urine except the unchanged sodium p-arsanilate. Arsenic ingested as an inorganic form appears in urine in both the inorganic and methylated forms. However, the proportions of the forms of arsenic in urine is species dependent. Tam et al. (1979a) found that the arsenic in human urine was 51% dimethylarsinic acid, 21% monomethylarsenic compound, and 27% inorganic arsenic after an oral dose of inorganic arsenic. Di methylarsinic acid was much more prevalent in urine of mice (about 80% total arsenic) given inorganic arsenic orally (Vahter, 1981) and in urine of dogs (about 90% total arsenic) given inorganic arsenic intravenously (Tam et al., 1978). Rats given an oral dose of arsenate excreted only 4% of the dose as dimethylarsinic acid and 0.3% as methylarsonic acid after 48 hours. About 13% of the dose was excreted as inorganic arsenic (Vahter, 1981).

Crecelius (1977) suggested that two different processes with different rates were involved in the removal of ingested arsenite. The first process involved the removal of arsenic as arsenite by the kidneys and began at 5 hours, diminished at 20 hours, and was not evident at 60 hours after ingestion. The second process involved the methylation of ingested arsenite to methylarsonic acid and dimethylarsinic acid. The excretion of methylated arsenic began within 5 hours after ingestion of arsenite but reached maximum levels much later than excretion of arsenic. Most of the arsenic ingested was excreted within 85 hours, and the apparent half-life was approximately 30 hours. The half-life increased slightly with increasing dosages of arsenite (Buchet et al., 1981). Crecelius (1977) found that excretion of ingested arsenate was difficult to assess but suggested that some unchanged arsenate was rapidly excreted from the body. Part of the ingested arsenate was methylated and excreted within several days.

Unlike marine animals, there is no doubt that mammals have the ability to methylate arsenic. The preceding described findings of methylated arsenic in urine, especially after the intravenous or intraperitoneal injection of inorganic arsenic, was strong evidence for this ability. However, because of the possibility that the methylation of arsenic was done by microorganisms in the gastrointestinal tract, or in urine, conclusive evidence remained to be shown by other studies. Vahter and Gustafsson (1980) found that germ-free mice were able to methylate inorganic arsenic to the same extent as conventional mice, thus showing intestinal bacteria were not a major source of methylation. Shirachi et al. (1981) found that, in vitro, rat liver methylated sodium arsenate. Furthermore, they suggested two different enzymes were involved in the methylation, because monomethylarsonic acid was formed in all subcellular fractions of the liver, whereas dimethylarsinic acid was formed mainly in the supernatant fraction. The liver apparently is the major site of biomethylation, because Tam et al. (1979b) found that human, or dog, urine, plasma, and red blood cells did not methylate arsenic in vitro.

When retention of arsenic by higher animals is discussed, it should be noted that rats, unlike other mammals, concentrate arsenic in their red blood cells (Cikrt et al., 1980). This accumulated arsenic was found to be mainly associated with the hemoglobin, or hematin (99%), fraction (Fuentes et al., 1981). Apparently, rats are unique in their metabolic management of arsenic. This was demonstrated by Marafante et al. (1982), who compared the metabolism of arsenic in rats and rabbits after an intraperitoneal injection of ^{74}As-labeled arsenite. The highest arsenic concentrations at 12 and 48 hr after injection were found in liver, kidney, and lung of rabbits. The arsenic in these tissues was rapidly cleared. In the corresponding tissues of rats, the rate of arsenic decline was significantly lower, because of the higher binding of arsenic to tissue constituents. Poor binding of arsenic to plasma proteins was seen in rabbits, while in rats it was totally bound to this fraction. The amount of dimethylarsinic acid in the tissues was lower in the rat than in the rabbit, reflecting the total amount of diffusible arsenic, which was also much lower in tissues of rats than in rabbits.

Vahter (1981) suggested that decreased methylation is the reason for the high retention of arsenic in the rat. This suggestion was supported by the finding of Stevens *et al.* (1977) that arsenic administered to rats in methylated form (dimethylarsinic acid) was readily excreted, whereas arsenic administered in inorganic form was substantially retained. Vahter (1981) found no specific affinity of arsenic by rat erythrocytes compared to mice erythrocytes *in vitro*, indicating further that the lack of methylation, and thus more inorganic arsenic to bind to tissue constituents, as the reason rats retain inorganic arsenic. The apparently poorer ability to methylate arsenic means that retention findings from rats would not be applicable to other species.

Pomroy *et al.* (1980) over a 103-day period studied arsenic retention in persons given oral ^{74}As as arsenic acid. They found that their data were best represented by a three-component exponential function, with the values of the coefficients being 65.9%, with a half-life of 2.09 days; 30.4%, with a half-life of 9.5 days; and 3.7%, with a half-life of 38.4 days. Their finding of very little arsenic being retained after 10 days was similar to the finding of Hoffmann *et al.* (1980) that in chickens and goats, 1% and 4%, respectively, of an oral dose of inorganic arsenic were retained after 96 hours. Thus homeostatic mechanisms, probably involving biomethylation, are present in most higher animals. These homeostatic mechanisms prevent unnecessary arsenic accumulation, or retention, of arsenic ingested at physiological levels. (The term *physiological* here refers to the amount of arsenic generally found in the diet.) Recent surveys indicate a dietary intake of usually <100 μg of arsenic/day for humans (Jelinek and Corneliussen, 1977; Pfannhauser and Woidich, 1980). Arsenic intakes higher than 100 μg/day might occur when diets are high in fish and seafood, which often contain relatively high levels of nontoxic forms of organic arsenic.

14.6 Arsenic Toxicity

The toxicity of a given arsenical depends upon its rate of excretion from the body and, thus, the degree to which it accumulates in tissues. Arsenicals that are excreted the slowest are the most toxic. The general pattern of toxicity is:

$$AsH_3 > As^{3+} > As^{5+} > R\text{-}As\text{-}X > As°$$

Inorganic arsenite is considerably more toxic to animals than arsenate. However, the physical state of the arsenic compound influences its toxicity. For example, the oral LD_{50} of sodium arsenite and arsenic trioxide in rats were reported to be, respectively, 24 and 293 mg/kg body weight (Done and Peart, 1971). Also, arsenic trioxide given orally in the solid form is much less toxic than the compound given in solution (Harrison *et al.*, 1958). In addition to physical state, there apparently are species-specific differences in susceptibility to arsenic poisoning. Vallee *et al.* (1960) estimated that the acute fatal dose of arsenic trioxide for humans is 70–180 mg (about 0.76–1.95 mg As/kg body weight of a 70-kg human). This dose is much less than that for rats.

The biochemical mechanism behind the toxicologic manifestations of inorganic trivalent arsenicals apparently is the reaction of arsenic with sulfhydryl groups of tissue proteins and enzymes. The formation of an arylbis(organylthio)arsine inhibits normal tissue protein and enzyme function. Generally, the activity of an enzyme inhibited by arsenite can be restored by adding an excess of monothiol. Toxicity of arsenate apparently stems from its ability to substitute competitively for inorganic phosphate in phosphorylation reactions to form unstable arsenyl esters, which then decompose spontaneously (Bencko and Symon, 1970).

Generally, organoarsenicals and elemental arsenic are less toxic than inorganic arsenicals. Also, the toxicity of the organoarsenicals is affected by the valence state of the arsenic component. Trivalent organoarsenicals produce signs of toxicity similar to those produced by inorganic arsenite, indicating that they react with sulfhydryl groups in proteins. Pentavalent organoarsenicals are quite nontoxic. Elemental arsenic, being insoluble, is essentially nontoxic.

Numerous epidemiologic studies have suggested an association between chronic arsenic overexposure and certain diseases, such as cardiovascular disease or cancer. However, the role of arsenic in these disorders, especially carcinogenesis, remains controversial, because laboratory studies have not succeeded in producing tumors in animals. Nielsen *et al.* (1983) have examined this controversy and decided that there apparently is some influence of arsenic on carcinogenesis. However, this influence is not that of a direct carcinogenic action, but that of indirectly influencing other metabolic systems (i.e., immune system) or nutrients (i.e., zinc) that may have a more direct role in the carcinogenic process. Nielsen *et al.* (1983) stated that this indirect action means that, depending upon the form, method of administration, and dosage, arsenic may have antagonistic, synergistic, or no action in carcinogenesis. In other words, arsenic is similar to many other nutrients, such as vitamin A, selenium, or zinc, which do not initiate, but under certain conditions and with certain types of cancer, can positively or negatively modify later tumor development.

14.7 Summary

Knowledge of the biochemistry of arsenic is important not only because arsenicals can have toxic and pharmacologic actions, but also because arsenic has a physiologic role *in vivo*. Arsenic deprivation signs have been described for the chick, goat, minipig, and rat. For the mammals, these signs include depressed growth, impaired reproduction characterized by elevated perinatal mortality and depressed fertility, and sudden death (in goats) during lactation from myocardial damage. For the rat and chick, more than 25 ng of arsenic/g of diet apparently is required to prevent deficiency.

Arsenic is present in all living material. The form of arsenic most important in the biochemistry, nutrition, and physiology of living organisms is not known. However, organic arsenic compounds that contain methyl groups, and are closely related to some phospholipids, are most likely to be found of primary importance. Evidence to date indicates that the absorption, retention, and excretion of arsenic involves methylated compounds. In marine organisms, these compounds include trimethylarsoniumlactate, arsenobetaine and arsenocholine. In some bacteria, and in mammals studied to date, these compounds include dimethylarsinic acid and methylarsonic acid. Arsenic compounds other than those of the methylated form that are of interest in biochemistry are those formed when arsenate binds to biological molecules to form arsenate esters. Because arsenylated compounds may be the form of arsenic that performs an essential function they should not be ignored.

To date no specific biological function of arsenic has been found. Recent findings suggest that arsenic has a function that influences arginine and zinc metabolism in mammals. Arsenic may do this through a function in some enzyme system(s) because arsenic can either activate or inhibit enzymes in vitro. Another possibility is that arsenic has a structural role as part of a membrane phospholipid.

Although arsenic has been synonymous with poison for centuries, and a number of reports have associated arsenic with some forms of cancer, it is actually much less toxic than selenium, a trace element with a well-established nutritional value. Toxic quantities of arsenic generally are measured in milligrams and the ratio of the toxic to nutritional dose for rats apparently is near 1250. Thus any beliefs that any form or amount of arsenic is unnecessary, toxic, or carcinogenic are unrealistic and of concern. The findings described in the preceding show that there are safe and, most likely, necessary levels of arsenic intakes that would permit optimal health for animals and humans throughout a lifetime.

References

Almquist, H. J., and Grau, C. R., 1944. Interrelation of methionine, choline, betaine and arsenocholine in the chick, *J. Nutr.* 27:263–269.

Andreae, M. O., and Klumpp, D., 1979. Biosynthesis and release of organoarsenic compounds by marine algae, *Environ. Sci. Technol.* 13:738–741.

Anke, M., Grün, M., and Partschefeld, M., 1976. The essentiality of arsenic for animals, in *Trace Substances in Environmental Health*, Vol. 10, D. D. Hemphill (ed.), University of Missouri, Columbia, pp. 403–409.

Anke, M., Grün, M., Partschefeld, M., Groppel, B., and Hennig, A., 1978. Essentiality and function of arsenic, in *Trace Element Metabolism in Man and Animals—3*, M. Kirchgessner (ed.), Tech. Univ. München, Freising-Weihenstephan, BRD, pp. 248–252.

Anke, M., Groppel, B., Grün, M., Hennig, A., and Meissner, D., 1980. The influence of arsenic deficiency on growth, reproductiveness, life expectancy and health of goats, in *3. Spurenelement—Symposium Arsen*, M. Anke, H.-J. Schneider, and Chr. Bruckner (eds.), Friedrich-Schiller-Universität, Jena, GDR, pp. 25–32.

Bencko, V., and Symon, K., 1970. The cumulation dynamics in some tissue of hairless mice inhaling arsenic, *Atmos. Environ.* 4:157–161.

Benson, A. A., and Summons, R. E., 1981. Arsenic accumulation in great barrier reef invertebrates, *Science* 211:482–483.

Benson, A. A., Cooney, R. V., and Summons, R. E., 1980. Arsenic metabolism—a way of life in the sea, in *3. Spurenelement—Symposium Arsen*, M. Anke, H.-J. Schneider, and Chr. Bruckner (eds.), Friedrich-Schiller-Universitat, Jena, GDR, pp. 139–145.

Bruckner, Chr., and Dietze, B., 1980. Arsenic in history, in *3. Spurenelement—Symposium Arsen*, M. Anke, H.-J. Schneider, and Chr. Bruckner (eds.), Friedrich-Schiller-Universität, Jena, GDR, pp. 5–10.

Buchet, J. P., Lauwerys, R., and Roels, H., 1981. Urinary excretion of inorganic arsenic and its metabolites after repeated ingestion of sodium metaarsenite by volunteers, *Int. Arch. Occup. Environ. Health* 48:111–118.

Calesnick, B., Wase, A., and Overby, L. R., 1966. Availability during human consumption of the arsenic in tissues of chicks fed arsanilic-^{74}As acid, *Toxicol. Appl. Pharmacol.* 9:27–30.

Cannon, J. R., Edmonds, J. S., Francesconi, K. A., Raston, C. L., Saunders, J. B., Skelton, B. W., and White, A. H., 1981. Isolation, crystal structure and synthesis of arsenobetaine, a constituent of the western rock lobster, the dusky shark, and some samples of human urine, *Aust. J. Chem.* 34:787–798.

Challenger, F., 1978. Biosynthesis of organometallic and organometalloidal compounds, *ACS Symp. Ser.* 82:1–22.

Cikrt, M., Bencko, V., Tichy, M., and Benes, B., 1980. Bilary excretion of ^{74}As and its distribution in the golden hamster after administration of ^{74}As (III) and ^{74}As (V), *J. Hyg. Epidemiol. Microbiol. Immunol.* 24:384–388.

Cooney, R. V., and Benson, A. A., 1980. Arsenic metabolism in *Homarus americanus*, *Chemosphere* 9:335–341.

Crecelius, E. A., 1977. Changes in the chemical speciation of arsenic following ingestion by man, *Environ. Health Perspect.* 19:147–150.

Cristau, B., Chabas, E., and Placidi, M., 1975. L'acide *p*. arsanilique est-il biotransformé chez le rat et le cobaye, *Ann. Pharm. Fr.* 33:37–41.

Cullen, W. R., Froese, C. L., Lui, A., McBride, B. C., Patmore, D. J., and Reimer, M., 1977. The aerobic methylation of arsenic by microorganisms in the presence of L-methionine-methyl-d$_3$, *J. Organomet. Chem.* 139:61–69.

Cullen, W. R., McBride, B. C., and Pickett, A. W., 1979. The transformation of arsenicals by *Candida humicola*, *Can. J. Microbiol.* 25:1201–1205.

Diplock, A. T., and Mehlert, A., 1980. The effect of arsenic and vitamin E on the uptake and distribution in the rat of ^{75}Se-labelled sodium selenite, in *3. Spurenelemente—Symposium Arsen*, M. Anke, H.-J. Schneider, and Chr. Bruckner (eds.), Friedrich-Schiller-Universität, Jena, GDR, pp. 75–81.

Done, A. K., and Peart, A. J., 1971. Acute toxicities of arsenical herbicides, *Clin. Toxicol.* 4:343–355.

Dutkiewicz, T., 1977. Experimental studies on arsenic absorption routes in rats, *Environ. Health Perspect.* 19:173–177.

Edmonds, J. S., and Francesconi, K. A., 1981a. Isolation and identification of arsenobetaine from the American lobster *Homarus americanus*, *Chemosphere* 10:1041–1044.

Edmonds, J. S., and Francesconi, K. A., 1981b. Arseno-sugars from brown kelp (*Ecklonia radiata*) as intermediates in cycling of arsenic in a marine ecosystem, *Nature* 289:602–604.

Edmonds, J. S., Francesconi, K. A., and Hansen, J. A., 1982. Dimethyloxarsylethanol from anaerobic decomposition of brown kelp (*Ecklonia radiata*): A likely precursor of arsenobetaine in marine fauna, *Experientia* 38:643–644.

El-Begearmi, M. M., Ganther, H. E., and Sunde, M. L., 1982. Dietary interaction between methylmercury, selenium, arsenic, and sulfur amino acids in Japanese quail, *Poult. Sci.* 61: 272–279.

Freeman, H. C., Uthe, J. F., Fleming, R. B., Odense, P. H., Ackman, R. G., Landry, G., and Musial, C., 1979. Clearance of arsenic ingested by man from arsenic contaminated fish, *Bull. Environ. Contam. Toxicol.* 22:224–229.

Frost, D. V., 1970. Tolerances for arsenic and selenium: A psychodynamic problem, *World Rev. Pest Control* 9:6–28.

Fuentes, N., Zambiano, F., and Rosenmann, M., 1981. Arsenic contamination: Metabolic effects and localization in rats, *Comp. Biochem. Physiol. C* 70C:269–272.

Gresser, M. J., 1981. ADP-arsenate. Formation by submitochondrial particle under phosphorylating conditions, *J. Biol. Chem.* 256:5981–5983.

Harrison, J. W. E., Packman, E. W., and Abbott, D. D., 1958. Acute oral toxicity and chemical and physical properties of arsenic trioxides, *AMA Arch. Ind. Health* 17:118–123.

Herrera-Lasso, J. M., and Benson, A. A., 1982. Arsenic detoxication in *Macrocystis pyrifera*, *Environ. Sci. Res.* 23:501–505.

Hoffmann, G., Anke, M., Grün, M., Groppel, B., and Riedel, E., 1980. Absorption, distribution and excretion of [76]arsenic in hens and ruminants, in *3. Spurenelement—Symposium Arsen*, M. Anke, H.-J. Schneider, and Chr. Bruckner (eds.), Friedrich-Schiller-Universität, Jena, GDR, pp. 41–48.

Hove, E., Elvehjem, C. A., and Hart, E. B., 1938. Arsenic in the nutrition of the rat, *Am. J. Physiol.* 124:205–212.

Huang, C.-H., and Mitchell, R. A., 1971. Stimulation by arsenate of ATP-driven energy-linked reduction of NAD^+ by succinate, *Biochem. Biophys. Res. Comm.* 44:1102–1108.

Hwang, S. W., and Schanker, L. S., 1973. Absorption of organic arsenical compounds from the rat small intestine, *Xenobiotica* 3:351–355.

Irgolic, K. J., Woolson, E. A., Stockton, R. A., Newman, R. D., Bottino, N. R., Zingaro, R. A., Kearny, P. C., Pyles, R. A., Maeda, S., McShane, W. J., and Cox, E. R., 1977. Characterization of arsenic compounds formed by *Daphnia magna* and *Tetraselmis Chuii* from inorganic arsenate, *Environ. Health Perspect.* 19:61–66.

Jaffe, K., and Apitz-Castro, R., 1977. Studies on the mechanism by which inorganic arsenate facilitates the enzymatic reduction of dihydroxyacetone by α-glycerophosphate dehydrogenase, *FEBS Lett.* 80:115–118.

Jelinek, C. F., and Corneliussen, 1977. Levels of arsenic in the United States food supply, *Environ. Health Perspect.* 19:83–87.

Klassen, C. D., 1974. Biliary excretion of arsenic in rats, rabbits and dogs, *Toxicol. Appl. Pharmacol.* 29:447–457.

Klevay, L. M., 1976. Pharmacology and toxicology of heavy metals: Arsenic, *Pharmacol. Ther. A* 1:189–209.

Klumpp, D. W., and Peterson, P. J., 1981. Chemical characteristics of arsenic in a marine food chain, *Marine Biol.* 62:297–305.

Kurosawa, S., Yasuda, K., Taguchi, M., Yamazaki, S., Toda, S., Morita, M., Uehiro, T., and Fuwa, K., 1980. Identification of arsenobetaine, a water soluble organo-arsenic compound in muscle and liver of a shark, *Prionace glaucus*, *Agric. Biol. Chem.* 44:1993–1994.

Lagunas, R., 1980. Sugar-arsenate esters: Thermodynamics and biochemical behavior, *Arch. Biochem. Biophys.* 205:67–75.
Larsen, N. A., Pakkenberg, H., Damsgaard, E., and Heydorn, K., 1979. Topographical distribution of arsenic, manganese, and selenium in the normal human brain, *J. Neurol. Sci.* 42:407–416.
Levander, O. A., 1977. Metabolic interrelationships between arsenic and selenium, *Environ. Health Perspect.* 19:159–164.
Lindgren, A., and Dencker, L., 1980. Preliminary study on the long time retention of arsenite and arsenate in epididymis, thyroid and lens in mice, in *3. Spurenelement—Symposium Arsen*, M. Anke, H.-J. Schneider, and Chr. Bruckner (eds.), Friedrich-Schiller-Universität, Jena, GDR, pp. 57–63.
Lunde, G., 1977. Occurrence and transformation of arsenic in the marine environment, *Environ. Health Perspect.* 19:47–52.
Mangal, P. C., and Singh, G., 1980. Uptake of selenium-75 and its interaction with arsenic, cadmium and mercury in the rat, *Proc. Indian Nat. Sci. Acad.* B46:615–620.
Marafante, E., Bertolero, F., Edel, J., Pietra, R., and Sabbioni, E., 1982. Intracellular interaction and biotransformation of arsenite in rats and rabbits, *Sci. Total Environ.* 24:27–39.
McBride, B. C., and Wolfe, R. S., 1971. Biosynthesis of dimethylarsine by methanobacterium, *Biochemistry* 10:4312–4317.
McChesney, E. W., Hoppe, J. O., McAuliff, J. P., and Banks, W. F., Jr., 1962. Toxicity and physiological disposition of sodium p-N-glycolylarsanilate. I. Observations in the mouse, cat, rat and man. *Toxicol. Appl. Pharmacol.* 4:14–23.
Moxon, A. L., 1938. The effect of arsenic on the toxicity of seleniferous grains, *Science* 88:81.
National Academy of Sciences, 1977. *Arsenic*, Report of the Committee on Medical and Biologic Effects of Environmental Pollutants, Washington, D.C.
Nielsen, F. H., Givand, S. H., and Myron, D. R., 1975. Evidence of a possible requirement for arsenic by the rat, *Fed. Proc., Fed. Am. Soc. Exp. Biol.* 34:923 (abs).
Nielsen, F. H., Myron, D. R., and Uthus, E. O., 1978. Newer trace elements—vanadium (V) and arsenic (As) deficiency signs and possible metabolic roles, in *Trace Element Metabolism in Man and Animals-3*, M. Kirchgessner (ed.), Tech. Univ. München, Freising-Weihenstephan, BRD, pp. 244–247.
Nielsen, F. H., Uthus, E. O., and Cornatzer, W. E., 1983. Arsenic possibly influences carcinogenesis by affecting arginine and zinc metabolism, *Biol. Trace Element Res.* 5:389–397.
Nissen, P., and Benson, A. A., 1982. Arsenic metabolism in fresh water and terrestrial plants, *Physiol. Plant.* 54:446–450.
Norin, H., and Christakopoulos, A., 1982. Evidence for the presence of arsenobetaine and another organoarsenical in shrimps, *Chemosphere* 11:287–298.
Pfannhauser, W., and Woidich, H., 1980. Source and distribution of heavy metals in food, *Toxicol. Environ. Chem. Rev.* 3:131–144.
Pomroy, C., Charbonneau, S. M., McCullough, R. S., and Tam, G. K. H., 1980. Human retention studies with [74]As, *Toxicol. Appl. Pharmacol.* 53:550–556.
Pyles, R. A., and Woolson, E. A., 1982. Quantitation and characterization of arsenic compounds in vegetables grown in arsenic treated soil, *J. Agric. Food Chem.* 30:866–870.
Schroeder, H. A., and Balassa, J. J., 1966. Abnormal trace metals in man: Arsenic, *J. Chron. Dis.* 19:85–106.
Shirachi, D. Y., Lakso, J. U., and Rose, L. J., 1981. Methylation of sodium arsenate by rat liver in vitro, *Proc. West. Pharmacol. Soc.* 24:159–160.
Siewicki, T. C., and Sydlowski, J. S., 1981. Excretion of arsenic by rats fed witch flounder or cacodylic acid, *Nutr. Reports Int.* 24:121–127.
Stevens, J. T., Hall, L. L., Farmer, J. D., DiPasquale, L. C., Chernoff, N., and Durham, W. F., 1977. Disposition of ^{14}C and/or [74]As-cacodylic acid in rats after intravenous, intratracheal, or peroral administration, *Environ. Health Perspect.* 19:151–157.

Tam, G. K. H., Charbonneau, S. M., Bryce, F., and Lacroix, G., 1978. Separation of arsenic metabolites in dog plasma and urine following intravenous injection of ^{74}As, *Anal. Biochem.* 86:505–511.

Tam, G. K. H., Charbonneau, S. M., Bryce, F., Pomroy, C., and Sandi, E., 1979a. Metabolism of inorganic arsenic (^{74}As) in humans following oral ingestion, *Toxicol. Appl. Pharmacol.* 50:319–322.

Tam, G. K. H., Charbonneau, S. M., Lacroix, G., and Bryce, F., 1979b. *In vitro* methylation of ^{74}As in urine, plasma and red blood cells of human and dog, *Bull. Environ. Contam. Toxicol.* 22:69–71.

Tam, G. K. H., Charbonneau, S. M., Bryce, F., and Sandi, E., 1982. Excretion of a single oral dose of fresh-arsenic in man, *Bull. Environ. Contam. Toxicol.* 28:669–673.

Uthus, E. O., Cornatzer, W. E., and Nielsen, F. H., 1983. Consequences of arsenic deprivation in laboratory animals, in *Arsenic Symposium, Production and Use, Biomedical and Environmental Perspectives,* W. H. Lederer (ed.), Van Nostrand Reinhold, New York, pp. 173–189.

Vahter, M., 1981. Biotransformation of trivalent and pentavalent inorganic arsenic in mice and rats, *Environ. Res.* 25:286–293.

Vahter, M., and Gustafsson, B., 1980. Biotransformation of inorganic arsenic in germfree and conventional mice, in *3. Spurenelement—Symposium Arsen,* M. Anke, H.-J. Schneider, Chr. Bruckner (eds.), Friedrich-Schiller-Universität, Jena, GDR, pp. 123–129.

Vallee, B. L., Ulmer, D. D., and Wacker, W. E. C., 1960. Arsenic toxicology and biochemistry, *A.M.A. Arch. Indust. Health* 21:132–151.

Wrench, J. J., and Addison, R. F., 1981. Reduction, methylation, and incorporation of arsenic into lipids by the marine phytoplankton *Dunaliella tertiolecta, Can. J. Fish. Aquat. Sci.* 38:518–523.

Cadmium

15

Harry A. Smith

15.1 Introduction

15.1.1 Historical Perspectives and Properties of Cadmium

The existence of cadmium was appreciated as early as 1556, when Agricola warned miners to wear heavy clothing to protect against a "kind of cadmia" that "eats away the feet of the workmen and injures their lungs and eyes" (quoted in Nriagu, 1980). The element was first discovered by Strohmeyer in 1817 and named cadmium from the Greek *cadmia,* which actually referred to the zinc carbonate ore from which it was isolated. Not long after its initial discovery, cadmium was recognized by Sovet as a potential health hazard with cadmium poisoning being a serious occupational disease (Prodan, 1932). Cadmium has thus been recognized for a long time as being toxic, but only rather recently has attention been given to its possible role as an essential element in biological systems.

Cadmium is a silvery white, soft, ductile metal with a slight bluish tinge and a density of 8.65 g/cm^3 at 20°C. Like mercury, cadmium is highly volatile and the vapor is quite reactive, forming finely divided cadmium oxide. There are eight stable isotopes of the metal naturally occurring, giving rise to an atomic weight of 112.40. Several radioactive isotopes are known; the most important in biological studies are ^{109}Cd (gamma, $t_{1/2}$ = 450 d) and ^{115m}Cd (IT, beta $t_{1/2}$ = 34 d). The melting and boiling points are 321°C and 765°C, respectively. The bond strengths of cadmium in diatomic molecules range from 2.7 Kcal/mole in Cd–Cd to 73 Kcal/mole in Cd–F. Cd–Cl and Cd–S have intermediate bond strengths of about 50 Kcal/mole. The specific heat of the metal does not differ very much between the solid (0.055 cal/g°C at 25°C) and the liquid (0.062 cal/g°C between 321 and 700°C), whereas the vapor pressure of the liquid changes

---

Harry A. Smith • Department of Chemistry, Florida State University, Tallahassee, Florida 32306.

almost three orders of magnitude (1–760 mm Hg) over a temperature range of about 350°C (394–767°C).

Cadmium is relatively rare, being concentrated in the lithosphere at levels of 0.1–0.5 µg/g. It is more rare than both mercury and zinc, its comembers of the group IIB family. In addition to its scarcity, cadmium is not found free in nature; rather, it occurs in association with ores of zinc, iron, and copper. Industrially, the metal is obtained in large quantities primarily as a by-product of zinc and copper refining. A major consequence of the rise in industrial use of cadmium over the past few decades has been a significant increase in the deposition of both the free metal and its salts into the biosphere. For example, there are markedly elevated levels of cadmium in air and soil near sites where the metal is used industrially when compared to rural areas (Friberg et al., 1974). It is not surprising, then, that interest in the biochemistry of cadmium has focused on the toxicological implications of the metal in reference to chronic environmental contamination.

15.1.2 Metallothionein and Its Interactions with Cadmium

Vallee and coworkers (Margoshes and Vallee, 1957; Kägi and Vallee, 1960) were the first to describe a biochemical localization of cadmium definitively, isolating a low-molecular-weight protein from equine renal cortex that apparently bound the metal with high affinity and in high quantities. They named the protein metallothionein, in keeping with its high sulfhydryl and metal content. During the next few decades following this initial discovery a large proportion of research on the biological and toxicological effects of cadmium focused on the central role of metallothionein-like proteins in regulating cadmium distribution (both organismic and intracellular) with special emphasis on the possible protective role of these proteins in sublethal exposure to the metal. It is difficult to appreciate many of the cellular and systemic effects of cadmium without some knowledge of the biological properties of these proteins. For this reason a brief review of the salient features of the metallothioneins will conclude this section, to set the stage for the discussions that follow. No attempt is made to give complete coverage, as many excellent reviews have recently appeared (Foulkes, 1982; Brady, 1982; Kägi and Nordberg, 1979; Cherian and Goyer, 1978). The interested reader is directed to these sources for a more comprehensive survey.

Metallothioneins are low-molecular-weight (6000–10,000 daltons) proteins rich in cysteine and metal content. Approximately 30% of the amino acids that comprise the protein are cysteine, all of which are involved in metal bonds in Cd-thionein. There are thus no free sulfhydryl groups or disulfide linkages in native metallothionein. The protein also lacks aromatic residues and therefore shows a minimal absorbance at 280 nm, but has an appreciable absorbance near

250 nm when the metal bound is cadmium. (If cadmium is replaced by another metal, this absorption band disappears.) The native form of the protein contains approximately 7 g-atoms of metal/g-molecular weight and is heat stable, possibly because of this high metal content.

The protein binds a number of metals besides cadmium, including Zn, Cu, Hg, and Ag. The affinity of the protein for these metals is in the order Ag > Hg > Cu > Cd > Zn. The relatively high affinity of thionein for Hg has allowed the development of assays for detecting levels of metallothioneins in tissues, using mercury exchange techniques. This method is adequate for detecting elevated amounts of metallothionein but is difficult to use when the protein is in low concentrations, such as in control animals. More recent cadmium exchange methods that remove excess cadmium by gel filtration or sequestering with RBC lysates have permitted the determination of low levels of endogenous zinc-thionein in adult and fetal tissues (Kern *et al.*, 1981; Onosaka and Cherian, 1982; Brady and Webb, 1981).

The amino acid sequences of several metallothioneins have been elucidated, and it is generally concluded that as a whole, the protein seems to have been highly conserved in evolution. There is no definitive report on the three-dimensional structure of the protein, primarily because of difficulties in obtaining good crystals. From ^{135}Cd-NMR studies, there appear to be two distinct metal "clusters" in the protein containing only cadmium. One cluster has four metals and 11 cysteines; the second has three metals and 9 cysteines. In (Cd, Zn)-thionein there is a marked separation of the metals, with most of the cadmium found in the first cluster and most of the zinc in the second.

Exposure to metals such as Cd, Zn, and Hg induces the synthesis of thionein primarily in the liver, but to some extent in the kidney and other tissues as well. In terms of cadmium induction, the protein is synthesized *in vivo* until most of the exogenous metal is associated with it. The physical half-life of Cd-induced thionein is longer than that of Zn-thionein or Zn-induced thionein. In addition, cadmium remains bound to thionein fractions in the liver for a time much longer than the half-life of the molecule. This suggests that once induced, thionein synthesis is maintained at some level in the tissue and there is an effective exchange of cadmium to newly synthesized chains. The net biological effect is a retention of the metal in the liver and other tissues where the protein is synthesized.

The mechanism whereby cadmium elicits the synthesis of thionein is only poorly understood but appears to involve regulation at the transcriptional level (Bryan *et al.*, 1979; Mayo and Palmiter, 1982).

The protective role of metallothionein synthesis against cadmium toxicity has been amply demonstrated. For example, Webb and Verschoyle (1976) have shown that pretreatment of rats with cadmium protected the animal against a subsequent lethal dose. The protection was optimal 1–3 days following pretreatment. In Webb's study, elevated levels of zinc-thionein resulting from starvation did not protect against cadmium toxicity, leading Webb to question whether the protein itself had a protective role. However, in a similar study, Leber and Miya (1976) demonstrated that cadmium pretreatment as well as zinc pretreatment can nearly double the LD_{50} for cadmium. It is possible that Webb's results with elevated levels of zinc-thionein are due to other factors resulting from starvation that potentiate cadmium toxicity.

Although the intracellular appearance of these proteins has been implicated as a protective mechanism against cadmium toxicity in both whole animal and cell culture studies (Nordberg et al., 1971; Yoshida et al., 1979), plasma levels of metallothioneins have been found to contribute to the nephrotoxic effects of long-term cadmium exposure (Cherian et al., 1977; Goyer and Cherian, 1977), complicating the early notion that the biosynthesis of the protein was primarily an adaptive response on the part of the organism to cadmium toxicity. The natural biological role(s) of this protein appears to be mainly concerned with the regulation of zinc absorption and distribution (Richard and Cousins, 1975; Kern et al., 1981), suggesting that its involvement in the detoxification of cadmium may be a fortuitous use of a molecular mechanism that had been previously evolved to handle the metabolism of closely related but more essential metals. There is evidence, for example, that the protein is able to transfer zinc to certain apoenzymes under mild *in vitro* conditions. *Escherichia coli* alkaline phosphatase and bovine erythrocyte carbonic anhydrase can be reactivated following zinc depletion to the same extent by Zn-thionein as repletion with zinc salts (see Brady, 1982 for a review). Copper can also be transferred from Cu-thionein to apocarbonic anhydrase. It is not known whether metallothionein can or does perform these functions *in vivo*.

The presence of elevated levels of metallothioneins in fetal and neonatal tissues has led some investigators to suggest that the protein is present in tissues undergoing rapid growth and development that require zinc and perhaps copper for macromolecular biosynthesis. For example, the necessity of placental transfer of zinc for normal DNA synthesis in fetal liver has been recently demonstrated (Webb and Samarawickrama, 1981). There is little evidence, however, that Zn-thionein is able to transfer the metal to enzymes involved in DNA, RNA, or protein synthesis.

Besides metal exposure, a number of stressful conditions as well as glucocorticoid exposure can elicit the synthesis of metallothionein. The common feature of the latter two treatments may involve a mobilization of zinc from depots in the body, migration of zinc to the liver where pools of the metal increase, and subsequent initiation of thionein synthesis. It is not clear whether zinc also mediates the ability of cadmium and other toxic metals to induce the protein. The mechanism of induction by glucocorticoids appears to differ from that of cadmium, since the ability of glucocorticoids to induce the protein can be selectively lost following a number of genetic manipulations while cadmium inducibility is retained (Mayo and Palmiter, 1982).

In summary, metallothionein proteins are integrally involved in the metabolism of cadmium and a number of other trace elements. The physiological roles of the proteins are not completely understood but appear to relate most closely to zinc homeostasis. The protective role of the proteins against short-term cadmium exposure has been demonstrated, but following long-term exposure the protein itself may contribute significantly to the nephrotoxicity of the metal.

15.2 Chemistry of Cadmium: Biological Perspectives

15.2.1 General Chemical Properties of Cadmium

Cadmium is a member of the Group IIB family and shares many of its chemical properties with mercury and zinc. Unlike mercury, however, cadmium and zinc do not form strong metal–carbon bonds, which are so important in the geochemical cycles and toxicity of mercury. The normal valence of cadmium is $+2$, but it readily forms a number of ionic complexes, such as CdX_3^{-1} and CdX_4^{-2} (where X = halogen). The metal dissolves readily in both organic and inorganic acids, with maximum solubility in nitric acid, which is used experimentally to solubilize cadmium from organic materials. Cadmium is not amphoteric and does not dissolve in alkaline solutions. There are only a few minerals of cadmium known, with highest concentrations of the element found in the sulfides. None of these minerals is naturally abundant, however, nor contributes appreciably to commercial sources. Cadmium is found most abundantly in association with sulfide ores of zinc, lead, mercury, copper and iron.

15.2.2 Biological Implications

In terms of its electronic structure and reactivity to organic ligands, cadmium closely resembles zinc. However, cadmium has a much higher affinity for thiol groups than does zinc and a lower affinity for ligands containing oxygen and nitrogen (Venugopal and Luckey, 1978). Cadmium is able to replace zinc in a number of organic complexes, including certain zinc-requiring enzymes. It is not always possible, however, to correlate these data directly with cadmium toxicity *in vivo* as in some cases cadmium replacement leads to an enhanced activity of the enzyme and in others to deactivation (Vallee and Ulmer, 1972). Copper, lead, and mercury are the only trace metals able to displace cadmium from thiol ligands *in vitro,* although their role in determining the intracellular distribution of the metal is not clear.

Attempts have been made to take advantage of these chemical properties of cadmium in identifying therapeutic agents to protect against acute cadmium toxicity. A number of early studies (Gilman *et al.*, 1946; Teperman, 1947; Tobias *et al.*, 1946) demonstrated that treatment with BAL (2,3-dimercaptopropanol) was able to reduce, to some extent, the toxicity of a single acute challenge of $CdCl_2$ in rabbits. However, the BAL treatment also resulted in an increase in renal accumulation of the metal, with concomitant nephrotoxicity.

Cherian and Rodgers (1982) have recently demonstrated that animals treated with BAL or DTPA (diethylnitriamine pentaacetic acid) following chronic cadmium exposure (under conditions where metallothionein levels are significantly elevated over those found in acute exposure) showed an increase in fecal and biliary excretion of cadmium, thus lowering the body burden of the metal. Most important, treatment under these conditions did not result in increased amounts of cadmium in renal tissues; rather, it resulted in a decrease in both hepatic and renal levels of the metal. It is possible that the nephrotoxicity associated with immediate treatment of cadmium exposure with chelating agents may be overcome, in part, by delaying treatment until the metal has been sequestered with metallothioneins *in vivo*.

15.3 Evidence for the Possible Essentiality of Cadmium

It is safe to estimate that most research done on the biological effects of cadmium has focused in one way or another on the toxicity of the metal. This is perhaps to be expected in light of the many systemic pathologies produced by cadmium poisoning in mammals (Frieberg *et al.*, 1974; Nriagu, 1980; Venugopal and Luckey, 1978). It has been only within the last 10 years or so that there has been interest in the possibility that cadmium may be stimulatory or essential to growth in some mammalian systems. One must be careful, however, in distinguishing between stimulatory effects of the metal given in supplement to a normal diet and clearly defined deficiencies when the metal is eliminated from the diet. Only the latter can be taken as evidence that the metal is essential for normal growth. Stimulatory effects may be due to a variety of interactions that may be incidental to the metal's presence in the organism because of supplementation. For example, cadmium administration is followed by an increase in the activity of Elongation Factors 1 and 2, but no evidence is available that suggests that this is a normal physiological function of cadmium in nature. Unfortunately, most of the studies investigating possible physiological roles of cadmium have focused on stimulatory effects of the metal given in supplement to a metal-free or normal diet, and very little evidence exists on deficiencies resulting from elimination of the metal from the diet.

Early studies indicated that cadmium may be stimulatory to the growth of

Aspergillus niger and other algae (Scharrer and Schropp, 1934; Pirschle, 1934). When the algae were grown on a basal medium supplemented with cadmium, growth effects were noted at cadmium concentrations of 10^{-8}–10^{-5} M, with maximal stimulation (147–206% of control) occurring at 10^{-5} M. Interestingly, at 10^{-4} M there was a sharp reduction of growth, and no growth at all was detected at higher concentrations. For comparison, zinc was found to be stimulatory up to concentrations of 10^{-3} M, whereas mercury, which showed clear toxic effects at 10^{-6} M, was not stimulatory at any of the concentrations used in the study.

The best evidence that cadmium may be stimulatory to growth in mammals has come initially from the work of Klaus Schwarz, who pioneered the development of animal isolation techniques necessary to study deficiency and stimulatory effects of trace metals. It is to Schwarz's credit that within the last 20 years or so at least six elements (Se, Cr, Sn, V, F, Si) were first found to be stimulatory and subsequently essential.

In a report to *Federation Proceedings,* Schwarz and Spallholz (1976) described significant, if small, growth effects of cadmium under trace element controlled conditions (Table 15-1). The levels of cadmium producing these growth effects were below what is normally considered to be toxic levels and were similar to levels found in the normal diet of the animals. When standard diets (containing trace levels of cadmium) were fed and additional cadmium was added, no growth effects were seen.

Although the preceding data suggest that cadmium may be stimulatory to growth under certain conditions, there is no indication of growth deficiencies due to elimination of cadmium from the diet. Anke *et al.* (1978) have shown such deficiencies in the growth and sexual maturation of goats fed diets containing less than basal levels of cadmium. However, the data are preliminary and further study is needed to clarify these effects.

The history of trace metal research should serve as a warning against taking

Table 15-1. Growth Effects of Cadmium[a]

Cd level in diet (ppm)	Enhanced growth (% over control)	
	Cd-acetate	Cd-sulfate
0.0125	—	0
0.025	0	—
0.05	2.5	8.7
0.1	8.9	7.5
0.2	—	13
0.5	12	—

[a] Summarized from the work of Schwarz and Spallholz, 1976.

too pessimistic a view on the possible essentiality of cadmium, because a number of other elements (notably Se) whose toxicity alone was initially appreciated have been subsequently proved to be essential.

15.4 Metabolism of Cadmium

15.4.1 Absorption of Cadmium

Organisms are exposed to cadmium from a variety of sources, including air, food, and water. The sites of cadmium absorption include respiratory and alimentary surfaces, skin, and placental circulatory systems. In humans and many other mammals, the main route of absorption is via the diet, although pulmonary absorption becomes important primarily with air-breathing organisms that are in close proximity to industrial sources of the metal. Human workers in such industries have been reported to have elevated body burdens of the metal when compared to normal populations (Tohyama et al., 1981; Nordberg et al., 1982). Cigarette smoke is an additional source of pulmonary exposure (Menden et al., 1972).

Absorption from respiratory surfaces depends on several factors, including the physical form of the particle, its size, and its solubility (Friberg et al., 1974; Nordberg, 1974). Pulmonary absorption is apparently more efficient (25–50% of dose is adsorbed) than that of the gastrointestinal tract (4–6% absorbed). However, nonindustrial concentrations of airborne cadmium are so low that this route of absorption contributes little to the overall body burden (a maximal absorption of 0.26 μg/day has been calculated in the preceding reports). Cigarette smoke, on the other hand, has been estimated to add as much as 1.0 μg/pack to the body burden of the metal (Menden et al., 1972).

In humans not exposed to industrial levels of cadmium, the major route of absorption is via the diet. Absorption of orally ingested cadmium is generally low, with typical values of 4–6% reported (Shaikh and Smith, 1980). The average dietary intake in the United States has been estimated to be about 50 μg/day, approximately 3 μg of which is actually absorbed (Duggan and Corneliussen, 1972). There are a number of dietary factors that modulate the amount of cadmium that is absorbed. The most important are the chemical and physical forms of the metal as well as the general nutritional state of the animal.

The mechanism of cadmium absorption in rats (Foulkes et al., 1982) involves an initial uptake of the metal from the intestinal lumen mediated by a saturable membrane transport system. Following this uptake, the metal is transferred from the mucosa into the blood, but at a slower rate than the initial uptake.

Several factors affect intestinal uptake in these *ex vivo* preparations. For example, Zn is able to compete effectively with Cd for transport; both milk and bile salts also inhibit cadmium uptake, the former primarily because of its Ca^{2+} content, the latter because of micelle formation. Metallothionein does not seem to have a role in the retention of Cd in the mucosa under the *ex vivo* conditions employed in these experiments. This is in contrast to the work of Cousin's group, which suggests that metallothionein plays an important role in determining the uptake of zinc and cadmium in the intestine, as well as interactions between these metals. There is a measurable difference in absorption between the ilium and the jejunum, in that ilial absorption shows a much faster transport of cadmium from the mucosa into the blood. Copper and zinc have a much higher mucosal transport than cadmium, although cadmium transport can achieve rather high values in neonates. These results suggest that Cd may share a common absorption mechanism with other metals but that the efficiency of this mechanism is less for cadmium and may be regulated physiologically in an age-dependent manner.

Cutaneous absorption and placental transport are two additional, if less important, routes of absorption. Cutaneous absorption is probably of minor significance when compared to inhalation or digestion. Placental transport has been well established in a number of species (Tsevetkova, 1970; Webb and Samarawickrama, 1981; Waalkes *et al.*, 1982), but interestingly enough, there seems to be a singularly effective placental barrier in goat (Anke *et al.*, 1970).

Placental interactions with cadmium are important because the metal proves to be a powerful teratogen at a certain dose. Studies have focused on the effects of cadmium in preventing transport of other essential elements to the fetus as well as direct effects on the fetus itself. At teratogenic doses (1.25 mg Cd/kg body weight) given parenternally, the metal inhibits placental transport of zinc in rats for approximately 48 hours but does not seem to affect the transport of sugars, amino acids, or nucleic acid precursors (Webb and Samarawickrama, 1981). DNA synthesis in fetal tissue is also inhibited in a time course that parallels the reduction of zinc transport. The inhibition of DNA synthesis is apparently due to a decrease in the activity of thymidine kinase, which can be restored *in vitro* through the addition of exogenous zinc. This suggests that zinc depletion may be the fundamental mechanism for the decrease in DNA synthesis seen in fetal tissue, following exposure of the mother to cadmium. These effects of cadmium can be inhibited by simultaneous injection of zinc at a 2 : 1 ratio.

The fetal rat responds to subteratogenic doses of maternally injected cadmium with a decrease in metallothionein content in much the same fashion as with teratogenic doses (Waalkes and Bell, 1980). At nonlethal doses in rabbit significant fetal effects are seen, including a reduction in fetal body weight, liver weight, kidney weight, and crown rump length while no placental damage is apparent (Waalkes *et al.*, 1982). However, at a nonlethal dose of 0.50 mg Cd/kg body weight, fetal rabbits show an increase in metallothionein levels and an increase in the cytosolic zinc in the fetal hepatocytes appears to coincide with a redistribution of the metal from extrahepatic sites. At this dose, there is no change in the total fetal content of zinc.

It is apparent that maternal exposure to cadmium can affect fetal life at a number of levels, including (1) placental transport of zinc (and perhaps of other trace elements as well), (2) macromolecular metabolism in fetal tissue, and (3) fetal hepatic and extrahepatic distribution of key trace elements.

15.4.2 Transport of Cadmium in Blood

Following absorption of cadmium from alimentary or respiratory structures, the metal is transported primarily via the blood. In general, there is a higher level of cadmium in plasma than in erythrocytes, although this ratio can shift as cadmium exposure increases (Friberg et al., 1974). Within the plasma, cadmium binds readily to serum proteins such as albumin and is distributed to target tissues, where it is taken up by specific transport mechanisms. Under conditions of acute exposure, cadmium is rapidly cleared from the blood. However, during chronic exposure, there is a secondary slower decline in blood cadmium followed by an even later increase in concentration of the metal in both plasma and cells. This increase in blood cadmium during chronic exposure does not occur until metallothionein synthesis in hepatic and other tissues has been enhanced.

Blood levels and clearance of cadmium are affected by excess levels of zinc, with maximal inhibition of clearance seen at a Zn : Cd molar ratio of 17 : 1 (Frazier, 1981). The mechanism of this effect is not clear, but may be due to the displacement of cadmium from albumin by zinc and thus an alteration of its normal mode of transport and uptake by target tissues.

The intracellular distribution of cadmium in erythrocytes has been only cursorily investigated. Circulating erythrocytes have detectable levels of metallothionein-like proteins (Nordberg, 1972) to which the cadmium may be bound. The source of metallothioneins in erythrocytes following cadmium exposure is not known but is presumed to arise from extraerythrocytic sites because of the limited ability of these cells to initiate new protein synthesis. The possibility that a stable thionein message exists in these cells and is regulated at the translational level cannot be ruled out. Nevertheless, current thinking suggests that the metallothionein either is of hepatic origin or is induced in erythrocyte precursors. It is not clear how a transport of hepatic metallothionein to erythrocytes could occur, although under certain exposure conditions, circulating levels of the protein are detectable. A more attractive hypothesis is that the protein is synthesized in erythrocyte precursors and then accumulates in maturing erythrocytes. Very little work of any definitive character has been done to distinguish between these mechanisms.

15.4.3 Organ, Tissue, and Subcellular Distribution of Cadmium

The distribution of cadmium among various tissues following exposure has been studied extensively, and Cherry (1980) has provided an exhaustive review of the literature pertaining to human tissues. Under conditions of low-level exposure, the metal accumulates primarily in the liver and kidney. With higher levels of exposure and over longer periods of time, the liver content of cadmium increases and the metal appears in other organs, such as the pancreas, testes, lungs, spleen, and various endocrine organs, in particular the thyroid. However, as Cherry points out, there is a direct relationship found only between cadmium contents of kidney, liver, and lung under low exposure conditions. Little correlation can be made between cadmium contents of other tissues primarily because the low levels lead to analytical problems that increase the background "noise" in the data.

The distribution of cadmium in tissues following exposure depends on the form of the metal administered. When experimental animals are given sublethal doses of cadmium salts, most of the metal is deposited in the liver. However, if cadmium is administered in the form of Cd-thionein or (Cd, Zn)-thionein, the metal is preferentially accumulated by the kidney, with a concomitant increase in urinary excretion of both the protein and the metal (Cherian *et al.*, 1976). The presence of the Cd-thionein complex in renal tissue has been implicated in the morphological changes seen in the plasma membranes of renal tubular cells and damages to mitochondria of these cells following injection of Cd-thionein. Comparable doses of cadmium salts or Zn-thionein do not produce these nephrotoxic effects. It is not known whether the metallothionein molecule itself is the chief nephrotoxic agent or whether cadmium ions released intra- or extracellularly are causing the damage. There is some evidence to suggest that the protein–metal complex can be taken up intact by absorptive cells and may rupture cell membranes in the process (Valberg *et al.*, 1977). The Cd-thionein that has entered the cell may then alter energy metabolism by damaging the mitochondria sufficiently to cause cell death.

The subcellular distribution of cadmium in hepatocytes exposed to cadmium both *in vivo* and *in vitro* has been investigated by Bryan *et al.* (1979). Isolated cells take up the metal very rapidly after exposure by a diffusion-like process, metal uptake being linear with dose. Once inside hepatocytes, the metal distributes among subcellular components in a time-dependent fashion both *in vivo* and *in vitro*. After an acute exposure to cadmium the metal initially binds nonspecifically to high-molecular-weight proteins ($>12,000$ daltons). There is a gradual shift in the distribution of the metal in time to a protein fraction enriched in metallothionein-like proteins. After 24 hours following an i.p. injection of $CdCl_2$, more than 90% of the intrahepatic metal is associated with metallothioneins.

Early after cadmium exposure *in vivo,* cadmium enters the nucleus and establishes an equilibrium between cytosolic and nuclear components. The amount of cadmium detected in the nucleus decreases over the subsequent 24 hours and becomes sequestered with the metallothionein fraction. The rate of disappearance from the nucleus follows an exponential decay with a decay constant of approximately 0.11 hour. The cadmium associated with nuclei does not seem to be due to contamination from the cytosol, since isolated nuclei do not "pick up" any metal when incubated with cadmium-containing cytosol, nor do cadmium-containing nuclei release much of the metal when incubated with control or experimental cytosols.

A recent report (Banerjee *et al.*, 1982) using immunohistochemical techniques has yielded new information on the intracellular localization of Cd-containing metallothioneins following cadmium exposure. In control animals (not exposed to the metal) native metallothioneins were found to be primarily localized in the cytoplasms of hepatocytes, renal collecting duct epithelium, and distal convoluted tubule epithelium. Two weeks following daily cadmium treatment, metallothionein was found mainly in nuclei rather than in the cytoplasm. After 4–8 weeks of continuous cadmium treatment metallothioneins appeared both in cytoplasm and nuclei. Long-term exposure results in immunohistochemical staining of metallothionein in a number of other tissues. No staining was detected in vascular endothelial cells, fibroblasts, or leukocytes. The significance of nuclear-bound thionein is not yet understood; however, it is tempting to speculate on whether some of cadmium found in nuclei early after exposure might not be associated with these small levels of thionein found in nuclei of control cells.

Hayden *et al.* (1982) have devised a compartmental model for intracellular cadmium localization using data available on cadmium transport and distribution in CHO cells. Using Monte Carlo techniques they have generated a stochastic model that is more easily adaptable to computation and does not require numerical solutions of differential equations which arise from most deterministic models. The model adequately describes the intracellular accumulation and localization of cadmium including the time dependent sequestering of the metal with metallothionein-like proteins. Such models offer a unique opportunity for investigating possible relationships between variables that might not be obvious from more traditional lines of inquiry. Care must be taken, however, in extrapolating any correlations seen in these models to *in vivo* situations. Such models should be used to suggest novel experimental approaches rather than to replace them.

15.4.4 Cadmium Excretion

Cadmium is excreted in man and many other mammals primarily via the urinary and gastrointestinal tracts. The bulk of ingested cadmium is excreted in the feces, primarily because of the low absorption from the intestinal tract. Renal excretion is normally low under conditions of low-level exposure, but urinary output increases slowly with age and in the presence of renal dysfunction. Chronic, low-level exposure results in a cadmium excretion of about 1–2 μg/day. This level is elevated in cadmium-treated individuals, and there appears to be a relationship between urinary excretion of the metal and the total body burden (Friberg et al., 1974). Biliary excretion has been demonstrated in experimental animals, and evidence suggests that the metal is transported from plasma into bile against a concentration gradient (Klaasen, 1976). Apparently, cadmium is excreted in bile in association with a low-molecular-weight substance, possibly as a cadmium–glutathione complex (Cherian and Vostal, 1977).

Shaikh and Smith (1980) have investigated carefully the excretion of orally ingested cadmium in humans using ^{109}Cd and ^{115m}Cd isotopes. They found that a large part (40%) of a single oral dose of 50 μg Cd (5–10 μCi ^{115m}Cd or ^{109}Cd) was excreted in the feces within 7 days. As a consequence of the low-level exposure used, urinary excretion was negligible relative to fecal excretion. Biliary excretion was not investigated. Analysis of whole body retention of the metal yielded data that was best fit using three exponentials having half-times of 1.58, 33.7, and 9605 days. The slowest component accounted for 2.5% of the initial dose and is significantly longer than values previously reported (half-times of 100–200 days). If the retention data are analyzed to day 200 rather than to day 800 of the study, the best fit analysis gives a long-term half-time of only 270 days. This discrepancy illustrates the necessity for long-term monitoring in analyzing body retention of cadmium. The other two components can be identified with the elimination of most of the unabsorbed dose (half-time = 1.58 days), and with a continued elimination of both unabsorbed and absorbed metal (half-time = 33.7 days).

Humans exposed chronically to high levels of cadmium (primarily through occupational exposure) have detectable levels of metallothionein and cadmium in their plasma and urine (Tohyama et al., 1981; Nordberg et al., 1982). There is an apparent linear relationship between log (metallothionein) present in urine and log (cadmium) in the kidney and liver. However, there appears to be less of a correlation between urinary levels of metallothionein and blood cadmium. Hence levels of urinary metallothionein may be a better indicator of the extent of cadmium exposure than tests that are more commonly used (such as blood levels of cadmium).

The important features of cadmium metabolism may be summarized as follows: (1) The metal has a relatively long biological half-life. (2) There is no apparent homeostatic mechanism for the metal, resulting in high retention in certain tissues. (3) Furthermore, there is an uneven distribution among tissues with preferential localization in liver and kidney. (4) There is low absorption from the diet because of retention in mucosal cells. (5) There may be signifcant interactions of Cd with other essential metals in their absorption and distribution both systemically and intracellularly.

15.5 Biochemical Effects of Cadmium

Cadmium exerts a number of toxic effects on living organisms at the cellular and systemic level. In only a few cases have the primary biochemical lesions responsible for the effects been elucidated. In addition, the nontoxic and possible stimulatory effects of cadmium under conditions of low level exposure have received relatively little attention in the literature. This section will summarize a limited number of the biochemical effects of cadmium with little distinction as to their definitive role(s) in the toxic or stimulatory consequences of exposure to the metal. Indeed, it is difficult to extrapolate effects cadmium may have in one tissue (the lung, for example) under experimental conditions, to the pathological effects the metal may have in other tissues under more natural exposure conditions. Nevertheless, an understanding of the possible biochemical effects of the metal is important in evaluating its role in toxicity or essentiality.

15.5.1 Nucleic Acid and Protein Synthesis

The interactions of metals such as zinc, copper, and cadmium with nucleic acids have been reviewed (Sissöeff *et al.*, 1976). The interactions of metals with chromatin *in vivo* is not well understood, since the sites of these interactions have not been clearly defined. *In vitro,* however, metals can bind to either the phosphate groups or nitrogenous bases in DNA. When bound to phosphate groups, metals tend to stabilize the double helix, but when bound to nucleotide bases, they cause a destabilization of the double helix. When such metals associate with the phosphate groups in RNA, they bring about a depolymerization through cleavage of the phosphodiester bond. This interaction, at least, appears to be fairly specific, since the tendency to cleave a given bond varies with the nature of the base adjacent to the bond. Polyribosomes are destabilized in the presence of excess amounts of zinc and copper (McGown *et al.*, 1971), an effect that is also seen following cadmium exposure (Gamulin *et al.*, 1977; Car *et al.*, 1979). In only a few cases are the functional or pathological significance of these interactions appreciated.

Stoll *et al.* (1967) have investigated the effects of cadmium on nucleic acid and protein synthesis under a variety of conditions. The metal inhibits thymidine incorporation into the DNA of regenerating liver cells *in vivo*. The metal also increases *in vitro* the absorbance at 260 nm of the melting profile of DNA without affecting its T_m. It is not clear whether this effect is realized by an actual unwinding of the DNA or by some other mechanism. The effects of cadmium on *in vivo* RNA metabolism are less easily understood. At low doses (0.34 mg/kg) the metal apparently stimulates *in vivo* orotic acid incorporation into RNA from 1 to 72 hours after treatment. A similar effect was noted by Enger *et al.* (1977) using mammalian cell cultures. At higher levels of exposure (3.41 mg/kg) the metal initially stimulates RNA synthesis at 1 hour, but subsequently causes a decrease in orotic acid incorporation. When isolated nuclei are exposed to the metal, UTP incorporation into RNA was inhibited at all levels of cadmium used (5×10^{-8} M to $1 \times 10^{-4} M$) and was not dose dependent. The level of incorporation was reduced by one third at all concentrations except the highest (1×10^{-4} M), where incorporation was reduced by one tenth. It is interesting that *in vitro* cadmium is able to inhibit RNA synthesis at concentrations much lower than those realized in rat liver (10^{-5}–10^{-6} M) after typical doses of the metal are given. Care must be taken in interpreting these results, however, since *in vivo* the metal may not be as free to interact with nuclear components as it is in an *in vitro* reaction mixture.

Protein synthesis is inhibited following cadmium exposure in rats (Hidalgo *et al.*, 1976). In the liver, there is a biphasic response with maximal inhibition occurring at 1 hour following challenge and levels of synthesis returning to that of control by 16 hours. This and other observations on the deleterious effects of cadmium on protein synthesis have led several groups to investigate the possible mechanisms of this action. Hidalgo *et al.* (1976) reported an early observation that RNA polymerase activity is inhibited following cadmium treatment, which inhibits maximally 1 hour after exposure. These results suggest that the decrease in protein synthesis might be due to a decrease in the synthesis of translatable mRNAs during this time period. This suggestion is difficult to reconcile with the observation by Stoll *et al.* (1976) and Enger *et al.* (1977) that net RNA synthesis may be stimulated early after cadmium challenge. Of course net RNA synthesis may be increased without producing any significant translatable mRNA's because of alterations in the processing of the message, but this seems unlikely in view of the fact that cadmium exposure elicits the synthesis of specific new messages coding for metallothioneins soon after cadmium treatment. Consequently, the mechanism of cadmium inhibition of protein synthesis must be looked for elsewhere.

The decrease in protein synthesis noted by Hidalgo *et al.* (1970) can be correlated quite nicely with a similar biphasic change in the state of polysomal aggregation noted by Gamulin *et al.* (1977) and Car *et al.* (1979). Polysomal disaggregation was maximal 1 hour following cadmium exposure and polymeric profiles resembling that from control animals were observable at the end of 12 hours. These studies have also demonstrated that the effect of cadmium seems to be directly at the level of the ribosomes itself rather than being due to some cytoplasmic component interfering with the ribosomes' function. Stoll *et al.* (1976) have carefully investigated the effects of cadmium on protein synthesis *in vivo* and *in vitro*. Liver microsomes isolated from animals pretreated with cadmium were assayed for their ability to incorporate phenylalanine into peptides. When the microsomes were preincubated to remove any endogenous messages and tested with the synthetic message polyuridylic acid, cadmium pretreatment seemed to inhibit the ribosome's ability to direct incorporation of phenylalanine. However, when the microsomes were not preincubated, thus allowing the translation of proteins already initiated, cadmium pretreatment did not seem to inhibit protein synthesis by the ribosomes; in fact, there was a slight trend toward enhanced synthesis. When the metal was added *in vitro* a similar effect was seen, with inhibition occuring at all concentrations in the preincubated preparations, but only at the highest concentrations with ribosomes having endogenous message. The significant difference between these two sets of experiments lies in the fact that preincubated ribosomes must undergo reinitiation in order to begin translation of the synthetic messages, whereas ribosomes containing endogenous message do not have to undergo this process in order for incorporation to proceed. Hence, assuming that cadmium affects either the rate of elongation of protein synthesis or the ability of ribosomes to initiate new rounds of synthesis, these experiments suggest that the effect may be at the level of initiation. Polysomal disaggregation could thus be explained by invoking an inability on the part of the ribosomes to initiate protein synthesis, and hence polysome formation. A recent report by Rakhra and Nicholls (1982) has shed further light on this problem by clearly demonstrating that, following cadmium treatment, peptide elongation was enhanced. This effect was apparently due to an increase in the activity of both Elongation Factors 1 and 2. Some investigators now feel that cadmium affects protein synthesis primarily through inhibiting the ribosome's ability to reinitiate protein synthesis (Car *et al.*, 1979; Smith *et al.*, 1982).

In none of the preceding studies was the association of cadmium with ribosomes or polysomes demonstrated. Smith *et al.* (1982) have recently studied the localization of cadmium, zinc, and copper with polysomal fractions at various times following cadmium challenge. Although cadmium was not normally detectable in control polysomes, it did associate with this subcellular fraction following challenge. The level of cadmium found in polysomes decreased with time in an apparent exponential decay, having the same time course as the disappearance of cadmium from nuclei. The cadmium associated with the polysomes was not due to contamination from cytoplasmic cadmium-containing thionein, as addition of ^{115}Cd-labeled thionein to supernatants from which polysomes were isolated did not contribute significantly to the amount of metal that was found when compared to levels noted when cadmium salts alone were added *in vitro* or *in vivo*. The precise localization of the metal with ribosomal subunits was not investigated in this report, although preliminary results of protease and ribonuclease digestion of polysomes suggest that the metal may be associated with proteins. Protease digestion released 70% more cadmium from polysomes than buffer treatment, whereas ribonuclease digestion released no more cadmium than the buffer control. These results, however, do not rule out the association of cadmium with RNA sequences in the interior of the ribosomes that are insensitive to RNase cleavage. It should be emphasized that the polysomal fractions used in these experiments are not a highly purified fraction and further studies on purified subunits are necessary to clarify the precise site(s) of cadmium's interactions with ribosomes.

Zinc and copper were detectable in polysomes isolated from control animals. Cadmium was not detected in these fractions. The presence of total zinc or copper in the cytosol associated with polysomes changed upon treatment. Early after exposure to cadmium (1–3 hours) there was a rise in both zinc and copper levels in polysomes. Elevated levels returned to nearly that of control by 16 hours. Hence the rise of total metal associated with polysomes paralleled both the inhibition of protein synthesis and the disaggregation of polysomes seen following exposure to cadmium. Since polysomes isolated in the presence of excess zinc or copper are destabilized and tend to depolymerize (McGown *et al.*, 1971), it is possible that the excess metals associated with this function after cadmium treatment may contribute to their disaggregation *in vivo*.

15.5.2 Induction of Thionein by Cadmium

The classical hallmark of cadmium exposure is the appearance of intracellular metallothioneins in certain target tissues. This section will address more carefully the evidence that cadmium, in some manner, is able to elicit the synthesis of new mRNA's specific for metallothioneins. The evidence for this can be divided into two types: (1) classical studies demonstrating that transcription is required for metallothionein synthesis using various inhibitors and (2) modern studies using newly developed techniques in molecular biology to demonstrate directly the production of new mRNA molecules coding for the protein. Although the early studies were important in demonstrating that a transcriptional event was required, the newer evidence has provided important insights bearing on the mechanism of the induction. For this reason more emphasis is placed here on the later studies, particularly those from Palmiter's group that have, literally, ushered in a new age in the study of metallothionein induction by cadmium.

The synthesis of metallothionein-like proteins can be elicited by a number of causative agents besides cadmium exposure, including stress, bacterial infections, alkylating agents, dexamethazone, endotoxins, hepatectomy, adrenalectomy, inflammatory reactions and agents, and alterations of zinc levels (Maitani and Suzuki, 1981). It is difficult, thus, to dissect out the mechanism whereby cadmium influences metallothionein synthesis *in vivo* apart from a number of other whole body effects. Several investigators have utilized isolated cell suspensions or cultures to focus directly on the cellular effects of the metal in the absence of systematic hormonal or organ interactions. A series of such studies on isolated hepatocytes and hepatic tumor cells has been summarized by Bryan *et al.* (1979). The isolated hepatocyte retains the ability to synthesize thionein (as does the hepatic tumor cell) and without the apparent lag period normally seen *in vivo*. The early appearance of thionein following exposure to cadmium is due to new message and protein synthesis and not to displacement of zinc from mature thioneins, as demonstrated by the following experiments.

Cyclohexamide and Actinomycin-D were added either at zero time following cadmium exposure or 2 hours after exposure. Cyclohexamide inhibited thionein synthesis at both time points, whereas Actinomycin-D inhibited completely at 0 hour but only partially at 2 hours. The authors interpret these results as indicating that enough mRNAs were synthesized between 0 and 2 hours to initiate a limited synthesis of the protein after inhibition of transcription. Hence, cadmium appears to act at the transcriptional level in isolated cells in the absence of hormonal effects, although intracellular metal interactions are not ruled out.

Although Palmiter's group was not the first to demonstrate that metals were able to elicit the synthesis of new mRNA molecules coding for thionein (Squibb and Cousins, 1977), they certainly have led the field in applying state of the art techniques in molecular biology to address the mechanisms of induction of the protein. In 1980 Durnam *et al.* reported the isolation of cDNA and genomic clones coding for metallothionein-I. Use of these molecular probes allowed Palmiter's group to monitor directly changes in gene transcription, gene organization, and mRNA accumulation in response to both cadmium and glucocorticoid exposure.

Durnam and Palmiter (1981) have investigated the transcriptional regulation of the metallothionein-I gene by a variety of metals in various tissues. In liver and kidney, cadmium administration resulted in a 25- and 17-fold increase, respectively, in the rate of RNA synthesis directed by the metallothionein I gene. The rate of transcription of the gene remained high in liver up to 9 hours after exposure. In the kidney, the rate dropped rather rapidly over the first 6 hours, after which it reached levels similar to that in control tissues. The levels of mRNA rose rapidly following an initial 30-minute lag and peaked at 4–5 hours following injection of the metal. At higher doses (5 and 25 mg/kg body weight), induction of the protein was noted in several other tissues, including intestine, heart muscle, brain, and spleen. Induction was not noted in testes, however. There seems to be a dose response in that tissues that absorbed more cadmium generally synthesized more mRNA for the protein. However, various tissues did not respond to the metal in the same fashion, as total message produced per unit of cadmium absorbed differed from tissue to tissue. When comparable doses of Cd, Zn, Hg, and Cu were compared for effectiveness in inducing the protein, the following series were noted: liver: Cd > Zn > Cu > Hg; kidney: Cd ~ Hg > Zn > Cu. However, no determination was made of the relative accumulation of these metals in the tissues that might explain the series. (For example, Hg is known to accumulate preferentially in the kidney rather than liver.) The general conclusions from these data are that (1) several heavy metals are able to regulate the expression of the metallothionein-I gene at the level of transcription and (2) transcriptional regulation occurs in both liver and kidney in disagreement with earlier suggestions that cadmium affected only translation in kidney (Shaikh and Smith, 1977).

The mouse W7 thymoma cell line does not express the metallothionein-I gene in response to either cadmium or glucocorticoid treatment. Compere and Palmiter (1981) have analyzed the methylation patterns of the metallothionein-I gene in these cells and have found that specific restriction sites in the vicinity of the metallothionein-I gene are methylated. In contrast, similar sites in cells that were able to express the gene were undermethylated. Interestingly, when W7 cells were treated with 5-azacytidine, which prevents methylation following DNA synthesis, the metallothionein gene was then found to be induced by both cadmium and glucocorticoids, and the restriction sites were found to be less methylated than in untreated cases.

In the experiments cited earlier, both cadmium and glucocorticoid induction were affected by DNA methylation. This suggests that the mechanisms of induction might be the same in both cases. Further work by Palmiter's group has shown this is not the case. Brinster *et al.* (1982) have developed a fusion plasmid of the promoter–regulator region of the metallothionein genes with the structural unit of the viral thymidine kinase gene. When injected into mouse ova, the metallothionein promoter in the fusion plasmid was equally as effective in eliciting the synthesis of functional thymidine kinase as the normal viral promoter. Furthermore, the metallothionein promoter responded to cadmium in a regulatory fashion and thymidine kinase synthesis was stimulated by exposure of the ova to cadmium. When the ova were fertilized and reinserted into foster mothers, some of the offspring contained the fusion plasmid and were able to respond to metals with increased thymidine kinase activity (Palmiter *et al.*, 1982). Glucocorticoids were singularly ineffective in inducing the protein in all transgeneic mice tested. The inheritance pattern of the fusion plasmids was quite variable, with the plasmid gene being expressed in some cases and repressed in others. In only some of the offspring was there a correlation between methylation patterns and expressibility of the gene, where loss of all methyl groups resulted in increased thymidine kinase activity. Evidently, some change occurs during development that modified the gene in such a way that regulation was significantly altered.

The inability of glucocorticoids to induce the synthesis of thymidine kinase from the fusion plasmids suggests that some genetic factor may be lost during the manipulatory phases of the experiments that is necessary for glucocorticoid regulation. Further evidence that some extragenetic factor is required for glucocorticoid regulation that is not necessary for cadmium induction can be found in the paper by Mayo and Palmiter (1982). In certain cadmium-resistant variants of the mouse sarcoma cell line (S180), the metallothionein-I gene was found to be amplified 10-fold. The genes were regulated by cadmium in the same fashion as the normal gene in nonresistant cells but were essentially nonresponsive to glucocorticoids, even though more than 18 kilobases of DNA flanking the 5' side of the gene were also amplified. Apparently, the genetic unit responsible for glucocorticoid regulation may have been lost during amplification or modified so that it was nonfunctional. In either case, it is clear that metal-induced and glucocorticoid-induced synthesis of thionein can occur somewhat independently. This may be extremely important to future experimental work in elucidating a mechanism whereby a metal ion can elicit the synthesis of specific proteins.

Other future research possibilities that appear to be promising in this area include an analysis of methylation patterns in fetal and adult metallothionein genes, and whether the induction by cadmium requires the participation of other ions or molecules or whether cadmium acts by itself. For example, one can ask whether zinc induction is possible in cell lines that retain cadmium inducibility but do not respond to glucocorticoids. If not, this would suggest that different metals have different specific mechanisms of induction. Such a finding would be extremely significant in pointing out unique specificities metal ions may have in fundamental biological processes.

15.5.3 Other Biochemical Effects of Cadmium

A large literature exists on the effects of cadmium on various other biochemical processes and has been reviewed previously (Friberg *et al.*, 1974; Webb, 1979; Doyle, 1977; Neathery, 1981). Most of the experimental studies on the biochemical effects of cadmium have used laboratory animals and procedures with widely different dosage levels and times of exposure. It is difficult to correlate these studies and, more important, to extrapolate these studies to questions of human concern. It should be clear from the previous discussions on human exposure and metabolism that laboratory studies in which animals are exposed to low levels of cadmium for long periods of time are the most relevant to human toxicity. Unfortunately, few of these long-term studies have focused on biochemical changes, and most of our information on this subject arises from acute short-term exposures or *in vitro* studies.

Some general conclusions from this vast literature are summarized in Table

Table 15-2. Selected Biochemical Effects of Cadmium

Biochemical site	General effects of cadmium
1. Hepatic drug metabolizing systems	Generally there is inhibition in both content and activity of enzymes of this system, although in a few cases activation can be noted.
2. RNA synthesis	Inhibition *in vitro*, stimulation *in vivo*
3. Protein synthesis	Early inhibition *in vivo*, some stimulation may occur; *in vitro* translations inhibited by both *in vivo* exposure and *in vitro* addition of metal
4. cAMP metabolism	General increase in levels of cAMP and activity of synthetic enzymes from *in vivo* exposure, but inhibition seen *in vitro*
5. Oxidative phosphorylation	*In vitro*, uncoupling is seen with an acceleration of state IV respiration, Similar results are seen from *in vivo* exposure
6. H_2O and CO_2 associated enzymes	Catalase and carbonic anhydrase are inhibited
7. Phosphatases	Generally inhibited, but to a small extent
8. Vitamin D metabolism	Low levels inhibit 1,25-dihydroxy Vitamin D_3
9. Phagocytic activity of macrophages	Generally inhibited
10. General effect on carbohydrate metabolism	Increased aldolase activity Increased phosphorylase a activity Increased pyruvate carboxylase activity Increased PEP carboxykinase activity Increased fructose 6-di-phosphatase activity Inhibition of insulin secretion

15-2. The interested reader is directed to the preceding references, which treat separate aspects of metal toxicity in a comprehensive manner.

A number of points should be made concerning the effects listed in Table 15-2. First of all, most effects of the metal *in vitro* are inhibitory, although *in vivo* stimulatory effects may be seen. This raises the point alluded to earlier that the biochemical handling of the metal *in vivo* may present a totally different paradigm for the metal's effect than that seen when the metal is added exogenously to *in vitro* assays. Thus, one must be cautious in interpreting any *in vitro* effects with respect to a biologically meaningful response *in vivo*. Second, the carbohydrate metabolism effects all give rise to some general symptoms, which include a tendency to increase glucose synthesis without utilization. Hence, hyperglycemia is a predictable and observable result from significant cadmium exposure. The *in vivo* effects of cadmium on cAMP activity are in agreement with the overall glycogenolytic, gluconeogenic, and diabetogenic effect of the metal. Finally, many of the effects of cadmium on these specific processes can be protected against or reversed by selenium or zinc administration.

It should be clear from the preceding discussions that the toxic effects of cadmium are due, in part, to alterations in the metabolism, distribution, and ultimately the physiological function of a number of trace elements, including Zn, Cu, Fe, Se, Ca, and Mn. Specific mechanisms whereby cadmium interacts with these elements have been reviewed (Weatherby, 1981; Friberg *et al.*, 1974). Only the essential features are given here. When animals are deficient in these metals, the toxicity of cadmium can be markedly potentiated. On the other hand, when the metals are given in excess, some of the toxic effects of cadmium may be lessened or even reversed. For example, protection against cadmium toxicity by zinc has been demonstrated in a number of cases. Copper, on the other hand, seems to potentiate cadmium toxicity when administered simulataneously (Irons and Smith, 1976; Bremner and Young, 1976), presumably by a partial aggregation of metallothionein. In a general sense, cadmium may affect the metabolism of these metals at a number of sites, including (1) absorption of the metal, (2) transport of the metal to target tissues, (3) incorporation of the metal into key biomolecules, and (4) turnover of the metal in tissues in the body as a whole. It is tempting to suggest that cadmium affects these processes primarily by competing for binding sites on molecules that transport the metal in question or utilize it for some physiological function. Care must be taken in offering such an explanation as a general mechanism of cadmium toxicity, since in a number of enzymes, direct replacement of zinc by cadmium *in vitro* can lead to activation rather than inactivation (Vallee and Ulmer, 1972), and at least one well documented case of *in vivo* activation has been reported (Rakhara and Nicholls, 1982). In addition, in the case of calcium, cadmium exerts part of its effects by interfering with the normal functions of vitamine D_3 apart from direct effects on the biochemical distribution of calcium.

15.6 Summary

Cadmium is a relatively toxic metal whose essentiality has only recently been investigated. The toxicity of the metal appears to be related to its ability to replace endogenous metal ions from cellular constituents, particularly proteins. Exposure to cadmium elicits, in target tissues, the synthesis of a group of specific proteins (the metallothioneins), which bind the metal with high affinity. The expression of the metallothionein gene is being actively studied at the molecular level and has proven to be a model system par excellence for investigating general aspects of gene regulation as well as the possible mechanisms whereby a metal may exert its specific biological effects.

The metabolism of cadmium is characterized by a lack of homeostatic regulation in mammals, with high body retention in certain tissues (such as liver and kidney) being characteristic of exposure to the metal. The nature of the distribution of cadmium in body tissues depends on a number of factors, but most significantly on the form of the metal given. Cadmium administered as cadmium salts or chelates accumulates preferentially in the liver after acute exposure. On the other hand, when administered as Cd-thionein, the metal accumulates primarily in the kidney.

The biochemical effects of cadmium are wide and varied (see Table 15-2). The metal does have inhibitory effects on several basic cellular processes, but these effects are variable and depend on whether the experiments are carried out *in vivo* or *in vitro*. Interactions of cadmium with the metabolism of various other metals (Zn, Cu, Fe, Se) have been extensively studied, and it is apparent that cadmium can interfere at almost any stage of a given metal's metabolism. It is not clear, in many cases, whether these interferences are direct in the competitive sense of the word or indirect, via an alteration of biological activities that influence the metal in question but do not directly use it.

Cadmium was shown early in this century to be stimulatory to the growth of algae. More recently, the metal was shown to stimulate the growth of rats kept on a highly purified diet in metal-free environs. At least one report has appeared that suggests that cadmium deficiency may lead to a constellation of biological impairments. Although these reports are not sufficient to establish definitively a specific function for cadmium, the metal is a good candidate for essentiality.

References

Agricola, G. 1556. *De Re Metallica,* Basel. English translation H. C. Hoover and L. H. Hoover, Dover, New York, 1950.

Anke, M., Hennig, A., Groppel, B., Partschfeld, M., and Grun, M., 1978. The biochemical role of cadmium, in *Proceedings of the Third International Symposium on Trace Element Metabolism in Man and Animals,* M. Kirchgessner (ed.), Freising-Weihenstephan, Germany, pp. 450–548.

Anke, M., Hennig, A., Schneider, H. J., Ludke, H., Von Cagern, W., and Schlegal, H., 1970. The interrelations between cadmium, zinc, copper and iron in metabolism of hens, ruminants, and man, in *Trace Element Metabolism in Animals,* C. F. Mills (ed.), Livingstone, Edinburgh, p. 317.

Banerjee, P., Onosaka, S., and Cherian, G., 1982. Immunohistochemical localization of metallothionein in cell nucleus and cytoplasm of rat liver and kidney, *Toxicology* 24:95.

Brady, F. O., 1982. The physiological function of metallothionein, *Trends Biochem. Sci.* 7:143.

Brady, F. O., and Webb, M., 1981. Metabolism of zinc and copper in the neonate-(zinc, copper)-thionein in the developing rat kidney and testes, *J. Biol. Chem.* 256:3931.

Bremner, I., and Young, B. W., 1976. Isolation of (copper, zinc)-thionein from pig liver, *Biochem. J.* 155:631.

Brinster, R. L., Chen, H. Y., Warren, R., Sarthy, A., and Palmiter, R. D., 1982. Regulation of metallothionein-thymidine kinase fusion plasmids injected into mouse eggs, *Nature* 296:39.

Bryan, S. E., Hidalgo, H. A., Koppa, V., and Smith, H. A., 1979. Cadmium, an effector in the synthesis of thionein, *Environ. Health Persp.* 28:281.

Car, N., Narancsik, P., and Gamulin, S., 1979. Effects of cadmium on polyribosome sedimentation pattern in mouse liver, *Exp. Cell Biol.* 47:250.

Cherian, G. M., and Goyer, R. A., 1978. Metallothioneins and their role in the metabolism and toxicity of metals, *Life Sci.* 23:1.

Cherian, M. G., Goyer, R. A., and Delaquerriere-Richardson, L., 1976. Cadmium-metallothionein-induced nephropathy, *Toxicol. Appl. Pharmacol.* 38:399.

Cherian, M. G., Goyer, R. A., and Richardson, L. D., 1977. Relationship between plasma cadmium-thionein and cadmium-induced nephropathy (Abstr.), *Appl. Pharmacol.* 41:145.

Cherian, G. M., and Rodgers, K., 1982. Chelation of cadmium from metallothionein *in vivo* and its excretion in rats repeatedly injected with cadmium chloride, *J. Pharmacol. Exp. Ther.* 222:699.

Cherian, M. G., and Vostal, J. J., 1977. Biliary excretion of cadmium in the rat. I. Dose-dependent biliary excretion and the form of cadmium in the bile, *J. Toxicol. Environ. Health* 2:945.

Cherry, W. H., 1980. Distribution of cadmium in human tissues, in *Cadmium in the Environment Part II: Health Effects*, J. O. Nriagu (ed.), Wiley, New York, 980 pp.

Compere, S. J., and Palmiter, R. D., 1981. DNA methylation controls the inductibility of the mouse metallothionein-I gene in lymphoid cells, *Cell* 25:233.

Doyle, J. J., 1977. Effects of low levels of dietary cadmium in animals—a review, *J. Environ. Qual.* 6:111.

Durnam, D. M., and Palmiter, R. D., 1981. Transcriptional regulations of the mouse metallothionein-I gene by heavy metals, *Biol. Chem.* 256:5712.

Durnam, D. M., Perrin, F., Gannon, E., and Palmiter, R. D., 1980. Isolation and characterization of the mouse metallothionein-I gene, *Proc. Nat. Acad. Sci. USA* 77:6511.

Duggan, R. E., and Corneliussen, P. E., 1972. Dietary intake of pesticide chemicals in the United States (III), June 1968–April 1970 *Pestic. Monit. J.* 5:331.

Enger, M. D., Hildebrand, C. E., Jones, M., and Barrington, H. L., 1977. Altered RNA metabolism in cultured mammalian cells exposed to low levels of cadmium (2^+) ion: Correlation of the effects with cadmium (2^+) ion uptake and intracellular distribution. DOE Symp. Ser. 47:37.

Evans, G. W., Grace, C. J., and Hahn, C., 1974. The effect of copper and cadmium on ^{65}Zn absorption in zinc-deficient and zinc-supplemented rats, *Bioinorg. Chem.* 3:115.

Foulkes, E. C., Johnson, D. R., Suyawara, N., Bonewitz, R. F., and Voner, C., 1982. Mechanisms of cadmium absorption in rats, *Gov. Rep. Announce. Index* 82:52.

Foulkes, E. C., 1982. *Biological Roles of Metallothionein*, Elsevier/North-Holland, New York.

Frazier, J. M., 1981. Effects of excess zinc on cadmium disappearance from rat plasma following intravenous injection, *Fundam. Appl. Toxicol.* 1(6):452.

Friberg, L., Piscator, M., Nordberg, G. F., and Kjellstrom, T., 1974. *Cadmium in the Environment*, Cleveland, C.R.C. Press, 248 pp.

Gamulin, S., Car, N., and Narancsik, P., 1977. Effect of cadmium on polysome structure and function in mouse liver, *Experientia* 33:1144.

Gilman, A., Philips, F. S., Allen, R. P., and Koelle, E. S., 1946. The treatment of acute cadmium intoxication with 2,3-dimercaptopropanol and other mercaptans, *J. Pharmacol. Exp. Ther.* 87:85.

Goyer, R. A., and Cherian, M. G., 1977. *Clinical Chemistry and Chemical Toxicology of Metals*, Elsevier/North-Holland Biomedical Press, Amsterdam, pp. 89–103.

Hayden, T. L., Turner, J. E., Williams, M. W., Cook, J. S., and Hsie, A. W., 1982. A model for cadmium transport and distribution in CHO cells, *Comput. Biomed. Res.* 15:97.

Hidalgo, H., Koppa, V., and Bryan, S. E., 1976. Effect of cadmium on RNA-polymerase and protein synthesis in rat liver, *FEBS Lett.* 64:159.
Irons, R. D., and Smith, J. C., 1976. Prevention by copper of cadmium sequestration by metallothionein in liver, *Chem. Biol. Interact.* 15:289.
Kägi, J. H. R., and Nordberg, M. (eds.), 1979. *Metallothionein,* Birkhäuser, Verlag, Basel.
Kägi, J. H. R., and Vallee, B. L., 1960. Metallothionein: A cadmium- and zinc-containing protein from equine renal cortex, *J. Biol. Chem.* 235:3460.
Kern, Sidney R., Smith, Harry A., Fontaine, David, and Bryan, Sara E., 1981. Partitioning of zinc and copper in fetal liver subfractions: Appearance of metallothionein-like proteins during development, *Toxicol. Appl. Pharmacol.* 59:346.
Klaasen, C. D., 1976. Biliary excretion of metals, *Drug. Metab. Rev.* 5:165.
Leber, A. P., and Miya, T. S., 1976. A mechanism for cadmium- and zinc-induced tolerance to cadmium toxicity: Involvement of metallothionein, *Toxicol. Appl. Pharmacol.* 37:430.
Maitani, T., and Suzuki, K. T., 1981. Alterations of essential metal levels and induction of metallothionein by carrageenan injection, *Bioch. Pharmacol.* 30:2353.
Margoshes, M., and Vallee, B. L., 1957. A cadmium protein from equine kidney cortex, *J. Am. Chem. Soc.* 79:4813.
Mayo, K. E., and Palmiter, R. D., 1982. Glucocorticoid regulation of the mouse metallothionein-I gene is selectively lost following amplification of the gene, *J. Biol. Chem.* 257:3061.
McGown, E., Richardson, A., Henderson, L. M., and Swan, P. B., 1971. Anomalies in polysome profiles caused by contamination of the gradients with Cu^{+2} or Zn^{+2}, *Bioch. Biophys. Acta* 241:765.
Menden, E. E., Elia, V. J., Michael, L. W., and Petering, H. G., 1972. Distribution of cadmium and nickel of tobacco during cigarette smoking, *Environ. Sci. Technol.* 6:830.
Neathery, M. W., 1981. Metabolism and toxicity of cadmium in animals, in *Cadmium in the Environment Part II: Health Effects,* J. O. Nriagu (ed.), Wiley, New York. 908 pp.
Nordberg, G. F., 1974. Health hazards of environmental cadmium pollution, *Ambio* 3:55.
Nordberg, G. F., 1978. Studies on metallothionein and cadmium, *Environ.Res.* 15:381.
Nordberg, G. F., Garvey, J. S., and Chang, C. C., 1982. Metallothionein in plasma and urine of cadmium workers, *Environ. Res.* 28:179.
Nordberg, G. F., Piscator, M., and Lind, B., 1971. Distribution of cadmium among protein fractions in mouse liver, *Acta. Pharmacol. Toxicol.* 29:456.
Nriagu, J. O., 1980. Production, uses, and properties of cadmium, in *Cadmium in the Environment, Part I: Ecological Cycling,* J. O. Nriagu (ed.), Wiley, New York, pp. 35–70.
Onosaka, S., and Cherian, G. M., 1982. Comparison of metallothionein determination by polarographic and cadmium-saturation methods, *Toxicol. Appl. Pharmac.* 63:270.
Palmiter, R. D., Chen, H. Y., and Brinster, R. L., 1982. Differential regulation of metallothionein thymidine kinase fusion genes in transgenic mice and their offspring, *Cell* 29:701.
Pirschle, Karl, 1934. Ver Gleichende unter Suchugen über die Physiologische Wirkung der Elemente nacht Wachstoms-versuchen mit *Aspergillus Niger* (Stimulation und Toxizität), *Planta* 23:177.
Prodan, L., 1932. Cadmium poisoning I: The history of cadmium poisoning and uses of cadmium, *J. Ind. Hyg.* 14:132.
Rakhra, G. S., and Nicholls, D. M., 1982. Does cadmium administration change peptide elongation in rat liver? *Environ. Res.* 27:36.
Richards, M. P., and Cousins, R. J., 1975. Mammalian zinc homeostasis: Requirement for RNA and metallothionein synthesis, *Biochem. Biophys. Res. Comm.* 64:1215.
Scharrer, K., and Schropp, W., 1934. Sand-und Wasserkulturversuche über die Wirkung des Zink und Kadmium-Ions, *Zeits. Pflanz Dung A.* 34:14.
Schwarz, K., and Spallholz, T., 1976. Growth effects of small cadmium supplements in rats maintained under trace-element controlled conditions, *Abst. Fed. Proc.* 35:255.

Schroeder, H. A., and Nason, A. P., 1974. Interactions of trace metals in rat tissues: Cadmium and nickel with zinc, chromium, copper, manganese, *J. Nutr.* 104:168.

Shaikh, Z. A., and Smith, J. C., 1980. Metabolism of orally ingested cadmium in humans, in *Mechanism of Toxicity and Hazard Evaluation,* B. Holmstedt, R. Lauweryes, M. Mercier, and M. Roberfroid (eds.), Elsevier/North-Holland Biomedical Press, Amsterdam, pp. 569–574.

Shaikh, Z. A., and Smith, J. C., 1977. The mechanisms of hepatic and renal metallothionein biosynthesis in cadmium exposed rats, *Chem. Biol. Interact.* 19:161.

Sissoeff, I., Grisvard, J., and Guillé, E., 1976. Studies on metal ions–DNA interactions: Specific behaviour of reiterative DNA sequences, *Prog. Biophys. Molec. Biol.* 31:165.

Squibb, K. S., and Cousins, R. J., 1977. Synthesis of metallothionein in a polysomal cell-free system, *Biochem. Biophys. Res. Commun.* 75:806.

Smith, Harry A., Hidalgo, Humberto A., and Bryan, Sara E., 1982. Heavy metal composition of polysomal fractions following cadmium challenge. *Biol. Trace Elements Res.* 4:57.

Stoll, R. E., White, J. F., Miya, T. S., and Bousquet, W. F., 1976. Effects of cadmium on nucleic acid and protein synthesis in rat liver, *Toxicol. Appl. Pharmacol.* 37:61.

Teperman, H. M., 1974. The effect of BAL and BAL-glucoside therapy on the excretion and tissue distribution of injected cadmium, *J. Pharmacol. Exp. Ther.* 89:343.

Tohyama, C., Shaikh, Z. A., Ellis, K. J., and Cohn, S. H., 1981. Metallothionein excretion in urine upon cadmium exposure: Its relationship with liver and kidney cadmium, *Toxicology* 22:181.

Tobias, J., Lusbaugh, C., Patt, H., Postell, S., Swift, M., and Gerhard, R., 1946. The pathology and therapy with 2,3-dimercaptopropanol (BAL) of experimental Cd poisoning, *J. Pharmacol. Exp. Ther.* 87:102.

Tsevetkova, R. P., 1970. Influence of cadmium compounds on the generative function, *Gig. Tr. Prof. Gabol.* 14:31.

Valberg, L. S., Haist, J., Cherian, M. G., Richardson, L. D., and Goyer, A., 1977. Cadmium induced enteropathy: Comparative toxicity of cadmium chloride and cadmium thionein, *J. Toxicol. Environ. Health* 2:964.

Vallee, B. L., and Ulmer, D. D., 1972. Biochemical effects of mercury, cadmium, and lead 1972. *Ann. Rev. Biochem.* 41:91.

Venugopal, B., and Luckey, D., 1978. *Metal Toxicity in Mammals, Vol. 2. Chemical Toxicity of Metals and Metalloids,* Plenum Press, New York, 409 pp.

Waalkes, M. P., and Bell, J. V., 1980. Depression of metallothionein in fetal rat liver following maternal cadmium exposure, *Toxicology* 18:103.

Waalkes, M. P., Thomas, J. A., and Bell, J. V., 1982. Induction of hepatic metallothionein in the rabbit fetus following maternal cadmium exposure, *Toxicol. Appl. Pharmacol.* 62:211.

Webb, M., and Samarawickrama, G. P., 1981. Placental transport and embryonic utilization of essential metabolites in the rat at the teratogenic dose of cadmium, *J. Appl. Toxicol.* 1(5):270.

Webb, M., and Verschoyle, R. D., 1976. An investigation of the role of metallothionein in protection against the acute toxicity of the cadmium ion, *Biochem. Pharmacol.* 25:673.

Yoshida, A., Kaplan, B. E., and Kimura, M., 1979. Metal-binding and detoxification effect of synthetic oligo peptides containing three cysteinyl residues, *Proc. Natl. Acad. Sci. USA* 76:486.

Lead 16

A. M. Reichlmayr-Lais and M. Kirchgessner

16.1 Introduction and History

The lead content of the uppermost layer of the earth's crust (16 km thick) amounts to 0.0016%. Accumulations of the metal occur in lead-specific deposits, most commonly as the sulfide mineral galena, which was known to the Egyptians 5000 years ago. The atomic weight of lead is 207.2. There are several isotopes. Chemically, lead is bivalent and quadrivalent, the Pb(II) salts being the most common and forming the most stable compounds. The properties of lead, especially its ductility and high resistance to erosion and chemical reagents, stimulated a diversified application of this metal, especially in alloys with other metals. Because of its widespread use, lead intoxication is common and has been known as "saturnism" or "plumbism" since ancient times. Hippocrates (370 B.C.) connected lead exposure for the first time with subsequent clinical signs. In the second century B.C. Nicander compiled the pathological effects on lead workers. Epidemic occurrences of lead intoxication are known from the sixteenth century, specifically in Amsterdam because of contaminated drinking water and in Devon because of Pb-containing apple wine (see Morgan *et al.*, 1966).

Anemia caused by lead was mentioned for the first time by Laennec (1831). In 1881 Charcot and Gomboult (cited by Aub *et al.*, 1925) characterized the pathogenesis of chronic lead nephropathy, while acute effects of lead on the kidney were differentiated later from chronic nephropathy by Oliver (1914). An early clinical study (Thomas, 1904) pointed to a peripheral neuropathy in the case of lead poisoning. In 1937 the first classical description of Pb encephalopathy was published (Blackman, 1937).

Recently, scientific interest has concentrated on the biochemical changes caused by an increased lead intake in relation to alteration of Pb metabolism.

A. M. Reichlmayr-Lais and M. Kirchgessner • Institute of Nutrition Physiology, Technical University München D-8050 Freising-Weihenstephan, FRG.

Detailed knowledge of the mechanisms of lead toxicity is important in therapeutic applications, particularly with respect to preventing Pb intoxication or enabling early diagnosis and initiation of expedient therapy. Latent forms of lead toxicity are especially difficult to evaluate clinically, without knowledge of the metal's toxicity at the molecular level.

Despite its toxicity, lead had been classically administered by many physicians as a treatment for a variety of illnesses. Altogether, the notable effects were of a pharmacological nature. Indications of a stimulatory effect of lead on metabolic processes and the high affinity of lead toward biological ligands gave rise to the proposition in recent years that lead might also be an essential element.

In the present report we attempt to review the current status of knowledge about major aspects of the metabolism, essentiality, and toxicity of lead.

16.2 Metabolism of Lead

16.2.1 Occurrence and Intake

In nature, lead occurs ubiquitously and, as far as is known, primarily in inorganic form. The metal can also be detected in the animal and human body under conditions of natural low-level exposure. Humans and animals take in lead with food, water, and air. Under normal conditions the lead content in foodstuffs, as recorded in the literature, ranges from 0.02 to 3 mg/kg fresh weight, in drinking water from 0.01 to 0.03 mg/L, and in air from 0.03 to 0.1 $\mu g/m^3$. These concentrations often are higher because of increasing environmental pollution. This is especially so in the areas surrounding the emission sources, such as automobile and industrial exhausts; lead-containing utensils and paints also contribute to increasing levels of lead in the environment. Lead is added to gasoline as Pb tetraethyl or Pb tetramethyl, together with organic halogen compounds and o-tricresyl-phosphate, to serve as an antiknock agent. In the engine, Pb alkylene is combusted to PbO_2, which then reacts with the halogen compounds to form volatile Pb halogens. In part, $Pb(SO_4)_2$ and Pb phosphate are also formed. Fifty to 70% of these Pb compounds are emitted with the exhaust fumes as a fine mist. Consequently, the lead content of the vegetation near busy streets may rise above 300 ppm in extreme cases (e.g., Vetter and Mählhop, 1971). The lead concentration of the air may reach levels between 2 and 20 $\mu g/m^3$, depending on traffic density and climatic conditions (see Reichel et al., 1970).

The daily lead intake of humans has been estimated at 0.1–2 mg (e.g., Kehoe et al., 1974; Goyer et al., 1973), intake is predominantly oral, and a minor amount is taken in through the respiratory tract. Occupational Pb intoxications very often result from the inspiration of Pb vapors or Pb-containing dust. Tetraethyl lead and related compounds, being lipid soluble, may also be absorbed through the skin.

16.2.2 Absorption of Lead

Pulmonary Absorption

Pulmonary lead absorption depends on the state of aggregation (gas, solid particles), particle size of the lead-containing dust, respiratory volume, concentration in the air, and distribution within the respiratory tract. Particles of <0.5% μm in the average mass median equivalent diameter especially are retained in the nasopharynx and tracheobronchial tree, including the terminal bronchioles. Larger particles are removed by the activity of the ciliated cells of the respiratory epithelium. Lead retained by the alveolae is not totally absorbed. A portion of inhaled lead may be cleared by pulmonary macrophages (Bingham *et al.*, 1968) but both the mechanism and the quantity of this clearance are largely unknown.

Gastrointestinal Absorption

The gastrointestinal absorption of lead depends on many factors, such as the amount of intake, chemical form, dietary composition, intraluminal interactions with other dietary constituents, presence of bile acids, age, and so on. According to the literature, the absorbability of lead falls into the range of 5–15% (see Goyer and Mahaffey, 1972).

The absorbability depends critically on the solubility of lead compounds. Lead is readily absorbed as acetate, chloride, oxide, and tetraethyl. Less soluble yet fairly absorbable are chromates, sulfides, sulfates, and carbonates. Delwaide *et al.* (1968) suspect that part of the lead is converted by the action of the digestive secretions to Pb chloride and complexes with bile acids and then absorbed in these forms. Conrad and Barton (1978) also point out that bile stimulates the transport of lead across the intestinal mucosal cells and, subsequently, the transport of this mucosal lead into the body. Ascorbic acid (see Goyer and Mahaffey, 1972) and amino acids with sulfhydryl groups (Conrad and Barton, 1978) improve the solubility and hence the absorbability of Pb compounds. Lead absorption is increased in the absence of food or during food restriction (Rabinowitz *et al.*, 1980). The intake of calcium, iron, magnesium, phosphate, ethanol, and high-fat diets lower the absorption of lead (Fine *et al.*, 1966; Barltrop and Kehoe, 1975; Ragan, 1977). This decrease in lead absorption by various elements is attributed mainly to their competition for carrier systems in the intestinal epithelium.

In the case of children and young animals, lead absorption is increased

(Kostíal *et al.*, 1971). This may, at least partially, explain their high sensitivity to larger amounts of lead. This phenomenon has also been shown to apply to other elements and might be explained by an as yet immature absorption system.

At present, little is known about the transport mechanism of lead. Newer studies implicate an active transport of lead. Results by Conrad and Barton (1978) point to a saturable mechanism of an acceptor affecting the inverse relationship between dose and relative absorption of lead. In the case of high lead doses, diffusion processes might predominate.

16.2.3 Excretion of Lead

Lead can be excreted via feces, urine, sweat, and saliva (Schroeder and Tipton, 1968). Fecal excretion is particularly high because of the high percentage of unabsorbed lead and may average around 0.2 mg/day. By comparison, urinary excretion amounts to an average of 30 μg/day. In the kidneys lead is excreted usually by glomerular filtration. In the case of high blood lead concentration tubular secretion also occurs (Vostal and Heller, 1968). The excretion through sweat, reported by Schroeder and Tipton (1968) to average 60μg/day, appears to be sizable.

16.2.4 Transport and Distribution

A total lead content of an adult person has been estimated to be around 130 mg under normal conditions. Disappearance curves of radiolabeled lead from tissues and organs in kinetic studies indicate the presence of roughly three compartments. Blood shows the shortest half-life for lead compounds. The soft tissues, including the skeletal muscle, represent a pool of medium half-life (weeks), and the skeleton represents a pool of very long half-life (months up to years) (e.g., Castellino and Aloj, 1964). Goyer and Mahaffey (1972) distinguish between a diffusible or mobile form and a nondiffusible or fixed form of lead in the body. The diffusible part is defined as the transport forms of lead in blood and intracellular lead that can be mobilized and transported through membranes. Lead in blood is to a large extent exchangeable with that in tissues and organs. The distribution between tissues and organs is a function of time, dose, supply state, and turnover rates of the particular compounds. During normal exposure, relatively high Pb concentrations were found in bones, and somewhat lower levels were found in liver and kidneys (Kehoe, 1964).

In blood, more than 90% of the lead is bound to erythrocytes (e.g., Castellino and Aloj, 1964). This erythrocyte-bound lead probably represents, in part, a transport form, since lead disappears from red cells at a rate that is not explained by their life span (Conrad and Barton, 1978). Studies by Barltrop and Smith (1971) and by Kaplan *et al.*, (1975) indicate that lead occurs mainly in the cellular constituents of the erythrocytes and to a lesser extent in the stroma. It is assumed that lead is bound mainly to hemoglobin. Lead may, however, also be present in low-molecular compounds. Raghavan and Gonick (1977) discovered a lead-binding protein with a molecular weight of 10,000 in erythrocytes of lead exposed but not in normal individuals. This protein may have a protective function.

The plasma lead concentration appears to remain constant over a wide range of whole-blood concentration (Rosen *et al.*, 1974). This constancy also holds true for lead in the hematocrit (Rosen *et al.*, 1974) and the lead content of the erythrocytes (Clakson and Kench, 1958). The plasma lead—as the studies by Kochen and Greener (1975) have shown—is bound predominantly to transferrin and at the same binding sites as iron. Saturation of the total iron-binding capacity suppresses the uptake of plasma lead by red cells and increases the uptake of lead by the liver.

The relationship between dose and effect is very often evaluated on the basis of the blood Pb content. It must, however, be realized that the blood lead level is influenced by numerous factors and that any value represents just a snapshot of dynamic processes of distribution. A lead concentration of <30–40 µg/100 ml is normally recorded in persons without serious lead exposure. Concentrations between 60 and 80 µg/100 ml reflect a Pb absorption high enough to induce biochemical changes without, however, manifestation of clinical symptoms. The critical value for the occurrence of clinical symptoms is suggested as 80 µg/100 ml (e.g., Goyer and Rhyne, 1973). These limits might be lower in the case of children or women.

With respect to soft tissues, the subcellular distribution of lead has been studied especially in liver and kidneys (e.g., Castellino and Aloj, 1964; Sabbioni and Marafante, 1976). Fractionations of organ homogenates into nuclei, mitochondria, lysosomes, microsomes, and soluble fraction have shown that lead is bound to all fractions. Its distribution between the individual fractions, however, largely depends on the state of supply and the dose applied. The major part of lead doses is recovered in the nuclei and in the soluble fraction. In the nuclear fraction lead seems to be bound in particular to nuclear membrane components. Sabbioni and Marafante (1976) could not detect lead in high-molecular-weight components, comprising RNA-bound membranes combined with nuclear pore complexes, nor in low-molecular-weight phospholipids of the cell nuclei. The binding of lead in the cell nuclei does not seem to be restricted to membranes, but may also involve intranuclear bulk chromatin components, predominantly a histone fraction. The intramitochondrial distribution of lead shows a marked localization in the heavy subfractions consisting of inner membranes and part of the matrix. There lead is bound to proteins of the membranes. Walton (1973) and Barltrop et al. (1971) showed that pretreatment of experimental animals with lead diminishes mitochondrial Pb uptake, indicating saturation of binding sites. High lead concentrations result in morphological and functional changes of the mitochondria. They show impaired respiratory and phosphorylative abilities and, accordingly, impaired active transport. This may explain several pathological effects of lead. In the endoplasmic reticulum lead is associated with both membranous and ribosomal components. In the cytosol, lead may be bound to high-molecular-weight components.

Rüssel (1970) demonstrated that, in the case of toxic doses, lead in kidneys, liver, and spleen is partially bound to ferritin and partially to an insoluble ferric hydroxide. Whether this holds true also for normal Pb metabolism needs to be investigated.

In the brain, lead concentration differs between various regions. The highest concentrations are found in cortical gray matter and basal ganglia (Klein et al., 1970). Subcellularly, lead is localized especially within the neural mitochondria, perhaps at the Ca-binding sites.

More than 90% of the lead in the body is located in bones even under normal conditions. Seventy percent of this falls to the share of cortical bones (Barry, 1978). In these dense cortical bones lead concentration is higher than in the spongy hemopoietic trabecular bone (Barry, 1978). In bones, lead is deposited at first in labile form, later as triphosphate (Ligeois et al., 1961). Lead metabolism in bones is very similar to that of calcium. In certain metabolic situations bringing about demineralization, lead may be mobilized again from the bones and can even cause a lead crisis under certain circumstances (see Six and Goyer, 1970).

The lead concentration in teeth is higher than that in bone. Lead in teeth appears to be firmly bound (see Barry, 1978).

Generally, the binding sites for lead seem to be macromolecules, mainly proteins. It is known that proteins with several free SH-groups very tightly bind lead both *in vitro* and *in vivo*. Metallothionein, which has a high cysteine content, can also bind lead both *in vitro* and *in vivo* (Ulmer and Vallee, 1969). It is, however, not yet clear whether metallothionein plays a role in the detoxification of lead and hence in its metabolic regulation or whether lead may induce the *de novo* synthesis of metallothionein, as demonstrated, for example, in the case of zinc, copper, and cadmium (Bremmer, 1974). Lead forms not only mercaptides with SH-groups of cysteine but also much less stable complexes with other side chains of amino acids (Vallee and Ulmer, 1972), for example, with the ε-amino group of lysine, the carboxyl group of glutamic and aspartic acids, the phenoxyl group of tyrosine and with imidazole residues (see Wong et al., 1979).

16.2.5 Interactions

Elements with similar physicochemical properties can influence each other during absorption, intermediary metabolism, and excretion. In the case of lead such interactions are known to occur especially with calcium and iron, but also with zinc, copper, magnesium, and cadmium.

Lead and Calcium, Phosphorus, Vitamin D

A decrease in Pb retention in response to increased Ca intake and, conversely, an increase in Pb retention in response to reduced Ca intake have been demonstrated repeatedly (e.g., Shields and Morrison, 1975). These effects are attributed mainly to interactions at the site of absorption (e.g., Quarterman and Morrison, 1975; Barton *et al.*, 1978). This explains why a very high Pb content in the diet may impair the absorption of calcium (Toraason *et al.*, 1981). Lead administration reduces, for example, the serum calcium level after a transient elevation (Six and Goyer, 1970). Lead may also be attached to the Ca-binding sites—for example, of the erythrocytes (Ong and Lee, 1980) and insulin (Blundell and Jenkins, 1977). Ca ions, above certain concentrations, can again displace lead from its binding sites in erythrocytes (Rosen and Haymovits, 1973). This displacement, however, does not become effective until Pb concentrations reach a certain level (see Rosen and Trinidad, 1974), an indication of different binding forms of lead in erythrocytes being dependent on concentrations. Pounds *et al.* (1982) showed that Pb acetate affects calcium homeostasis in isolated hepatocytes and, consequently, may impair the function of calcium as a second messenger. A rise in the tissue Pb content due to low dietary calcium content has also been observed in the case of soft tissues and, to a lesser extent, in bone (e.g., Mahaffey *et al.*, 1973). Similarly, a reduction of the dietary phosphorus level below requirements causes a greater Pb retention (Quarterman and Morrison, 1975). With respect to additive effects of reducing the contents of calcium and phosphorus in the diet, different results have been found (Quarterman *et al.*, 1978).

Vitamin D supplementation induces a rise in the concentration of Ca-binding proteins in the intestinal mucosa and, therefore, an increase in Pb absorption (Wasserman and Carradino, 1973). Smith *et al.* (1978) have also demonstrated these relationships both *in vivo* and *in vitro*. In addition, they showed that Pb absorption increases only up to the state of saturation of the mucosa with vitamin D. Lead toxicity, in turn, influences the formation of metabolites as well as several functions of vitamin D. Accordingly, it has been reported that the 1, 25-$(OH)_2D_3$ content of plasma decreases and that the intestinal response to 1, 25-$(OH)_2D_3$, 25-$(OH)D_3$, and vitamin D_3 is inhibited when lead intake is very high (Smith *et al.*, 1981). These effects of high lead exposure may be reduced by a high Ca content or by a low P content of the diet.

Lead and Iron

A correlation between lead poisoning and iron deficiency, especially in children, has been known for a long time. The prevalence of Fe deficiency in children certainly is a major factor increasing the risk of Pb intoxication. A series of experimental studies with different species confirms the increased Pb retention in response to deficient Fe supply (Six and Goyer, 1972) and the association with increased Pb toxicity (e.g., Mahaffey, 1974). Increased Pb retention because of Fe deficiency results in elevated Pb contents in both soft tissues and bones (Six and Goyer, 1972). The major reason for the increased Pb retention might be the enforced Pb absorption because of deficient Fe supply, as demonstrated both *in vivo* and *in vitro* (e.g., Barton *et al.*, 1978; Hamilton, 1978). The studies of Barton *et al.* (1978) show that the intraluminal presence of iron diminishes lead absorption. These findings point to a competition between lead and iron for common mucosal acceptors.

Conversely, the absorption or retention of lead may be lowered by Fe supplementation (Mahaffey *et al.*, 1978; Barton *et al.*, 1978). Mahaffey *et al.* (1978) showed that a threefold increase of the Fe supply lowered Pb concentrations in kidneys, femur, and blood.

Very high molar ratios of lead to iron may, on the other hand, diminish Fe retention (Hamilton, 1978). The usual environmental lead concentrations may, however, not yet be so high as to bring about this effect. Rather, they induce disturbances in iron utilization by inhibiting nearly all enzymes involved in heme synthesis and may, accordingly, even bring about an accumulation of iron in the reticulocytes (see Section 16.4.1). Studies that failed to find any effect of lead on iron absorption (Dobbins *et al.*, 1978) may be explained by these relationships between the two elements.

In the case of experimental Pb deficiency, there was a decreased Fe retention in the progeny of Pb-depleted rats (see Section 16.3).

Lead and Zinc

The zinc status seems to affect the accumulation of lead and also the biochemical changes resulting from toxic lead exposure. Investigations by Cerklewski and Forbes (1976) showed that the Pb content of tissues in rats and, accordingly, the severity of Pb toxicity, decreased in response to increasing dietary zinc concentrations (8, 35, 200 ppm). Zinc can, for example, counteract the inhibitory effect of lead on the δ-ALAD activity (Meredith *et al.*, 1974). Willoughby *et al.* (1972) found that Pb distribution between different organs of the horse is altered by increasing zinc intake. While the lead content in bones decreased, the hepatic and renal lead concentrations rose. Simultaneously, the sensitivity toward toxic lead doses was reduced, which led Bremner (1974) to speculate that higher zinc doses induce the synthesis of metallothionein, which then would bring about an increased binding and hence detoxification of lead. A higher dietary Zn content (120 ppm) was also found to reduce the lead content in blood and liver of rat fetuses in mothers that were fed diets with 500 ppm lead.

16.3 Lead Deficiency

Evidence of a possible essentiality of lead is scarce in the literature. Pecora *et al.* (1968) reported that very low lead concentrations stimulated heme synthesis. In the microorganism *Sarcina flava,* growth was observed to be enhanced by lead (Devigne, 1968). Also, the study of Schwarz (1974) may be seen as a mere indication of the possible essentiality of lead. He found the growth of rats— which was, however, very low overall, even in the case of the control animals— to be slightly improved by 0.1–0.2 g (10–16%) when a diet with 1 or 2.5 ppm as compared to 0.5 ppm lead was fed. These findings led Reichlmayr-Lais and Kirchgessner (1981) to investigate the essentiality of lead. The experimental approach of their studies was basically different from that of Schwarz (1974). Schwarz studied the growth-promoting effect of supplemental lead added to a basal diet that was inadequate to meet the nutritional requirements of the experimental animals and reported growth responses that appear relatively large compared to the overall poor growth rate of the controls. Reichlmayr-Lais and Kirchgessner designed experiments to investigate the essentiality of lead by inducing growth depressions resulting from a deficiency, along with biochemical changes and possibly clinical symptoms that could either be prevented or alleviated by lead supplementation. Depletion was attained by compounding a low-lead diet from purified components. The lead content of this diet (<45 ppb) was less than one-tenth that of the diet used by Schwarz. At first, no negative effects were observed in comparison with a control diet containing 1 ppm lead. It was not until the mother animals had already been fed the same low-lead diet that physiological changes could be demonstrated for the first time in their progeny (F_1 generation) (Reichlmayr-Lais and Kirchgessner, 1981). In the young rats from the lead-depleted mothers, hematocrit, hemoglobin, and MCV were reduced by about 10–15%.

By further reduction of the dietary lead content (<20 ppb), it was possible to reproduce the hematological changes to an even greater degree. In the 20-day-old progeny of depleted mothers the hematocrit was reduced by 22%; hemoglobin by 23%; and MCV by 28%, compared with the control values. The erythrocyte count, however, was not affected (Reichlmayr-Lais and Kirchgessner, 1981). In this experiment the deficiency was so pronounced that the growth of the progeny from the lead-depleted rats was already depressed during the nursing period. At the time of weaning, the weight depression amounted to 11%, which could be extended to 20% by continued depletion.

The anemia induced by Pb depletion led to the hypothesis that Pb deficiency is tangent to iron metabolism. Therefore, criteria of iron metabolism were investigated. In all deficient animals, the blood iron level was markedly reduced, whereas the Fe-binding capacity was increased (Kirchgessner and Reichlmayr-Lais, 1981). In older depleted animals of the F_1 generation (38 days), the iron contents in the liver, spleen, and carcass were also analyzed and found to be reduced. This implies that lead deficiency brings about diminished Fe retention (Kirchgessner and Reichlmayr-Lais, 1981). The Fe retention was already reduced in newborns of severely depleted mothers (Reichlmayr-Lais and Kirchgessner, 1981). In the case of a more pronounced deficiency, the activity of catalase was also impaired (Reichlmayr-Lais and Kirchgessner, 1981). Recent results from *in vitro* absorption studies are also indicative of a secondarily induced Fe deficiency (Reichlmayr-Lais and Kirchgessner, 1983). Further studies are necessary to elucidate the cause of these relationships.

In other experiments it was established that the hematological changes and abnormalities of Fe metabolism occurring in response to lead deficiency could be alleviated or prevented by lead supplementation (Reichlmayr-Lais and Kirchgessner, 1981; Kirchgessner and Reichlmayr-Lais, 1981).

Additional physiological changes observed in older deficient animals (38 days) of the F_1 generation included altered enzyme activities (alkaline phosphatase, glutamate oxaloacetate transaminase, glutamate pyruvate transaminase, coeruloplasmin in the liver and alkaline phosphatase in the femur) and metabolite concentrations (glucose, triglycerides, cholesterol, phospholipids, low-density lipoproteins in the liver) (Kirchgessner and Reichlmayr-Lais, 1981; Reichlmayr-Lais and Kirchgessner, 1981). These changes may represent secondary effects indicative of disturbances in metabolism but do not yet allow any conclusions with regard to possible functions.

In summary, these results show that deficiency symptoms can be induced in the progeny of rats provided with very low levels of lead and that lead is needed for the full functioning of metabolism. Accordingly, the essential nature of lead seems to be established; knowledge of its possible functions is still lacking.

16.4 Toxicity of Lead

Lead intoxications are generally divided into acute and chronic forms. Chronic poisoning describes a prolonged exposure. Manifestations, however, are often acute. Depending on duration and severity of exposure, the effects range from lead colic to encephalopathy and death. Early symptoms of intoxication are usually headache, anorexia, fatigue, nervousness, tremor, and constipation. Later, lead colics occur at repeated intervals. Frequently, a marked fatigue of the extensor muscles may accompany the distinct state of tremor.

All clinical symptoms result from the toxic effects of lead, which are manifested mainly in the blood (anemia), nervous system (encephalopathy and neuropathy), and kidneys (renal dysfunction), the three major sites of lead intoxication. Recently, there has been increasing discussion of the mitogenic, mutagenic, and teratogenic effects of toxic Pb doses.

16.4.1 Hematologic Effects of Toxic Lead Doses

Acute and chronic lead intoxications cause anemia associated with reticulocytosis and basophilic stippling of the erythroblastic cells (Goyer and Rhyne, 1973). The anemia results from diminished hemoglobin synthesis, hemolysis of immature erythrocytes, and direct hemolysis of mature erythrocytes in conjunction with a shortened life span.

The reduced hemoglobin synthesis following Pb intoxication may be attributed primarily to an inhibition of heme synthesis (Goldberg, 1968). Toxic Pb doses adversely affect several enzymes of the heme pathway: δ-aminolevulinic acid synthetase (δ-ALAS), δ-aminolevulinic acid dehydratase (δ-ALAD) (Dresel and Faulk, 1956), heme synthetase (= ferrochelatase) (Goldberg *et al.*, 1956), and uroporphyrinogen I synthetase (Piper and Tephly, 1974).

Inhibition of δ-ALAD, depending on blood Pb levels (e.g., Hernberg and Nikkanen, 1970) and duration of the intoxication, has been observed *in vivo* in both humans and animals even in the case of very small Pb concentrations (De Bruin, 1968; Nakao *et al.*, 1968). Gibson *et al.* (1955) also demonstrated the inhibitory effect of lead on the isolated enzyme. Inhibition of the δ-ALAD occurs at small Pb doses and can be detected before the manifestation of clinical symptoms (De Bruin, 1968). Above a Pb concentration of 35 μg/100 ml blood, there is a linear correlation between the inhibition of the enzyme and the blood Pb concentration (Wada, 1976). δ-ALAD is present not only in erythrocytes, but also in all cells with aerobic metabolism (Gibson *et al.*, 1955), the highest activities occurring in the liver, kidneys, and bone marrow. In the case of experimental intoxication, inhibition of this enzyme was observed in erythrocytes as well as in the brain, liver, kidneys, and bone marrow (Goldberg *et al.*, 1956; Gibson and Goldberg, 1970).

The mechanism of the δ-ALAD inhibition *in vitro* or *in vivo* seems to be different in nature (Hernberg, 1972). While the inhibition under *in vitro* conditions and possibly also in the case of acute experimental intoxication results from a direct and competitive blocking of the sulfhydryl groups, the inactivation of sulfhydryl groups seems to be of minor importance under *in vivo* circumstances. Vergans *et al.* (1969) postulated that lead inactivates the enzyme via a thermolabile component. Likely candidates for thermolabile components are intermediates in the heme synthesis, for example, protoporphyrin or protoporphyrin IX (Calissano *et al.*, 1966). Hernberg (1972), however, assumes a more direct effect of lead via an allosteric inhibition.

Although δ-ALAD is the enzyme most responsive to lead, it is not certain whether its inhibition is the major cause of the defects in heme synthesis. Nevertheless, under defined conditions its activity is the most sensitive and reliable test of lead exposure.

The rate-limiting enzyme for heme formation is δ-ALAS (Goldberg *et al.*, 1956). Several *in vivo* studies show that the activity of this enzyme increases in response to lead intoxication (e.g., Wada, 1976; Kusell *et al.*, 1978). This is attributed to a feedback effect of the reduced synthesis of the end product heme. Kussel *et al.* (1978) assume on the basis of their studies that the effect of lead on δ-ALAS results more likely from an induction of its synthesis than from an activation of the enzyme. *In vitro,* however, Goldberg *et al.* (1956) noted an inactivation of the isolated enzyme by lead.

Ferrochelatase is also inhibited by toxic lead doses (Otrzonsek, 1967). It catalyzes the incorporation of ferrous iron into the porphyrin ring structure, with the result that iron accumulates in the mitochondria of bone marrow reticulocytes in the form of ferritin and ferrunginous micelles (Bessis and Jensen, 1965). Sideroblasts in bone marrow are consistently associated with lead poisoning anemia. Higher amounts of zinc protoporphyrin, which combines with globin, are synthesized in place of heme (Lamola *et al.*, 1975).

The defects in the pathway of heme synthesis due to toxic lead levels bring about an accumulation of intermediary products, especially aminolevulinic acid and coproporphyrin, in blood and bone marrow, and an increase of their urinary excretion (Nakao *et al.*, 1968), even in the absence of clinical symptoms (Chisholm, 1964). Consequently, their concentrations in urine represent very sensitive diagnostic parameters.

Because of the defective heme formation, disturbances other than the impaired synthesis of hemoglobin must be expected also in the case of enzymes possessing heme as the prosthetic group. For example, the activity of the cytochrome P-450 complexes has been shown to be reduced, especially in acute lead poisoning, while there was only a mild inhibition in chronic cases (Alwares *et al.*, 1976; Fischbein *et al.*, 1977).

Apart from the defects in the heme pathway, the synthesis of hemoglobin may also be diminished by a reduced globin synthesis following lead poisoning (White and Harvey, 1972). Piddington and White (1974) postulated that the *in vivo* inhibition of the globin synthesis is caused secondarily by the inhibition of the heme synthesis. This relates to theories of the control of heme and globin synthesis. It has been observed both *in vivo* with children (White and Harvey,

1972) and *in vitro* (Piddington and White, 1974) that the synthesis of the α-chains of hemoglobin is inhibited to a greater extent than that of the β-chains.

The anemia due to lead poisoning is caused not only by impaired hemoglobin synthesis, but also by a shortened life span of the erythrocytes. Toxic lead levels exert a direct hemolytic effect on mature erythrocytes (Berk *et al.*, 1970). The cause may be found in the high affinity of erythrocytes for lead, especially of their membranes; this results in an increase of the osmotic resistance and mechanical fragility (Griggs, 1964). The consequence is that potassium is released from the erythrocytes into the plasma (e.g., Hasan *et al.*, 1967). This could also be associated with a reduced activity of the Na^+/K^+-ATPase (Raghavan *et al.*, 1981).

Another phenomenon, which can be observed soon after the first days of lead exposure, is the basophilic stippling of the polychromatic erythroblasts and reticulocytes in the cytoplasm because of changes in the ultrastructure of these cells (see Albahary, 1972). This change concerns the mitochondria and, especially, the ribosomes of the erythroblasts and reticuloblasts. Furthermore, lesions of the nuclear membrane and swelling of the Golgi apparatus occur. The basophilic stippling represents essentially aggregations of undegraded and partially degraded ribosomes (Jansen *et al.*, 1965). The cause might be an inhibition of the erythrocyte pyrimidine-5'-nucleotidase, with the consequence that large amounts of pyrimidine nucleotides accumulate intracellularly and bring about retardation of ribosomal RNA degradation, similar to hereditary nucleotidase deficiency. These defects ultimately result in a premature erythrocyte hemolysis.

In summary, it may be concluded that lead affects the erythroid precursor cells and the mature circulating cells, resulting in impaired erythroid maturation, defective hemoglobin synthesis, sideroblastosis, stippling of the erythroid cells, and distinct hemolytic tendencies.

16.4.2 Neurotoxic Effects of Lead

Among all organ symptoms, the brain reacts most sensitively to higher lead doses (Silbergeld, 1982). Depending on the exposure level, lead toxicity may result in lead encephalopathy and neuropathy. Children and young animals appear to be especially sensitive, according to the ontogenic differences in neural tissue susceptibility or because of a greater accessibility to lead of the nervous system in the young (Keller and Doherty, 1980).

Clinical Manifestation

Lead affects the functions of both the central and peripheral nervous system. Early lead exposure of children correlates with decreased IQ, symptoms of hyperkinesis or minimal brain dysfunction, poor learning, or defects in specific neuromotor tasks (Landgrin et al., 1980; Needleman, 1980). Higher loads cause depressed behavior, lethargy, mental regression, coma, and seizure (see Silbergeld, 1982).

In adults, changes in mood and other effects after low-level exposure have been described, with mania, psychosis, and seizure noted after high-level exposure (see Silbergeld, 1982). A peripheral neuropathy manifests itself by paralysis and decreased nerve conduction. All clinical symptoms described have been confirmed by experimental studies.

Neuroanatomic Pathology of Lead

Cerebral edema resulting from lead poisoning is often associated with a rise of the cerebrospinal fluid pressure. Other changes observed include proliferation and swelling of the endothelial cells accompanied by dilatation of the arterioles and capillaries, proliferation of the glial cells, focal necrosis, and neural degeneration. Severe cases manifest themselves as acute communicating hydrocephalus because of inflammation of the pia arachnoid or choroid plexus (Mirando and Ranasinghe, 1970). Another symptom of lead encephalopathy is a diffuse astrocytic proliferation in the gray and white matter (Pentshew, 1965). Peripheral signs are segmental demyelination resulting from the degeneration of Schwann cells and axons (Fullerton, 1966).

Biochemical Mechanisms of the Neurotoxicity of Lead

Little is known about the mechanisms responsible for the neurotoxic effects of lead. On the basis of biochemical changes evident from *in vitro* and *in vivo* models, only hypotheses may be forwarded (see Silbergeld, 1982).

At present, studies concentrate on the effects of lead on neurotransmitter functions concerning especially acetylcholine, catecholamines, and γ-aminobutyric acid. Results of *in vivo* studies indicate that lead exposure reduces cholinergic neurotransmission perhaps as the result of an altered reaction in the course of the presynaptic uptake of precursors, or the synthesis and release of the transmitter. Results concerning the effect of lead on the catecholamines are conflicting. However, it has been reported frequently that there is an increase in catecholaminergic neurotransmission accompanied by an increased turnover of norepinephrine, decreased levels of dopamine and norepinephrine, increased activity of tyrosine hydroxylase, decreased reuptake of dopamine, and increased levels of catecholamine metabolites. *In vitro,* lead can enhance the resting release of dopamine and inhibit synaptosomal reuptake of dopamine. Inhibition of the dopamine-sensitive adenylcyclase has also been noted *in vitro*. Concerning γ-aminobutyric acid, findings of *in vivo* studies imply a decreased uptake, a decreased release, and a supersensitivity to this neurotransmitter.

At the molecular level, the biochemical changes of lead poisoning may be explained by interactions with other elements, for example, with Ca, Na, and Mg. Calcium is involved in neurotransmitter function (e.g., the stimulus-coupled release of transmitters, the regulation of some enzymes in neurotransmitter synthesis, the storage of transmitter, and the regulation of hormone-sensitive cyclases) (Rubin, 1970). Higher lead concentrations reduce the calcium concentrations in bathing or superfusing media, possibly because of interactions at the site of the transmembrane calcium transport leading to defects in calcium functions (see Kober and Cooper, 1976). Lead–sodium interactions may explain the accumulation of calcium and the enhanced dopamine release from CNS synaptosomes (Silbergeld and Adler, 1978). Interactions between Pb and Mg might play a role in mitochondrial binding and in the oxidative phosphorylation. Another cause of the neurotoxicity of lead may be related to a direct inhibition of the activity of myelin-synthesizing enzymes, as demonstrated by *in vitro* studies. Also, inhibition of heme synthesis resulting in an increase of γ-aminolevulinic acid and porphobilinogens may be involved as a cause of the encephalopathy. *In vitro*, γ-aminolevulinic acid can displace γ-aminobutyric acid in its binding to the synaptic membranes and block its reuptake by the synaptosomes because of their structural similarity.

16.4.3 Renal Effects of Toxic Lead

Functional, metabolic, and morphological abnormalities of the kidneys have been reported in the case of lead intoxication. The vessels and tubules are affected especially. In acute poisoning, tubular lesions dominate, while in the case of chronic poisoning vascular lesions also occur and may sometimes be progressive, with the ultimate development of nephrosclerosis (see Haeger-Aronson, 1960).

These morphological changes bring about impairment of the reabsorptive mechanisms, with the effect of aminoaciduria, glycosuria, hyperphosphaturia, and acidosis (see Morgan, 1976). Newer experimental studies also report on an increased urinary excretion of K, Na, and Ca (Mouw *et al.*, 1978). These functional defects of the kidney caused by lead poisoning may be explained by the influence of lead on the mitochondria (Goyer, 1968). Accordingly, active transport mechanisms may be impaired because of the resulting deficiency of ATP. Other explanations concern changes in the tubular permeability or interference with carrier molecules.

Apart from the toxic effects of lead on the tubular lining cells, there are also indications that excessive doses of lead interfere with the metabolism of the juxtaglomerular apparatus, with the consequence of a transient decrease in the synthesis or release of renin (McAllister *et al.*, 1971), followed by a prolonged increase (Mouw *et al.*, 1978). The effects of lead on the renin–angiotensin–aldosterone system might also explain the hypertension encountered during lead exposure (Beevers *et al.*, 1976).

16.4.4 Intranuclear Inclusion Body

Another characteristic phenomenon of lead poisoning is the occurrence of intranuclear inclusion bodies that were detected first in the liver cord cells and proximal tubular cells of the kidney and, in recent times, in the osteoclasts and brain glial cells. They occur mainly in the nucleus and only partly in the cytoplasm. There is evidence that they are formed in the cytoplasm and then migrate into the nucleus. The structure of the inclusion bodies consists of a dense central core and an outer fibrillary zone (see Moore and Goyer, 1974). They represent insoluble lead-protein complexes containing also small amounts of lipids. The presence of nucleic acids in the inclusion bodies has also been observed. It is noteworthy that the Ca content of inclusion bodies is relatively high (Six and Goyer, 1970).

The formation of these inclusion bodies is considered to be a protective mechanism of the body, which maintains a relatively low lead concentration in the cytoplasm, preventing deleterious effects. The lead bound in the inclusion bodies is designated as nondiffusible as opposed to diffusible lead.

16.4.5 Mutagenic, Mitogenic, and Teratogenic Effects of Lead

With respect to mitogenic, mutagenic, and teratogenic effects of excessive lead doses, divergent findings have been reported, which may be largely related to differences in experimental procedures and the chosen levels of lead exposure.

Early reports on lead intoxication described nuclear polyploidy and abnormalities in the mitosis of bone marrow cells (see Waldron, 1966). Newer studies, both *in vitro* and *in vivo*, implicate structural and quantitative aberrations of chromosomes, depending on the lead doses (e.g., Sirouer and Volb, 1976; Jachimezek and Skotarczak, 1978; Nordenson *et al.*, 1978). Further research is necessary to elucidate the mode of action of lead in bringing about these changes.

A direct influence of lead on nucleic acids has been suspected for a long time. Russek (1970) showed that lead is able to form complexes with the phosphate groups of nucleotides and nucleic acids and catalyze the nonenzymatic hydrolysis of nucleoside triphosphates.

Depending on the concentration of lead there is also experimental evidence that this metal stimulates DNA replication in the kidney, which requires *de novo* synthesis of RNA and protein before the onset of DNA synthesis (Choie and Richter, 1974a,b; Russel, 1970). This might explain the elevated cell proliferation following lead exposure, as observed by Choie and Richter (1974).

Very high lead doses, however, bring about a disaggregation of polyribosomes (Waxman and Rabinowitz, 1966), shown primarily in reticulocytes, where lead degrades particularly those regions of the polynucleotides with little secondary structure. This might be caused by an activation of ribonuclease by lead (Farkas *et al.*, 1972).

Ferm and Carpenter (1967) demonstrated teratogenic activities of different Pb salts in mammalian embryos. There are maternal and paternal effects possible. The teratogenic changes bring about reduced birth weights (Nordström *et al.*, 1978a), spontaneous abortions (Nordström *et al.*, 1978b), fetal malformation (Ferm and Carpenter, 1967), and premature deliveries (Fakim *et al.*, 1976). The causes for these effects are multifaceted, one certainly being related to the aforementioned changes in the chromosomes (Muro and Carpenter, 1969). Besides, biochemical studies on lead-intoxicated embryos found a depressed activity of δ-ALAD, increased porphyrins, and diminished incorporation of iron into heme (Jaquet *et al.*, 1977; Gerber and Mals, 1978). However, a lowered placental blood flow (Gerber *et al.*, 1978) may play a role as well as elevated lead concentrations in the fetuses (see Gall, 1978; Ryü *et al.*, 1978).

16.5 Conclusions

Despite the remarkably high toxicity of lead, very small amounts of this metal have been proved to be essential. The essentiality of lead can be demonstrated in model studies by growth depressions, biochemical changes, and clinical symptoms occurring in the progeny of rats fed a lead-depleted diet. Now, the essential nature of lead needs to be investigated with regard to its biological functions in the body and to the mode of interaction between lead and iron.

With respect to the toxicity of lead, numerous studies exist on clinical symptoms and biochemical changes; the primary causes of Pb toxicity at the molecular level are, however, still largely unknown. This is particularly true for the mitogenic, mutagenic, and teratogenic effects. The manifestation of toxic symptoms is affected not only by the dose and duration of exposure, but also by numerous other factors (e.g., dietary composition, supply status, physiological state). Interactions with other trace elements during absorption, excretion, and intermediary metabolism play a key role in this regard. Newer results indicate that such interactions may even be the primary reason for the toxic symptoms.

Thus, high Pb doses can displace other elements (calcium, for example) from their binding sites in proteins and, accordingly, interfere with the function of these elements.

Results from studies on lead metabolism should be interpreted with care, because they were often obtained after the application of toxic or subtoxic amounts of lead and, therefore, cannot be inferred to apply to normal metabolic states without reservations. This is especially true for the mechanisms of absorption, distribution, and excretion and for their regulation. Also, the specific compounds of lead present in the body are largely unknown. Basically, it must be realized that, over the entire range of lead intake (deficient to toxic), the modes of action of lead may often be very different between *in vitro* and *in vivo* situations, because *in vivo* responses, in contrast to isolated systems *in vitro,* may be affected by many factors.

Also, analytical problems arise in the range of usual and deficient lead intake, since the concentrations of lead in the body or in organs and tissues lie near or below the limits of detection of the common analytical procedures, so that reproducibility and comparability of the results are impeded.

The numerous influences on lead metabolism resulting from the close interrelationships in metabolism, the differences in experimental methods, and the analytical problems make it evident that results are valid only under strictly defined conditions. Accordingly, the setting of thresholds—for practical matters, particularly important the setting of tolerance thresholds—is problematic. This problem concerns particularly the diagnostic parameters so that several criteria should be used for an unequivocal diagnosis. Basically, biochemical changes are suitable criteria since they become evident much earlier than clinical symptoms.

References

Albahary, C., 1972. *Am. J. Med.* 50:367.
Alwares, A. P., Fischbein, A., Sassa, S., Anderson, K. E., and Kappas, A., 1976. *Clin. Pharmacol. Ther.* 19:193.
Barltrop, D., Barrett, A. J., and Dingle, J. T., 1971. *J. Lab. Clin. Med.* 77:705.
Barltrop, D., and Kehoe, H. E., 1975. *Post-grad. Med. J.* 51:795.
Barton, J. C., Conrad, M. E., Nuby, S., and Harrison, L., 1978. *J. Lab. Clin. Med.* 92:536.
Beevers, D. G., Erskine, E., and Robertson, M., 1976. *Lancet* 2:1.
Berk, P. D., Tschudy, D. P., Shepley, L. A., Waggoner, J. G., and Berlin, N. I., 1970. *Am. J. Med.* 48:137.
Bessis, M. C., and Jensen, W. N., 1965. *Br. Haematol.* 11:49.
Blackman, S. S., 1937. *Bull. Johns Hopkins Hospital* 61:1.
Blundell, B. T., and Jenkins, J. A., 1977. *Chem. Soc. Rev.* 6:139.
Bingham, E., Pfitzer, E. A., Barkley, W., and Radford, E. P., 1968. *Science* 162:1297.
de Bruin, A., 1968. *Med. Lavora* 59:411.

Calissano, P., Bonsignore, D., and Carasegna, G., 1966. *Biochem. J.* 191:550.
Castellino, N., and Aloj, S., 1964. *Br. J. Industr. Med.* 21:308.
Cerklewski, F. L., and Forbes, R. M., 1976. *J. Nutr.* 106:689.
Chisholm, J., 1964. *J. Pediatr.* 64:174.
Choie, D. D., and Richter, G. W., 1974. *Lab. Invest.* 30:652.
Conrad, M. E., and Barton, J. C., 1978. *Gastroenterology* 74:731.
Delwaide, P. C., Hensghem, C., and Noirfalise, A., 1968. *Ann. Biol. Clin.* 26:987.
Devigne, J. P., 1968. *CR Acad. Sci. D.* 267:935.
Devigne, J. P., 1968. *Arch. Inst. Pasteur Tunis* 45:341.
Dobbins, A., Johnson, D. R., and Nathan, P., 1978. *J. Toxicol. Environ. Health* 4:541.
Dresel, E. I. B., and Faulk, J. E., 1956. *Biochem. J.* 63:72.
Farkas, W. R., Hewins, S., and Welch, J. W., 1972. *Chem. Biol. Interact.* 5:191.
Ferm, V. H., and Carpenter, S. J., 1967. *J. Exp. Mol. Path.* 7:208.
Ferm, V. H., 1972. *Adv. Teratol.* 5:51.
Fine, B., Barth, A., Sheffet, A., and Levenhar, M., 1976. *Environ. Res.* 12:224.
Fischbein, A., Alvares, A. P., Anderson, K. E., Sassa, S., and Kappas, A., 1977. *J. Toxicol. Environ. Health* 3:431.
Fullerton, P. M., 1966. *J. Neuropathol. Exptl. Neurol.* 25:214.
Gale, Th. F., 1978. *Environ. Res.* 17:325.
Gerber, G., Maes, J., and Deroo, J., 1978. *Arch. Toxicol.* 41:125.
Gerber, G., and Maes, J., 1978. *Toxicology* 9:173.
Gibson, K. D., Neuberger, A., and Scott, J. J., 1955. *Biochem. J.* 61:618.
Gibson, K. D., and Goldberg, A., 1970. *Clin. Sci.* 38:63.
Goldberg, A., Ashenbrucker, H., Cartwright, G. E., and Wintrobe, M. M., 1956, *Blood* 11:821.
Goldberg, A., 1968. *Seminars of Hematology* 9:424.
Goyer, R. A., 1968. *Lab. Invest.* 19:71.
Goyer, R. A., and Mahaffey, K. R., 1972. *Environ. Health Perspect.* 2:73.
Goyer, R. A., and Rhyne, B. C., 1973. In *Int. Rev. Exp. Path.* 12, G. W. Richter and E. A. Epstein (eds.), Academic Press, New York, pp. 1.
Griggs, R. C., 1964. *Prog. Hematol.* 4:117.
Haeger-Aronson, B., 1960. *J. Clin. Lab. Invest.* 12:1.
Hamilton, D. L., 1978. *Toxicol. Appl. Pharmacol.* 46:651.
Hassan, J., Hernberg, S. L., Metsälä, P., et al., 1967. *Arch. Environ. Health* 14:309.
Hernberg, S., and Nikkanen, J., 1970. *Lancet* 1:63.
Hernberg, S., Nikkanen, J., Mellin, G., et al., 1970. *Arch. Environ. Health* 21:140.
Hernberg, S., 1972. *Prac. Lek.* 24:77.
Jachimczak, D., and Skotarczak, B., 1978. *Genet. Pol.* 19:353.
Jaquet, P., Gerber, G. B., and Maes, J., 1977. *Bull. Environ. Contam. Toxicol.* 16:271.
Jansen, W. N., Moreno, G. D., and Bessis, M. C., 1965. *Blood* 25:933.
Kaplan, M. L., Jones, A. G., Davis, M. A., and Kopito, L., 1975. *Life Sci.* 16:1545.
Kehoe, R. A., 1964. *Environ. Health* 8:232.
Keller, C. A., and Doharty, R. A., 1980. *Environ. Res.* 21:217.
Kirchgessner, M., and Reichlmayr-Lais, A. M., 1981. *Int. J. Vit. Nutr. Res.* 51:421.
Kirchgessner, M., and Reichlmayr-Lais, A. M., 1981. *Biol. Trace Elements Res.* 3:279.
Kirchgessner, M., and Reichlmayr-Lais, A. M., 1981. *Ann. Nutr. Metab.* 26:50.
Klein, M., Namer, R., Harpur, E., and Corbin, R., 1970. *N. Engl. J. Med.* 283:669.
Kober, T. E., and Cooper, G. P., 1976. *Nature* 262:704.
Kochen, J., and Greener, Y., 1975. *Ped. Res.* 9:323.
Kostial, K., Simonovic, I., and Pisonic, M., 1971. *Nature* 233:564.
Kusell, M., Kake, L., Anderson, M., and Gerschenson, L. E., 1978. *J. Toxicol. Environ. Health* 4:515.

Lamola, A. A., Piomelli, S., Poh-Fritzpatrick, M. B., Yamane, T., and Harber, L. C., 1975. *J. Clin. Invest.* 56:1528.
Landgrin, P. J., Baker, E., Whitworth, R., and Feldman, R. G., 1980. In *Low Lead Exposure. The Clinical Implementation of Current Research*, H. L. Needleman (ed.), p. 17.
Ligeois, F. J., Derivaux, J., and Depelchin, A., 1961. *Ann. Med. Vet.* 2:57.
Mahaffey, K. R., Goyer, R. A., and Haseman, J. K., 1973. *J. Lab. Clin. Med.* 82:92.
Mahaffey, K. R., 1974. *Environ. Health Perspect.* 7:107.
Mahaffey, K. R., Stone, C. L., Banks, T. A., Reed, G., 1978. In *Trace Element Metabolism in Man and Animals*, Vol. 3, M. Kirchgessner (ed.), ATW Freising-Weihenstephan, FRG.
McAllister, R. G., Michelakis, A. M., and Sandstead, H. H., 1971. *Arch. Intern. Med.* 127:919.
Meredith, P. A., Moore, M. R., Goldberg, A., 1974. *Biochem. Soc. Trans.* 2:1243.
Mirando, E. H., and Ranasinghe, L., 1970. *Med. J. Aust.* 2:966.
Moore, J. F., and Goyer, R. A., 1974. *Environ. Health Respect. Exp. Issue* 7:121.
Morgan, J. M., 1976. *South. Med. J.* 69:881.
Mouw, D. R., Vander, A. J., Cox, J., Fleischer, N., 1978. *Toxicol. Appl. Pharmacol.* 46:435.
Muro, L. A., and Goyer, R. A., 1969. *Arch. Pathol.* 87:660.
Nakao, K., Wada, O., and Yano, Y., 1968. *Chim. Acta* 19:319.
Needleman, H. L., 1980. In *Lead Toxicity*, R. Singhal and J. A. Thomas (eds.), Urban Schwarzenberg, Baltimore.
Nordenson, J., Beckman, G., Beckman, I., and Nordström, S., 1978. *Hereditas* 88:263.
Nordström, S., Beckman, I., and Nordenson, I., 1978. *Hereditas* 88:43.
Oliver, T., 1914. *Lead Poisoning*. Lewis, London.
Ong, C. N., and Lee, W. R., 1980. *Br. J. Ind. Med.* 37:70.
Otrzonsek, N., 1967. *Int. Arch. Gewebepathol. Gewebehyg.* 24:66.
Pecora, L., Fati, S., Mole, R., and Pesaresi, C., 1965. *Folia Med.* 48:33.
Pentschew, A., 1965. *Acta Neuropathol.* 5:133.
Piddington, S. K., White, J. M., 1974. *Br. J. Haematol.* 27:415.
Piper, W. N., Tephly, T. R., 1974. *Life Sci.* 14:873.
Pounds, J. G., Wright, R., Morrison, D., and Casciano, D. A., 1982. *Toxicol. Appl. Pharmacol.* 63:389.
Quarterman, J., and Morrison, J. N., 1975. *Br. J. Nutr.* 34:351.
Quarterman, J., Morrison, J. N., and Humphries, W. R., 1978. *Environ. Res.* 17:60.
Rabinowitz, M. B., Kopple, J. D., and Wetherill, G. W., 1980. *Am. J. Clin. Nutr.* 33:1784.
Ragan, H., 1977. *J. Lab. Clin. Med.* 90:700.
Raghavan, S. R. V., Culver, B. D., Gonick, H. C., 1981. *J. Toxicol. Environ. Health* 7:561.
Reichel, G., Wobith, F., and Ulmer, W. T., 1970. *Int. Arch. Arbeits-med.* 26:84.
Reichlmayr-Lais, A. M., and Kirchgessner, M., 1981. *Z. Tierphysiol., Tierernährg. Futtermittelkde.* 46:1.
Reichlmayr-Lais, A. M., and Kirchgessner, M., 1981. *Ann. Nutr. Metab.* 25:281.
Reichlmayr-Lais, A. M., and Kirchgessner, M., 1981. *Z. Tierphysiol., Tierernährg. Futtermittelkde.* 46:8.
Reichlmayr-Lais, A. M., and Kirchgessner, M., 1981. *Z. Tierphysiol., Tierernährg. Futtermittelkde.* 46:145.
Reichlmayr-Lais, A. M., and Kirchgessner, M., 1981. *Zbl. Vet. Med. A.* 28:410.
Reichlmayr-Lais, A. M., and Kirchgessner, M., 1983. In preparation.
Rosen, J. F., and Haymovits, A., 1973. *Pediatr. Res.* 7:393.
Rosen, J. F., and Trinidad, E. E., 1974. *Environ. Health Perspect. Exp. Issue* 7:139.
Rosen, J. F., Zarate-Salvador, C., and Trinidad, E. E., 1974. *J. Pediatr.* 84:45.
Rubin, R. P., 1970. *Pharmacol. Rev.* 22:289.
Ruessel, H. A., 1970. *Bull. Environ. Contam. Toxicol.* 5:115.

Ryn, J. E., Ziegler, E. E., and Foman, S. J., 1978. *J. Pediatr.* 93:476.
Sabbioni, E., and Marafante, E., 1976. *Chem. Biol. Interactions* 15:1.
Schroeder, H. A., Tipton, I. H., 1968. *Arch. Environ. Health* 17:965.
Schwarz, K., 1974. In *Trace Element metabolism in Animals*, Vol. 2, W. G. Hoekstra, J. W. Suttie, H. E. Ganther, and W. Mertz (eds.), University Park Press, Baltimore, p. 335.
Shields, J. B., and Mitchell, H. H., 1940. *J. Nutr.* 21:541.
Silbergeld, E. A., Adler, H. S., 1978. *Brain Res.* 148:451.
Silbergeld, E. K., 1982. In *Mechanism of Actions in Neurotoxic Substances*, K. N. Prasad and A. Verhadakis (eds.), Raven Press, New York.
Six, K. M., and Goyer, R. A., 1970. *J. Lab. Clin. Med.* 76:933.
Six, K. M., and Goyer, R. A., 1972. *J. Lab. Clin. Med.* 79:128.
Sirover, M. A., and Loeb, L. A., 1976. *Science* 194/4272:1434.
Smith, C. M., De Luca, H. F., Tanaca, Y., and Mahaffey, K. R., 1978. *J. Nutr.* 108:843.
Smith, C. M., De Luca, H. F., Tanaca, Y., and Mahaffey, K. R., 1981. *J. Nutr.* 111:1321.
Thomas, H. M., 1904. *Johns Hopkins Hosp. Bull.* 15:209.
Toraason, M. A., Barbe, J. S., and Knecht, E. A., 1981. *Toxicol. Appl. Pharmacol.* 60:62.
Ulmer, D. D., and Vallee, B. L., 1969. Proc. 2nd. Missouri Conf. Trace Substances Environ. Health 1968, p. 7.
Vallee, B. L., and Ulmer, D. D., 1972. *Ann. Rev. Bioch.* 41:91.
Vergano, C., Cartasegna, C., and Ardoina, V., 1969. *Med. Lavora.* 60:505.
Vetter, H., and Mählhop, R., 1971. *Landwirtsch. Forsch.* 24:294.
Vostal, J., and Heller, J., 1968. *Environ. Res.* 2:1.
Wada, S., 1976. In *Effect and Dose-Response Relationships of Toxic Metals*, G. F. Nordberg (ed.), Elsevier Scientific Publishing Company, Amsterdam.
Walton, 1973. *Nature* 243:100.
Wasserman, R. H., and Corradino, R. A., 1973. *Vitam. Horm.* 31:43.
Waxman, H. S., and Rabinovitz, M., 1966. *Biochim. Biophys. Acta* 129:369.
White, J. M., and Harvey, D. R., 1972. *Nature* 236:71.
Wong, P. R. S., Silverberg, B. A., Chan, Y. K., and Hodson, P. V., 1978. In *The Biogeochemistry of Lead in the Environment*, J. O. Nriagu (ed.), Elsevier/North Holland, Amsterdam, p. 279.

Boron 17

Carol J. Lovatt and W. M. Dugger

17.1 Boron in Biology

17.1.1 Introduction

Boron is the only nonmetal in a family otherwise comprised of active metals, Group IIIA of the periodic table. As expected, boron exhibits bonding and structural characteristics intermediate to both. Like carbon (atomic number 6), boron (atomic number 5) has a tendency to form double bonds and macromolecules. In addition, there are several features that are more or less unique to boron and this group of elements. These include electron-deficient molecules (such as boron trifluoride) and bridge bonds (such as those in diborane, B_2H_6). These tendencies have formed the basis for the many hypotheses attempting to predict the mode of action of boron as a nutrient essential to the metabolism of vascular plants (Section 17.2.1), as a toxicant to animals (Section 17.1.5), and for achieving boron accumulation in cancer cells (Section 17.1.8).

In this chapter the biochemistry of boron is reviewed predominantly in vascular plants. The discussion is organized around those metabolic processes that boron nutrition repeatedly has been shown to influence: carbohydrate metabolism, hormone action, membrane structure and function, and nucleic acid biosynthesis.

17.1.2 Criteria for Essentiality

The biochemistry of boron encompasses the history of almost 75 years of research seeking to elucidate the primary role of boron in the metabolism of

---

Carol J. Lovatt and W. M. Dugger • Department of Botany and Plant Sciences, University of California, Riverside, California 92521.

vascular plants. Few organisms, other than tracheophytes, have been demonstrated to require boron for growth. For plants, the essential nature of an element is established according to a set of criteria defined by Arnon and Stout (1939): (1) the element must be essential to the completion of the life cycle of the plant; (2) the element cannot be substituted for or replaced by any other element; (3) the element must have a distinct function (e.g., enzyme cofactor, as in the case of zinc, which is a cofactor of alcohol dehydrogenase; molecular component, such as magnesium in chlorophyll; or structural component, such as silicon in the frustules of diatoms). Two early articles demonstrated that boron met these criteria: (1) Sommer and Lipman (1926) provided evidence that boron was essential for the completion of the life cycle of a number of mono- and dicotyledonous plants; and (2) Warington (1923) demonstrated that none of the 52 elements she tested could alleviate the symptoms of boron deprivation for the several species of leguminous plants with which she worked. Thus, by 1926, the essential nature of boron to the growth of angiosperms had been established in accordance with the criteria of Arnon and Stout except for demonstration of a distinct function for boron in a plant. Although there is universal agreement that boron is essential to vascular plants, over 70 years have passed and a specific biochemical role for boron in the metabolism of a higher plant remains to be elucidated.

17.1.3 Effect of Boron on the Growth of Organisms

There is minimal evidence that boron is essential to organisms other than vascular plants. In the Monera, the only bacterium tested for a boron requirement was *Azotobacter chroococcum*, a free-living nitrogen-fixing bacterium. Boron was shown to stimulate nitrogen fixation in this prokaryote, but it could not be demonstrated that boron was essential to this process (Anderson and Jordan, 1961). Nitrogen fixation also was stimulated in three species of Cyanophyta: *Calothrix parietina, Anabaena cylindrica,* and *Nostoc muscorum*. In addition, boron stimulated the growth rate of these three species and the growth of the nonnitrogen-fixing blue-green alga, *Microcystis aeruginosa*, when nitrate was omitted from the growth medium. The increase in growth rate was not as dramatic when nitrate was provided (Gerloff, 1968). Only *Nostoc muscorum* was shown to require boron for growth. The cells deprived of boron became chlorotic in the third week of culture and were completely white by the end of 8 weeks. At this time the population of *N. muscorum* determined by cell count was only 39% of the population of the cultures supplied with boron (Eyster, 1952).

Boron is not required for the fungi *Neurospora crassa, Saccharomyces cerevisiae, Aspergillus niger,* or *Penicillium chrysogenum* (Bowen and Gauch, 1966; Gerloff, 1968). However, the morphology of the spore-forming organs has been shown to be sensitive to variations in boron concentrations (Davis et al., 1928). Species of fungi exhibited effects of boron toxicity, resulting in the aborted growth of mycelia, perithecia, and ascospores (Bowen and Gauch, 1966) and in the failure of gametes to cleave (Zittle, 1951). Foster (1949) demonstrated that penicillin production is stimulated by boron in both *Penicillium chrysogenum* and *P. notatum*, and Davis et al. (1928) reported that boron enhanced the growth of *P. italicum*.

Although boron was shown to enhance the growth of *Chlorella vulgaris*, with 0.5 mg boron per liter being sufficient for optimal growth (McIlrath and Skok, 1958), no requirement for boron could be demonstrated among the many species of chlorophyta tested (Gerloff, 1968). In fact, boron was found to be toxic to four species of *Chlorella* at the concentrations indicated: (1) *C. vulgaris*, greater than 0.5 mg boron per liter; (2) *C. vannielei*, 50 mg boron per liter; (3) *C. emersonii*, 100 mg boron per liter; and (4) *C. protothicoides*, 100 mg boron per liter (Bowen et al., 1965). Lewin (1966a) reported an absolute requirement for boron in the marine pennate diatom, *Cylindrotheca fusiformis*, in both the light and dark. She subsequently reported an absolute requirement for boron in 16 species of marine diatoms and eight species of freshwater diatoms (Lewin, 1966b). For marine diatom species, the requirement for boron was demonstrated upon the first transfer to a boron-free culture medium. However, the boron requirement for freshwater diatoms could be shown only after a number of transfers to boron-free medium had been carried out (Lewin, 1966a,b). Preliminary studies with other phytoplankton indicated that a boron requirement is not unique to diatoms; species of Dinophyceae and Prasinophyceae do not multiply under boron-deficient conditions (Lewin, 1966b).

A requirement for boron in bryophytes has not been reported (Lewis, 1980a). In the Lycopsida *Selaginella apoda*, boron seems to be essential to the formation of spore-producing organs (Bowen and Gauch, 1965). This effect appears to be true for the fern *Dryopteris dentata*, which, when grown under boron-deficient conditions, produced no visible sori or incomplete indusia with aborted sporangia (Bowen and Gauch, 1965).

In the gymnosperms, tissue cultures derived from *Ginkgo biloba* pollen have an absolute boron requirement (Yih et al., 1966); and several reports of a boron requirement in conifers are found in the literature (Ludbrook, 1942; Walker et al., 1955; Blaser et al., 1967). Boron deficiency in *Thuja plicata* resulted in restricted growth in meristematic regions; needles were closely bunched, giving a "rosette" appearance to shoots, and roots were short (Walker et al., 1955). The symptoms of boron deficiency were most obvious during rapid growth and periods of meristematic activity (Blaser et al., 1967).

Boron is considered an "apparently" nonessential inorganic constituent in the dietary requirements of the rat (McCoy, 1967). There is only one study on record that attempts to determine whether or not boron is an essential nutrient for animals. Using the laboratory rat as the test animal, Hove *et al.* (1939) concluded that if boron was needed by the growing rat, 0.8 mg per day satisfied this requirement.

17.1.4 Plant Evolution and an Essential Role for Boron

Any unified theory attempting to explain the biochemical function of boron in plant metabolism must account for the fact that some taxonomic groups have an absolute requirement for boron while others do not. Thus, two conclusions can be drawn from the previous section: (1) there must be specific metabolic pathways common to both vascular plants and the species of diatoms, Dinophyceae and Prasinophyceae, requiring boron that are generally lacking among the other taxonomic groups; or (2) there must be at least two unique pathways requiring boron: one in tracheophytes and the other in the phytoplankton listed previously.

It has been proposed that a study of the evolution of vascular plants may provide a possible key to the role of boron in plant metabolism. Lewis (1980a) proposed that the development of an essential role for boron was a prerequisite to the evolution of vascular plants. He contends that the primary role of boron is in lignin biosynthesis and, in conjunction with auxin (indole-3-acetic acid, IAA), in xylem differentiation. This hypothesis is based on (1) the many reports in the literature that phenols accumulate under conditions of boron deficiency (Spurr, 1952; Perkins and Aronoff, 1956; Watanabe *et al.*, 1961; Watanabe *et al.*, 1964; Troitskaya *et al.*, 1970); (2) the suggestion in the literature that there is a concomitant decrease in lignin biosynthesis in boron-deficient tissue (Skok, 1958); (3) the fact that hyperauxiny is a frequent symptom of boron deficiency (Odhnoff, 1957; Neales, 1960; Jaweed and Scott, 1967; Coke and Whittington, 1968; Bohnsack and Albert, 1977); and (4) the fact that monocots, which have a much lower requirement for boron, possess an additional lignin biosynthetic pathway not possessed by dicots.

Despite this evidence, there is room to doubt Lewis's hypothesis that the primary role of boron in vascular plants is in lignin biosynthesis. In fact, lignin synthesis may be enhanced under conditions of boron deficiency (Neales, 1960); increased availability of lignin precursors and cell wall thickening are common features of boron-deficient tissues (Kouchi and Kumazawa, 1975b; Hirsch and Torrey, 1980). The relationship between boron and IAA may also be considered a secondary one. Bohnsack and Albert (1977) and Hirsh *et al.* (1982) have shown that at the biochemical and ultrastructural levels, boron deficiency and hyperauxiny are different (Section 17.3).

Lewin (1966a) proposed that the role of boron in diatoms is due to its structure as B_2O_3, which closely mimics SiO_2. This property results in the incorporation of boron into the diatom shell. Based on the silicon to borate ratio, she concluded that boron was an important structural component of the diatom cell wall and essential to normal wall morphogenesis. In the evolution of plants, it is probable that a requirement for boron arose more than once.

17.1.5 Boron Toxicity

Economic losses due to boron toxicity are common in irrigated regions of the world; such losses arise not only through a reduction in plant productivity, but also through boron intoxication of livestock that feed on plants that have accumulated toxic levels of boron. Boron inhibits many enzymic activities (Section 17.2.9) with concomitant deleterious effects on both livestock and humans.

In vascular plants, boron is carried passively in the transpiration stream and accumulates where the transpiration stream ends (Kohl and Oertli, 1961). Because boron is relatively immobile in the phloem, very little of the accumulating boron moves out of these tissues (Oertli and Richardson, 1970; Raven, 1980). For these reasons, leaves usually exhibit the first symptoms of boron toxicity: yellowing of the leaf tip, with the chlorosis subsequently progressing along the leaf margin and then spreading into the blade. Necrosis of the chlorotic tissue follows, followed by leaf abscission (Wilcox, 1960). Thus, boron toxicity results in a loss of plant productivity.

While there are many reports in the literature relating the development of leaf symptoms to elevated levels of boron in leaves, research attempting to determine the actual manner by which boron is toxic to plants has been minimal. It is assumed that the chlorosis and subsequent necrosis of leaf tissue resulting from the accumulation of boron to a toxic level result in a loss of photosynthetic capacity that accounts for the subsequent loss in plant productivity.

Since the uptake of boron is passive through the transpiration stream in tracheophytes, boron intoxication is a function of the concentration of boron to which the plant is exposed, the length of exposure, and the rate of transpiration. Because of these phenomena, Kohl and Oertli (1961) predicted that if a leaf lived long enough, it would show symptoms of boron toxicity even when grown in the presence of a concentration of boron considered optimal. Boron concentrations that are toxic to a particular species or cultivar have proved to be nearly constant, regardless of the stage of growth (El-Sheikh *et al.*, 1971).

Some vascular plants grow well under conditions of excess boron. In some boron-tolerant species, boron fails to accumulate in the leaves, or does so at a reduced rate. In most cases, this is due to reduced boron uptake. The mechanism is unknown; reduced boron uptake may simply be due to a lower rate of transpiration in the boron-tolerant species. The sequestering of boron along the root, which occurs in some species tolerant to sodium, has not been demonstrated in plants tolerant to boron. In other boron-tolerant species, a high concentration of boron accumulates, but the leaves do not exhibit the symptoms associated with boron toxicity (El-Sheikh et al., 1971).

17.1.6 Problems Associated with Studies of Boron Metabolism

Two significant factors contribute to the difficulty of elucidating boron's mode of action. The first is the lack of a radioactive or heavy isotope of boron that would facilitate localization and transport studies. Without a radioisotope of boron, all metabolic investigations of a role for boron in plant metabolism have been comparative studies employing boron-sufficient and boron-deficient tissues.

The second is the difficulty of establishing a zero boron concentration. Boron is a common component of "hard" glass and has been shown to leach from it in a relatively short period of time. Boron is also found as a contaminant of chemicals and distilled water. Furthermore, tissue derived from boron-sufficient plants contains endogenous boron. In light of the fact that 0.1 mg boron per liter provides adequate boron for most monocots and dicots, boron sufficiency is often attained by contamination. A further complication arises from the fact that optimal and toxic levels of boron are extremely close to one another. For example, it has been demonstrated that 0.1 mg boron per liter is optimal for the growth of tomato plants (*Lycopersicum esculentum*,) while a concentration greater than 1.0 mg boron per liter is toxic to them (Davies and Addo, 1957). The micronutrient status of boron results form the fact that very low concentrations—0.01–4.0 mg boron per liter—are adequate for most gymnosperms and angiosperms grown in solution culture (Wilcox, 1960). Since boron is transported passively in the transpiration stream (Kohl and Oertli, 1961), tissue levels of boron are a function not only of the concentration of available boron but of the length of exposure to boron and the rate of transpiration. Thus, boron concentrations within the optimum range, or slightly higher, may prove toxic to some species (Kohl and Oertli, 1961; Mengel and Kirby, 1978). Unless one is careful, varying degrees of boron contamination may make it difficult to know whether one is observing the effects of boron deficiency, sufficiency, or toxicity.

17.1.7 ^{10}B (n, α) ^{7}Li Nuclear Reaction

Recently, technical progress has been made in the use of a (n, α) nuclear reaction with the stable isotope of boron—^{10}B. Initial studies on the distribution and compartmentalization of boron are beginning to appear in the literature

(Martini and Thellier, 1975; Thellier et al., 1979). The distribution of boron was associated with four classical compartments: the free space (including easily dissociable borate monoesters), the cytoplasm, the vacuole, and the cell wall (as stable borate diesters) (Thellier et al., 1979). The use of the ^{10}B (n, α) ^{7}Li nuclear reaction for *in vivo* studies on the localization and redistribution of boron between the cellular compartments under different physiological conditions might shed an important new light on our understanding of the biological role of boron.

17.1.8 Therapeutic Uses for Boron and Organoborates in Medicine

The idea that boron or organoborate compounds may prove efficacious in cancer therapy originated from the following observations: (1) ^{10}B is present in nature at about 20% of the concentration of ^{11}B and was not toxic to living tissues; (2) slow neutron bombardment (an average speed of 2200 m/sec with an energy of only 0.025 eV) was not harmful to living tissues; and (3) used together, the nucleus of ^{10}B absorbed slow neutrons undergoing fission to yield ^{7}Li and an α particle that share an average energy of 2.4 MeV.

The ^{10}B isotope is uniquely suited for the treatment of cancer by thermal neutron capture for several reasons. It has a large cross-section capture, 3850 barns (1 barn = 1×10^{-24} cm^2). This is significantly greater than the cross-section capture of ^{11}B and of all the common elements of normal cells (Soloway, 1964). Thus, if the malignant tissue can be enriched in ^{10}B or simply in boron compounds (which, in nature, are 20% ^{10}B), the destruction of malignant cells is possible with a minimal number of nuclear reactions occurring with those elements that are usually found in greatest abundance within a cell. Second, the size of the particles (lithium and helium nuclei) and the energy (2.4 MeV) released in the ^{10}B (n, α) ^{7}Li reaction limits the radius of action to 9 μm from the site of ^{10}B disintegration. For comparison, the sphere of travel for these particles would be about the size of a red blood cell. Thus, the radiation would be limited to the one cell containing ^{10}B and its immediate neighbor cells (Soloway, 1964). This is also important for another reason. If boron enrichment of tissue could not be achieved by an organ- or tissue-specific process, but only through the general enrichment of all tissues in a patient, this would not be a severe limitation to the use of ^{10}B-neutron capture therapy. This is because the area irradiated can be highly localized, and the reaction itself is of limited sphere (Kliegel, 1972). This makes ^{10}B more suitable than several other nonradioactive isotopes that have larger cross-section capture values than ^{10}B but produce gamma radiation when they undergo neutron capture (Soloway, 1964).

With neutron therapy in mind, researchers have synthesized and tested a number of organic and inorganic boron compounds in addition to borax and boric acid (Kliegel, 1972). The necessary prerequisites for use of any of these compounds in cancer therapy in human patients are paraphrased from Soloway's (1964) guidelines:

1. To permit injection, the boron compound must have sufficient solubility in water, in the ideal case at pH 7.4, so that the physiology of the system is minimally disturbed.
2. The boron compound must be stable so that *in vivo* no oxidation or hydrolysis occurs.
3. Because 50 mg of boron per kg of tumor tissue is necessary for the ^{10}B (n, α) ^7Li reaction to be effective, the boron compound must have low toxicity or high boron content.
4. In order to obtain maximum destruction of tumor cells, it is desirable for the tumor cells to have a greater concentration of boron than the normal cells. Since cancerous cells grow at a more rapid rate than normal cells, intracellular or intranuclear enrichment can be achieved by using boron-containing pyrimidines, purines, or amino acids. Such enrichment would maximize the destruction of the chromosomes of cancer cells.
5. Because the dosage of boron or boron-containing compounds must be high to be effective, it is desirable that the source of boron or resulting complexes be colorless.

From studies employing animals, nine compounds have been identified that have low toxicity and give the desired enrichment of cancer cells with boron. One is a boric acid ester with triisopropanolamine, six are arylboric acid derivatives, and two are borohydrides of the $B_{10}H_{10}$ type (Kliegel, 1972). The latter, with their great stability, relatively low toxicity, and high boron content, are especially attractive for use in the treatment of cancer by ^{10}B-thermal neutron capture (Kliegel, 1972). One of the borohydrides, $Na_2B_{10}H_{10}$, is excreted chemically unaltered, demonstrating that this compound cannot be metabolized by mammals—a property that makes it almost perfect for ^{10}B-neutron irradiation therapy.

The idea of using boron-containing substrate analogs in cancer therapy, either in ^{10}B-thermal neutron capture or as metabolic inhibitors, has also been explored with some success. In clinical tests, the replication of herpes simplex virus type 1 was inhibited 57 and 97% by 5-dihydroxyboryl-2'-deoxyuridine, respectively, at concentrations of 200 and 800 μM (Schinazi and Prusoff, 1978).

17.2 Carbohydrate Metabolism

Of the various roles that have been assigned to boron in plant growth and development, the ones that have been studied most are those that involve the

Boron

element in carbohydrate synthesis, translocation, transformation, and utilization. For vascular plants it has been proposed that boron has a role in starch synthesis, respiration, photosynthesis, carbohydrate translocation, cellulose synthesis, phenol biosynthesis, sugar-phosphate metabolism, and a number of carbohydrate-involved growth and developmental processes. In this section the various roles of boron in carbohydrate metabolism are briefly reviewed. For additional information see Dugger (1973, 1983) and Augsten and Eichhorn (1976).

17.2.1 Boron Complexes

Boron is one of the chemical elements, along with aluminum, germanium, and several others, that will form complexes with certain organic compounds. The ability of boron to complex with compounds is dependent on their having adjacent OH-groups in the cis position. Thus, sugars, sugar alcohols, and other compounds in plants having this configuration form complexes with boron; however, attempts to isolate such boron complexes from plants have failed. The type of complexes formed between boron and polyhydroxy compounds depends on the ratio of borate to the diol as well as the pH. In plants with the normal physiological level of boron, it has been proposed that the type of complex formed when the diol to borate ratio is high is the BD_2 type, while the form that exists under a low diol to borate ratio is the BD type (see diagrams).

$$\begin{bmatrix} =C-O & OH \\ & \diagdown \diagup \\ & B \\ & \diagup \diagdown \\ =C-O & OH \end{bmatrix}^- \quad H^+ \quad \begin{bmatrix} =C-O & O-C= \\ & \diagdown \diagup \\ & B \\ & \diagup \diagdown \\ =C-O & O-C= \end{bmatrix}^-$$

$$\text{(BD)} \qquad\qquad\qquad \text{(BD}_2\text{)}$$

Compounds that have ring configurations other than cis may also form complexes with boron. For example, boron may complex with compounds having trans 1,2-diol groups; the important factor seems to be the angle between the OH-groups relative to the carbon axis (Zittle, 1951).

This ability of boron to complex with a large number of biologically important substances may alter the involvement of those substances in the metabolic reactions of plants. Boron might complex with substrate molecules, end products, or enzymes themselves, thereby inhibiting or stimulating a metabolic pathway involving one or more of these modified molecules. In turn, this altered metabolic pathway may cause a change in the level of a particular metabolite that could alter a subsequent metabolic reaction. This, in turn, may alter a more obvious developmental process or the growth of the plant.

17.2.2 Sugar Translocation

Since many of the polyhydroxy compounds in plants that have the proper configuration of OH-groups are carbohydrates, much of the early research investigated the possible roles of boron in carbohydrate metabolism. There is evidence that boron deficiency in plants leads to an accumulation of sugars and starch in leaves (Gauch and Dugger, 1954; Dugger, 1973). It was pointed out that (1) such deficiency also altered the types of carbohydrates that accumulated; (2) such accumulation resulted from a breakdown of the phloem; and (3) in some fashion, boron was involved with the translocation of carbohydrates in plants. All of these proposals have been substantiated by research in a number of laboratories and over a number of years.

A primary question involves whether boron has a direct or indirect effect on sugar transport in the phloem. Skok (1958) believed that there was "some relationship between boron and sugar translocation," but since sugar-boron complexes have not been isolated from phloem sap or plant tissues, perhaps the relationship is "indirect." However, a number of reports indicate a possible direct and early effect of boron on sugar movement (Gauch and Dugger, 1954; Sisler *et al.*, 1956) and on the sugar-dependent movement of plant hormones from leaves of bean plants (*Phaseolus vulgaris*) to the hypocotyl (Mitchell *et al.*, 1953).

17.2.3 Photosynthesis

Earlier reviews on boron point out that there is little data to support a direct role for boron in photosynthesis. More recent work demonstrated that a suboptimal level of boron caused an increase in the photosynthetic rate of *Wolffia arrhiza* (Eichhorn and Augsten, 1974). It has also been reported that an increase in photosynthetic rate occurred in the marine alga *Cylindiotheca fusiformis* when the cultures contained no added boron (Smyth and Dugger, 1980). In both of these cases, there was an increase in the level of potassium in the boron-deficient cells. Subba Rao (1981) observed that nanoplankton responded to boron. During the winter months, when the temperature was low and there were more nitrates, phosphates, and silicates available, the addition of boron caused an increase in the rate of carbon assimilation. When these nutrients were lower and the temperature higher during the warmer part of the year, added boron caused no positive response in the photosynthetic rate.

Augsten and Eichhorn (1976) pointed out that boron stimulated noncyclic photophosphorylation, but boron deficiency induced a reduced level of photophosphorylation in isolated chloroplasts.

17.2.4 Respiration

The earlier literature regarding the direct influence of boron on respiration is contradictory. (For review of this literature see Gauch and Dugger, 1954; Dugger, 1973, 1983.) Since boron deficiency results in the accumulation of starch and other carbohydrates, an increase in tissue respiration would be expected to result. However, when boron deficiency intensifies, respiration decreases to values below that of the controls. Augsten and Eichhorn (1976) pointed out that under boron deprivation, there is an increase in the level of phenolic compounds and an activation of phenol oxidases as the ratio of substrate metabolized via the glucose-monophosphate pathway increases.

When bean leaf tissue was infiltrated with glucose and various levels of boron from 5 to 100 μM, there was an increased level of O_2 uptake after 24 hours. In some way, boron enhanced the utilization of glucose supplied to the leaf tissue (Dugger et al., 1957).

17.2.5 Starch

Winfield (1945) reported a slight inhibition of starch phosphorylase by boron. He proposed that boron reacted with sugars in the same way as phosphorus did; however, because of the low boron-to-phosphorus ratio in plants, he did not believe that boron influenced the starch ⇌ glucose-1-phosphate equilibrium. On the other hand, others (Dugger et al., 1957; Scott, 1960) reported that boron inhibited the *in vitro* conversion of glucose-1-phosphate to starch catalyzed by starch phosphorylase. It was proposed that this enzyme was not inhibited in boron-deficient plants, and therefore starch accumulated. Scott (1960) proposed that "boron performs a protective function in plants, in that it prevents excessive polymerization of sugars at the site of sugar synthesis. . . ."

Augsten and Eichhorn (1976) proposed that starch synthesis is stimulated in boron-deficient plants by stimulation of the enzyme that converts glucose-6-phosphate to glucose-1-phosphate and by concomitant inhibition of the enzyme that synthesizes UDP-glucose from glucose-1-phosphate. Considering that the pathway for starch synthesis is now thought to involve the sugar nucleotide ADP-glucose, this proposal seems more feasible. The authors also pointed out that steric hindrance would make the formation of a complex between boric acid and glucose-1-phosphate impossible. Regardless of the exact mechanisms involved, it is generally agreed that starch does accumulate in boron-deficient plants.

17.2.6 Cellulose and Cell Wall Glucans

Cellulose and other wall polymers constitute the largest fraction of carbohydrates in plants. If boron has a role in carbohydrate metabolism, it would seem that the synthesis of these polysaccharides would be affected. Spurr (1957) reported that the structure of the cell wall of boron-deficient celery plants was altered. He concluded that boron affected not only the rate but also the actual process by which carbohydrates were incorporated into cell walls. During *in vitro* culture of parenchyma cells from tobacco pith, boron deficiency caused a two-fold increase in the cell wall content; the largest change was noted in the galactan fraction (Wilson, 1961).

^{3}H-*myo*-inositol was incorporated at a higher rate into the D-galacturonosyl and L-arabinosyl components of pectin material in pollen tubes when the germinating medium contained boron (Stanley and Loewus, 1964). The authors suggested that boron has a role in the synthesis of pectin materials of germinating pollen, possibly related to the synthesis of D-galacturonosyl units. Others have observed (1) an increase in cell wall materials in *Lemna minor* and *Ginkgo* pollen-derived tissue cultured *in vitro,* and (2) disorganized microfibrillar structure of the cell walls of boron-deficient sunflower plants (Dugger, 1983).

In oil palm (*Elaeis guineensis*) leaves, and sunflower (*Helianthus annuus*) and soybean (*Glycine max*) plants, the level of pectic and hemicellulose substances was higher when plants were subjected to boron deficiency (Yamanouchi, 1973; Rajaratnam and Lowry, 1974). Callose, a β-1,3 glucan, accumulated in bean (*Phaseolus vulgaris*) and cotton (*Gossypium hirsutum*) plants subjected to boron deficiency (Van de venter and Currier, 1977). In addition, Timashov (1977) observed that boron-deficient sunflower root tips incorporated more ^{14}C-glucose into the total cell wall fraction than control root tissue.

Morphological changes were observed at the ultrastructural level in cell walls and cellular organelles of boron-deficient roots (Kouchi and Kumazawa, 1976). They reported an increase in the pectin and hemicellulose fractions but a decrease in the cellulose fraction under boron deficiency. It was proposed that abnormal activity of the Golgi apparatus caused the alteration in cell wall synthesis. The hydroxyproline level in the cell wall fraction of sunflower roots was observed to increase when plants were cultured with a low boron level (Shive and Barnett, 1973); others, however, have not observed this change (Troitskaya et al., 1975).

When UDP[^{14}C]-glucose was supplied to cotton fibers attached to ovules cultured *in vitro*, a large fraction of the substrate was incorporated into cell wall glucans. Although most of these glucans were not β-1,4 linked, culturing the ovules with boron in the medium did cause more of the label to appear in the cellulose fraction (Dugger and Palmer, 1980).

Augsten and Eichhorn (1976) pointed out that the growth-promoting activity of phenylboric acid derivatives results from the complexing of organic borate with cell polysaccharides. They proposed that boron controls the stabilization of the cell by forming these complexes. A correlation was observed between the complexing ability of organoborate compounds and their growth-promoting ability: the BD_2 type links the polyhydroxy-containing molecules that exist between the fibrillar chains and thus reduces elasticity, whereas the BD type forms between boron, or phenylboric acid derivatives, and wall polysaccharides and does not reduce elasticity.

17.2.7 Phenols

Boron deficiency in plants generally leads to a buildup of phenolic compounds (see reviews by Dugger, 1973, 1983; Augsten and Eichhorn, 1976). However, there is no consistent agreement as to whether this response is a primary or secondary one. Boron deficiency in celery (*Apium* sp.) resulted in an induced fluorescence of the stem tissue, a response that Spurr (1952) suggested was due to an increased level of caffeic and chlorogenic acids. Others (Dear and Aronoff, 1965) suggested that the increase in caffeic acid caused tissue necrosis; caffeic acid or its metabolite caused a breakdown of conductive tissue in the plant, bringing about death of the tissue. Others have also observed the accumulation of fluorescent products, presumably phenols, in boron-deficient plants (Watanabe et al., 1961, 1964; Shkol'nik, 1974; Rajaratnam and Lowry, 1974).

This observed increase in the level of phenolic compounds in boron-deficient tissue has led Shkol'nik to propose that such accumulation is the "original cause of plant death during boron starvation" (Shkol'nik, 1974). In his hypothesis he pointed out that monocots have a lower requirement for boron than dicots and did not accumulate phenols to the same degree as boron-deficient dicots. The subsequent accumulation of auxin (IAA) in boron-deficient dicots resulted from the accumulation of phenols that inhibited IAA-oxidase. While others have observed hyperauxiny in boron-deficient tissue, some investigators (Bohnsack, 1974; Hirsch and Torrey, 1980; Hirsch et al., 1982) do not agree with the hypothesis that boron deficiency is accounted for by altered IAA levels (Section 17.3).

17.2.8 Lignin

Because in plants boron is thought to regulate the metabolic pathways leading to phenol synthesis, it would seem appropriate that the element be involved in the subsequent pathway leading to the synthesis of lignin. In fact, Lewis (1980a) proposed that boron's primary role in plants is in the biosynthesis of lignin and the differentiation of xylem tissue. In his interesting hypothesis, he points out that p-coumarate and ferulate, branch points in the biosynthesis of lignin, can be either directly reduced to two lignin precursors (p-coumaryl alcohol and coniferyl alcohol) or hydroxylated to the corresponding O-diphenols (caffeate and hydroxyferulate). With sufficient boron in plant tissues, the element complexes with these intermediates and subsequent methylation occurs in the process of lignin synthesis. However, under boron deficiency, utilization of the intermediates of the pathway is reduced, so they accumulate. This hypothesis generally agrees with published results (see reviews by Dugger, 1973, 1983; Augsten and Eichhorn, 1976). Other related compounds, such as leucoanthocyanins, flavonols, flavonones, and flavonol-3-glucosides, have been reported to accumulate in boron-deficient plant tissue (Rajaratnam and Lowry, 1974; Shkol'nik and Abysheva, 1975).

Lewis (1980a) discussed the interrelationships among boron, lignification, peroxidase enzymes, and auxin in plants. In addition, he pointed out the similarities by which wounding a vascular plant, microbial infection of the plant, and boron deficiency bring about changes in the levels of phenolic compounds and in IAA metabolism (Lewis, 1980b).

17.2.9 Boron in Enzymic Reactions

As pointed out in earlier reviews (Dugger, 1973, 1983; Augsten and Eichhorn, 1976), boron has not been shown to be a cofactor or specific component of any enzymic reaction. However, many reports, in both *in vivo* and *in vitro* systems, have shown that the element, or lack of it, alters the rates of many enzymic reactions. Reed (1947) reported that boron deficiency caused an increase in the activity of catechol oxidase. The products of this oxidation, quinones, resulted in the accumulation of phenols in boron-deficient celery plants.

It has also been shown that boron-deficient tissue had a more active polyphenol oxidase than the control tissue (MacVicar and Burris, 1948) and that boron was a competitive inhibitor of xanthine oxidase (Roush and Norris, 1950).

Later it was reported that the element also inhibited the oxidation of tyrosine (Yasunobu and Norris, 1957), was a competitive inhibitor of alcohol dehydrogenase (Roush and Gowdy, 1961), and was an inhibitor of alkaline phosphatase activity in milk and intestinal mucosa (Zittle and Della Monica, 1950).

More recent work has shown that catechol oxidase, polyphenol oxidase, tyrosinase, peroxidase, and IAA oxidase are inhibited by boron (Odhnoff, 1957; Parish, 1968, 1969; Shive and Barnett, 1973; Eichhorn and Augsten, 1974). In a technique developed by Alvarado and Sols (1957), boron was used to complex with fructose-6-phosphate, inhibiting phosphoglucose isomerase and making it possible to assay for phosphomannose isomerase. It has also been reported that boron will inhibit starch synthesis by inhibiting starch phosphorylase (Winfield, 1945; Dugger et al., 1957; Scott, 1960), and Loughman (1961) reported that phosphoglucomutase is inhibited by boron.

Dugger and Humphreys (1960) reported that boron stimulated the synthesis of UDP-glucose catalyzed by UDP-glucose pyrophosphorylase. The authors suggested that the latter result was probably because boron stimulated the conversion of UTP and glucose-1-phosphate to UDP-glucose. In cell-free preparations from pea seeds (*Pisum sativa*) and sugarcane seedlings (*Saccharum officiarum,*) which contain several of the enzymes leading to sucrose synthesis, it was observed that boron in the reaction mixture stimulated the synthesis of sucrose when the substrates UDP-glucose and fructose were provided (Dugger and Humphreys, 1960). Teare (1974) has reported that the level of UDP-glucose in the roots of boron-deficient bean plants is much lower than that in the roots of boron-sufficient plants. For the growth of cotton fibers on ovules cultured in a defined medium, boron is an absolute requirement. Without boron in the medium, growth was typified by a rapid proliferation of undifferentiated cells (callus) compared to the normal growth of ovule epidermal cells into the fiber cells when boron was provided. Boron deficiency in such a system reduced the activity of the pyrimidine pathway; however, there was no reduction in nucleic acid biosynthesis, since callus production occurred. It was suggested that UDP-glucose synthesis was reduced in boron-deficient tissue; therefore cellulose synthesis from this substrate was also inhibited (Birnbaum et al., 1977). The incorporation of ^{14}C-orotic acid into UDP-glucose by cotton fibers growing on ovules cultured at a suboptimal boron level was less than that in fibers cultured at an optimal level. A larger fraction of the ^{14}C-orotate was incorporated into the RNA fraction of the fibers grown at the lower level of boron (Wainwright et al., 1980).

Cresswell and Nelson (1973) have reported that the level of α-amylase activity in germinating seeds of *Themeda traindra* was higher if the germinating medium contained boron, and the level of β-amylase in leaves and stems of sugarcane plants was lower if the plants were grown under a low-boron regime (Zapata, 1973).

Lee and Aronoff (1967) demonstrated that the substrate 6-phosphogluconate complexed with boron, thereby inhibiting 6-phosphogluconate dehydrogenase. Without boron in the *in vitro* reaction, the oxidative decarboxylation of 6-phosphogluconate to ribulose-5-phosphate occurred; subsequent reactions of the pentose phosphate pathway were also stimulated. The increase in the amount of substrate metabolized via this pathway and the decrease in that metabolized via the TCA cycle resulted in the synthesis of more phenolic acids when boron was absent. Lee and Aronoff also proposed that the phenolic compounds may in turn complex with boron. This decreased the boron available for regulating 6-phosphogluconate dehydrogenase even more, thereby setting in motion the autocatalytic synthesis of phenolic acids. Shkol'nik and Il'inskaya (1975) reported an increase in the amount of this enzyme in several boron-deficient dicots. Lewis (1980a) proposed that as a result of the complexing of borate with caffeate, the caffeate ⇌ caffeoyl *o*-quinone reaction is prevented, and the substrate electron donor reduces the enzyme involved in the shuttle directly, thereby inhibiting catechol oxidase activity. As Lewis previously pointed out, the complexing of borate with intermediates of the lignin biosynthetic pathway regulates the conversion of p-coumarate to lignin precursors.

Smith and Johnson (1976) reported that borate inhibited alcohol dehydrogenase from yeast; borate as $B(OH)_4^-$ was competitive with NAD^+ rather than NADH. Although boron is not generally thought to be a constitutive part of any specific enzyme, acid phosphatase purified from sweet potato tubers contained boron along with manganese, magnesium, and silicon (Uehara *et al.*, 1974).

Other plant enzymes were reported to be regulated in some fashion by borate: (1) β-glucosidase activity in several dicot species was increased by boron deficiency (Maevskaya *et al.*, 1974, 1975, 1977), although not observed to increase in boron-deficient diatom cells (Smyth and Dugger, 1981); (2) the activities of ATPase and acid phosphatase were higher in boron-deficient tissue with a decrease in alkaline pyrophosphatase activity (Hinde and Finch, 1966) and reduced amino acid-dependent ATP pyrophosphate exchange (Hinde *et al.*, 1966); (3) RNAase increased due to boron deficiency (Chapman and Jackson, 1974; Sherstnev, 1974; Dave and Kannan, 1980); and (4) KCl-stimulated ATPase activity from *Z. mays* was decreased by boron deficiency (Pollard *et al.*, 1977).

17.2.10 Pollen Germination

In a previous review of the earlier literature, Gauch and Dugger (1954) referenced a large number of papers on this subject as well as on the flowering and fruiting of plants. Schmucker (1932) was one of the earlier investigators to report the several interesting effects resulting from the omission of boron from the artificial medium for germinating pollen: (1) the number of pollen grains that germinated was reduced; (2) pollen tubes that did form were short and malformed; and (3) a high proportion of the pollen tubes burst. Schmucker proposed that the ability of boron to complex with the hydroxyl-rich compounds of the pollen tube wall (cellulose and pectic materials) in some way regulated the synthesis of wall materials in the growth of pollen tubes. Reference has also been made to the possibility that boron may regulate the uptake of water by germinating pollen (Skok, 1958; Dugger, 1973, 1983; Augsten and Eichhorn, 1976). If so, regulation may be at the cellular membrane level rather than in the development of the pollen tube wall.

O'Kelley (1957) observed that the absorption of sugars and the uptake of oxygen by germinating pollen were stimulated by boron. However, he proposed that the effect of boron on pollen tube elongation was not related to the absorption of sugar or the increase in respiration.

Kumar and Hecht (1970) reported that the addition of boron to the styles of *Oenothera organensis*, which is self-incompatible, enhanced pollen tube growth. Utilization of endogenous sugar and a decrease in callose in the styles were both influenced by boron. Stanley and Loewus (1964) reported that boron played a "definite role in pectin synthesis in germinating pollen" of *Pyrus communis;* Samorodov and Golubinskii (1978) also observed that boron stimulated pear pollen tube growth, and Dickinson (1978) observed that boron stimulated growth of *Lilium longiflorum* pollen tubes.

Plant hormones such as IAA and gibberellic acid did not enhance pollen germination or pollen tube growth. Vaughan (1977) reported that the tassels on boron-deficient *Zea mays* plants did not produce viable pollen, and the silks of these plants were not receptive to compatible pollen.

17.2.11 Conclusions

The wide array of observed plant responses to boron deficiency indicates that the element is probably involved in a number of metabolic pathways or a cascade effect; therefore, regulating metabolic processes somewhat as has been proposed for plant hormones. Thus, regulation by boron occurs because of the ability of this element to complex with the large number of OH-rich compounds in plants and not because the element is involved directly in a specific metabolic reaction (Augsten and Eichhorn, 1976; Dugger, 1983).

17.3 Hormone Action

Although the exact role of boron has not been elucidated, there persists a common premise on which many investigators agree—boron is essential for the normal growth and functioning of apical meristems (a partial list includes Chapman and Jackson, 1974; Kouchi and Kumazawa, 1975a; Bohnsack and Albert, 1977; Cohen and Lepper, 1977; Hirsch and Torrey, 1980; Lovatt et al., 1981; also see reviews by Dugger, 1973, 1983; Augsten and Eichhorn, 1976). This led to experiments that examined the relationship between boron and plant growth regulators. The preponderance of this work has emphasized the relationship between boron and auxins for the following reasons:

1. Auxin, as IAA, is the principal growth hormone of higher plants.
2. Meristems have long been established as sources of IAA. The shoot apical meristem is the major site of free auxin formation from bound auxin precursors (free IAA is considered the metabolically active form).
3. Boron is essential for the normal growth and functioning of apical meristems.
4. Growth and geotropism have been shown to correlate with free auxin levels.
5. Normal growth and geotropic response are impaired in boron-deficient plants.
6. Adventitious root development is stimulated synergistically by the combined application of boron and auxin.

In addition to reduced growth at the root and shoot apices, boron-deficient broccoli (*Brassica oleracea*) and squash (*Cucurbita pepo*) plants exhibited a loss in geotropic response (Alexander, 1942; Bohnsack and Albert, 1977). Eaton (1940) suggested that boron-deficiency symptoms were those that would be expected in plants deficient in auxin. He successfully restored growth to boron-deficient cotton seedlings by spraying them with IAA.

A relationship between boron and auxin has also been demonstrated by other workers. Weiser (1959) reported that boron or auxin stimulated the rooting of two-node sections of *Clematis* sp., which was consistent with Eaton's hypothesis that boron-deficient plants suffered from hypoauxiny. Weiser also observed a synergistic effect when these two compounds were used together. Not only were more roots produced, but the rate of rooting was accelerated. Cuttings were soaked for 12 hours in boron (50 mg/liter) or in indolebutyric acid (IBA) (50 mg/liter) or in both, and then transferred to vermiculite. At the end of 32 days, cuttings treated with both boron and IBA had three times more roots than those treated with boron or IBA alone. It took an additional 22 days for the latter cuttings to produce as many roots. Synergism between IAA and boron on the rooting of *Phaseolus* cuttings was demonstrated by Gorter (1958). A boron concentration of 10^{-6} M was found to be optimal; and at this concentration, rooting increased proportionately with increased concentrations of auxin.

More recent work on adventitious root development in cuttings of *Phaseolus aureus* has provided some clarification of these earlier observations. No roots developed in cuttings of *P. aureus* without exogenous boron even when cuttings were soaked for 24 hours in 10^{-4} M IBA. Provision of boron at suboptimal concentrations severely limited both root number and root length (Middleton et al., 1978). Cuttings pretreated by soaking in 10^{-4} M IBA without added boron for 24 hours accumulated soluble sugars (sucrose, glucose, and fructose) in the hypocotyl, but no roots developed. IBA was shown to stimulate the translocation of ^{14}C-sucrose from leaves to the hypocotyls within 24 hours. Since root development does not occur without boron, the translocation of sucrose was independent of a root sink (Middleton et al., 1980).

Other workers have also demonstrated that auxin stimulates sugar translocation to shoot and root apices. Using *P. vulgaris* L., var. "Black Valentine" bean plants, Dyar and Webb (1961) demonstrated that auxin applied to the apical bud restored sugar translocation to a level greater than that in the boron-sufficient plants. Auxin applied to roots increased sugar translocation to buds and roots to rates 2 and 10 times greater, respectively, than those observed for these organs in boron-sufficient plants. However, it was shown that sucrose cannot fulfill the requirement for boron in normal root growth (Whittington, 1959; Neales, 1960; Albert and Wilson, 1961).

From this work it appears that in the absence of roots, auxin maintains the polar transport of carbohydrates to the cut end of the hypocotyl. It is not clear if auxin alone, boron alone, or both are required for initiation of meristematic activity during the early stages of formation of root primordia; but it is clear that boron is required for their subsequent organization and further development.

Shkol'nik et al. (1964) measured the level of IAA in corn and sunflower tissues. They found a decrease in the content of free auxin in the boron-deficient shoot and root apices that paralleled the decrease in growth. In both species there was an increase in the bound auxin content of the boron-deficient plants. Bound auxin is not usually found in apices, while free auxin is localized predominantly in this region. Young leaves provide the bound form as a precursor to free auxin that is formed in the apices. Shkol'nik's results suggest that this conversion is not occurring in boron-deficient apices. The authors investigated the possibility that the low level of free auxin in boron-deficient plants was due to its destruction by increased IAA oxidase activity. However, the activity of this enzyme was found to decrease under conditions of boron deprivation.

Crisp et al. (1976) found no difference in the content of IAA in the leaf margin of lettuce (*Letuca sativa*) plants grown in vermiculite at 0.001 mg boron per liter relative to control plants grown at 1.0 mg boron per liter, until the plants were 66 days old.

In a very thorough investigation, Smirnov et al. (1977) measured both free and bound IAA in roots and shoots of sunflower, bean, corn, and wheat (*Triticum vulgare*) hydroponically cultured in Knop's nutrient solution with boron (0.5 mg per liter) and without. Plants were harvested for analysis when initial symptoms of boron deficiency appeared for a given species. The results of this study clearly demonstrated that the influence of boron deprivation on IAA metabolism was different in the various tissues of the same plant and for the same tissues in the various species studied. Boron deficiency led to decreased levels of free IAA in shoots of sunflower, bean, and corn, but not wheat. In boron-deprived roots, free IAA decreased in corn but increased in bean and wheat. Under boron deprivation, bound IAA decreased in the shoots of sunflower, corn, and wheat, but increased in shoots of bean. Roots from the boron-deprived bean and wheat had significantly more bound IAA than the control plants.

Bohnsack and Albert (1977) demonstrated that the symptoms of boron deficiency can be induced by hyperauxiny in boron-sufficient squash plants (*Cucurbita pepo*) by the addition of 10^{-6} M IAA. Under conditions of boron deprivation or hyperauxiny, root elongation was inhibited, accompanied by swelling of the root tip; lateral root primordia were initiated near the root apex; a loss in geotropic response occurred; and IAA oxidase activity in apical and subapical root sections increased approximately 6–9 hours after boron was withheld, or 3–6 hours after IAA was added to the boron-sufficient medium. Thus, these authors concluded that boron-deficient plants may actually suffer from toxic effects of hyperauxiny. In addition, Bohnsack (1974) tested the possibility that boron-deficiency symptoms may actually be the result of toxic levels of ethylene synthesized in response to the state of hyperauxiny. Ethylene evolution in boron-deficient *C. pepo* roots decreased to 25% of the level in the boron-sufficient control within 24 hours after transfer of the 5-day-old plants to medium without added boron. On the other hand, boron-sufficient tissue of the same age treated with 10^{-6} M IAA showed a 63% increase in ethylene evolution over that in the boron-sufficient control. Thus, boron-deficiency symptoms are not due to increased ethylene biosynthesis, while auxin-induced boron deficiency-like symptoms may very well be a result of increased ethylene production. Boron deficiency and hyperauxiny both cause inhibition of root elongation, but apparently not for the same physiological reasons.

This conclusion was confirmed by Hirsch *et al.* (1982), who used electron microscopy to show that changes in sunflower root tissue characteristic of boron deprivation were not the same as those induced by IAA. The only apparent change in auxin-treated root cells was an increase in electron-dense material within the vacuoles; there was no increase in cell wall thickness, which is characteristic of boron-deficient roots. Consistent with the ultrastructural results, the levels of free IAA determined by use of a very sensitive radioimmunoassay were similar in both the boron-deprived and boron-sufficient root tips. Although boron deficiency and hyperauxiny elicit many similar responses in roots, it appears that they are not the same.

In light of the many and varied responses of auxin metabolism to boron deprivation, a single unifying hypothesis concerning a regulatory role for boron relative to the metabolism of this hormone seems unlikely.

Jackson, 1974). Thus, altered nucleic acid biosynthesis appears to be a primary effect of boron deprivation and not a secondary effect resulting from inhibition of either protein synthesis or respiration, or increased ribonuclease activity. Taken together, these results strongly suggest that the availability and/or utilization of purine or pyrimidine nucleotides is altered by boron deprivation.

Of particular interest is the observation that plants growing in the absence of boron can be protected from developing the symptoms of boron deficiency by the addition of a hydrolysate of yeast RNA to the nutrient solution (Shkol'nik and Soloviyova-Troitskaya, 1961). Several workers (Johnson and Albert, 1967; Johnson, 1971; Birnbaum *et al.*, 1974, 1977) tested the effects of both purine and pyrimidine bases on plant growth to determine which component(s) of the RNA hydrolysate afforded this protection. Both intact plants and isolated organs cultured in the absence of boron were protected to varying degrees from developing the symptoms of boron deficiency when pyrimidine bases were added to the culture medium.

This result was taken as evidence that the state of boron deficiency may, in fact, be a case of pyrimidine deprivation. Such an interpretation was supported by the observations that both barbituric acid and 6-azauracil, known inhibitors of pyrimidine biosynthesis (Handschumacher and Pasternack, 1958; Ross, 1964; Potvin *et al.*, 1978; Lovatt *et al.*, 1979), produced symptoms identical to those of boron deprivation (Johnson and Albert, 1967; Albert, 1968; Birnbaum *et al.*, 1977). The accumulating evidence prompted various investigators (Lewin, 1976; Birnbaum *et al.*, 1977; Wainwright *et al.*, 1980; Lovatt *et al.*, 1981) to propose that boron deficiency results in impaired *de novo* biosynthesis of pyrimidine nucleotides. Transferring 5-day-old squash plants (*Cucurbita pepo*) to boron-deficient nutrient solution resulted in cessation of root elongation within 18 hours. However, withholding boron for up to 30 hours did not result in either impaired *de novo* pyrimidine biosynthesis or a change in the sensitivity of the *de novo* pathway to regulation by end-product inhibition (Lovatt *et al.*, 1981). A shortage in available pyrimidine nucleotides could also result from an inability of boron-deficient plants to salvage or reutilize pyrimidine bases or nucleosides, or from an acceleration of pyrimidine catabolism. Boron deprivation had no significant effect on pyrimidine salvage; catabolism was slightly increased (Lovatt *et al.*, 1981). Whether a slight increase in catabolism would cause significant perturbations in the pool size of specific pyrimidine nucleotides is not known but emphasizes the need to determine the levels of the various pyrimidine nucleotides available during these two states of boron nutrition. These results argue against the hypothesis that boron-deficient plants are lacking in uridine nucleotides collectively but leave open the possibility that boron may be essential for maintaining adequate levels of one or more specific pyrimidine nucleotide species.

Two valid areas for further investigation are the possibility that boron de-

Kouchi and Kumazawa, 1975a; Bohnsack and Albert, 1977; Cohen and Lepper, 1977; Hirsch and Torrey, 1980; Krueger et al., 1979; Lovatt et al., 1981) and gymnosperms (Walker et al., 1955; Blaser et al., 1967). To date, some of the earliest effects of boron deficiency on plant metabolism reported in the literature were observed in root meristem tissues (Chapman and Jackson, 1974; Kouchi and Kumazawa, 1975a,b, 1976; Bohnsack and Albert, 1977; Cohen and Lepper, 1977; Krueger et al., 1979; Hirsch and Torrey, 1980; Lovatt et al., 1981). These effects were associated with the cessation of cell division and altered nucleic acid biosynthesis.

Determining the earliest effect of boron deficiency may be a possible key to elucidating its role in plant metabolism; the earliest symptom of boron deprivation is the one most likely to be associated with the primary role of boron in the metabolism of higher plants. The earliest effects of boron deprivation are on nucleic acid biosynthesis. The incorporation of [^{3}H]thymidine into the acid-insoluble fraction of root tips of intact 5-day-old squash plants (*Cucurbita pepo*) was decreased significantly after only 6 hours of boron deprivation and was reduced 66% when boron was withheld for an additional 6 hours. Decreased DNA synthesis correlated temporally with inhibition of root elongation (Krueger et al., 1979). Similar experiments using autoradiography revealed that the incorporation of [^{3}H]thymidine into root apical meristems ceased after 20 hours of boron deprivation (Cohen and Albert, 1974). When these plants were returned to a boron-sufficient medium for 12 hours, autoradiographs showed that their incorporation was indistinguishable from that of boron-sufficient root tips. These observations suggest that DNA synthesis is impaired under conditions of boron deficiency (Cohen and Albert, 1974).

Measurements of the incorporation of radiolabeled precursors into RNA provide evidence that RNA synthesis is also impaired when boron is withheld. For example, Sherstnev and Razumova (1965) reported decreased incorporation of [^{14}C]adenine into RNA of boron-deficient sunflower leaves and roots, while other workers have demonstrated that increased incorporation of [^{14}C]orotic acid (Wainwright et al., 1980) and [^{14}C]uridine (Chapman and Jackson, 1974) into RNA were early effects of boron deficiency in cotton ovules (*Gossypium hirsutum*) and mung bean (*Phaseolus aureus*) root apices, respectively.

Several workers (Albert, 1965; Jaweed and Scott, 1967; Johnson and Albert, 1967; Chapman and Jackson, 1974) have reported decreased RNA content in boron-deficient roots. When nucleic acid biosynthesis, protein biosynthesis, and respiration were measured in the same system, changes in nucleic acid biosynthesis preceded (1) a reduction in protein content (Johnson, 1971); (2) a decrease in the incorporation of [^{14}C]leucine into protein and a reduction in respiration (Krueger et al., 1979); and (3) a decrease in RNA content and the observed increase in RNAase activity that accompanies boron deprivation (Chapman and

stimulated ATPase, a membrane-bound enzyme, in boron-deficient Z. *mays* root tissue. This activity could also be restored by the 1-hour pretreatment with boron. The authors suggested that the results support the view that boron is required for "efficient operation of the membrane transport system" and that the influence is direct, rather than an effect of the accumulation of metabolic intermediates.

In more direct studies on the influence of boron on membrane functions, Tanada (1974, 1978, 1982) reported that boron affected the bioelectric field potential of excised mung bean (*Phaseolus aureus*) hypocotyls from seedlings grown in the dark and irradiated with red light. The author suggested that this observed effect was induced by a modification of some membrane component when boron was present. Further studies also showed that boron stimulated the translocation of fluorescein in mung bean hypocotyl sections following gravitational or red irradiation stimulation. In addition, boron was necessary for the induction of the delaying action that irradiation at 710 nm has on the nyctinastic closing of *Albizzia julibrissin* pinnules. Tanada has suggested that boron was required to stabilize the positive electrostatic charge in the plasma membrane that was generated by the action of gravity and phytochrome.

It has been shown in diatom cells (*Cylindrotheca fusiformis*) (Smyth and Dugger, 1980) that a change in the influx and efflux of ^{86}Rb was an early effect of boron deficiency. After 5 hours of culture in boron-deficient medium, the ^{86}Rb influx rate was reduced 20%. However, there was an accumulation of ^{86}Rb by the boron-deficient diatoms brought about by a lower efflux rate in these cells. After 24 hours the boron-deficient cells had 30% more ^{86}Rb than the control cells. These changes could not be reversed by preincubation of the diatom cells with 50 μM boron for 3 hours. This observed effect of boron deficiency on membrane function in diatoms preceded the inhibition of cell division by 10 hours.

Roth-Bejerano and Itai (1981) reported that boron enhanced KCl-induced stomatal opening in epidermal strips of *Commelina communis*. They propose that boron combines with membrane-polyhydroxy compounds, influencing the potassium influx or efflux through guard cell membranes.

These observations led many researchers (Pollard *et al.*, 1977; Hirsch and Torrey, 1980; Smyth and Dugger, 1980; Roth-Bejerano and Itai, 1981; Hirsch *et al.*, 1982) to conclude that boron may play an important role in membrane transport or in maintaining membrane integrity.

17.5 Nucleic Acid Biosynthesis

The dependence on boron for normal meristematic activity is well documented for both angiosperms (a partial list includes Chapman and Jackson, 1974;

17.4 Membrane Structure and Function

The hypothesis that was presented earlier to explain the observed effect of boron on sugar translocation in plants suggested that boron was perhaps associated with cell membranes. As a constituent of membranes, the element reacted chemically with the sugar molecules, and the resulting complex was transported across the membrane; a subsequent reaction on the inside of the membrane released the sugar into the cytoplasm (Gauch and Dugger, 1953). No experimental evidence was provided to substantiate the presence of boron in cell membranes. However, Shkol'nik and Kopman (1970) reported that the apical meristems of boron-deficient sunflower plants contained a lower level of phospholipids than control tissues. They proposed that since phospholipids occur in intracellular membranes, boron was apparently involved in the ultrastructural organization of meristem cells in roots. Hirsch and Torrey (1980) observed ultrastructural changes in the cellular membranes of sunflower roots after 6 hours in a boron-deficient medium.

Dave and Kannan (1980) observed that RNAase was enhanced in *Phaseolus vulgaris* under boron deficiency. They suggested that since RNAase is located in cellular membranes, such an enhancement indicated an alteration or disruption of membrane permeability.

Robertson and Loughman (1973, 1974a,b) carried out a series of experiments to study the effects of boron deficiency on inorganic ion flux by *Vicia faba* roots. In addition to reducing root elongation, removing boron from the growth medium reduced the uptake of ^{86}Rb. In fact, the terminal centimeter of roots showed a marked reduction in ^{86}Rb uptake. Older root tissue (2–3 cm from tip) from boron-deficient plants showed a marked increase in absorption when compared to the control root tissue. Adding boron to the growth medium before the start of the ^{86}Rb absorption period did not restore the capacity to absorb ^{86}Rb to the terminal centimeter of root tissue. In a similar study measuring the uptake of ^{32}P-labeled phosphate, these authors showed that boron-deficient root tissue exhibited a reduction in ^{32}P uptake, as had also been observed for the ^{86}Rb studies. However, the effect of boron deficiency on phosphate absorption was reversed by adding boron 1–2 hours before the ^{32}P-phosphate uptake was initiated. The authors suggested that the observed effect may relate to a rapid restoration of the phosphate carrier system in the membranes, with boron involved as a component of the system. A later study from the same laboratory (Pollard et al., 1977) showed that in both *Vicia faba* and *Zea mays*, the rate of absorption of ^{32}P-phosphate was decreased by boron deficiency; but it was rapidly restored to approximately the same rate of the control by the addition of 10^{-5} M boron 1 hour before the absorption study began. In *Z. mays*, the boron-deficient reduction of Cl^- and Rb^+ ion absorption could also be restored by the 1-hour pretreatment with boron. There was a one-third reduction in the activity of KCl-

privation leads to (1) an impairment of the utilization of uridine nucleotides for the synthesis of cytidine or thymidine nucleotides or (2) an alteration in the interconversion of nucleotides for the provision of adequate levels of pyrimidine mono-, di-, and triphosphates. Preliminary measurements have been made using HPLC to determine the pool size of available nucleotides in root apical meristems excised from boron-sufficient and -deficient squash roots. Within 24 hours after 5-day-old *C. pepo* plants were transferred to boron-deficient nutrient solution, the quantity of available UMP and UDP was less, while that of UTP was greater in the boron-deficient root apices than in the boron-sufficient controls. The pool of available CMP also was less. The latter observation is consistent with the loss of available CMP in 9-day-old chick pea roots (*Cicer arietinum*) deprived of boron for 6 days (Mamedova and Rasulov, 1977).

The shift in availability of specific pyrimidine nucleotides could be due to two possibilities that remain to be investigated: (1) that the interconversions of UMP, UDP, UTP, and UDP-glucose, or synthesis of ribonucleotides or deoxyribonucleotides of cytidine or thymidine are impaired under conditions of boron deficiency, or (2) that the utilization of a particular pyrimidine nucleotide or pyrimidine nucleotide-sugar in a specific metabolic process might be impaired under conditions of boron deficiency.

The hypothesis that boron has a fundamental role in pyrimidine metabolism, through either of the two possibilities listed previously, unifies the seemingly separate roles of boron in two very different systems: the first includes dividing cells without concomitant maturation, such as meristematic cells of root and shoot apices (Chapman and Jackson, 1974; Kouchi and Kumazawa, 1975a; Bohnsack and Albert, 1977; Cohen and Lepper, 1977; Hirsch and Torrey, 1980; Lovatt *et al.*, 1981) and DNA repair in the generative cell of pollen grains (Jackson and Linskens, 1979). The second comprises elongating cells that do not undergo cell division, such as *in vitro* cotton fiber development (Birnbaum *et al.*, 1974, 1977) and pollen tube growth (Stanley and Loewus, 1964; Yih *et al.*, 1966; Vaughn, 1977). Boron deprivation has also been shown to have a marked influence on carbohydrate metabolism (Section 17.2). Dugger and co-workers (Dugger and Humphreys, 1960; Birnbaum *et al.*, 1977; Wainwright *et al.*, 1980) have provided evidence that suggests that boron deprivation results in reduced UDP-glucose formation. This would result in reduced sucrose synthesis (Dugger and Humphreys, 1960), increased starch formation (Dugger *et al.*, 1957; Scott, 1960) (Section 17.2.5), and interference with normal cell wall formation (Kouchi and Kumazawa, 1975b, 1976; Dugger and Palmer, 1980) (Section 17.2.6), all of which typify boron deficiency. Pyrimidine nucleotides are involved directly in the biosynthesis of UDP-glucose and other nucleotide sugars. Thus, the hypothesis that boron is essential to the maintenance of adequate levels of one or more specific pyrimidine nucleotide species, or to utilization of a particular pyrimidine nucleotide in a specific metabolic process, also unifies the seemingly disparate roles of boron in nucleic acid biosynthesis and carbohydrate metabolism.

17.6 Summary

Formation of electron-deficient molecules or bridge bonds, chemical features more or less unique to boron, have been considered the possible basis for the roles that boron has in biology. Most major taxa have been screened for boron-requiring organisms, but the only evidence demonstrating that boron is essential to a taxon is limited to the Tracheophyta and a number of taxonomically distinct marine and freshwater phytoplankton species. Although there is universal agreement that boron is essential to vascular plants, over 60 years have passed since boron was demonstrated to be essential to this group, and a specific biochemical role for boron in their metabolism still remains to be elucidated. There are, however, a number of metabolic processes in vascular plants that have been shown repeatedly to be influenced by boron nutrition: carbohydrate metabolism, hormone action, membrane function, and nucleic acid synthesis.

Several researchers have interpreted the wide array of observed plant responses to boron deficiency as evidence that the element is involved in a number of metabolic pathways or in a "cascade" effect. They contend that metabolic regulation occurs because boron complexes with a large number of hydroxyl-rich compounds and not because the element is involved directly in a specific metabolic reaction (Augsten and Eichhorn, 1976; Dugger, 1983). Other researchers argue that boron does have a single biochemical role in the metabolism of vascular plants and that the primary role of boron in plant metabolism is most likely to be associated with the earliest symptom of boron deprivation occurring at the molecular level (Johnson and Albert, 1967; Krueger et al., 1979; Lovatt et al., 1981).

Despite the considerable amount of research designed to determine the role(s) of boron in vascular plant metabolism, definitive evidence to support a specific role has not been obtained. Most investigations, because of the lack of a radioactive or heavy isotope of boron, have been comparative studies employing boron-sufficient and boron-deficient tissues. Some of this research is compromised by failure to establish true zero boron conditions and by failure of the researchers to distinguish between early events resulting from a primary effect of boron deprivation and the later events, which are more likely to be secondary effects. Thus the controversy surrounding the role of boron in vascular plant metabolism still awaits resolution.

With the increased use of irrigated land for food and forage crop production, a concomitant increase in the instances of boron toxicity must be expected.

Boron complexes with a number of enzymes with resulting deleterious effects to plants and the animals that consume them. Finally, the unique biochemical properties of boron are being exploited for their therapeutic potential, principally through the use of organoborate compounds in chemotherapy.

References

Albert, L. S., 1965. Ribonucleic acid content, boron deficiency symptoms and elongation of tomato root tips, *Plant Physiol.* 40:649–652.

Albert, L. S., and Wilson, C. M., 1961. Effect of boron on elongation of tomato root tips, *Plant Physiol.* 36:244–251.

Alexander, T. R., 1942. Anatomical and physiological responses of squash to various levels of boron supply, *Bot. Gaz.* 103:475–491.

Alvarado, F., and Sols, A., 1957. Borate and phosphoglucose isomerase in the assay of phosphomannose isomerase, *Biochim. Biophys. Acta* 25:75–77.

Anderson, G. R., and Jordan, J. V., 1961. Boron: a non-essential growth factor for *Azotobacter chroococcum*, *Soil Sci.* 92:113–116.

Arnon, D. I., and Stout, P. R., 1939. The essentiality of certain elements in minute quantities for plants with special reference to copper, *Plant Physiol.* 14:371–375.

Augsten, H., and Eichhorn, B., 1976. Biochemistry and physiology of the effect of boron in plants, *Biol. Rundsch.* 14:268–285.

Birnbaum, E. H., Beasley, C. A., and Dugger, W. M., 1974. Boron deficiency in unfertilized cotton (*Gossypium hirsutum*) ovules grown *in vitro*, *Plant Physiol.* 54:931–935.

Birnbaum, E. H., Dugger, W. M., and Beasley, C. A., 1977. Interaction of boron with components of nucleic acid metabolism in cotton ovules cultured *in vitro*, *Plant Physiol.* 59:1034–1038.

Blaser, H. W., Marr, C., and Takahashi, D., 1967. Anatomy of boron-deficient *Thuja plicata*, *Am. J. Bot.* 54:1107–1113.

Bohnsack, C. W., 1974. Early effects of boron deficiency on indoleacetic acid oxidase levels, peroxidase localization and ethylene evolution of root tips of squash, Ph.D. thesis, University of Rhode Island, Kingston.

Bohnsack, C. W., and Albert, L. S., 1977. Early effects of boron deficiency on indoleacetic acid oxidase levels of squash root tips, *Plant Physiol.* 59:1047–1050.

Bowen, J. E., and Gauch, H. G., 1965. Essentiality of boron for *Dryopteris dentata* and *Selaginella apoda*, *Am. Fern Journal* 55:67–73.

Bowen, J. E., and Gauch, H. G., 1966. Non-essentiality of boron in fungi and the nature of its toxicity, *Plant Physiol.* 41:319–324.

Bowen, J. E., Gauch, H. G., Krauss, R. W., and Galloway, R. A., 1965. The nonessentiality of boron for *Chlorella*, *J. Phycol.* 1:151–154.

Chapman, K. S. R., and Jackson, J. F., 1974. Increased RNA labelling in boron-deficient root-tip segments, *Phytochemistry* 13:1311–1318.

Cohen, M. S., and Albert, L. S., 1974. Autoradiographic examination of meristems of intact boron-deficient squash roots treated with tritiated thymidine, *Plant Physiol.* 54:766–768.

Cohen, M. S., and Lepper, R., 1977. Effect of boron on cell elongation and division in squash roots, *Plant Physiol.* 59:884–887.

Coke, L., and Whittington, W. J., 1968. The role of boron in plant growth. IV. Interrelationships between boron and indol-3yl-acetic acid in the metabolism of bean radicles, *J. Exp. Bot.* 19:295–308.

Cresswell, C. F., and Nelson, H., 1973. The influence of boron on the RNA level, α amylase activity, and level of sugars in germinating *Themeda triandra* Forsk seed, *Ann. Bot.* 37:427–438.

Crisp, P., Collier, G. F., and Thomas, T. H., 1976. The effect of boron on tip-burn and auxin activity in lettuce, *Scientia Hort.* 5:215–226.

Dave, I. C., and Kannan, S., 1980. Boron deficiency and its associated enhancement of RNAase activity in bean plants, *Z. Pflanzenphysiol.* 97:261–263.

Davis, A. R., Marloth, R. H., and Bishop, C. J., 1928. The inorganic nutrition of the fungi: I. The relation of calcium and boron to growth and spore formation, *Phytopathology* 18:949.

Davis, D., and Addo, P. E. A., 1957. Boron toxicity in *Lycopersicon esculentum* L., *Plant Physiol.* 32(Suppl.):23.

Dear, J., and Aronoff, S., 1965. Relative kinetics of chlorogenic and caffeic acids during the onset of boron deficiency in sunflower, *Plant Physiol.* 40:458–459.

Dickinson, D. B., 1978. Influence of borate and pentaerythritol concentrations on germination and tube growth of *Lilium longiflorum* pollen, *J. Am. Soc. Hort. Sci.* 103:413–416.

Dugger, W. M., 1973. Functional aspects of boron in plants, in *Trace Elements in the Environment*, E. Kothny (ed.), Advances in Chemistry Series, Vol. 123, pp. 112–129.

Dugger, W. M., 1983. Boron in plant metabolism, in *Inorganic Plant Nutrition, Encyclopedia of Plant Physiology*, Vol. 15, A. Läuchli and R. L. Bieleski (eds.), Springer-Verlag, Heidelberg, pp. 626–650.

Dugger, W. M., Jr., and Humphreys, T. E., 1960. Influence of boron on enzymatic reactions associated with biosynthesis of sucrose, *Plant Physiol.* 35:523–530.

Dugger, W. M., Humphreys, T. E., and Calhoun, B., 1957. The influence of boron on starch phosphorylase and its significance in translocation of sugars in plants, *Plant Physiol.* 32:364–370.

Dugger, W. M., and Palmer, R. L., 1980. Effect of boron on incorporation of glucose from UDPGlucose into cotton fibers grown *in vitro*, *Plant Physiol.* 65:266–273.

Dutta, J. J., and McIlrath, W. J., 1964. Effects of boron on growth and lignification in sunflower tissue and organ cultures, *Bot. Gaz.* 125:89–96.

Dyar, J. J., and Webb, K. L., 1961. A relationship between boron and auxin in ^{14}C translocation in bean plants, *Plant Physiol.* 36:672–676.

Eaton, F. M., 1940. Interrelations in the effects of boron and indoleacetic acid on plant growth, *Bot. Gaz.* 101:700–705.

Eichhorn, V. M., and Augsten, H., 1974. Influence of boron on the populations of *Wolffia arrhiza* (L.) Wimm. of different ages cultivated in chemostat, *Biochem. Physiol. Pflanzen.* 165:371–385.

El-Sheikh, A. M., Ulrich, A., Awad, S. K., and Mawardy, A. E., 1971. Boron tolerance of squash, melon, cucumber, and corn, *J. Am. Soc. Hort. Sci.* 96:53–57.

Eyster, C., 1952. Necessity of boron for *Nostoc muscorum*, *Nature* 170:755.

Foster, J. W., 1949. *Chemical Activities of Fungi*, Academic Press, New York.

Gauch, H. G., and Dugger, W. M., Jr., 1953. The role of boron in the translocation of sucrose, *Plant Physiol.* 28:457–466.

Gauch, H. G., and Dugger, W. M., Jr., 1954. The Physiological Action of Boron in Higher Plants: A Review and Interpretation, Univ. of Maryland Agr. Exp. Stat. Bulletin #A-80 (Technical):1–43.

Gerloff, G. C., 1968. The comparative boron nutrition of several green and blue-green algae, *Physiol. Plant.* 21:369–377.

Gorter, C. J., 1958. Synergism of indole and indole-3-acetic acid in root production of *Phaseolus* cuttings, *Physiol. Plant.* 11:1–9.

Handschumacher, R. E., and Pasternack, C. A., 1958. Inhibition of orotidylic acid decarboxylase, a primary site of carcinostasis by 6-azauracil, *Biochim. Biophys. Acta* 30:451–452.

Hinde, R. W., and Finch, L. R., 1966. The activities of phosphatases, pyrophosphatases and adenosine triphosphatases from normal and boron-deficient bean roots *Phytochemistry* 5:619–623.

Smyth, D. A., and Dugger, W. M., 1980. Effects of boron deficiency on ^{86}Rb uptake and photosynthesis in the diatom *Cylindrotheca fusiformis, Plant Physiol.* 66:692–695.

Smyth, D. A., and Dugger, W. M., 1981. Cellular changes during boron-deficient culture of the diatom *Cylindrotheca fusiformis, Physiol. Plant.* 51:111–117.

Soloway, A. H., 1964. Boron compounds in cancer therapy, in *Progress in Boron Chemistry*, Vol. 1, H. Steinberg and A. L. McCloskey (eds.), Macmillan, New York, pp. 203–234.

Sommer, A. L., and Lipman, C. B., 1926. Evidence of the indispensable nature of zinc and boron for higher green plants, *Plant Physiol.* 1:231–249.

Spurr, A. R., 1952. Fluorescence in ultraviolet light in the study of boron deficiency in celery, *Science* 116:421–423.

Spurr, A. R., 1957. The effect of boron on cell-wall structure in celery, *Am. J. Bot.* 44:637–650.

Stanley, R. G., and Loewus, F. A., 1964, Boron and myo-inositol in pollen pectin biosynthesis, in *Pollen Physiology and Fertilization*, H. F. Linskens (ed.), North-Holland, Amsterdam, pp. 128–136.

Subba Rao, D. V., 1981. Effect of boron on primary production of nanoplankton, *Can. J. Fish. Aquat. Sci.* 38:52–58.

Tanada, T., 1974. Boron-induced bioelectric field change in mung bean hypocotyl, *Plant Physiol.* 53:775–776.

Tanada, T., 1978. Boron—key element in the actions of phytochrome and gravity? *Planta* 143:109–111.

Tanada, T., 1982. Role of boron in the far-red delay of nyctinastic closure of *Albizzia* pinnules, *Plant Physiol.* 70:320–321.

Teare, I. D., 1974. Boron nutrition and acid-soluble phosphorus compounds in bean roots, *HortScience* 9:236–238.

Thellier, M., Duval, Y., and DeMarty, M., 1979. Borate exchanges of *Lemna minor* L. as studied with the help of the enriched stable isotopes of a (n, α) nuclear reaction, *Plant Physiol.* 63:283–288.

Timashov, N. D., 1977. Effect of a boron deficiency on the incorporation of glucose-^{14}C into fractions of polysaccharides of sunflower organ cell walls, *Vestn. Khar'k. Univ.* 158:36–38.

Troitskaya, E. A., Dranik, L. I., and Shkol'nik, M. Y., 1970. Phenol composition of sunflower leaves during boron deficiency, *Soviet Plant Physiol.* 18:328–330.

Troitskaya, E. A., Maevskaya, A. N., and Témp, G. A., 1975. Hydroxyproline content in cell walls of plants with different boron requirements, *Soviet Plant Physiol.* 22:854–857.

Uehara, K., Fujimoto, S., and Taniguchi, T., 1974. Studies on violet-colored acid phosphatase of sweet potato. I. Purification and some physical properties, *J. Biochem.* 75:627–638.

Van de venter, H. A., and Currier, H. B., 1977. The effect of boron deficiency on callose formation and ^{14}C translocation in bean (*Phaseolus vulgaris* L.) and cotton (*Gossypium hirsutum* L.), *Am. J. Bot.* 64:861–865.

Vaughan, A. K. F., 1977. The relation between the concentration of boron in the reproductive and vegetative organs of maize plants and their development, *Rhod. J. Agric. Res.* 15:163–170.

Wainwright, I. M., Palmer, R. L., and Dugger, W. M., 1980. The pyrimidine pathway in boron-deficient cotton fiber, *Plant Physiol.* 65:893–896.

Walker, R. B., Gessel, S. P., and Haddock, P. G., 1955. Greenhouse studies in mineral requirements of conifers: Western red cedar, *Forest Sci.* 1:51–60.

Warington, K., 1923. The effect of boric acid and borax on the broad bean and certain other plants, *Ann. Bot.* 37:629–672.

Watanabe, R., Chorney, W., Skok, J., and Wender, S. H., 1964. Effect of boron deficiency on polyphenol production in the sunflower, *Phytochemistry* 3:391–394.

Watanabe, R., McIlrath, W. J., Skok, J., Chorney, W., and Wender, S. H., 1961. Accumulating of scopoletin glucoside in boron deficient tobacco leaves, *Arch. Biochim. Biophys.* 94:241–243.

Weiser, C. J., 1959. Effect of boron on the rooting of clematis cuttings, *Nature* 183:559–560.

Robertson, G. A., and Loughman, B. C., 1974a. Response to boron deficiency: A comparison with responses produced by chemical methods of retarding root elongation, *New Phytol.* 73:821–832.

Robertson, G. A., and Loughman, B. C., 1974b. Reversible effects of boron on the absorption and incorporation of phosphate in *Vicia faba* L., *New Phytol.* 73:291–298.

Ross, C., 1964. Influence of 6-azauracil on pyrimidine metabolism of cocklebur leaf discs, *Biochem. Biophys. Acta* 87:564–573.

Roth-Bejerano, N., and Itai, C., 1981. Effect of boron on stomatal opening in epidermal strips of *Commelina communis, Physiol. Plant.* 52:302–304.

Roush, A., and Norris, E. R., 1950. The inhibition of xanthine oxidase by borates, *Arch. Biochem.* 29:344–347.

Roush, A. H., and Gowdy, B. B., 1961. The inhibition of yeast alcohol dehydrogenase by borate, *Biochim. Biophys. Acta* 52:200–202.

Samorodov, V. N., and Golubinskii, I. N., 1978. Stimulation of pear pollen germination under the effect of physiologically active substances, *Ukr. Bot. Zh.* 35:401–406.

Schinazi, R. F., and Prusoff, W. H., 1978. Synthesis of 5-dihydroxyboryl-2'-deoxyuridine and related boron-containing pyrimidines, *Am. Chem. Soc.* 176(Suppl.):Abstr. 38.

Schmucker, T., 1932. Bor als physiologisch entischeidendes element, *Naturwissenchaften* 20:839.

Scott, E. G., 1960. Effect of supra-optimal boron levels on respiration and carbohydrate metabolism of *Helianthus annuus, Plant Physiol.* 35:653–661.

Sherstnev, E. A., 1974. Protein and nucleic acid metabolism in plants during a boron deficiency, *Biol. Rol Mikroelem. IKH Primen. Sel'sk. KOHZ. MED.* 263–272.

Sherstnev, E. A., and Razumova, M. V., 1965. The effect of boron deficiency on the ribonuclease activity in young leaves of sunflower plants, *Agrochimica* 9:348–350.

Shive, J. B., Jr., and Barnett, N. M., 1973. Boron deficiency effects of peroxidase, hydroxyproline, and boron in cells walls and cytoplasm of *Helianthus annuus* L. hypoctyls, *Plant and Cell Physiol.* 14:573–583.

Shkol'nik, M. Y., 1974. General conception of the physiological role of boron in plants, *Soviet Plant Physiol.* 21:140–150.

Shkol'nik, M. Y., and Abysheva, L. N., 1975. Effect of boron deficiency on the level of the growth inhibitor flavonol-3-glycoside and other flavonoids in tomatoes, *Fiziol. Biokhim. Kul't. Rast.* 7:291–297.

Shkol'nik, M. Y., and Il'inskaya, N. L., 1975. Effect of a boron deficiency on activity of glucose-6-phosphate dehydrogenase in plants with different boron requirements, *Soviet Plant Physiol.* 22:695–699.

Shkol'nik, M. Y., and Kopman, I. V., 1970. Effect of boron on the phospholipid level in the sunflower, and the possible role of this element in the structural organization of a cell, *Tr. Bot. Inst., Akad. Nauk USSR* 20:108–113.

Shkol'nik, M. Y., Krupnikova, T. A., and Dmitrieva, N. N., 1964. Influence of boron deficiency on some aspects of auxin metabolism in the sunflower and corn, *Soviet Plant Physiol.* 11:164–169.

Shkol'nik, M. Y., and Soloviyova-Troitskaya, E. A., 1961. The physiological role of boron: I. Treatment of boron deficiency by nucleic acid, *Bot. Zh.* 46:161–173 *Biol. Abstr.* (1962) 39:3694.

Sisler, E. D., Dugger, W. M., Jr., and Gauch, H. G., 1956. The role of boron in the translocation of organic compounds in plants, *Plant Physiol.* 31:11–17.

Skok, J., 1958. The role of boron in the plant cell, in *Trace Elements*, C. A. Lamb, O. G. Bentley, and J. M. Beattie (eds.), Academic Press, New York, pp. 227–243.

Smirnov, Y. S., Krupnikova, T. A., and Shkol'nik, M. Y., 1977. Content of IAA in plants with different sensitivity to boron deficits, *Soviet Plant Physiol.* 24:270–276.

Smith, K. W., and Johnson, S. L., 1976. Borate inhibition of yeast alcohol dehydrogenase, *Biochemistry* 15:560–564.

Ludbrook, W. V., 1942. Effects of various concentrations of boron on the growth of pine seedlings, *J. Aust. Inst. Agr. Sci.* 8:112–114.

MacVicar, R., and Burris, R. H., 1948. The relation of boron to certain plant oxidases, *Arch. Biochem.* 17:31–39.

Maevskaya, A. N., Troitskaya, E. A., and Témp, G. A., 1974. Effect of boron deficits on activity of β-glucosidase in sunflower, *Soviet Plant Physiol.* 21:505–507.

Maevskaya, A. N., Troitskaya, E. A., Témp, G. A., and Andreeva, E. N., 1975. Activity of β-glucosidase in plants with different boron requirements, *Soviet Plant Physiol.* 22:476–480.

Maevskaya, A. N., Troitskaya, E. A., and Yakovleva, N. S., 1977. Effect of boron starvation on activity of β-glucosidase in plants of the families leguminosae and gramineae, *Soviet Plant Physiol.* 23:1073–1076.

Mamedova, T. K., and Rasulov, F. A., 1977. Changes of free nucleotides in chick pea roots during growth under conditions of boron deficiency, *Soviet Plant Physiol.* 24:521–524.

Martini, F., and Thellier, M., 1975. Study with the help of the ^{10}B (n, α) ^{7}Li nuclear reaction on the redistribution of boron in white clover after foliar application, *Newslett. Applic. Nuclear Meth. Biol. Agr.* 4:26–29.

McCoy, R. H., 1967, Dietary requirements of the rat, in *The Rat in Laboratory Investigation*, R. J. Farris and J. Q. Griffeth (eds.), Hafner Publishing Co., New York, pp. 68–103.

McIlrath, W. J., and Skok, J., 1958. Boron requirement of *Chlorella vulgaris*, *Bot. Gaz.* 119:231–233.

Mengel, K., and Kirby, E. A. (eds.), 1978. *Principles of Plant Nutrition*, International Potash Institute, Bern, Switzerland, pp. 483–494.

Middleton, W., Jarvis, B. C., and Booth, A., 1978. The boron requirement for root development in stem cuttings of *Phaseolus aureus* Roxb., *New Phytol.* 81:287–297.

Middleton, W., Jarvis, B. C., and Booth, A., 1980. The role of leaves in auxin and boron-dependent rooting of stem cuttings of *Phaseolus aureus* Roxb., *New Phytol.* 84:251–259.

Mitchell, J. W., Dugger, W. M., Jr., and Gauch, H. G., 1953. Increased translocation of plant-growth-modifying substances due to application of boron, *Science* 118:354–355.

Neales, T. F., 1960. Some effects of boron on root growth, *Aust. J. Biol. Sci.* 13:232–248.

Odhnoff, C., 1957. Boron deficiency and growth, *Physiol. Plant.* 10:984–1000.

Oertli, J. J., and Richardson, W. F., 1970. The mechanism of boron immobility in plants, *Physiol. Plant.* 23:108–116.

O'Kelley, J. C., 1957. Boron effects on growth, oxygen uptake and sugar absorption by germinating pollen, *Am. J. Bot.* 44:239–244.

Parish, R. W., 1968. *In vitro* studies on the relations between boron and peroxidase, *Enzymologia* 35:239–252.

Parish, R. W., 1969. Studies on the effect of calcium and boron on peroxidases of plant cell walls, *Z. Pflanzenphysiol.* 60:211–216.

Perkins, H. J., and Aronoff, S., 1956. Identification of the blue fluorescent compounds in boron-deficient plants, *Arch. Biochem. Biophys.* 64:506–507.

Pollard, A. S., Parr, A. J., and Loughman, B. C., 1977. Boron in relation to membrane function in higher plants, *Exp. Bot.* 28:831–841.

Potvin, B. W., Stern, H. J., May, S. R., Lam, G. F., and Krooth, R. S., 1978. Inhibition by barbituric acid and its derivatives of the enzymes in rat brain which participate in the synthesis of pyrimidine nucleotides, *Biochem. Pharmac.* 27:655–665.

Rajaratnam, J. A., and Lowry, J. B., 1974. The role of boron in the oil-palm (*Elaeis guineensis*), *Ann. Bot.* 38:193–200.

Raven, J. A., 1980. Short- and long-distance transport of boric acid in plants, *New Phytol.* 84:231–249.

Reed, H. S., 1947. A physiological study of boron deficiency in plants, *Hilgardia* 17:377–411.

Robertson, G. A., and Loughman, B. C., 1973. Rubidium uptake and boron deficiency in *Vicia faba* L., *J. Exp. Bot.* 24:1046–1052.

Hinde, R. W., Finch, L. R., and Cory, S., 1966. Amino acid-dependent ATP-pyrophosphate exchange in normal and boron deficient bean roots, *Phytochemistry* 5:609–618.

Hirsch, A. M., Pengelly, W. L., and Torrey, J. G., 1982. Endogenous IAA levels in boron-deficient and control root tips of sunflower, *Bot. Gaz.* 143:15–19.

Hirsch, A. M., and Torrey, J. G., 1980. Ultrastructural changes in sunflower root cells in relation to boron deficiency and added auxin, *Can. J. Bot.* 58:856–866.

Hove, E., Elvehjem, C. A., and Hart, E. B., 1939. Boron in animal nutrition, *Am. J. Physiol.* 127:689–701.

Jackson, J. F., and Linskens, H. F., 1979. Pollen DNA repair after treatment with the mutagens 4-nitroquinoline-1-oxide, ultraviolet and near-ultraviolet irradiation, and boron dependence of repair, *Molec. Gen. Genet.* 176:11–16.

Jaweed, M. M., and Scott, E. G., 1967. Effect of boron on ribonucleic acid metabolism in the apical meristems of sunflower plants, *Proc. W. Virginia Acad. Sci.* 39:186–193.

Johnson, D. L., 1971. Nucleic acid and protein contents of root tip cells from squash plants during development of and recovery from boron-deficiency and mitotic cell cycle changes in boron-deficient root tip cells, Ph.D. thesis, University of Rhode Island, Kingston.

Johnson, D. L., and Albert, L. S., 1967. Effect of selected nitrogen-bases and boron on the ribonucleic acid content, elongation and visible deficiency symptoms of tomato root tips, *Plant Physiol.* 42:1307–1309.

Kliegel, W., 1972. Bor-Verbindungen aus pharmazeutisch-chemischer Sicht, *Pharmazie*, 1:1–14.

Kohl, H. C., and Oertli, J. J., 1961. Distribution of boron in leaves, *Plant Physiol.* 36:420–424.

Kouchi, H., and Kumazawa, K., 1975a. Anatomical responses of root tips to boron deficiency. I. Effects of boron deficiency on elongation of root tips and their morphological characteristics, *Soil Sci. Plant Nutr.* 21:21–28.

Kouchi, H., and Kumazawa, K., 1975b. Anatomical responses of root tips to boron deficiency. II. Effect of boron deficiency on the cellular growth and development in root tips, *Soil Sci. Plant Nutr.* 21:137–150.

Kouchi, H., and Kumazawa, K., 1976. Anatomical responses of root tips to boron deficiency. III. Effect of boron deficiency on sub-cellular structure of root tips, particularly on morphology of cell wall and its related organelles, *Soil Sci. Plant Nutr.* 22:53–71.

Krueger, R. W., Lovatt, C. J., Tremblay, G. C., and Albert, L. S., 1979. The metabolic requirement of *Cucurbita pepo* for boron, *Plant Physiol.* 63(Suppl.):115.

Kumar, S., and Hecht, A., 1970. Studies on growth and utilization of stylar carbohydrate by pollen tubes and callose development in self-incompatible *Oenothera organensis*, *Biol. Plant.* 12:41–46.

Lee, S., and Aronoff, S., 1967. Boron in plants: A biochemical role, *Science* 158:798–799.

Lewin, J. C., 1966a. Physiological studies of the boron requirement of the diatom, *Cylindrotheca fusiformis* Reimann and Lewin, *J. Exp. Bot.* 17:473–479.

Lewin, J. C., 1966b. Boron as a growth requirement for diatoms, *J. Phycol.* 2:160–163.

Lewin, J., and Chen, C. H., 1976. Effects of boron deficiency on the chemical composition of a marine diatom, *J. Exp. Bot.* 27:916–921.

Lewis, D. H., 1980a. Boron lignification and the origin of vascular plants—a unified hypothesis, *New Phytol.* 84:209–229.

Lewis, D. H., 1980b. Are there inter-relations between the metabolic role of boron, synthesis of phenolic phytoalexins and the germination of pollen? *New Phytol.* 84:261–270.

Loughman, B. C., 1961. Effect of boric acid on the phosphoglucomutase of pea seeds, *Nature* 191:1399–1400.

Lovatt, C. J., Albert, L. S., and Tremblay, G. C., 1979. Regulation of pyrimidine biosynthesis in intact cells of *Cucurbita pepo*, *Plant Physiol.* 64:562–569.

Lovatt, C. J., Albert, L. S., and Tremblay, G. C., 1981. Synthesis, salvage, and catabolism of uridine nucleotides in boron-deficient squash roots, *Plant Physiol.* 68:1389–1394.

Whittington, W. J., 1959. The role of boron in plant growth. II. The effect on the growth of the radicle, *J. Exp. Bot.* 10:93–103.

Wilcox, L. V., 1960. *Boron Injury to Plants*, USDA Agriculture Information Bulletin No. 211.

Wilson, C. M., 1961. Cell wall carbohydrates in tobacco pith parenchyma as affected by boron deficiency and by growth in tissue culture, *Plant Physiol.* 36:336–341.

Winfield, M. E., 1945. The role of boron in plant metabolism. III. The influence of boron on certain enzyme systems, *Aust. J. Exp. Biol. Med. Sci.* 23:267–273.

Yamanouchi, M., 1973. The role of boron in higher plants (Part 2), The influence of boron on the formation of pectic substances, *Bull. Fac. Agric. Gottori*, 25:21–27.

Yasunobu, K. T., and Norris, E. R., 1957. Mechanism of borate inhibition of diphenol oxidation by tyrosinase, *J. Biol. Chem.* 227:473–482.

Yih, R. Y., Hille, F. K., and Clark, H. E., 1966. Requirement of *Ginkgo* pollen-derived tissue cultures for boron and the effects of boron deficiency, *Plant Physiol.* 41:815–820.

Zapata, R. M., 1973. Studies on the roles of boron in growth and sugar-transport processes of sugarcane, *J. Agric. Univ. Puerto Rico.* LVII:9–23.

Zittle, C. A., 1951. Reaction of borate with substances of biological interest, in *Advances in Enzymology and Related Subject of Biochemistry*, F. F. Nord (ed.), Vol. 12, Interscience, New York, pp. 493–527.

Zittle, C. A., and Della Monica, E. S., 1950. Effects of borate and other ions on the alkaline phosphatase of bovine milk and intestinal mucosa, *Arch. Biochem.* 26:112–122.

Index

Abundance of elements in ocean, 4
Actinides, 13
Alkali metals, 9, 10
Alkaline earth metals, 11
Aluminum, 11
   and Alzheimer's disease, 11
Anemia and iron, 19
Animals, growth and survival, 2
Antimony, 3, 8
Antioxidant action, 225–227
Arsenic, 26, 319–340
   compounds of, 320–323
   deficiency of, 323
   functions of, 327
   history of, 26, 319–320
   interaction with other nutrients, 323
   metabolism of, 328–334
   toxicity of, 334–335

Barium, 11
Beryllium, 11
Bismuth, 6, 9
Blood, thyroid hormones in, 42
Boron, 8, 389–415
   and auxins, 406
   carbohydrate metabolism, 396–399
   cellulose biosynthesis, 400–401
   complexes of, 397–398
   effect on growth, 390–392
   effect on hormone action, 405–408
   effect on membrane structure and function, 409–410
   effect on nucleic acid biosynthesis, 410–414
   effect on plant metabolites, 401–402
   essentiality of, 389–390
   enzyme reactions of, 402–404
   history of, 389

Boron (*cont.*)
   lignins, phenols, 401–402
   metabolism of, 394
   plant evolution, 392
   in pollen germination, 404–405
   toxicity of, 393
   uses for boron compounds, 395–396
Bromine, 6–8
Bulk elements, 5, 18

Cadmium, 26, 341–366
   absorption of, 348–350
   chemistry of, 345
   distribution of, 351–352
   effects on nucleic acid and protein synthesis, 354–357
   essentiality of, 346–348
   excretion of, 353–354
   history of, 26, 341–342
   induction of thionein, 357
   metabolism of, 348–352
   other biochemical effects, 360–362
   relation to metallothionein, 342–345
   transport of, 351
Calcium, 6, 18
Cancer induction, 4
Carbon, 5
Cesium, 6, 7
Chromium, 24, 175–201
   absorption, transport of, 177
   biologically active, 176
   concentrations in blood, hair, tissue, 178–182
   deficiency signs, 184
   dietary requirements, 189–192
   excretion of, 183
   history of, 24

Chromium (cont.)
    metabolism of, 187–188
    properties of, 175
    toxicity of, 193
Cobalt, 21, 22, 133–148
    in cells and tissues, 134
    deficiency of, 138
    effects of cobalt, 143
    function in animal and human nutrition, 14, 138–140
    history of, 21, 22, 133
    metabolism and toxicity, 141, 144
    in soils, plants and animals, 134–136
    in vitamin B_{12}, 140
Copper, 6, 20
    antagonism to molybdenum, 27
    antagonism to zinc, 20
    in ceruloplasmin, 12

Dental caries, 83
Dose response range
    for fluorine, 2
    for selenium, 2

Elements, 7, 19, 1–16
    essential, history of, 17–28
    factors involved in essentiality, 4–5
    mechanism of action of essential, 13–14
Evolution of essential trace elements, 4, 5

Fluorine, 8, 14, 25, 55–88
    absorption, 69
    acute toxicity of, 80
    in cells and soft tissues, 59
    chronic toxicity of, 81
    deficiency, 61
    dose-response of, 2
    effect on cells, 78
    effect on enzymes, 75
    excretion, 72
    in extracellular fluids, 56
    functions, 63
    history of, 25
    inhibition of mineral dissolution, 64
    metabolism of, 67
    in mineralized tissues, 56
    in secretions, 74
    toxicity of, 75
    transport, 70
Francium, 9

Gallium, 3
Germanium, 6, 7
Gold, 12

Hormetins, 3
Hydrogen, 5
Hypertension, related to cadmium, 351

Iodine, 8, 14, 19, 33–54
    biochemical reactions, 35
    deficiency, 44
    iodothyronines, 37
    iodotyrosines, 37
    isotopes, 34
    metabolism of, 39
    properties of, 34
    toxicity, 45
Iron, 6, 19
    mobilization by ceruloplasmin, 12
    prevalence of, 5

Lanthanides, 9, 13
Lead, 26, 367–387
    absorption of, 369–370
    deficiency of, 375–376
    history of, 26, 367–368
    inhibition of enzymes of heme biosynthesis, 377–379
    metabolism of, 368–369
    mutagenic, teratogenic effects, 382–383
    relation to calcium, phosphorous and vitamin D, 373
    relation to iron and zinc, 374–375
    renal and other toxic effects, 381–383
    toxicity of, 376–383
        hematological effects, 377–379
        neurotoxic effects, 379–381
    transport of, 370–373
Lithium, 3, 6
    effect on manic psychosis, 10
    possible essentiality, 9, 10, 27
    salt balance, 10
    toxicity, 10

Macrominerals, 6
Manganese, 20, 21, 89–132
    absorption, 92–94
    biochemistry of, 97
    and brain function, 110
    in cancer and other diseases, 121

Index

Manganese (cont.)
  in carbohydrate metabolism, 104
  as a coenzyme, 98
  concentration in animal tissues, 89
  deficiency of, 112
  excretion, 96
  genetic interactions, 115
  in humans, 114, 117
  in lipid metabolism, 108
  in metal-activated reactions, 98
  in metalloenzymes, 101
  toxicity of, 117
  transport and distribution, 94

Mercury, 12
Metalloenzymes, 13, 14
Metallothionein, relation to cadmium, 342–345
Methylation
  of mercury, 12, 13
  of selenium, 207
Molybdenum, 22, 149–174
  antagonism to copper, 27
  deficiency of, 168
  enzymes containing, 151, 152
    action of molybdenum enzymes, 165–167
    aldehyde oxidase, 156
    carbon monoxide oxidase, 162
    as a cofactor, 162–164
    formate dehydrogenase, 161
    nicotinic acid hydrolase, 161
    nitrate reductase, 159
    nitrogenase, 151
    sulfite oxidase, 157
    xanthine oxidase, 153
  history of, 22, 149, 159
  toxicity of, 169
Molybdopterin, 164

Nickel, 26, 293–308
  in cells and tissues, 293–296
  deficiency of, 296–300
  functions of, 300
  history, 26, 27
  interactions with other elements, 301–305
  interactions with zinc, 303
  metabolism and toxicity, 304–306
Nickeloplasmin, 295–296
Nitrogen, 5, 11
Nonmetals, 6, 8

Oxygen, 5, 11

Peroxidase-catalyzed iodination, 36
Phosphorus, 6
Platinum, 12
Posttransition metals, 12, 13
Pretransition metals, 9
Protein synthesis effects by cadmium, 354

Radioiodination of biological molecules, 37
Redox and substitution reactions, 35–36
Rubidium, 6, 7, 9, 10, 27

Selenium, 22, 201–238
  amino acid derivatives of, 204–206
  antioxidant action of, 225–228
  deficiency diseases, 213–215
  dose-response of, 2
  effect of organic compounds, 227
  exudative diathesis, 217
  formate dehydrogenase, 207
  glutathione peroxidase, 210–212
  glycine reductase, 208
  growth effects of, 220
  history of, 23, 24, 201
  interaction with other metal ions, 228–231
  isotopes of, 202
  liver and muscle diseases, 216, 217
  liver neurosis, 213–215
  macromolecular weight compounds, 207–213
  metabolism of, 222–225
  nicotinic acid hydrolase, 209
  other low-molecular-weight compounds, 206–207
  oxyacids of, 203
  relation to vitamin E, 214, 215
  reproductive diseases, 218–220
  seleno-t-RNAs, 212
  selenoproteins, 212
  specific selenium diseases, 220–222
  toxicity of, 231–237
    blind staggers, 232
    in man, 234–235
    xanthine dehydrogenase, 209
Silicon, 25, 257–292
  in aging, 275
  in bone formation, 267
  chemistry of, 258

Silicon (*cont.*)
　in connective tissue, 269–272
　deficiency of, 264–266
　essentiality, 259
　history of, 25, 26, 257
　metabolism, of, 277
　silicosis, asbestosis, and other diseases, 281–206
　in structure, 273–274
　in tissues, 259–264
　toxicity of, 281
Silver, 12
Stimulatory metals, 3–4, 346
Strontium, 6, 7, 11
Sulfur, 6, 11, 13

Teeth, 56
Thyroid gland, 39
Thyroxine, 39
Titanium, 6, 7, 12
Trace elements, 6
Transition metals, 9–12
Triiodothyronine, 39
Tungsten, 9, 10, 12
Thyronine, 34
Tyrosine, 34
Thyroid hormones, 39
　in blood, 42
　　mechanism of action, 46
　　structure-activity relationships, 47
　　in tissues, 43

Tin, 24, 309–318
　chemical properties, 309
　deficiency, 311
　distribution in tissues, 311
　history of, 24
　metabolism of, 312
　organotin compounds, 315
　toxicity of, 314

Urease, nickel content, 295–296

Vanadium, 25, 239–256
　ATPases, 247
　biological effects, 244–246
　deficiency of, 243
　distribution of, 240–243
　effect of hormones on, 249
　essentiality, 240
　functions of, 243
　history of, 25, 239
　metabolism of, 248
　on metabolism of glucose, lipids, 246
　toxicity in humans and other mammals, birds, 250–252
Vitamin D and lead toxicity, 373
Vitamin E, effects of, 226
　in selenium deficiency, 214–226

Zinc, 2, 19, 21, 26
　as an active site in metalloenzymes, 5, 12
Zirconium, 12